Handbook of Research on Digital Research Methods and Architectural Tools in Urban Planning and Design

Hisham Abusaada
Housing and Building National Research Center, Egypt

Carsten Vellguth
German Academic Exchange Service, Germany

Abeer Elshater
Ain Shams University, Egypt

A volume in the Advances in Civil and Industrial Engineering (ACIE) Book Series

IGI Global
DISSEMINATOR OF KNOWLEDGE

Published in the United States of America by
IGI Global
Engineering Science Reference (an imprint of IGI Global)
701 E. Chocolate Avenue
Hershey PA, USA 17033
Tel: 717-533-8845
Fax: 717-533-8661
E-mail: cust@igi-global.com
Web site: http://www.igi-global.com

Library of Congress Cataloging-in-Publication Data

Names: Abusaada, Hisham, 1956- editor. | Vellguth, Carsten, 1960- editor. | Elshater, Abeer, 1976- editor.
Title: Handbook of research on digital research methods and architectural tools in urban planning and design / Hisham Abusaada, Carsten Vellguth, and Abeer Elshater, editors.
Description: Hershey, PA : Engineering Science Reference, 2020.
Identifiers: LCCN 2018060053| ISBN 9781522592389 (hardcover) | ISBN 9781522592402 (ebook)
Subjects: LCSH: City planning--Technological innovations. | Architectural design--Technological innovations. | Architecture and technology.
Classification: LCC NA9012 .D54 2020 | DDC 720/.47--dc23 LC record available at https://lccn.loc.gov/2018060053

This book is published in the IGI Global book series Advances in Civil and Industrial Engineering (ACIE) (ISSN: 2326-6139; eISSN: 2326-6155)

British Cataloguing in Publication Data
A Cataloguing in Publication record for this book is available from the British Library.

All work contributed to this book is new, previously-unpublished material. The views expressed in this book are those of the authors, but not necessarily of the publisher.

For electronic access to this publication, please contact: eresources@igi-global.com.

Advances in Civil and Industrial Engineering (ACIE) Book Series

Ioan Constantin Dima
University Valahia of Târgovişte, Romania

ISSN:2326-6139
EISSN:2326-6155

Mission

Private and public sector infrastructures begin to age, or require change in the face of developing technologies, the fields of civil and industrial engineering have become increasingly important as a method to mitigate and manage these changes. As governments and the public at large begin to grapple with climate change and growing populations, civil engineering has become more interdisciplinary and the need for publications that discuss the rapid changes and advancements in the field have become more in-demand. Additionally, private corporations and companies are facing similar changes and challenges, with the pressure for new and innovative methods being placed on those involved in industrial engineering.

The **Advances in Civil and Industrial Engineering (ACIE) Book Series** aims to present research and methodology that will provide solutions and discussions to meet such needs. The latest methodologies, applications, tools, and analysis will be published through the books included in **ACIE** in order to keep the available research in civil and industrial engineering as current and timely as possible.

Coverage

- Hydraulic Engineering
- Operations research
- Quality Engineering
- Transportation Engineering
- Engineering Economics
- Optimization Techniques
- Construction Engineering
- Structural Engineering
- Productivity
- Urban Engineering

The Advances in Civil and Industrial Engineering (ACIE) Book Series (ISSN 2326-6139) is published by IGI Global, 701 E. Chocolate Avenue, Hershey, PA 17033-1240, USA, www.igi-global.com. This series is composed of titles available for purchase individually; each title is edited to be contextually exclusive from any other title within the series. For pricing and ordering information please visit http://www.igi-global.com/book-series/advances-civil-industrial-engineering/73673. Postmaster: Send all address changes to above address.

Titles in this Series

For a list of additional titles in this series, please visit: www.igi-global.com/book-series

Ambient Urbanities as the Intersection Between the IoT and the IoP in Smrt Cities
H. Patricia McKenna (AmbientEase, Canada)
Engineering Science Reference • ©2019 • 333pp • H/C (ISBN: 9781522578826) • US $195.00

Optimizing Current Strategies and Applications in Industrial Engineering
Prasanta Sahoo (Jadavpur University, India)
Engineering Science Reference • ©2019 • 382pp • H/C (ISBN: 9781522582236) • US $245.00

Big Data Analytics in Traffic and Transportation Engineering Emerging Research and Opportunities
Sara Moridpour (RMIT University, Australia)
Engineering Science Reference • ©2019 • 188pp • H/C (ISBN: 9781522579434) • US $165.00

Optimization of Design for Better Structural Capacity
Mourad Belgasmia (Setif 1 University, Algeria)
Engineering Science Reference • ©2019 • 283pp • H/C (ISBN: 9781522570592) • US $215.00

Reusable and Sustainable Building Materials in Modern Architecture
Gülşah Koç (Yildiz Technical University, Turkey) and Bryan Christiansen (Global Research Society, LLC, USA)
Engineering Science Reference • ©2019 • 302pp • H/C (ISBN: 9781522569954) • US $195.00

Measuring Maturity in Complex Engineering Projects
João Carlos Araújo da Silva Neto (Magnesita SA, Brazil) Ítalo Coutinho (Saletto Engenharia de Serviços, Brazil) Gustavo Teixeira (PM BASIS, Brazil) and Alexandro Avila de Moura (Paranapanema SA, Brazil)
Engineering Science Reference • ©2019 • 277pp • H/C (ISBN: 9781522558644) • US $245.00

Big Data Analytics for Smart and Connected Cities
Nilanjan Dey (Techno India College of Technology, India) and Sharvari Tamane (Jawaharlal Nehru Engineering College, India)
Engineering Science Reference • ©2019 • 348pp • H/C (ISBN: 9781522562078) • US $225.00

Contemporary Strategies and Approaches in 3-D Information Modeling
Bimal Kumar (Glasgow Caledonian University, UK)
Engineering Science Reference • ©2018 • 313pp • H/C (ISBN: 9781522556251) • US $205.00

IGI Global
DISSEMINATOR OF KNOWLEDGE

701 East Chocolate Avenue, Hershey, PA 17033, USA
Tel: 717-533-8845 x100 • Fax: 717-533-8661
E-Mail: cust@igi-global.com • www.igi-global.com

Editorial Advisory Board

List of Contributors

Table of Contents

Detailed Table of Contents

Section 1
Newly-Developed Digital Methods

This section includes 12 chapters. These chapters provide many cases of the newly digital methods in the fields of urban planning and design. The example for the methods is given to a crowdsourcing approach, BigGIS with Hadoop, the sensation of thermal comfort by using the digital simulation, internet of things, GeoDesign, digital simulation for the outdoor thermal comfort assessment, noise measurement sensors by using a device "Q sensor," building information modelling (BIM) with GIS, spatial decision simulator "SD-SIM," a review of models of land-use change, and space syntax.

This chapter examines the problem of excessive similarity when designing new cities. It focuses on the generating of innovative ideas through urban design paradigms. The purpose of this work is to support the efforts of planners and designers toward the creation of new cities based on the concept of cities of singularity. This chapter is a bibliographic review of some conventional Western paradigms in urban planning and design. Based on this work, the three initial singularities of cities can be sketched as being architecturally singular (artwork-like/artistic and organic), societally singular (social, economic, and transcultural), or technologically and informationally singular (smart) in nature. The analytical reading depends on content analysis—which follows the potentiality of exploring the meaning of singularity and its characteristics, indicators, and principles. It collects the interrelationships of the old and new paradigms. The outcomes provide a framework for creating 'cities of singularity' based on a crowdsourcing approach.

BigGIS is a new product that resulted from developing GIS in the "Big Data" area, which is used in storing and processing big geographical data and helps in solving its issues. This chapter describes an optimized Big GIS framework in Map Reduce Environment M2BG. The suggested framework has been integrated into Map Reduce Environment in order to solve the storage issues and get the benefit of the Hadoop environment. M2BG include two steps: Big GIS warehouse and Big GIS Map Reduce. The first step contains three main layers: Data Source and Storage Layer (DSSL), Data Processing Layer (DPL), and Data Analysis Layer (DAL). The second layer is responsible for clustering using swarms as inputs for the Hadoop phase. Then it is scheduled in the mapping part with the use of a preempted priority scheduling algorithm; some data types are classified as critical and some others are ordinary data type; the reduce part used, merge sort algorithm M2BG, should solve security and be implemented with real data in the simulated environment and later in the real world.

Creating a successful public outdoor space for the city people is significantly affected by the sensation of thermal comfort, which is influenced by the local microclimate. With the increase in the complexity of the urban environment and the excessive change in the climate, there is a great need to quantitatively assess the thermal comfort in outdoor spaces to help designers and decision makers to establish the optimum scenario from the first stages of design. The traditional methods of assessment have been criticized for dependency on people's subjective perceptions and responses to different thermal environments. This chapter aims to explore the potentials and limitations of the "Digital Simulation" as one of the methods that allow objective assessment. This chapter tackles several related issues and discusses researchers' recent attempts and contributions, which would help the readers to understand the pros and cons of using the digital simulation method in outdoor thermal comfort assessment and proposes possible solutions for several challenges.

The livability standard still has not considered the chaos city that may stem from or lead to cities of hardship. This chapter rectifies this by making the phenomena of chaos and hardship the centerpiece of the analysis. It depends on the internally displaced persons (IDPs) to display the characteristics of liability and the hardship of living and be the indicators of chaos city. This chapter addresses the non-perceptible processes of the IDPs from outside and inside Cairo in Egypt. This internal displacement supposes the lead-in to chaotic changes in the lifestyles of the cities; it can even be said that they become cities of hardship. The theoretical reading depends on conventional and digital methods (content analysis and the internet of things) to follow these changes, which occur not only due to migrations but also due to ignoring decentralization. The outcomes provide an action plan to create cities free from hardship, displacement, and chaos.

explaining how the theory succeeded in presenting numerical and concrete measures to reveal what have been intangible values of urban settlements such as legibility and accessibility.

This chapter sets out to provide a detailed description of the relationship between space and society. It begins by discussing how people co-live in spaces and how such spaces co-live as communities. Understanding the relationship between space and society requires shedding light on how (1) communities emerge and work and (2) people build their social network. The chapter's main premise is that spatial configuration is the container of activities and the way we construct our cities influences our social life. Therefore, the urban environment should be analyzed mathematically using urban models in order to evaluate and predict future urban policies. The chapter reviews a space-people paradigm, Space Syntax. It defines, elaborates, and interprets its main concepts and tools, showing how urban space is modelled and described in terms of various spatial measures including connectivity, integration, depth, choice, and isovist properties.

Street vending is a growing controversial phenomenon in urban environments. It is a survival strategy and an economic opportunity for countless numbers of marginalized vendors. However, the temporal presence of vendors is portrayed as the source of substantial urban issues, which detract from the quality of the urban public space and the public life of individuals. This chapter aims to propose a practical approach to understand the impact of vendors' temporal presence on the quality of urban space and social life. By space syntax theory, this study utilizes pragmatic methods in the fields of social and human sciences; to analyze the socio-spatial and temporal attributes of the vending phenomenon in relation to urban users' movement in a case study street market at Cairo. The findings introduce a syntactic methodology that highlights the profound relationship between users and informal urban markets to be applied in diverse contexts.

Section 2
Conventional Methods

This section includes four chapters that examine the effectiveness of introducing terms as challenges in urban planning and design. These terms describe the characteristics of traditional cities – for instance, cities of singularity and cities of hardship. The term cities of hardship expresses the idea that the quality of the city is unmatched. The cities of singularity was presented to distinguish cities. The last four chapters deal with these two challenges through an integrative review about urban morphology, energy-efficient mobility, urban policies, and economic diagnosis in historical areas.

Urban settlements are recognized in literature to be a significant energy consumer. Among the primary contributors of such consumption is the transportation system. Recent literature that discusses the relation between energy efficient mobility and urban form has been continually branching and growing. However, as it grew, it becomes more specialized, less self-conscious, and to some extent contradicting. After adopting Conzen's approach to urban morphology, this chapter discusses through an argumentative review the relationship between energy efficient mobility and buildings density, land use, and streets networks. In addition to providing a snapshot to critical studies in the last three decades, two arguments are refuted. First, the residents' behavior resulting from socio-economic factors or individuals' choice has more impact than urban morphology characteristics on energy-efficient mobility. Second, morphological characteristics of urban settlements that lead to energy efficient mobility do not vary with the variation of context.

Alexandria's old town center is like passing into an extensive gallery include various architectural forms of several ancient buildings. Since theses listed buildings in Egypt are not profitable from the financial perspective, the owners have plans to destroy them and locate an option for their monetary needs. The author, in the research objective of this chapter, tries to discover the best approach to raise the financial estimation of some listed villas in Alexandria until they have the same estimation of or more than the building if decimated and replaces by a tall structure. The results provide solutions for the owners who are ready to keep it and safeguard the cultural heritage of the city. The concluded remarks provide answers to the primary research question raises of how to achieve a win-win situation between listed buildings in Egypt and the economic needs of the users.

This chapter aims at understanding and interpreting the informal urban growth in Cairo as a physical objection act against the failing radical ways in managing the city. Accordingly, this chapter tries to analyze both informal physical actions and the formal policies and approaches adopted, and reach reasons for the successes and failures. The chapter first investigates briefly the evolution of Cairo's informal settlements and the hardship conditions behind. Then it examines these settlements through analyzing their two main types. Afterward, it discusses the reasons behind the success of these informal actions. Then it reviews policies and approaches. Consequently, it tries to discover the reasons behind the failure of formal policies and approaches. Thereafter, it discusses briefly using geospatial digital research methods in Cairo's informal settlements. Finally, the chapter ends with a concluded discussion tries to figure out how to reach the right path in dealing with Cairo's informal settlements.

Cairo's downtown, through sociopolitical conditions, had been moved from a single city to a hardship one. The authors attempt using a method of multiple readings to provide a new comprehension for the city by using the historical review side by side with several examples from Egyptian literature that describes this dramatic evolution. For this neighborhood, it is considered an active part of Cairo. The period they suggest for scanning the literature begins from Cairo Great Fire in 1952 and its consequences. The year of 1952 and the constitution of the first republic until the dramatic fall of this last by the revolution of January 2011. This chapter aims to articulate the evolution of Downtown Cairo from the singularity to the hardship.

Foreword

Handbook of Research on Digital Research Methods and Architectural Tools in Urban Planning and Design is a timely book. Many city planners, urban designers and architects recognize that there is a paradigm shift occurring that alters the relationships between the profession, the city and its multiple, individual inhabitants. Many professionals welcome this digital turn as a "Smart City" of digitally organized, corporate service systems. Others decry the digital systems potential for abuse and invasion of privacy. Digital Research Methods and Architectural Tools in Urban Planning and Design is more subtle and complex, side stepping the marketing push by the big tech companies. The authors examine the implications of the digital revolution from a wide variety of points of view, with a humanistic emphasis on the value culture and society.

Behind the "Smart City" corporate promotion lies an informational city, a new paradigm that contains within it three earlier paradigms that still exist in the contemporary city. The oldest is the metropolis, the mother city giving birth to empires, with a singular, privileged center of power, often with a religious base (powered by bio-power, wind, water, animals and slaves). Later, industrial versions of this metropolitan and colonial system created cities of 8 million by 1900, based on coal, steel, machinery, the telegraph and printing presses. In the 1950's a second paradigm based on oil, plastics and accelerated telecommunications linked huge corridors of development with 32 million inhabitants into the megalopolis. First identified on the East Coast of the USA in 1961, this new paradigm spread to Europe, Asia and Latin America, based on car and truck transportation, containers, air travel and satellite communications. A further development of communication systems allowed the megalopolis to fragment into a third paradigm, the fragmented metropolis with is multiple sub centers, sometimes historic, forming a complex poly centric, networked system over a large territory.

The Metacity revolution studied in Digital Research Methods and Architectural Tools in Urban Planning and Design subsumes these three previous paradigms within a far larger digital framework creating a new form of urban territory. Much of this volume concerns the creation and processing of information to document and improve lives in the informational city. A word cloud analysis of the chapter headings makes this very clear including; GIS, GPS, BIM, simulation labs, machine learning, algorithms and urban spatial analyses like Space Syntax (generative rules based on network analysis and traffic flows). Here the role of information is clear in the metacity, as meta-Informational space and time tags make the analysis of data files easier and faster, theoretically helping speed city processes and responses.

Such informational systems are only as good as the data they receive. There has been the scale jump in the capacity to process information resulting in the conceptual modeling of much wider city territories, including the megacity. This territorial concept is an artifact of the informational revolution, pioneered by the United Nations in the 1990's. Satellite imagery enables the mapping and modeling of fast expanding, informal cities. Still there have been embarrassing "Smart City" failures as in the recent Rio Olympics. It seems megacities, sometimes with 60% informality as in Rio, are chaotic systems beyond the comprehension of current informational systems. Human ground testing becomes of utmost importance to check the accuracy of this imagery, and this leads back to the human inhabitants migrating to the city in unprecedented numbers who often operate outside official systems.

The editors of Digital Research Methods and Architectural Tools in Urban Planning and Design keep this fleeting, human dimension always in mind. They include more traditional, human elements from earlier paradigms in their metacity model. Climate change information and increased heat energy flows highlight the future impact of the urban heat island effect, potentially making cities uninhabitable. Rising sea levels will affect Alexandria, where one chapter describes how the historic urban cultural heritage is in dire need of protection from development. Another chapter details the poor living conditions in an informal settlement crowded with new immigrants from the countryside outside Cairo. Historical memory, built into urban form as urban information and set in stone is one theme of the editors, also found in contemporary typo-morphological studies of both rich and poor settlements. These studies are not stuck in the metropolitan past but take on a life in contemporary fiction, making downtown Cairo into a mytho-poetic object, a mixture of fact and fiction in the last chapter. Digital Research Methods and Architectural Tools in Urban Planning and Design thus provides a true representation of the contemporary metacity and a welcome addition to the literature.

David Grahame Shane
Columbia University, USA

Preface

INTRODUCTION

Unquestionably, the realm of architecture today should benefit from the new ways that achieve more realistic results, principally given that the areas of this research focus on their closely related fields. Not only are the physical dimensions of the traditional and new cities addressed, but the authors also discuss the non-physical aspects and applied models that are currently under development in new settlements, such as sustainable cities, smart cities, creative cities, and intercultural cities are covered.

This book proposes to investigate and promote conventional and new digital methods concerning the impacts of contemporary digital technology in the field of architecture, mainly in urban planning and design disciplines. The efficient usage of these methods is becoming more critical in the present day (Mossberger, Tolbert, & Franko, 2013; Silva, 2018). This book explains how these methods can represent an innovative approach that exists side by side with the traditional research methods that are still predominant in the fields of social and human sciences.

The concepts of cities of singularity and cities of hardships are the critical two terms in this book. These terms are emerging in literature in relation to how an urban planner and designer are dealing with the problems of similarity in architectural forms that affect the city identity and inability to live in a convenient good city form(Abusaada, 2017; Abusaada & Elshater, 2019). Singularity appeared due to an acceptance of new urban design paradigms instead of the prevailing one that makes each city indistinct from the rest. Singularity means that each city, to be unparalleled (a haploid organism), should have a point of distinction that challenges ideas (Rossi, 1966). Each city should be a different entity and be able to improve itself autonomously over time. The aim of this concept is to make the city unmatched, unprecedented, and second to none. In this concept, in the preliminary stages of urban planning and design a new city. There is a need to emphasize the way of positive thinking about how this way of thinking enables one to begin a new vision before establishing of any modern city to become singular.

AIMS AND OBJECTIVES OF THE BOOK

The aim of this book is to explore and develop the new paradigms in the discipline of urban planning and design in the realm of architecture. There is a need to create prevailing traditional paradigms to achieve the real city's story, as appropriately adapted with its inhabitants, dwellers, and visitors. It is worth investigating ways to select the relevant parts to this singularity. In some cities of Global South, the hardship seems to be found in aspects of life and emerges by using the indicators of the degrees of inferior living. These cities of difficulties in convenient living—like cities of Global North—meet all the needs of urbanization in integrated environments. However, these cities lack the characteristics that are contrary to the requirements of the validity of life. Those cities suffer from urban chaos, which relates to the absence of order, whether in an engineering system or a behavioral setting.

The novelty of this book emerges in how the urban planner and designer use the digital method to investigate the problems of singularity and hardships. The authors provide a deep discussion of the modern and conventional methods in urban planning and design by using the digital new methods. The contribution depends on the beginning uses of the new digital methods in architecture, particularly in urban planning and design. Additionally, this book offers a theoretical and empirical viewpoint for applying the conventional and digital new methods and tools.

The proposed publication is noteworthy for three reasons: (a) It collects information about everything between the concepts and the processes of more than ten new methods in both planning and design in the field of architecture. (b) It includes the theoretical and practical works in exploring the three concepts, namely the unknown cities, hardship, and singularity. (c) It aims to explain the ways for researchers to apply these methods in their works.

TARGET AUDIENCE

The primary audience of this book consists of professionals working in the fields of architecture and urban planning and design, particularly researchers, academicians, practitioners, and advanced-level students (Master's/PhD), technology developers, and policymakers. Moreover, this book will be of interest to support executives who are concerned with recent technologies that have attracted attention for new methods and tools in scientific research, such as data analytics for sustainable and smart cities. Principally, the target audience can acquire further insight on how to use new methods in research and analytical work; they can use the various technologies contributed in this book for their research applications. This book is mainly for beginners.

ORGANIZATION OF THE BOOK

The contributed authors of the entire chapter aim to explore the potentials and limitations of "digital technology" as one of the global challenges that allow critical thinking. This book tackles several issues related to digital technology and discusses researchers' recent attempts and findings that would help the readers understand the pros and cons of using digital methods in the field of architecture. This book includes two parts, which have been partitioned into 16 chapters.

Section 1

Section 1 of the book focuses on ten new digital methods according to its chronology of their emergence, from the 2016 until 1970s. This part includes 12 chapters. In the first chapter, Hisham Abusaada and Abeer Elshater examine the problem of excessive similarity when designing new cities. This work is based on a bibliographic review of conventional Western paradigms in urban planning and design. The authors used two conventional and digital analytical research methods; the first is content analysis and the second is a crowdsourcing approach. This work helps unnamed stakeholders gather information on ways to make cities unique. These method and tools support the efforts of planners and designers in employing this approach professionally.

The next chapter, written by Nada Alhakkak, presents "BigGIS" as a new product that resulted from developing GIS in the "Big Data" area, which is used in storing and processing big geographical data and helping in solving its issues. This chapter describes an optimized Big GIS framework in Map Reduce Environment M2BG. The suggested framework has been integrated into the Map Reduce environment to solve its storage issues and benefit from the Hadoop environment. The authors acknowledge that M2BG should implemented with real data in the simulated environment and later in the real world.

The third chapter, written by Alshimaa Farag, Sama Badawi, and Rahma Doheim, aims to explore the potentials and limitations of the "Digital Simulation" as a method to allow for objective assessment. In this chapter, the authors examine the high possibility of digital simulation in invistagting the built environment. The authors recommend that urban designers and municipalities find in their decisions, in the initial stages of the design process. To use the simulated model and examine the thermal comfort in the outdoor space, the authors use the traditional methods in the post-occupied scene to provide socially successful open spaces in livable cities.

The fourth chapter, written by Hisham Abusaada and Abeer Elshater, focuses on the liability standard which still has not considered the chaos city can stem from or lead cities experiencing hardship. Internally displaced persons (IDPs) seem to display the characteristics of liability and the hardship of living and be the indicators of chaos city. The analytical methods used in this chapter are content analysis and internet of things (IoT) to follow the non-perceptible processes of the IDPs from outside and inside Cairo, Egypt. The outcomes provide an action plan to create cities free from hardship, displacement, and chaos. The authors use the method of IoT to analyze an example from the city of Lugano in Switzerland.

The fifth chapter, written by Mohammed Alfiky and Hans-Georg Schwarz-v Raumer, presents Geographic Information Systems (GIS) as a tool to digitalize and analyze reality in urban planning, particularly in geo design fields. In addition, this chapter includes how practitioners, urban planners, developers, designers, and scientists handle this method. The author introduces GIS-based modeling and geo design and analyzes an example of the "Filder" area in Stuttgart city, based on a set of pre-identified parameters. This case study demonstrates how GIS can be used to avoid the employment of sophisticated methods and tools.

The sixth chapter, written by Abdalrahman Alashi and Turgay Koramaz, presents the method Building Information Modeling (BIM), which is based on the integration and coordination of different disciplines and professions. BIM offers a platform to provide weighted combined data for reasonable assumptions, allowing all related disciplinaries to generate data of cities. These data will be coordinated and integrated under the supervision of urban planners. Many cases use to indicate how to integrate BIM with GIS-based on to comparatively analyze to show how integration between different disciplines will be illustrated.

The seventh chapter, written by Sana Layeb and Mohsen Ben Hadj Salem, discusses urban atmospheres that evoke several sensory registers that participate in our perception of singular tonalities of our daily situations, as in the case of Tunis. The authors quantify users' feelings by objective measures of electrodermal activity, which is conducted *in situ* in metrological work on the sound signal. These measurements were taken using a device called "Q sensor." This device quantifies emotional arousal by measuring electrodermal activity (EDA).

The eighth chapter, written by Ahmed Toimah and Samy Afifi, presents a conceptual framework for Spatial Decision Simulator (SD-SIM). SD-SIM is intended to be a platform that supports spatial decisions taken by different stakeholders to provide the capability for integrated modeling of socio-economic, man-made, and natural environmental impacts. This framework contains four components for expressing the evolution of spatial issues and reflecting into simulator. These four components are Districts Sub-system, Property Price, and Living Cost Simulator, Interventions Sub-System, and Development Scenarios Sub-System.

The ninth chapter, written by Eda Ustaoglu and Arif Çagdaş Aydınoglu, provides a framework for the study of land-use change processes and for making projections for the future land-use/cover patterns. This chapter focuses on various modeling tools and practices that range from pattern-based methods, such as machine learning and GIS-based approaches, to process-based methods, such as structural economic or agent-based models. The authors discuss the progress that is gained by geographers and natural and economic scientists in developing these models of spatial land-use change and indicates to the necessity needed improvement in the future.

The tenth chapter, written by Ahmed M. Refaat, presents the theory of space syntax and its contributions to urban design. The author demonstrates the space syntax theory at both conceptual and methodological levels, types of space syntax, importance of space syntax, and it potential for application. The author provides links between the measurements of connectivity, integration, choice, and intelligibility to discuss and relate the traditional urban values. This chapter provides a potential mechanism for explaining how the theory succeeded in presenting statistical and concrete measures to reveal what have been intangible values of urban settlements, such as legibility and accessibility.

The eleventh chapter, written by Abdelbaseer A. Mohamed, reviews a space-people paradigm, called space syntax. This chapter defines, elaborates, and interprets the main concepts and tools of space syntax, showing how urban space is modelled and described in terms of various spatial measures including connectivity, integration, depth, choice, and Isovist properties. the author presents a detailed description of the relationship between space and society, and how people co-live in spaces and how such spaces co-live as communities. The author explains that the spatial configuration is the container of activities and that the way we construct our cities influences our social life. Therefore, the urban environment should be analyzed mathematically, using urban models to evaluate and predict future urban policies.

In the twelfth chapter, written Amany Arisha and Nancy Abd El-Moneim, the authors use a syntactic methodology as a practical approach to understand the impact of vendors' temporal presence on the quality of urban space and social life. The authors use space syntax theory as a new digital method to analyzes the socio-spatial and temporal attributes. The authors extract criteria that interrelate socio-economic and tempo-spatial attributes of vending activity as a step towards developing a practical approach for managing public trading in public spaces and upgrading the vitality of public life. Moreover, the authors recommend using the digital methods side-by-side with traditional research tools for a more effective outcome.

Section 2

Section 2 of the book presents four chapters, each of which provides fresh insights and concepts on the terms related to cities of singularity and cities of hardship. As revisiting terms, cities of singularity and cities of hardship express the idea that sees the quality of the city is unmatched. These chapters discuss how to use conventional methods to explore the problems that affect cities of today through these two terms. The authors of the last three chapters deal with this problem through an integrative review of urban morphology, energy-efficient mobility, urban policies, and economic diagnosis.

The thirteenth chapter, written by Omar Galal, discusses how urban settlements are recognized as a significant energy consumer. This chapter focuses on the impacts of the transportation system. Through an integrative review, urban morphology, based on Conzen's approach, is the critical factor to explore the relationships between energy-efficient mobility, building density, land use, and street networks. Two issues in this work have impact on energy-efficient mobility: the residents' behavior or individuals' choices, and morphological characteristics of urban settlements.

The fourteenth chapter, written by Heba Aggour, discusses the fact that the listed buildings in Egypt are not profitable from a financial perspective; the owners are willing to destroy it and locate an option for their monetary needs. The question explored in this chapter is how to achieve a win-win situation between listed buildings in Egypt and the economic needs of the users. The author presents an approach to raise the financial estimation of the listed villa in Alexandria until it has the same estimation or more than the building if decimated and replaced by a tall structure so that the owners are willing to keep it and safeguard it.

The fifteenth chapter, written by Ahmed Hassan Abayazeed, discusses informal urban growth for exploring the radical ways of managing the city. This chapter focuses on the interactions between informal physical actions and formal policies and approaches. The author focuses on the policies and approaches for identifying reasons behind the success of the procedural actions and the failure of the formal systems. This chapter investigates how to reach the convenient path in dealing with Cairo's informal settlements.

In the last chapter in this book, Mougibelrahman Aboamer and Dalia Abdelfattah highlight the case of Cairo's downtown through sociopolitical transformation. The authors believes that these transformations lead to Cairo's downtown from the singularity to the hardship. The authors focus on the core cause of the development and its influence on the social, urban, political and economic domains. Authors interpret Cairo downtown morphisms through multiple reading approaches of novels and literature to de-code the transformation from a new perception of fact and fiction.

ABOUT THE KEY CONCEPTS USED IN THE BOOK

Many of the key concepts employed in this book are innovative and convey unique meanings for the authors, publishers, and the audience. The authors have proposed interpretations that have been readjusted to their contexts. The key concepts used in this book relate to the new digital method in the fields of architecture, particularly in the fields of urban planning and design. In addition, there are various concepts highlights on singularity, hardship, and displacement to explain the situations of some traditional cities in many developing countries. Moreover, the authors present their visions of other innovative in the key concepts that follow:

1. **Aspect Ratio (H/W):** A parameter usually used to quantify a built environment's density. Density is equivalent to the ratio between a street's height to width.
2. **Chaos City:** A phenomenon that embraces the political, social, and economic practices and refers to the imbalance of the performance of tasks and responsibilities; a state of utter confusion; and an absence of arrangement or order.
3. **Cities of Hardship:** A term to describe the negative concept that contradicts liveability as a positive concept. Both terms represent the two aspects of explaining the living of people in the city.

4. **Cities of Singularity:** A term that expresses the idea that the quality of the city is unequalled. This singularity emerged due to an acceptance of new urban design paradigms instead of the prevailing ones that make each city indistinct from the rest.
5. **Crowdsourcing Approach:** A term that expresses the idea of an open-call publication and online collaborative for problem-solving and focuses on the building of internet research and development platforms.
6. **City as Artwork:** Results from the efforts of the previous generations. This has a beginning and does not have a single inevitable end. The urban developers puts the origins of their drawings, the people involved in the construction according to the initial plan, and the character of the individuals living in this city emerges.
7. **Cold Air Flow:** Formed at night in only slightly inclined terrain in case of general wind calm. Cold Air Flow results from the tendency of cold air to move downwards in terrain by gravity. Cold air production that feeds cold air flow is dependent on thermal behavior of land surface.
8. **Concave Space:** A space in which a line connecting two points goes outside the boundary of that space.
9. **Culture Heritage:** Includes archaeological, historical, or scientific sites, structures, or other features of historical, scientific, artistic or architectural value, and ethnological structures of past cultures still extant invalid form.
10. **Electrodermal Activity:** The physiological reaction of the human body detectable by the surface of the human skin.
11. **Energy Efficient Mobility:** A form of mobility within the urban environment that consumes no energy (walking and cycling) or relatively low amounts of energy (means of public transports).
12. **Genotype:** In space syntax literature, Genotype is defined as the abstract and generative rules underlying spatial forms and guides its transformation process.
13. **Geo-design:** The process of planning or developing a spatial entity within a geographical context. In a rather narrow definition, Geo-design describes a design and planning method that, informed by geographic context, tightly couples the creation of a design proposal with impact simulations and supports iterative, sequential, and participative planning.
14. **Geographical Information System (GIS):** A computer program for gathering, managing, analyzing, and visualizing spatial data to make better-informed decisions. GIS analyzes geographical information regarding spatial location, co-incidence, distance, and neighborhood, and it organizes them into visualizations by maps or 3-D scenes. GIS also reveals insights into geographical patterns, relationships, and complexity of situations and configurations.
15. **Global Measure:** This considers the relationships between a space and all other spaces within a system.
16. **Graph:** A type of representation that consists of a set of nodes (vertices) and links. In space syntax theory, nodes represent axial lines or segment lines or convex spaces, while links commonly represent the intersection between pairs of spaces.
17. **Heritage Conservation (Ethnocentric):** A series of actions undertaken to slow the rate of decay of both cultural and natural heritage. This term also aims to communicate the messages held in a cultural heritage object.
18. **Iterative:** A process in which a quick and closed planning loop is established between sketching an intervention, followed by GIS-based impact evaluation of the intervention that is the basis of alterations in a next sketch.

19. **Local Measure:** The measure that confines the syntactic analysis to only spaces that are located at a distance from the selected root space.
20. **Locational Utility:** A measure of the utility of a place or region, which is determined by the individuals' total time spent in transportation modes to reach that place or region – in terms of cost or effort – to fulfill social, economic, or functional needs in a broader spatial system.
21. **Metapopulation:** Several distinct populations considered together with areas of suitable habitat that are unoccupied. In classical metapopulation theory, each population cycles in relative independence of the other populations and eventually goes extinct as a consequence of demographic stochasticity (fluctuations in population size due to random demographic events); the smaller the population, the more chances of inbreeding depression and the more prone the population is to extinction.
22. **Multiple Reading**: A method that consists of readings from various categories of data. By dominating all readings, one would develop a language of the city.
23. **Normalisation/ Standardisation Formula:** This term aims to remove the effect of the number of lines (or segments) in the graph from the global properties' calculations.
24. **Phenotype:** In space syntax literature, this term refers to the spatial forms of urban and architectural composition themselves.
25. **Step Depth/Point Depth:** This term reflects the extent to which each space is away from the selected root space (or spaces) in terms of topological or geometric distance.
26. **Semiotics Challenges**: Explores the study of signs and symbols as a significant part of communications.
27. **The semantic content of the city:** This phrase means that each city has a rhythm between its functional elements and its significant elements (between the significance and the non-significance). This rhythm might be based on the opposition, alternation, or juxtaposition. These are the forms of the relationship between the city's reality and its signification that physical maps can never explain.
28. **Singular Urbanism:** This term creates "a city of singularity" and makes it "a city imprint". Singular Urbanism should be through activation the practice that is based on an understanding of the function of the city and the nature of its inhabitant.
29. **Socio-Spatial Analysis:** The sociologic aspect of a configured spatial space is captured at the level of topological category description, where individual spatial units can be assigned to different social or cultural groups, individuals, or activities with a diverse behavior or pattern.
30. **Sound Effect:** A tool specific to the sound environment that connects perception and action.
31. **Space Syntax:** A theory and method to analyze the relationship between human societies and the structure of the inhabited space in all its diverse forms: cities, settlements, buildings, or even landscapes, and quantify the various levels of topological relationships within a layout.
32. **Spatial Configuration:** A set of interdependent relationships between various discrete spaces, including urban areas, buildings, or temporary structures, each of which is determined by the topological connection of pair-wise connections concerning the overall emerging pattern. The arrangement of solids and voids of urban layout. This is is a more complex concept than spatial relation, which commonly refers to the reciprocal relationship between a pair of spaces.
33. **Street Vending:** A pattern of socio-spatial interactions and behaviors, embedded in urban public life through a network of multi-functional and multi-layered dynamics. Street Vending goes beyond the classic stereotypes of destitution-informality, legality-illegality, and regularity-irregularity.
34. **Syntactic Measures:** This refers to the measures produced through the space syntax analysis, such as integration, connectivity, and intelligibility measures.

35. **Transit Oriented Development (TOD):** Dense, mixed-use urban environments that enjoy an interconnected streets network connecting transit stops, houses, jobs, commercial and leisure activities.
36. **Urban Heat Island:** The area of an urban landscape which reflects significant differences in surface and air temperature compared to the surrounding non-urban area.
37. **Urban Displacement:** This includes refugees and IDPs. Refugees flee from a nation due to natural and human-made disasters, while keeping of personal rights and international stability. IDPs are people who are move within the borders of their country and are dependent the legal protection of their governments.
38. **Urban Artefacts:** Material constructions that link with the term of the spirit of architecture. Architecture and urban artefacts both appear in cities as artistic works and emerge in the dynamic of the city.
39. **Urban Market:** A basic public setting for upgrading cities' functioning and integrating urban communities. Urban Markets accommodate a variety of economic, socio-cultural, and political activities.
40. **Western Paradigms:** A holistic intellectual base in the field of urban design throughout history, which encompasses theories, movements, schools, trends, and approach.

FUTURE DIRECTIONS

As researchers, academics and urban designers of toady cities, we find ourselves in a permanent debate between the logic of scientifically obtained data, recognized methodologies and 'other' facts that we discriminate against in the formulation of our findings. In our reviewing process for this book, we found a wide range of information from data to interpretation or behavioral profiles containing valid arguments that seem to have been voluntarily censored. It is of concern that the formulation of contributions to the discourse are, in some instances, fraught with difficulties in realms other than the topical. The editors find themselves not in a position to discuss the investment of meaning into the terminology used. We leave this for the readers to decipher. In international editions this is, of course, to be expected. We feel the importance to raise this since only through a discussion beyond the factual the reason behind the argument can truly be revealed. Current discourses will only ever be able to detect or reflect the arguments that are actually made. Any –even partial-omission as perceived by the audience will inevitably draw criticism and possibly devaluate or debase the argument entirely.

Ultimately, through introducing future parts in line with this area of specialization, editors present new digital methods in the field of architecture, mainly in urban planning and design. The contribution recommended new research methods that are (1) the methods of quantitative analysis of urban spaces. (2) Way-finding methods and techniques. (3) Isovist analysis. (4) Topological, angular, and metric radii of space syntax analysis (Depthmap X). (5) Human Sensory Assessment. (6) Urban Sensing: people as sensors. (8) Visual Graph Analysis, or VGA. (9) Vision-guided Agent-Based Modelling. (10) Segment-angular spatial network analysis. (11) Origin-destination weighted network analysis to advance the methods of spatial accessibility analysis. (12) Axial integration map. (13) The multi-criteria analysis (MCA). (14) Wearable Technologies: Concepts, Methodologies, Tools, and Applications. (15) CiteSpace: as visualizing patterns and trends in scientific literature.

We feel compelled to point out that the much-debated paradigm shifts of course, are happening. One of the challenges is that their actual (local) 'significance is only communicated to the extent that is perceived to be tolerable. If these situations persist, we might need to further explore the codes. But that would be a topic for another book.

Hisham Abusaada
Housing and Building National Research Center, Egypt

Carsten Vellguth
German Academic Exchange Service, Germany

Abeer Elshater
Ain Shams University, Egypt

REFERENCES

Abusaada, H. (2017). The untold story of the urban design practice of new cities in Egypt. In H. Elsharkawy, S. Zahiri, & J. Clough (Ed.), *International conference for sustainable design of the built environment* (pp. 787-798). London, UK: SDBE.

Abusaada, H., & Elshater, A. (2019). From Chaos to Order: Articulating the Urban Policies for Cities of Hardship. In *Industrial and Urban Growth Policies at the Sub-National, National, and Global Levels*. IGI Global. doi:10.4018/978-1-5225-7625-9.ch003

Mossberger, K., Tolbert, C. J., & Franko, W. W. (2013). *Digital Cities: The Internet and the Geography of Opportunity*. Oxford, UK: Oxford University Press.

Rossi, A. (1966). *The Architecture of the City*. Cambridge, MA: Oppositions Book, The MIT Press.

Silva, C. N. (2018). *New Approaches, Methods, and Tools in Urban E-Planning*. IGI Global. doi:10.4018/978-1-5225-5999-3

Acknowledgment

The editors should like to express their gratitude to everyone who participated in the progress of this publication. Special thanks to the chapters' authors of the current book for their enthusiastic contribution this to project and to the Editorial Advisory Board for their constructive recommendations in the reviewing process. Their conscientiousness is very much appreciated. We extend our special gratitude to David Grahame Shane who, in his foreword, put our publication into the wider context of current discourses.

We would also like to thank Jan Travers, the Director of Intellectual Property and Contracts, for her untiring support throughout the development of this book, to Amanda Fanton and Josephine Dadeboe from IGI Global for their assistance with the editing process. We are indebted to the Publisher IGI for giving us a platform to inform our professional audience of the innovative views and interesting arguments held and developed by the contributors of our publication.

Section 1

Newly-Developed Digital Methods

This section includes 12 chapters. These chapters provide many cases of the newly digital methods in the fields of urban planning and design. The example for the methods is given to a crowdsourcing approach, BigGIS with Hadoop, the sensation of thermal comfort by using the digital simulation, internet of things, GeoDesign, digital simulation for the outdoor thermal comfort assessment, noise measurement sensors by using a device "Q sensor," building information modelling (BIM) with GIS, spatial decision simulator "SD-SIM," a review of models of land-use change, and space syntax.

Chapter 1
A Crowdsourcing Approach in Urban Design:
A Bibliographic Review of Cities of Singularity

Hisham Abusaada
https://orcid.org/0000-0001-6530-7714
Housing and Building National Research Center (HBRC), Egypt

Abeer Elshater
https://orcid.org/0000-0002-5061-6861
Ain Shams University, Egypt

ABSTRACT

This chapter examines the problem of excessive similarity when designing new cities. It focuses on the generating of innovative ideas through urban design paradigms. The purpose of this work is to support the efforts of planners and designers toward the creation of new cities based on the concept of cities of singularity. This chapter is a bibliographic review of some conventional Western paradigms in urban planning and design. Based on this work, the three initial singularities of cities can be sketched as being architecturally singular (artwork-like/artistic and organic), societally singular (social, economic, and transcultural), or technologically and informationally singular (smart) in nature. The analytical reading depends on content analysis—which follows the potentiality of exploring the meaning of singularity and its characteristics, indicators, and principles. It collects the interrelationships of the old and new paradigms. The outcomes provide a framework for creating 'cities of singularity' based on a crowdsourcing approach.

DOI: 10.4018/978-1-5225-9238-9.ch001

INTRODUCTION

This chapter offers an attempt to rediscover a relatively undiscovered aspects of the term *singularity*, which has emerged since the 1960s in the field of architecture of the city (Gehlawat, 2011). That aspect of singularity believes that the architecture of the city should answer the crucial issue: Is it necessary that the design of any new city differ from existing ones? Many theorists have directly adopted this course of thought regarding urban singularity (Lynch, 1960; Lefebvre, 1984; Sitte, 1986; Baudrillard & Nouvel, 2002). Others have implemented their theoretical work in a practical way. For instance, Le Corbusier (1947) established his principles of design by using these ideas to create the northern Indian city of Chandigarh (Gehlawat, 2011). In addition, this concept emerges when Christopher Alexander (1965) identified the discrepancy between *natural* and *artificial* cities (Harary & Rockey, 1976). First, this was done by Le Corbusier based on the ideas of India's first Prime Minister Jawaharlal Nehru regarding the design of the northern Indian city of Chandigarh in 1947. In his view, the city should serve as a symbol of the new India via creating two forms of singularity: the singularity of the "historic moment and an aesthetic singularity" (Gehlawat, 2011, p. 353). In another vein, Alexander sees natural cities as having arisen spontaneously—and *artificial cities* as having been *planned from scratch* (Harary & Rockey, 1976, p. 375). As well, he saw a *real city* as being a *complex human institution* (p. 384 including social, economic, and physical—interacting—phenomena. This endeavor revolved around the efforts of Western thinkers to propose new architectural products based on innovative ideas—ideas which would make the new cities not alike.

Research into the planning and design of new cities has a long tradition dating from the 19th century until today, as is illustrated in Figure 1. For instance, Linear city (Soria mata, 1844), Garden city (Ebenezer Howard 1898), Industrial city (Tony Garnier, 1917), Contemporary city (Le Corbusier, 1922), Radiant city (Le Corbusier, 1933), Broadacre City (Frank Lloyd Wright, 1935), Eco-city (Register, Richard (1975), Collage City (Colin Rowe and Fred Koetter, 1978), Informational city, (Manuel Castells, 1989), A global, or alpha city or world centre, in the global economic system (Saskia Sassen, 1991),The Organic City (Caroline van Eck, 1994), The Generic city, (Rem Koolhaas, 1995), The Mechanic City (Ash Amin and Nigel Thrift, 2002), City branding, place marketing and place promotion (Graham Hankinson, 2004), An intercultural city (Phil Wood and Charles Landry, 2008), Ubiquitous city or u-City (J. Hwang, 2009), Cosmic, Machine And Organic Cities (Martin Johncox, 2009), Great city (Gary Warnaby, 2009).

The leading problem in over-similarity of planning and design is that the continuous employment of single ideas derived from some Western paradigms in city planning and design inevitably leads to similarity. Currently, this similarity is not only occurring between the layouts of those cities but also in their urban form. This is particularly so if these layouts are generated artificially and the clustering of their districts and neighborhoods follows strict hierarchies. Arguably, the new Egyptian cities that have emerged since the 1970s have built their basic planning ideas on a single Western intellectual orientation. This employment of single ideas was based on three essential determinants as follows: Urban hierarchy, planning grids, and standardization (as determined from profiling models of housing units).

Thus, the current problem, particularly during the transition from the 1970s to the present, is that the planning and design of the new cities need to enhance the efforts of architects in order not only to apply the latest innovative ideas but also to offer differences in plans and designs continuously. These issues discuss the ability of planners and designers adapted to the requirements of the *contemporary city*, which should be related to the concept of a city's singularity.

Figure 1. The intellectual visions of designing a city according to the Western Vanguards

This work examines the fact that the potentiality to be ascribed to these ideas lies in their ability to offer an analysis of the Western paradigms of urban planning and design. Interpretations of the previous problem can produce a presenting framework which enables the creation of *cities of singularity*—with all the parties concerned committing to resolve the critical global issues that most affect the design of the new cities relative to them each having their own *singularity*.

Therefore, this chapter needed an innovative approach helpful for accessing acquired knowledge. Such an approach would include searching for content related to the ideas of planning and designing the new cities in ways related to the concept of singularity. This would occur based on reviewing the Western paradigms of the city. As far as we know, recently, no researchers have performed an investigation designed to discuss the concept of singularity to create new cities; only a few studies have shown these effects (Baudrillard & Nouvel, 2002; Gehlawat, 2011; Abusaada, 2016; 2017). One way to overcome these problems is to investigate a relatively new area which has emerged from the question of how to use the ideas of urban planning and design to prevent the similarity of new cities. In order to illuminate this uncharted area, we examined the new digital methods that may help to identify proper methods by which to acquire knowledge, as well as help in development of creative solutions. Notwithstanding, alongside the effort of the talented architects that the conventional technique to get the innovative ideas still was the architectural competitions. In another vein, a crowdsourcing approach is widely considered to be an exceptional means of knowledge acquisition (Papadopoulou & Giaoutzi, 2014).

Thus, in order to reveal new views regarding the concept of singularity, we have developed a methodology which focuses on a crowdsourcing approach in four aspects—which are as follows:

Firstly, it involves establishing a bibliographic review of scientific production in the fields of urban planning and design.

Secondly, it involves identifying and classifying the main theoretical contributions related to the concept of singularity.

Thirdly, it involves establishing a theoretical framework for knowledge acquisition regarding *cities of singularity*.

Fourthly, it involves applying this framework in a real example.

The purpose of this work is to support the efforts of planners and designers toward the creation of new cities based on the concept of *cities of singularity*. The objective is to explore the singularity of the cities in six Western paradigms:

- *(1) artistic* and (2) *organic* architectural similarities;
- (3) *social*, (4) *economic*, and (5) *transcultural* social similarities; and
- (6) *smart* technological and informational singularities.

This work searches for the reasons behind the repetitive models that have appeared in the designing of some new cities around the world. In addition, it provides a framework for how to apply a crowdsourcing approach to urban design. The structure of this chapter includes five sections subsequent to the introduction and background, as follows: The first section presents a crowdsourcing approach. The second section highlights singularity in urban design. The third section presents singularity related to six Western paradigms. The fourth section discusses how the researchers used a crowdsourcing approach in relation to their exploration of *cities of singularity*. The concluding section gives the research outcomes.

BACKGROUND

After the Industrial Revolution, since the first decade of the last century, the theories of urban planning emerged, and the ideas of spatial and functional planning have dominated. These theories started with the residential neighborhood of Clarence Berry (1910), the *Super Block* of Henry Wright and Clarence Stein (1929), the *City of Tomorrow*, the *Radiant City*, the layout of the city of Chandigarh by Le Corbusier (1922; 1924; 1951), and the layout of Islamabad by Constantinos Doxiadis (1959). The planning of all those cities follows the same abstract structural model: to be like a tree. But according to the viewpoint of Christopher Alexander (1965), "a city is not a tree consists of consistent and sequential deployments such as army camps—as he opined when criticizing some of those kinds of cities, such as Greenbelt, Maryland by Clarence Stein (1924), Chandigarh by Le Corbusier (1951), and A plan by Ludwig Hilberseimer (1955), which looked similar to certain Roman towns based on the army camp archetype.

In conventional 19th-century Western paradigms regarding ancient cities, the term *cities of singularity* focused on the principles of design—principles such as individuality, locus, and memory—as well as on the concept of urban artifacts (Rossi, 1982). In addition, from the perspective of conventional and well-known planning and design paradigms, urban form configured the layout of the city, which appeared in the notion of cities themselves being works of art (Sitte, 1945; PULSA, 1972; Lynch, 1984; Olsen, 1986; Sitte, 1986)—an idea which clearly emerged in conceptualizations of the city of Venice (Lefebver, 1991).

Today, the concept of singularity developed implicitly and emerged in the cities of today based on a variety of technological developments. Some examples of these cities are Auckland, New Zealand; Dublin, Ireland; Chattanooga, southeastern Tennessee, USA; and Belgium's Flemish region (World Economic Forum, 2015). Moreover, new cities such as Clark Green City, Philippines; Colombo Port City, Sri Lanka; Gujarat International Finance Tec-City (GIFT), India; and King Abdullah Economic City, Saudi Arabia have emerged differently than have other cities (New Cities Foundation, 2015). In this regard, urban form does not only seem to be the single element that controls singularity (making each city—like human fingerprints—different). The sociocultural, economic, and technological impacts also have a significant effect on the singularity of a city.

In the new era, *cities of singularity* are established. New world cities display a diverse range of strengths, such as: An urban environment and visitor appeal (Auckland), culture and tourism (Barcelona), tech and human capital (Boston), culture and innovation (Berlin), town attractiveness and visitor appeal (Cape town, Scuth Africa), quality of life and talent (Copenhagen), tech and lifestyle (Denver), livability and commercial innovation (Hamburg), human capital (Melbourne), business friendliness (Miami), human capital and reputation (Montreal), Brand, innovation, and productivity (Munich), Infrastructure platform (Oslo), openness to investment (Santiago), digital technology (Tel Aviv), talent and innovation (Toronto), sustainability and quality of life (Vancouver), and lifestyle, visitor appeal, and reputation (Vienna) (Clark, Couturier, Kelly, & Moonen, 2015, p. 32).

Arguably, all the paradigms of those cities above are aimed at formulating *cities of singularity*. Now, the questions are: What is meant by the phrase *cities are not alike?* and the term *cities of singularity*?

This chapter focuses on the word *singularity* based on two issues:

- **The first issue** is that the term *singularity* appeared in the literature when Western paradigms conceptualized cities as being works of art.
- **The second issue** is that the experts invented—in dealing with cities of singularity— many new applications in urban planning and design, which evolved to match several needs. Sometimes, these needs targeted informational, economic, transcultural, or technical issues (each of which is applied individually). One of the key purposes in this work is how to create the singularity through involving more than one application (e.g. economic and technical issues or informational and financial, etc.) to finds the notion of the distinction between the cities without any problems.

CROWDSOURCING APPROACH

The *crowdsourcing* approach was developed by Jeff Howe and Mark Robinson in June 2006 (Brabham, 2008, p. 76). It started as a digital method of facilitating the gathering of information from individuals (as possible) in exchange for an optional reward. It depends on the cognitive assessments performed by means of Internet technologies and GIS. This approach was developed to be the use of tasks outsourced to a large group of non-expert individuals (Tarasov, Cullen, & Delany, 2010, p. 2). Crowdsourcing tools include "smartphone software, blogs, and wikis. Crowdsourcing tools include social bookmarking applications, social networks", and "peer-to-peer software" (Certoma, Corsini, & Rizzi, 2014, p. 93). Many web-based companies use this approach—companies such as Linux, Ubuntu, Threadless.com, iStock photo, Flickr.com, Amazon.com, and SurveyMonkey.com (p. 100).

Via voluntary, open-source collaboration across a network, from the expertise which exists amidst the public crowd, crowdsourcing provides "creative solutions" (Howe, 2008, p. 19). It is a modern, crowd-centered tool for problem-solving—emergent in the form of an open call to present proposals for an issue through a new web-based business model. It is an interactive, technology-based approach to public meetings and can include a large group of non-expert individuals, using virtual reality sites on the Internet and digital engagement with real (in-person) live streamed web events (Conroy & Gordon, 2004; Tarasov, Cullen, & Delany, 2010; Zhao & Zhu, 2012; Li, Hijazi, Reinhard Koenig, Zhong, & Schmitt, 2016).

Today, this approach offers ways to collect people's responses toward urban space which depends on the investigation and the collection of information by individuals, using virtual reality sites on the Internet. The assessment of essential qualities of urban space with emotional and visual data based on the model for measuring emotions is a process based on the concept of cognitive appraisal, which describes a relatively complex set of secondary emotions (Li, Hijazi, Reinhard Koenig, Zhong, & Schmitt, 2016).

In the field of urban design, the disadvantages revolved around how certain data which are collected in the web may provide a significant percentage of error. The reason is that some of the respondents who are involved do not have an uncommon ability to acknowledge the professionally developed questions. Thus, the advantages and disadvantages are as follows (Cheung, 2012; Brabham, 2008; Qualitest20, 2018):

The advantages (discussions of which focused on the sheer size of the crowd) include 1) the ability to get data from a substantial number of experts fast, and 2) the ability to do so via multiple worldwide destinations and not from a limited area.

The disadvantages include the possibility of 1) data stealing and the repeated re-circulation of that data by others; 2) a single person giving multiple responses (using different names and giving different responses each time) in order to get money; and 3) participants not caring to give accurate data.

Crowdsourcing provides practical ideas, comprehensive know-how, and bottom-up models. Likewise, its utility depends on there being a moral motivation to solve complex problems in the fields of interest, such as those related to rewards and scientific publishing. Now, the major questions of this work relate to how crowdsourcing should be utilized in the design of new cities with regard to the issue of singularity.

A THEORETICAL FRAMEWORK

According to the crowdsourcing approach, the process of acquiring knowledge includes four aspects to create cities of singularity, as follows: The first aspect focuses on a bibliographic review of cities of singularity to identify all the data related to the concept of singularity. These data include the notion, idea, concept, status, characteristics, indicators, elements, and principles. The second aspect focuses on the issue of sharing opinions concerning the data collected about the idea of singularity; it depends on continuous questionnaires via the web and social media. The third aspect focuses on the analysis, taking an indirect approach by hosting a forum in which the crowdsourcing transaction will take place, such as ResearchGate and Academia.edu. In this forum, competent authorities should develop something from web-based offerings regarding characteristics, indicators, and principles of cities of singularity. Furthermore, each researcher can present his or her own vision about the notion of singularity. The fourth aspect focuses on the documentation of the data. When the authorities are collecting the proposals and preparing a manual for designing the cities, each researcher must agree to use and distribute his or her proposals in the manual of cities of singularity. This aspect depends on experts in the field generating new knowledge that is appropriate to the purpose of city design to establish the final manual.

This framework requires large numbers of expert respondents. Further, many researchers are needed to collect, analyze, and verify the data and to document the outcomes. And, the approach requires substantial financial support sufficient for financing this process. Accordingly, this process requires convincing the higher authorities in the state to establish a database that includes aspects of *cities of singularity* inside the country. This process helps to establish a theoretical framework includes four directions as illustrated in Figure 2: Construct the web page, prepare the open call for participants, identify the scientific forum, and the techniques of GIS.

A search regarding the institution(s) that wish to fund the establishment of the database and follow a systematic process includes the following:

1. Select the cities,
2. Prepare the data, which includes photographs of the city and the checklist,
3. Investigate the types and classification of the crowd,
4. Prepare the open call for participants,
5. Collect the data,
6. Analyze and classify the results,
7. Present the results in conferences and workshops,
8. Prepare the manual for creating *cities of singularity*, and
9. Publish the data on the web for evaluation.

Thus, crowdsourcing points can be summarized in a theoretical framework in three phases—which are as follows (Figure 3):

Figure 2. Four directions to establish the theoretical framework

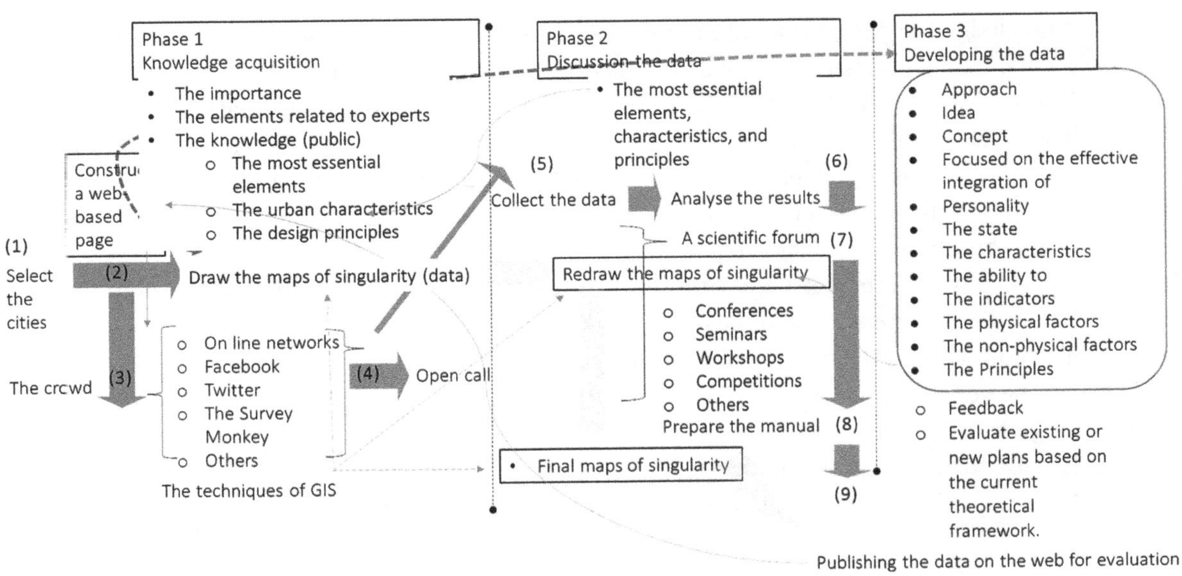

Figure 3. Theoretical framework of cities of singularity (based on a crowdsourcing approach)

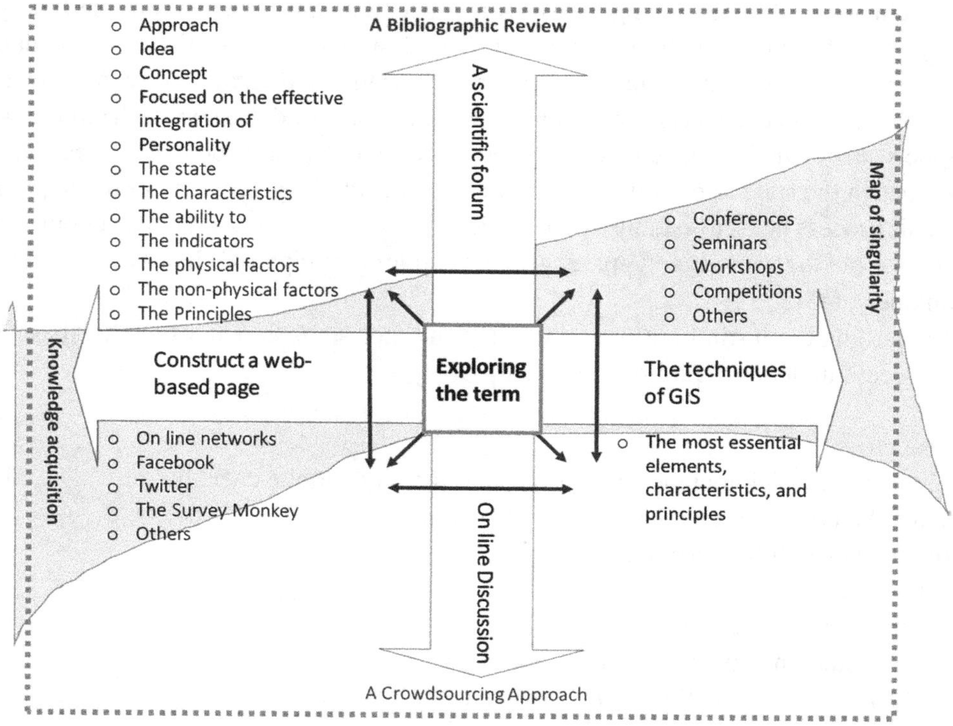

- Construct a Facebook webpage in order to follow the various beliefs of respondents that are suitable to exchange and share, which refer to gathering information regarding ideas that members of the crowd have regarding the ideal planning and design of their own urban center:
    - The importance of the concept of *cities of singularity*;
    - The physical and nonphysical elements of the singularity;
    - Knowledge the public (including both specialists and non-specialists) about the crucial elements, the urban characteristics, and the design principles of the singularity;
    - Maps of singularity based on the techniques of the Geographic Information System (GIS). These maps will be prepared by a group of specialists in the areas of field documentation and GIS. Non-specialists will participate—through sharing ideas as members of the online crowd (via the computer network)—in the selection of the most essential elements of singularity (buildings and urban artifacts).
- Try to involve the competent state authorities—encouraging them to contribute through:
    - Scientific conferences, seminars, and workshops;
    - The competitions of urban planning design.
- Develop the concepts, vision, principles, and practical applications of the project of city planning and design through:
    - Feedback from implementation test and integration;
    - Evaluate existing or new plans based on the current theoretical framework.

The argument that follows draws on extensive literature to examine the issues impacting on *cities of singularity* in greater depth based on Western thought—conventional and contemporary paradigms—and the overarching inquiry of city design as proposed in the theoretical framework.

THE REALM OF SINGULARITY

In the field of architecture, particularly in an urban design discipline, the word *singularity* in its contemporary sense back to many vanguards and Modernist in the Western thought, such as Camillo Sitte (1889), Le Corbusier (1952), Kevin Lunch (1960), Aldo Rossi (1982), Henri Lefebvre (1984). Figure 4 displays the singularity based on vanguard and modernist in the fields of urban design.

Sitte and Le Corbusier never used the word *singularity* explicitly.

Sitte believed that the term singularity should be used related to the urban context as an artistic principle and that it is derived from the temporal and spatial content. In addition, he believed the architectural styles "edifices produced by the Zealous along the Ringstrasse" does not make a singularity—being style "as something alien to the urban context; to the process of change; and something to be avoided because it was usually associated with a lack of humanity vis-a-vis the totality of the city"; (Collins, Sitte, & Collins, 2006, p. 19).

For Le Corbusier, based on Gehlawat (2011), the characteristics of singularity appeared in his conceptual design for the city of Chandigarh in India in 1946, referring to the unusual urban design and

Figure 4. Singularities based on the vanguard and modernist approaches
Source: Compiled by the author, based on (Lynch, 1960; Rossi, 1982; Sitte, 1986; Gehlawat, 2011).

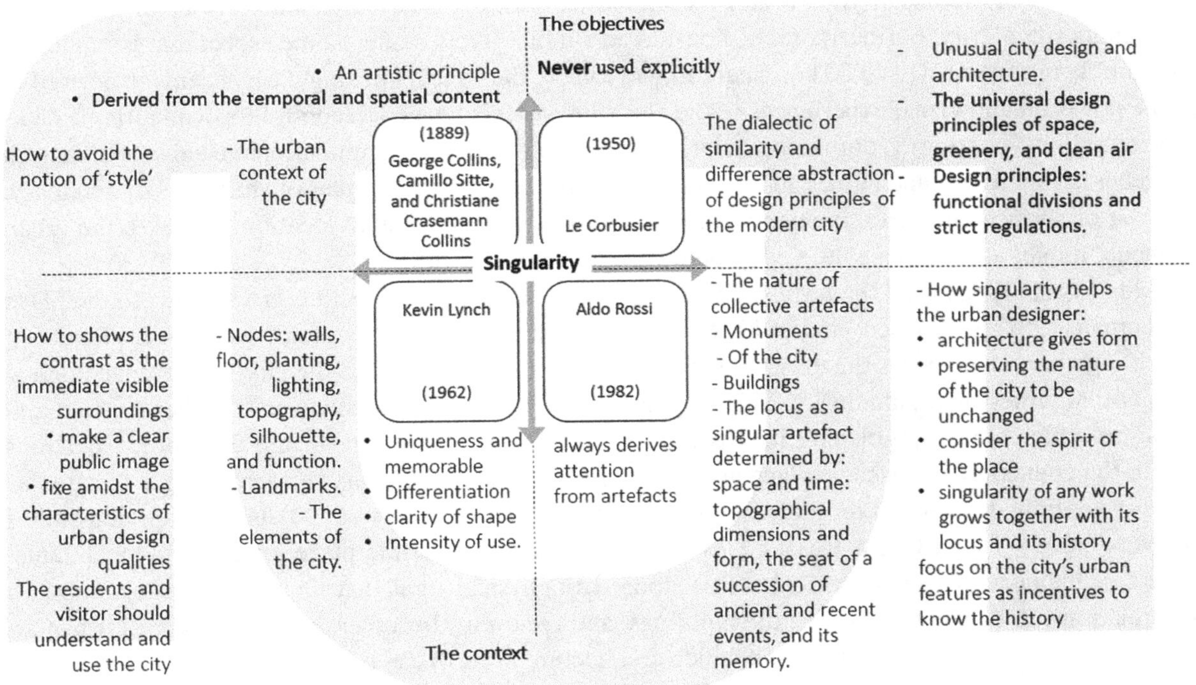

architecture. The singularity manifested every day in the city of Chandigarh based on the universal design principles of space, greenery, and clean air (Gehlawat, 2011, p. 354). Gehlawat sees Le Corbusier as following two guiding principles to create the singularity of the city of Chandigarh—functional divisions and strict regulations. In addition, the singularity has emerged in this project through the dialectic of similarity and difference abstraction of Le Corbusier's design principles regarding the modern city. Further, Chandigarh's singularity appeared related to the historical and cultural context; the intention is producing an "unexpected model" of what we might think of as the organic; this idea is based on "the production of space within the city" (p. 354). For Gehlawat, the discussions of the notion of the singularity in Le Corbusier's work produces three conceptual categories, which are:

- **Aesthetics:** The auratic potential of the original artwork;
- **Disciplinarily:** Which means the method, genre, form, discrete art, and critical practices; and
- **Subjectivity:** Which includes *the One* through authenticity, originality, uniqueness, or individuality; and *the Other*, which deals the work of identity and body politics, recognizing the differences of race, gender, and sexuality in order to recuperate the positions of marginalized individuals or groups (Gehlawat, 2011, pp. 349-350).

In another vein, a number of authors have explicitly recognized the term *singularity*—authors such as, for instance, Lynch (1960), Rossi (1982), and Lefebvre (1984).

Lynch (1960) used the word singularity to express on the continuity, and to describe how the possibility of gaining identity in a well-recognizable node. This singularity will be achieved through the elements of the outdoor areas of the city—such as the walls, floor, planting, lighting, topography, silhouette, function, clarity of shape, and intensity of use. In addition, for Lynch, singularity means characteristics of distinctiveness that give any place its identity; it is a major requirement of design principles for wayfinding. Lynch describes how to be defining physical characteristics of landmarks through its singularity. Furthermore, he considered singularity as being "some aspect that is unique or memorable in context" (p. 79). Thus, according to Lynch, the concept of singularity seems to be used to create powerful and visual experiences of the city's image. It is located to control form qualities such as form simplicity, continuity, dominance, clarity of joint, directional differentiation, visual scope, motion awareness, time series, and names and meanings. It depends on the principles of sharpness of boundary, contrast of surface and form, intensity, complexity, size, use, and spatial location. It makes the urban element "remarkable, noticeable, vivid, [and] recognizable" (p. 105).

Aldo Rossi (1982) used the term singularity to describe how "architecture gives form" to the place (Eisenman, 1982, p. 7). Locus and its history both enhance the singularity of any work—the identity of which together presupposes the existence of the architecture artifacts; and the city is an archaeological artifact that unites their singularity. This singularity (or uniqueness) depends on the phenomenon and the signs which are distinguishing it. Rossi discusses a specific location (locus) from other locations within the context of two issues: place and time. Moreover, the possibility of identifying a location as a site is not only determined through accommodating a series of events but "it also in itself constitutes an event" (p. 7). Accordingly, Rossi believes that the singularity of the place includes its inhabitants' collective memory of local historical events along with physical signs that such events occurred (p. 7). The relationship between the locus, the buildings, and the events becomes a characteristic of urban artifacts—wherein a locus is the place on which architecture or form can be imprinted. Also, architecture gives form to the singularity of the place. Further, Rossi's view is that the meaning of the architecture of the city is based on its broadest adaptability to multiple functions—which corresponds to an extreme precision of form (e.g. the city of Split in Yugoslavia)—wherein the architecture of the city has given new uses and new meanings to unchangeable forms. Concerning any work related to the design of the city, Rossi also sees that the presence of urban artifacts leads to this singularity—which often occurs through the understanding of how this singularity grows together with its locus and history (p. 106). Urban singularity always derives attention from those artifacts (pp. 107-113).

In 1984, Henri Lefebvre (1984) explored the "level of singularities" through everyday life of the city, when he wrote: "Let everyday life become a work of art! Let every technical means be employed for the transformation of everyday life!" (Lefebvre, p. 240). Lefebvre selects the city of Venice to serve as an example of artistic work; he gives it as the model of the richness of a city by virtue of its heritage and history. He sees that the city of Venice takes its singularity from its *architectural harmony*, which seems like *a monumental architectural unity*. This unity is according to the result of historical accumulations, the law of regulation, the privacy of the city, and its geographical location. This city combines between the practical, the symbolic, and the imaginary (Lefebvre, 1991, p. 75). Moreover, the first critical issue is the final product that "was not the result of the individual design of one person" (pp. 118-119).

CITIES OF SINGULARITY AND BEYOND

Diverse ideas have emerged throughout history in order to create a different new city. The ideas behind those cities have represented the backbone of the conventional paradigms in Western thought. Some of these selected paradigms started during the eighteenth and nineteenth centuries but are no longer influential today. Examples include the *Linear City*, the *Industrial City*, the *Radiant City*, and the *Broadacre City*. Meanwhile, ideas continued and evolved, and still have considerable influence. Such ideas include the *Garden City* movement, the *Neighborhood Unit*, and the *Organic City*. In this vein, in the third millennium, many other innovative ideas have had latency beyond the concept of singularities. *Such ideas include city branding, the smart city, and transcultural cities.*

The following is an attempt to discuss the first phase of the theoretical framework in order to discover how to identify the singularity of the city in several selected paradigms. These ideas follow the previous vision—ranging between:

1. Ideas that have emerged but whose effects are over;
2. Those whose effects are still evolving; and
3. Those which have emerged in the recent period.

Singularity Related to the Conventional Western Paradigms—1988 until 2000

The following displays three transformations of the conventional paradigms in order to show the approaches, concepts, and elements of singularity as follows:

The first is the *Linear City*—a longitudinal urban extension designed to be a "more human" city closer to nature. The element of singularity is the functional specialization of—and the infrastructure of—the city. It was invented by Soria Mata (1882). This concept was applied by Tony Garnier in the *Industrial City* (1918) and by Le Corbusier in the *Contemporary City* (1922) and the *Radiant City* (1935). The Linear City (*la Ciudad Lineal*) approach envisaged a 55-km-long open loop around the city of Madrid

(Furundzic & Furundzic, 2012, p. 722) is an example created by Soria Mata. The ten principles of the linear city are as follows (Boileau, 1959, pp. 230-231):

1. The layout of the town conforms with the linear path of the railway,
2. Town planning precedes town building,
3. There is a rectangular layout of roads and building blocks,
4. Twenty percent of the area of the site is built,
5. There is one house with a garden for each family,
6. Structures are set back away from the street to allow for road widening in future,
7. Nucleated towns are combined with linear towns to link the old towns together by means of the linear towns,
8. The path of linear development to the minimum distance required for double railway lines is reduced,
9. The goal is a return to nature and to "a civilized life" in the country, and
10. Land is divided in a way designed to benefit land owners and the general public.

The second is the *Broadacre City* (Wright, The Disappearing City, 1932). It develops the *Utopian* concept alongside the socio-political scheme. It is called a *Usonia*, which was designed to be an *embodied program for economic reform* as a *keystone of a model democracy* (Krohe, 2000, p. 28). The concept of its structural design was based on mobility (motor car) and electrical intercommunication (radio, telephone, and telegraph) factors (Wright, 1935, p. 346). The principles of *Broadacre City* are as follows:

1. Decentralization and mobility: speedy and efficient accessibility with regard to the institutions of advanced society, which are distributed throughout the whole city,
2. Single-family dwellings,
3. A tower building,
4. Land use planning related to function; harmonizing this with the existing landscape.

The third is the *Eco-city*—cities designed to be more sustainable, efficient, and livable (Fook, 2014, p. 8). It started from *Garden City*: *Urban complexity beyond spatialization* (Vernet & Coste, 2017, p. 57). The concept is self-sustaining societies surrounded by greenbelts to serve the evolution of the suburbs. This approach was presented by Ebenezer Howard (1898) and applied by Raymond Unwin (1933) and Frederic Osborn. The *City Beautiful Movement* for Daniel Burnham (1893), and *Eco-friendly cities* (*zero-carbon city*) for Richard Register 1975 the founder of the *Urban Ecology* organization in Berkeley, CA, USA in 1970 and which coined the term *Eco-city* (Fook, 2014, p. 9). This term represents an approach which built the city's singularity on creating a system of:

1. Compact urban planning,
2. Conserving energy,
3. Minimizing the use of other resources (Fook, 2014, p. 9), and
4. Enabling movement and free interaction on foot, bicycles, and mass transit vehicles without fear of traffic or toxins (Engwicht, 1992).

The ten principles of the *Eco-city* are (Roseland, 1977, p. 3):

1. Vital mixed use: a city which is compact, diverse, green, safe, and pleasant,
2. *Access by proximity*: commuting on foot, by bicycle, cart, and public transit taking precedence over automobiles,
3. Restore damaged urban environments,
4. Mixed housing: decent, affordable, safe, convenient, and racially and economically diverse,
5. Social justice: for women, people of color, and people with disabilities,
6. Local agriculture, urban greening projects, and community gardening,
7. Recycling, innovative technology, and resource conservation,
8. Ecologically sound economic activity, and the discouragement of pollution and of hazardous waste,
9. Voluntary simplicity and discouragement of the excessive consumption of material goods, and

Awareness of the local environmental bioregion and of ecological sustainability issues.

Singularity Related to the Contemporary Western Paradigms—2000 Until Today

Concerning the built environment, the *city of singularity* concept indicates that there are several terms which have emerged in the last two decades in urban design paradigms. These terms are used to interpret the relationship between people and place—and can help the designer to achieve the singularities of the cities as follows:

The first term is *technical information entity* [technology as a society]. It focuses on the economy, technology, and communication. It creates global and informational cities to explore and establish the notion of singularity. It does so through the physical element of singularity (ICT infrastructure) and the nonphysical elements (community participation and support localism). The principles are Variety of choices, and attractiveness (Castells, 1996; 1997; Lor & Britz, 2007; Stock, 2011; ATKearney, 2012).

The second term is *commodification* [making places famous]. The goals of city branding are to explore and to establish the notion of singularity as a distinctive place when compared to other places (Hankinson & Cowking, 1993; Greenberg, 2000; Ashworth, 2009; Anholt, 2010). In designing a *city of singularity*, the designer should take it as a good starting point for city marketing and as a framework to manage the city's image and "gain a competitive brand value" (Ashworth, 2009, p. 9). The idea focuses on how to make a unique combination of functional attributes and symbolic values (Hankinson & Cowking, Branding in Action, 1993, p. 10), based on the concept of the *real* material city compatible with the idealized image of the *brand-name city*. The visual symbols represent the physical elements; while the nonphysical elements are perception, value for money, customer focus, and affinity of the local residents. The visual symbols represent the physical elements (a strong and attractive positioning and image), while the nonphysical elements are perception, value for money, customer focus, and affinity of the 'local resident'. The principles are related to "management of the image of the city" (Castillo-Villar, 2018, p. 36) as follows: Visionary leadership, a brand oriented organizational culture, departmental coordination and process alignment, consistent communications across a wide range of stakeholders, and compatible partnerships (Hankinson, 2007, pp. 246-250).

The third term is *entrepreneurship* [a city as an object of renown]. A *smart city* aims to achieve differentiation between cities in terms of their history, economic and political makeup, cultural legacy, national boundaries, and indigenous government policies and laws (Begg, 1999; Hollands, 2008; Nam & Pardo, 2011; Bouton, et al., 2013; Susantia, Soetomoa, Buchoria, & Brotosunaryo, 2016). Singularity is enhanced based on the performance of *cities of singularity* and by promoting their *competitiveness* as compared to other cities. For instance, entrepreneurs can increase the prominence of the city by increasing its wealth. The concepts aimed to enhance the aspirations of the citizens, and to enhance specific layers of *smartness* in the overall infrastructure. The physical elements are those of digital technology of *smart community*, while the nonphysical elements are long-term outlook and the question of how to meet the needs of present and future generations (United Nations Economic Commission for Europe, 2017, p. 85). The principles are

1. The aspects of city life, achieves actual change,
2. A wide range of stakeholders and partners,
3. Vulnerability to external change,
4. Learning from successful approaches in other cities, and
5. Develop bottom-up approaches that have strong local ownership and buy-in (BSI Standards Publication, 2014, p. 9).

The fourth term is *trans-culturalism locus* [difference and diversity]—*cosmopolis/transcultural cities* (Hannerz, 1996; Bloomfield & Bianchini, 2002; Fainstein, 2005; Talen, 2006; Hadjicostandi, 2007; Landry & Wood, 2008; Manzo, 2011). The concepts in this paradigm are that people lead the economic and technological developments. Besides, people are also affecting the artistic and intellectual experiments. Moreover, foreigners play a key role as contributors in all these developments and experiments. Urban diversity creates singularity. The idea is that urban diversity involves a spatial assemblage between people and their environment. Allan Jacobs and Donald Appleyard (1987), Henri Lefebvre (1991), and David Harvey (2000) see diversity as being very essential for the urban life; the city is the locus of difference, diversity, and—ultimately—equality (Talen, 2006, p. 234). The physical elements focus on spatial mixed uses and places, while the nonphysical elements are city making, management, leadership, citizenship, tolerance, pluralism, and diversity in one's way of life. The principles are urban diversity, human experience, quality, and fairness.

1. The fifth term is *great city* (Sargent, 2000; Warnaby, 2009; Savitch, 2010; Bouton, et al., 2013; Herd, 2014). The idea is based on the availability of community services, and on quality of life. A *great city* is in the vanguard of the *zeitgeist* (spirit) of the times [ambiguity]. The concept is that urban life is essential for a more sustainable life. The city's greatness stems from its *economic boom*, based on technology and energy through the work of the metro system, which measures the efficiency of the city. The city's economy is fed by entrepreneurship, social media, mobile applications, and software. Economy of the city should be creating based on information storage systems, computer systems design, permanent publishing on the Internet, clean and dynamic technologies, and the performing of research based on the creative thinking. This aspects of dealing with the economy of the city together comprise makes up the singularity of the cities. The physical elements are the city's iconic buildings, infrastructure, and services. The nonphysical elements are economic balance and environmental responsiveness. The principles areFlexibility (Bouton, et al., 2013, p. 9),
2. The 4Cs: currency, cosmopolitanism, concentration, and charisma (Savitch, 2010, p. 43).

Finally, the sixth term is *empowerment* of people to play a role in the creation of their own city (which, in turn, attracts a certain demographic) [people creating the city: granting people participatory power in the creation of their own city]—an organic urban development creates the singularity of the cities (Schilders, 2010; Osterman, 2015). The idea is that people making up (comprising) the city—and the city being comprised of the people. The concept is that there is a necessity to eliminate failures in the way in that comprehensive planning (which is often just focusing on modernist thought) is performed. The physical elements are the city itself with all part can people involve creating, while the nonphysical elements are the civil initiatives and participatory power. The guiding principles is based on the management mission to facilitate the empowering people for "creating, keeping, and sustaining the city" (p. 29).

How to Apply the Framework

Now, based on a crowdsourcing approach, and after identifying that the goal of this work is how to get ideas for cities of singularity, we established the following strategy in four steps:

1. Review the data on urban planning and design paradigms to fill out the checklist, which includes the ten issues of an investigation into the idea of cities of singularity:
    a. **Approaches:** Any way of dealing with the issue of the singularity of each city is based on its ideas and concepts. It should follow one or more urban planning and design paradigms.
    b. **Idea:** The notions of the paradigms, such as technology for smart city or competitiveness for a branding city.
    c. **Concept:** Interpretation of the essential components of each idea,
    d. Effective integration between more than one idea to get an innovative solution to overcome the problem of singularity.
    e. **Personality:** The critical source of social meaning based on the integration of ethnic groups, cultural differences, and self-organised initiatives.
    f. **The State:** The main themes that help to create the singularity.
    g. **The Characteristics:** Characteristics are things that characterise a place or human, and they can be both apparent and formative (and can be determined by the five senses). It maybe exists only in this place, and it may be repeated by varying degrees and may appear elsewhere. The attributes are both visible and hidden, physical, and nonphysical (Cambridge Dictionary, 2018; Collins, 2018; English Oxford Living Dictionaries, 2018; Merriam Webster, 2018).
    h. **The indicators:** An indicator is a precise characteristic of an inevitable result that is observable and measurable (UNWomen, 2008).
    i. **Elements of Singularity:** These are associated with the urban form and people in the city, consisting of physical elements (visual symbols) and another intangibles (management).
    j. **Guiding Principles:** This is a statement of values which city leaders can use to steer business decision-making by multiple stakeholders over the long term and which focuses on outcomes, open data, incentives for innovation and collaboration, and avoidance of lock-in (BSI Standards Publication, 2014, pp. 4, 9).
2. Using the bibliometric analysis method by "feeding" the computer with ten basic pieces of information to monitor the identification of the interference between several data (Certoma, Corsini, & Rizzi, 2014). Figure 5 displays examples of how to identify the interferences between the data of ten issues of investigation (from the checklist in the Tables 1,2,3, and 4) for three cities (Certoma, Corsini, & Rizzi, 2014). Figure 6 show the relationships between the ten investigation issues of checklist.

Figure 5. The interferences between the ten investigation issues of checklist: bibliometric analysis

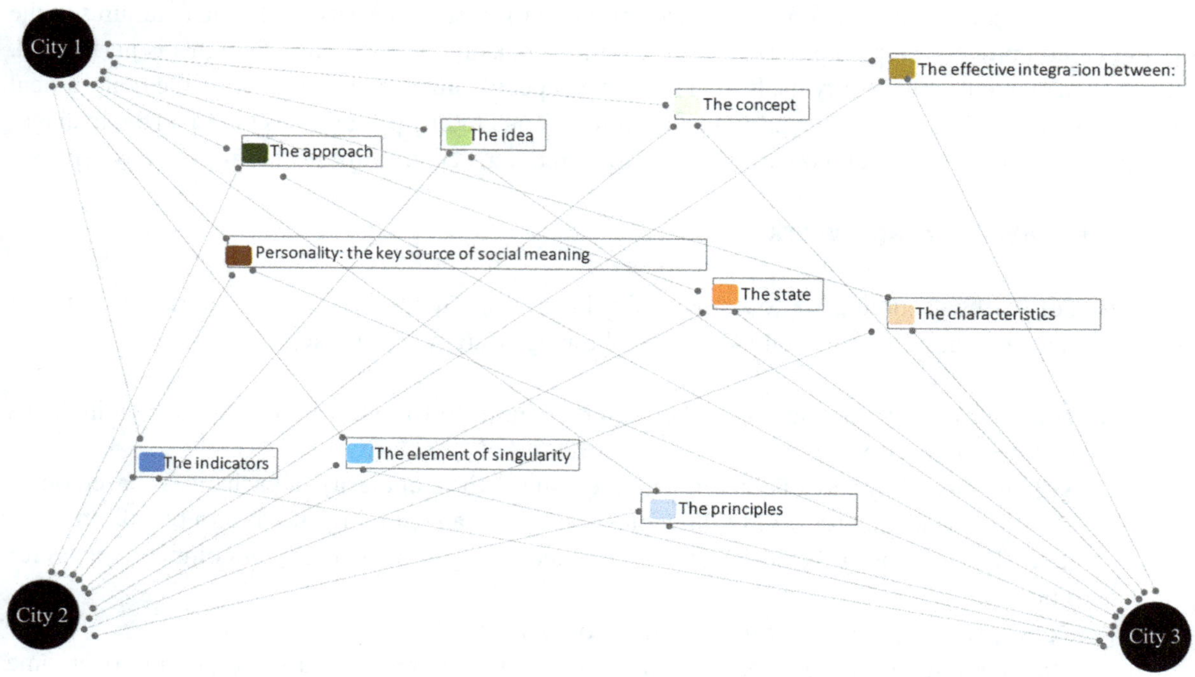

Figure 6. Example of the interferences between the three investigation issues of checklist

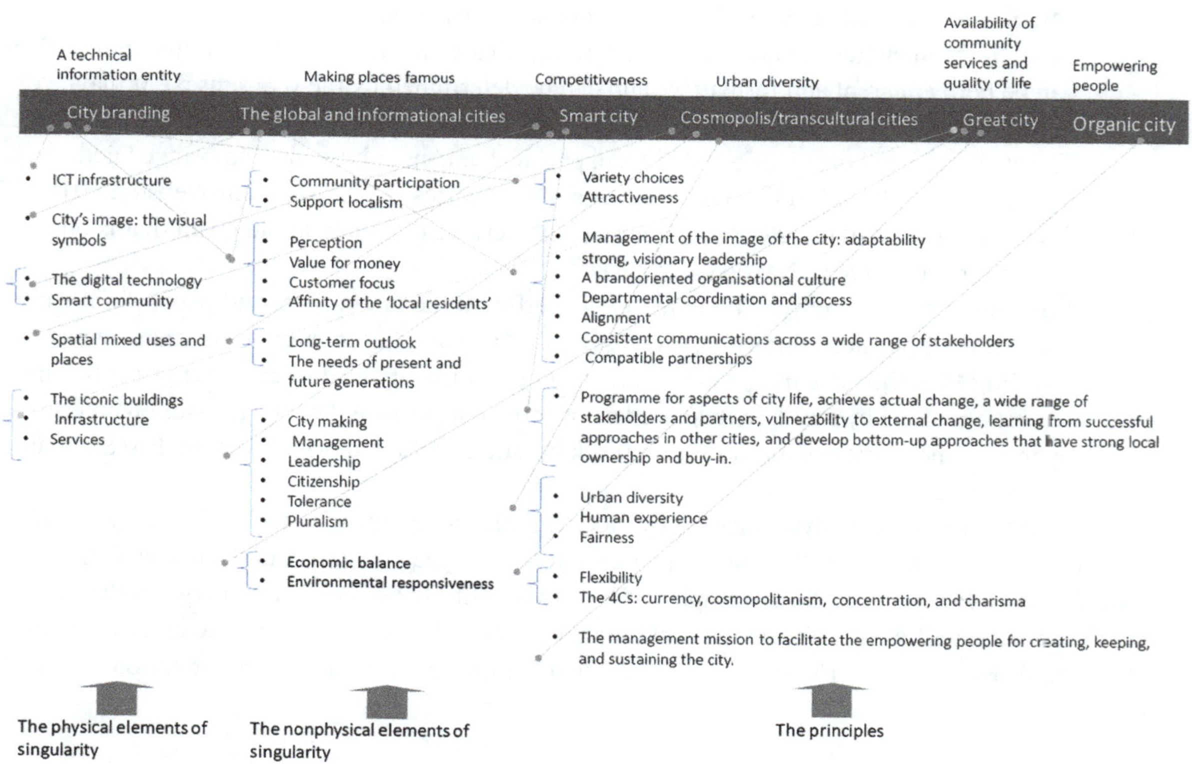

3. Identifying the comprise the targeted crowd, which includes people who have enough information about the topic and communicate with them through an open call online. This communication will be achieved by using scientific networks (such as Academia, Research Gate, and Survey Monkey) or via social media (such as Facebook or Twitter).
4. Developing and publishing the data.

Towards Example-Guided Crowdsourcing

The city of Lugano is one of the most famous Swiss tourist towns and is located on the shore of Lake Lugano. Its singularity seems to be in its urban form, the relationships between buildings and urban spaces, and urban artifacts. Besides, one of the unique characteristics and elements of a city of singularity comes from overlocking views of the structures built on the mountain. Figure 7 shows the features of the singularity of the traditional places of Lugano.

The empirical study revolves around the central theme, which aimed at identifying the factors that describe a city of singularity in the checklist of ten issues for the photo from the town of Lugano. This checklist includes the approach, idea, concept, effective integration, and personality of the city: the critical source of social meaning, the state, the characteristics, the indicators, the elements of singularity, and the principles of singularity. Each respondent selects the items that explain the singularity of each city well. Concerning the previous analysis, the current Tables, 1, 2, 3, and 4, show an analytical study based on the features of Lugano, Switzerland, regarding how the urban designers prepare the data for employing the crowdsourcing approach in the first phase of the framework.

The first phase is the definition of the problem: knowledge acquisition – the publicising of the issue through requesting an open call.

Figure 7. The city of Lugano

Figure 8. The checklist from the city of Lugano, Switzerland.
Source: The authors

The city of Lugano, Switzerland

1) The approach

- ☐ Process, not a place
- ☐ A bipolar opposition between the net and the self
- ☐ Strategies to create and market the cities
- ☐ Apply the new methods: the new generation of information technologies
- ☐ Attract human capital and encourages the innovation
- ☐ Ensure fairness and equal access to a variety of groups
- ☐ Promote the social and economic vitality
- ☐ Reinforce the notion of organicism
- ☐ Enhance participatory decision making

2) The idea

- ☐ Technology is society
- ☐ Unique combination of functional attributes and symbolic values
- ☐ Competitive urban economy
- ☐ Urban diversity: assembling between the people and functions spatially
- ☐ Availability of community services
- ☐ Quality of life
- ☐ People make the City

3) The concept

- ☐ Types of identity-building process lead to a different outcome in constituting society
- ☐ The 'real' city compatible with the idealized image of the brand-name city
- ☐ Enhance the aspirations of the citizens
- ☐ Promoting specific layers of 'smartness' in the overall infrastructure incrementally
- ☐ People lead to developments: economic, technological, artistic, and intellectual experiments
- ☐ Foreigners as contributors
- ☐ Urban life is essential for more sustainable life
- ☐ Eliminate the failure of the way of comprehensive planning, which is depending on the modernist thought

4) The effective integration:

- ☐ The economy, technology, and communication
- ☐ The successive and prominent commercial campaigns
- ☐ The functional, symbolic, and emotional needs
- ☐ Brand equity
- ☐ A registered trademark
- ☐ The economic, social, and environmental aspects
- ☐ New methods: the internet of things, cloud computing, big data and space geographical information integration
- ☐ Different ways of thought about the aesthetic sense and enjoyable experience
- ☐ Promoting the opportunities of the spatial expression of multiple cultural identities
- ☐ Enhancing the new cultural encounters and spatial practices
- ☐ The lovable and lovability not a livability
- ☐ The best amenities and resources
- ☐ The roles of the government, developers, citizens, and other parties involved
- ☐ People control a spatial and social sense individually as they see fit

5) Personality: the key source of social meaning

- ☐ Religious and ethnic identity
- ☐ Enhance the local identity related the citizens
- ☐ Adapt IT and engages society: people are a human capital;
- ☐ Respect the existing cultural differences: exceed the opportunities
- ☐ Respect the temper of the times
- ☐ Promote the self-organized initiatives

6) The state

- ☐ Prototypical cities of the know
- ☐ A world of global flows of wealth, power, and images
- ☐ High returns on investments by meeting the expectations of visitors and investors
- ☐ An interplay between the attributes of cities as locations and the strengths and weaknesses of the firms and the other economic agents
- ☐ The cultural boundaries in a state of flux and remaking
- Facilitate reciprocal understanding between people of different cultural backgrounds
- ☐ Great places to live and work
- ☐ The preservation of life, liberty, and property
- ☐ The interaction between planned and unplanned processes

1. Constructing a web-based page, discussing the data that exists in the first phase on one or more of these online networks, such as Facebook, Twitter, The Survey Monkey, or others.
2. Identifying the crowd, experts, academics, stockholders, or others.
3. Finding the interferences between the ten investigation issues of the checklist based on the notion of bibliometric analysis. This step is based on the data of the crowd.

The second phase is discussing the information – this phase includes two steps:

1. Discussing the problems by the users of those sites and rewriting the proposed data in the new checklist.
2. Addressing the issues and proposed solutions in a scientific forum: conferences, seminars, workshops, or competitions.

The third phase is writing down the results: developing the data – preparing the manual and publishing the data on the web for evaluation. This phase includes:

1. Collecting the new evaluated data.
2. Establishing the manual.
3. Releasing the information on the internet for assessment.

This application is an initial attempt to explain how crowdsourcing help to acquire knowledge related to a specific problem. The limitations of the present studies naturally include many aspects, such as, the nature of the crowd, the team that was collecting and analysing the data, and the cost of the study. Furthermore, the city of the singularity of each city needs big data analysis to identify the ten factors of the checklists before employing the method of crowdsourcing. This work needs applicable vision to show and try how to use this method in the fields of urban planning and design.

CONCLUSION

The topic of the research is how to help unnamed stakeholders gather information on a crowdsourcing method of making cities unique. This work provides support to the efforts of planners and designers who wish to employ this approach professionally. In the field of urban design, the crowdsourcing approach will be regarded in future research, particularly if the limitations are considered. These restrictions are expense, the factor of specialization, the results of the surveys to ensure the validity of the results, and the non-repetition of participation. Future research should consider the potential of new digital methods.

The summing up of this chapter can be defined as guiding the process of the way of thinking to create new cities. This process starts with a bibliographic review for all urban planning and design paradigms to identify the ten investigation issues about the notion of cities of singularity. These issues are 1) the approach, 2) the idea, 3) the concept, 4) the effective integration between, 5) the personality: the key source of social meaning, 6) the state, 7) the characteristics, 8) the indicators, 9) the elements of singularity, and 10) the guiding principles of singularity. The second step is how to acquire knowledge by using a crowdsourcing approach.

There are four dominant visions to inspire the importance of using the term of singularity in conventional methods through Le Corbusier, Kevin Lynch, Aldo Rossi, and Henri Lefebvre. This trend has continued until the present time to produce various visions providing outward singularities, such as informational cities, smart cities, city branding, transcultural cities, organic cities, and great cities. In brief, this research proposes that the concept of designing the city of today should be more than just an individual's work.

We presented cities of singularity as a concept to create unique cities, which have the collective advantages of technology and artwork on one side and community participation on the other. The suggestion of the current study is to review the contemporary and newest Western paradigms and combine them to create a character that suggests or is inspired by the singularity of the city. Therefore, the urban designer should commit to everything that achieves the integrations, not depending on his view but on the criteria presented by theorists over time. It then adheres to all the standards that govern the classification of the city as a place where people go to live the best time of their lives.

The designer should be dealing with the work of designing a city as not merely a comprehensive product done by one person using one paradigm. Sociocultural, economical, and technological aspects are essential to developing the singularity of any city, starting from the implementation of a local societal context to creating the cultural background, and explaining how the city works according to its components and the differences in its places.

Urban designers should respect the considerations of a city as a technical information entity [technology is a society]; a commodification [making places famous]; an entrepreneurship [a city is a renowned object]; a trans-culturalism locus [difference and diversity]; empowering [the city is made by the people and vice versa]. All these factors support cities in becoming great and liveable, as they can become cities of singularity.

Future study should estimate the inherent effects of the crowdsourcing approach more carefully. For example, all web pages should indicate the participants, the incentives, rewards, and justifications to profit from cooperation, whether in theoretical issues or applied projects.

FUTURE RESEARCH DIRECTIONS

This chapter suggests that the implementation of an empirical study depends on a crowdsourcing approach that is essential to gain knowledge relating to the topic of cities of singularity. This study proposes a redrawing of the maps of the new cities in some developing countries in order to identify the aspects of similarity in each city. A design manual in the field of generating knowledge should take into account the physical and nonphysical elements, characteristics, and principles of singularity — which will lead to the creation of unique cities in the future.

REFERENCES

Abusaada, H. (2016). An Approach of Making an Arabian City Branding: City Imprint (in Arabic). In 5th Jordan International Architectural Conference (pp. 1-13). Amman, Jordan: Jordanian Association of Engineers.

Abusaada, H. (2017). The untold story of the urban design practice of new cities in Egypt. In H. Elsharkawy, S. Zahiri, & J. Clough (Ed.), *International conference for sustainable design of the built environment* (pp. 787-798). London: SDBE.

Alexander, C. (1965). A City is Not a Tree. *The Architectural Forum*, *122*(1), 58–62.

Anholt, S. (2010). Definitions of place branding –Working towards a resolution. *Place Branding and Public Diplomacy*, *6*(1), 1–10. doi:10.1057/pb.2010.3

Ashworth, G. (2009). The Instrument of Place Branding: How is it done? *European Spatial Research and Policy*, *16*(1), 9–22. doi:10.2478/v10105-009-0001-9

ATKearney. (2012). *2012 Global Index and Emerging Cities Outlook*. Retrieved 7 10, 2017, from www.atkearney.com: https://www.atkearney.com/research-studies/global-cities-index/2012

Baudrillard, J., & Nouvel, J. (2002). *The Singular Objects of Architecture*. Minneapolis, MN: Minnesota Press.

Begg, I. (1999). Cities and Competitiveness. *Urban Studies (Edinburgh, Scotland)*, *36*(5-6), 795–809. doi:10.1080/0042098993222

Bloomfield, J., & Bianchini, F. (2002). *Planning for the Cosmopolitan City: A Research Report for Birmingham City Council*. Leicester, UK: Comedia, International Cultural Planning and Policy Unit.

Boileau, I. (1959). La Ciudad Lineal: A Critical Study of the Linear Suburb of Madrid. *The Town Planning Review*, *30*(3), 230–238. doi:10.3828/tpr.30.3.377v40725m3m0771

Bouton, S., Cis, D., Mendonca, L., Pohl, H., Remes, J., Ritchie, H., & Woetzel, J. (2013). *How to Make a City Great?* McKinsey & Company.

Brabham, D. C. (2008). Crowdsourcing as a Model for Problem Solving. *The International Journal of Research into New Media Technologies*, *14*(1), 75–90. doi:10.1177/1354856507084420

BSI Standards Publication. (2014). *Smart city framework – Guide customer service to establishing strategies for smart cities and communities*. BSI Standards Publication.

Cambridge Dictionary. (2018). *Meaning of "characteristic" in the English Dictionary*. Retrieved from https://dictionary.cambridge.org/dictionary/english/characteristic

Castells, M. (1996). *The Rise of the Network Society*. London: Blackwell.

Castells, M. (1997). *The Power of Identity, The Information Age: Economy, Society and Culture* (Vol. 2). Cambridge, MA: Blackwell.

Castillo-Villar, F. R. (2018). City branding and the theory of social representation. *Bitácora Urbano Territorial*, *28*(1), 33–38. doi:10.15446/bitacora.v28n1.52939

Certoma, C., Corsini, F., & Rizzi, F. (2014). Crowdsourcing urban sustainability. Data, people and technologies in participatory governance. *Futures*, *74*, 93–106. doi:10.1016/j.futures.2014.11.006

Cheung, S. (2012). How Companies Can Leverage Crowdsourcing. Cambridge, UK: Composite Information Systems Laboratory (CISL).

Clark, G., Couturier, J., Kelly, J., & Moonen, T. (2015). *Globalisation and Competition: The New World of Cities*. Cities Research Center.

Collins, G. R., Sitte, C., & Collins, C. C. (2006). *Camillo Sitte: The Birth of Modern City Planning*. Meniola, NY: Courier Corporation.

Collins. (2018). *Definition of 'characteristic'*. Retrieved from www.collinsdictionary.com: https://www.collinsdictionary.com/dictionary/english/characteristic

Conroy, M. M., & Gordon, S. I. (2004). Utility of interactive computer-based materials for enhancing public participation. *Environmental. Planning. Management*, *47*, 19–33.

Eisenman, P. (1982). Editor's Introduction by Peter Eisenman. In A. Rossi, *The Architecture of the city* (D. Ghirardo & J. Ockman, Trans.). Cambridge, MA: The MIT Press.

English Oxford Living Dictionaries. (2018). *Definition of characteristic in English*. Retrieved from https://en.oxforddictionaries.com/definition/characteristic

Engwicht, D. (1992). *Towards an Eco-City: Calming the Traffic*. Sydney, Australia: Envirobook.

Fainstein, S. S. (2005). Cities and Diversity: Should we want it? Can we Plan for it? *Urban Affairs Review*, *41*(1), 3–19. doi:10.1177/1078087405278968

Fook, L. L. (2014). Towards Eco-cities in Europe and Asia—Sharing of Best Practices and Experiences: An Introduction. In W. Hofmeister, P. Rueppel, & L. L. Fook (Eds.), Eco-Cities: Sharing European and Asian Best Practices and Experiences (pp. 7-18). Singapore: Konrad-Adenauer Stiftung and European Union.

Furundzic, D., & Furundzic, B. (2012). Infrastructure Corridor as Linear City. *1st International Conference on Architecture & Urban Design* (pp. 721-728). EPOKA University, Department of Architecture.

Gehlawat, M. (2011). Space, the Self, and Singularity in Le Corbusier and Christopher Isherwood. *Literary Imagination, 15*(3), 349–364. doi:10.1093/litimag/imr140

Greenberg, M. (2000). Branding Cities: A Social History of the Urban Lifestyle Magazine. *Urban Affairs Review, 36*(2), 228–263. doi:10.1177/10780870022184840

Hadjicostandi, J. (2007). Urbanism/Urban Culture. In G. Ritzer & G. Ritzer (Eds.), *The Blackwell Encyclopedia of Sociology* (pp. 5154–5159). Blackwell Publishing. doi:10.1002/9781405165518.wbeosu022

Hankinson, G., & Cowking, P. (1993). *Branding in Action.* London, UK: McGraw-Hill.

Hankinson, G. (2007). The management of destination brands: Five guiding principles based on recent developments in corporate branding theory. *Journal of Brand Management, 14*(3), 240–254. doi:10.1057/palgrave.bm.2550065

Hannerz, U. (1996). *Transnational Connections: Culture, People, Places.* London: Routledge.

Harary, F., & Rockey, J. (1976). A city is not a semilattice either. *Environment & Planning A, 8*(4), 375–384. doi:10.1068/a080375

Herd, M. (2014). *What makes your city so special?* Retrieved 7 23, 2015, from http://www.theguardian.com/cities/2014/jan/27/guardian-cities-website-welcome

Hollands, R. G. (2008). Will the real smart city please stand up? Intelligent, progressive or entrepreneurial? *City, 12*(3), 303–320. doi:10.1080/13604810802479126

Howe, J. (2008). *Crowdsourcing: Why the Power of the Crowd Is Driving the Future of Business.* Crown Business.

Jacobs, A., & Appleyard, D. (1987). Toward an Urban Design Manifesto. *Journal of the American Planning Association, 53*(1), 112–120. doi:10.1080/01944368708976642

Krohe, J. (2000). Return to Broadacre City. *Illinois Issues., 26*(4), 27–29.

Landry, C., & Wood, P. (2008). *The Intercultural City: Planning for Diversity Advantage.* Earthscan.

Lefebver, H. (1991). *The Production of Space* (D. Nicholson-Smith, Trans.). Malden, MN: Blackwell Publishers.

Lefebvre, H. (1984). *Everyday Life in the Modern World.* New Brunswick, NJ: Transaction Publishers.

Lefebvre, H. (1991). Critique of Everyday Life: Vol. I. *Introduction.* London: Verso.

Li, X., Hijazi, I., Reinhard Koenig, Z. L., Zhong, C., & Schmitt, G. (2016). Assessing Essential Qualities of Urban Space with Emotional and Visual Data Based on GIS Technique. *International Journal of Geo-Information*, *5*(11), 1–18. doi:10.3390/ijgi5110218

Lor, J., & Britz, J. J. (2007). Is a knowledge society possible without freedom of access to information? *Journal of Information Science*, *33*(4), 387–397. doi:10.1177/0165551506075327

Lynch, K. (1960). *The Image of the City*. Cambridge, MA: The MIT Press.

Lynch, K. (1984). The Immature Arts of City Design. *Places*, *1*(3), 10–21.

Manzo, L. C. (2011). 'We are the Fruit Bowl': Place, Cultural Identity and Social Ties among Immigrant Residents in Public Housing. In J. Hou, & J. L. Roberts (Ed.), *Transcultural Cities: Symposium Proceedings* (pp. 14-23). Washington, DC: Department of Landscape Architecture, University of Washington.

Merriam Webster. (2018). *Characteristic*. Retrieved from www.merriam-webster.com: https://www.merriam-webster.com/dictionary/characteristic

Nam, T., & Pardo, T. A. (2011). Conceptualizing Smart City with Dimensions of Technology, People, and Institutions. In *Proceedings of the 12th Annual International Conference on Digital Government Research* (pp. 282-291). Academic Press.

New Cities Foundation. (2015). *Building New Cities: Challenges, Opportunities and Recommendations: Summary and analysis of themes emerging from Cityquest*. KAEC Forum 2014. Retrieved from http://bit.ly/Cityquest2014

Olsen, D. J. (1986). *The City as a Work of Art: London, Paris, Vienna*. New Haven, CT: Yale University Press.

Osterman, A. (2015). The Empowerment of 'Self Power'. *Self-building City, 43*, 104-107. Retrieved 8 2, 2017, from http://volumeproject.org/the-empowerment-of-self-power/

Papadopoulou, C.-A., & Giaoutzi, M. (2014). Crowdsourcing as a Tool for Knowledge Acquisition in Spatial Planning. *Future Internet*, *6*(1), 109–125. doi:10.3390/fi6010109

PULSA. (1972). The city is an artwork. In G. Kepes (Ed.), *Arts of the Environment* (pp. 208–232). New York: George Braziller.

Qualitest20. (2018). *Advantages and Disadvantages of Crowdsourcing*. Retrieved from https://www.qualitestgroup.com/white-papers/advantages-disadvantages-crowdsourcing/

Roseland, M. (1977). *Eco-city Dimensions: Healthy Communities, Healthy Planet.* New Society Publishers.

Rossi, A. (1982). *The Architecture of the City* (D. Ghirardo & J. Ockman, Trans.). Cambridge, MA: The MIT Press.

Sargent, L. T. (2000). Utopian Traditions: Themes and Variations. In T. N. Schaer & G. C. Sargent (Eds.), *Utopia: The Search for the Ideal Society in the Western World* (pp. 8–17). New York: Oxford University Press in Association with The New York Public.

Savitch, H. (2010). What makes a great city great? An American perspective. *Cities (London, England)*, *27*(1), 42–49. doi:10.1016/j.cities.2009.11.012

Schilders, P. (2010). *The Organic City: Method or Metaphor? The Meaning of 'Organic' in Architecture and Urban Planning* (M. Provoost, W. Vanstiphout, Eds., & B. Annable, Trans.). International New Town Institute.

Sitte, C. (1945). *The Art of Building Cities: City Building According to Its Artistic Fundamentals*. New York: Reinhold Publishing Corporation.

Sitte, C. (1986). *City Planning According to Artistic Principles* (G. R. Collins & C. C. Collins, Trans.). Mineola, NY: Dover Publications, INC.

Stock, W. G. (2011). Informational Cities: Analysis and Construction of Cities in the Knowledge Society. *Journal of the American Society for Information Science and Technology*, *62*(5), 963–986. doi:10.1002/asi.21506

Susantia, R., Soetomoa, S., & Buchoria, I., & Brotosunaryo, P. (2016). Smart growth, smart city and density: In search of the appropriate indicator for residential density in Indonesia. *Social and Behavioral Sciences*, *227*, 194–201.

Talen, E. (2006). Design That Enables Diversity: The Complications of a Planning Ideal. *Journal of Planning Literature*, *20*(3), 233–249. doi:10.1177/0885412205283104

Tarasov, A., Cullen, C., & Delany, S. J. (2010). *Using crowdsourcing for labelling emotional speech assets*. W3C workshop on Emotion ML, Paris, France. doi:10.21427/D7RS4G

United Nations Economic Commission for Europe. (2017). *Country Profile of the Housing Sector: Kazakhstan*. New York: United Nation. Retrieved from http://www.unece.org/housing.html

Vernet, N., & Coste, A. (2017). Garden Cities of the 21st Century: A Sustainable Path to Suburban Reform. *Urban Planning*, *2*(4), 45–60. doi:10.17645/up.v2i4.1104

Warnaby, G. (2009). Towards a Service-Dominant Place Marketing Logic. *Marketing Theory*, *9*(4), 403–423. doi:10.1177/1470593109346898

World Economic Forum. (2015). *Inspiring Future Cities & Urban Services Shaping the Future of Urban Development & Services Initiative*. Industry Agenda, Cologny/Geneva Switzerland.

Wright, F. L. (1932). *The Disappearing City*. New York: Payson.

Wright, F. L. (1935). Broadacre City: A New Community Plan. *Architectural Record*, *77*(4), 345–349.

Zhao, Y., & Zhu, Q. (2012). Evaluation on crowdsourcing research: Current status and future direction. *Information Systems Frontiers*, *12*(3), 1–18. doi:10.100710796-012-9350-4

Chapter 2
BigGIS With Hadoop in MapReduce Environment:
Towards an M2BG Framework

Nada M. Alhakkak
https://orcid.org/0000-0002-5435-0730
Baghdad College of Economic Sciences University, Iraq

ABSTRACT

BigGIS is a new product that resulted from developing GIS in the "Big Data" area, which is used in storing and processing big geographical data and helps in solving its issues. This chapter describes an optimized Big GIS framework in Map Reduce Environment M2BG. The suggested framework has been integrated into Map Reduce Environment in order to solve the storage issues and get the benefit of the Hadoop environment. M2BG include two steps: Big GIS warehouse and Big GIS Map Reduce. The first step contains three main layers: Data Source and Storage Layer (DSSL), Data Processing Layer (DPL), and Data Analysis Layer (DAL). The second layer is responsible for clustering using swarms as inputs for the Hadoop phase. Then it is scheduled in the mapping part with the use of a preempted priority scheduling algorithm; some data types are classified as critical and some others are ordinary data type; the reduce part used, merge sort algorithm M2BG, should solve security and be implemented with real data in the simulated environment and later in the real world.

INTRODUCTION

Over the last years, big data got a great interest almost from all internet users, especially from the academic people, government and some organizations. Mainly, Big data includes multiple huge amount of unstructured data that made its storing process hard and complicated; when using the traditional techniques because it needs more real time analysis. According to the rapid growth of services presented by internet companies the big data increased, i.e. Google deals with data greater than a petabyte (PB).

Big data resources are cloud computing, mobile internet, and GIS services. Big data is categorized by 5 Vs, i.e. huge volume, high velocity, high variety, low veracity, high value. The high variety is one

DOI: 10.4018/978-1-5225-9238-9.ch002

of the main challenges in dealing with big data. Consequently, any publication would need to deal with Structured data and semi-structured data or unstructured data, like text, image, video, and voice. However, dealing with big data faced a great challenge, related to data: capturing, storing, searching, sharing, visualization, and analyzing that includes inconsistency and incompleteness, scalability, timeliness, and security. Moreover, in order to improve the quality of data and the used analysis tools; developers should understand how to preprocess their data in an efficient way. This work is organized as follows: a review on Big GIS and its relation to Hadoop to manage its storing and processing with the best warehousing, then the most related works that have been done in this area, to be ended with the proposed framework and the conclusion with future work (Giachetta, 2015; Pasha, 2016).

BACKGROUND

The background of the present research handles some key terms that are widely and newly discussed as urban tools. These tools are Big GIS, Warehousing, Hadoop, Cloud computing with big geo-information, Intelligent GIS services, Intelligent GIS services, Parallel GIS based on Hadoop Cluster.

Generally speaking, the term 'urban analysis' is related to collecting geographical information about cities (Elshater & Abusaada, 2016) and towns and analyzing it. GIS is one of the main tools used for storing the collected data in databases and analyzing it for better decision making. This improves the required work with big geospatial data generated from GIS applications (Yeh, 1999; Elshater, 2015).

According to the vision of data holders in the GIS area for big data in managing, processing, visualization, and analyzing; Big GIS has been introduced. Mostly, most traditional GIS software is limited in dealing with big data challenges, discussed before. As a result, for developing the traditional GIS software that is related to big data issues and challenges, BIG GIS have been introduced; this new term used to manage and process big geospatial data (Yue & Jiang, 2014). The faster growing of all data types used by nowadays applications requires more flexible data warehousing software with the focus on some factors, i.e. data format and volume, varieties in data sources, unstructured data's integration, and data analysis tool. With the faster change in Bigdata term, it's important to solve the issues and challenges related to data storing and its related warehouses and use more flexible tools, like Hadoop and Hive (Sebaa, Chick, Nouicer, & Tari, 2017)

For Hadoop, it's a Java-based programming framework with open source feature that supports managing big data in distributed an environment; solved the failure issues because of supporting clustering process in master-slave structure; which is all presented in Hadoop Distributed File System (HDFS). Meanwhile, there are two foremost programming tools that Hadoop depends on, i.e. Map Reduce and Spark. Each one has its own benefits and limitations. This work uses Map Reduce which covers three main factors. These factors are scheduling, monitoring, and re-executing failed tasks (Zhao, Chen, Ranjan, Choo, & He, 2016). However, Map Reduce leading idea can be summarized as follows:

- Divide the data into individual blocks, which are processed by map jobs in parallel.
- The output of the maps sorted by the framework is then inputted to the reduce tasks.

Cloud computing makes urban analysis possible to process geo-information in an effective and efficient way. SOLAP is one of its tools. However, adopting a cloud-based strategy improves the efficiency of geo-processing tasks. However, adopting those strategies means that the need is to rent commercial clouds as public ones or adopting special private clouds. Nonetheless, both types lead to a waste of resources in one way or another. This problem has been solved be developing new software trends that made full use of existing hardware techniques in geographic information system and remote sensing (GIS/RS) environment (Gao, Liu, Xie, & Tian, 2016).

In intelligent GIS services, GIS services are the fundamental technologies and methods in geospatial data science. Those services require developing existing traditional methodologies for data collection, data storage, data management, and data analysis, in order to get full advantage of special big data with huge volume, real-time restrictions, and device capabilities (Yue & Jiang, BigGIS: How big data can shape next-generation GIS, 2014; Yue, Baumann, Bugbee, & Jiang, 2015).

Parallel GIS is based on Hadoop which considered the main solution for spatial data processing. Hadoop GIS is a spatial warehousing tool which can achieve supporting the retrieval of huge data, this tool built from five parts: portioning the data, QL^{sp} query language, transforming the query, and query engine, as shown in Figure 1. This tool of analysis achieves a fast query of massive huge amounts of data by using various technologies; i.e. data pre-split, real-time special query engine (RESQUE), implicit parallel query based on Map-Reduce (Zhao, Chen, Ranjan, Choo, & He, 2016).

LITERATURE REVIEW

Research on big data has been developed rapidly. Subsequently, it is little surprise that big data in GIS has a big effect on how we manage spatial information. As we consider the way organizations are using geospatial data, one of the clearest methods is using geospatial data by government agencies that lead to the adoption of GIS, with multiple areas. However, some researchers were interested in collecting GIS data over the one on the internet (Pons & & Masó, 2016). This study did not make any obvious implementation to be compared with others. Meanwhile, others designed and integrated framework for GIS and hybrid cloud community in order to get better performance related to cost and storage capacity with the independent location so the GIS can be accessed from anywhere and anytime (Helmi, Farhan, & Nasr, 2018). This study did not represent in its framework how the data transferred to the cloud and back again in knowledge way.

At the same time, and during the search engine, an interesting study was found. This study presents the main idea of Big GIS, and its relation to data technology (Yue & Jiang, 2014). Essentially, this study

Figure 1. Simple Architecture of Hadoop-GIS
Source: (Cho & Choi, 2015; Aji, et al., 2013)

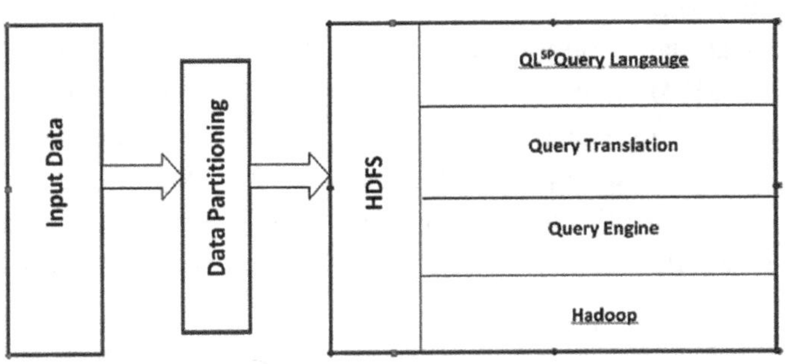

was the spark for my work, because it made me think about how we can manage all huge geospatial data generated from GIS sensors and applications. But the study has not been implemented for performance checking and comparison with others.

Likewise, other studies helped me to make the decision which type of methods or tools that could be used for managing a huge amount of data or Big Data. It's the Map Reduce, as have been proposed in the following studies by Giachetta (2015) proposed a framework for remote sensing data processing in a distributed environment. The framework has been based on Map Reduce Diagram and their toolkit got better results than current solutions. Similarly, Singh and Bawa (2016) used the Map Reduce for analyzing the big data as a parallel processing technique. They used two analysis methods; the first is Map Reduce and Arc GIS and the second is compared to the results. They figured out that using Map Reduce parallel programming model would improve the execution time of queries but simply they did not present any cleared architecture, framework or even an algorithm to represent their work. Conclusively, Sheshikala, Rao, and Prakash (2016) proposed a new algorithm (parallel join-less algorithm) for finding the pattern in a co-location. Their algorithm was developed to contain the followings: Grid-based Partitioning approach for neighbor-relations, filtering tool, and Map Reduce approach for finding special patterns to reduce searching time. However, the study cloud use swarm intelligent algorithm for finding the neighbors.

Another study presented a new platform (PAIRS) refers to Physical Analytics Integrated Repository and Services; which helps to discover rapid data processing, their (PAIRS) software was open source that manages many data processing techniques like download, automatic scalable and storage. PAIRS offers data pre-processing level with analysis step (Klein, et al., 2015). However, the study did not give any details about analyzing the collected data in the integration framework or any other part. As a result, the challenge of big data in GIS today is less about the hardware and more about finding people who can manage the huge volume of information and transform it into value for their organizations. Hence, many types of research have been done in order to develop Big GIS software or produce a new one, but none of them produced a new framework; like this study, for solving the big data generated from GIS tools.

INDUCTIVE LOGIC

Geographic Information System (GIS) is an important tool for urban planning; it's used for analysis, modeling, and visualization. As urban population grow and speed, the importance of GIS lies in its ability to solve complicated problems, such as optimizing new building placement. The ability to run a variety of queries and analytics on GIS data means experts can evaluate how new construction will fit in with existing infrastructure and meet regulatory demands. GIS technology empowers urban planners with enhanced visibility into data. The problem is how to manage the huge data generated from GIS tools and software (Yue, Baumann, Bugbee, & Jiang, 2015; Yue & Jiang, BigGIS: How big data can shape next-generation GIS, 2014) rather than classify it as Big GIS without introducing and management method or tool. As a result, and after reading and comparing search papers related to this area of study, it has been found that no other research proposed such a framework.

This work proposed an optimized framework to manage Big GIS data generated from GIS applications and sensors in the Hadoop framework with its tool Map Reduce, as a programming environment.

The proposed framework (M2BG), which stands for "Managing Big GIS in Map Reduce", as in Figure 2, contains two main parts. Part one is the Big GIS Warehouse and Part-two is the Big GIS Map Reduce

Predominantly, part one deals with collecting data and managing its storage for later forwarding to Map Reduce part. This part contains three main layers.

- The first layer is the Data Source and Storage Layer (DSS Layer). It contains GIS-data sources that are represented as an unstructured database.
- The second is the Data Processing Layer (DP Layer). This layer deals with the pre-processing step that contains two separated types of databases. It also contains an important part that Hadoop relies on it, it's the classifying or clustering pas, in this work the classifying depends on the k-means for a small amount of GIS data, otherwise using swarm algorithms as a clustering tool will bring better results. The output of this layer would be input for the Map Reduce part. At the same time, it may be stored in the databases with some specific pattern.

Figure 2. M2BG Framework as a proposed Framework
Source: The author

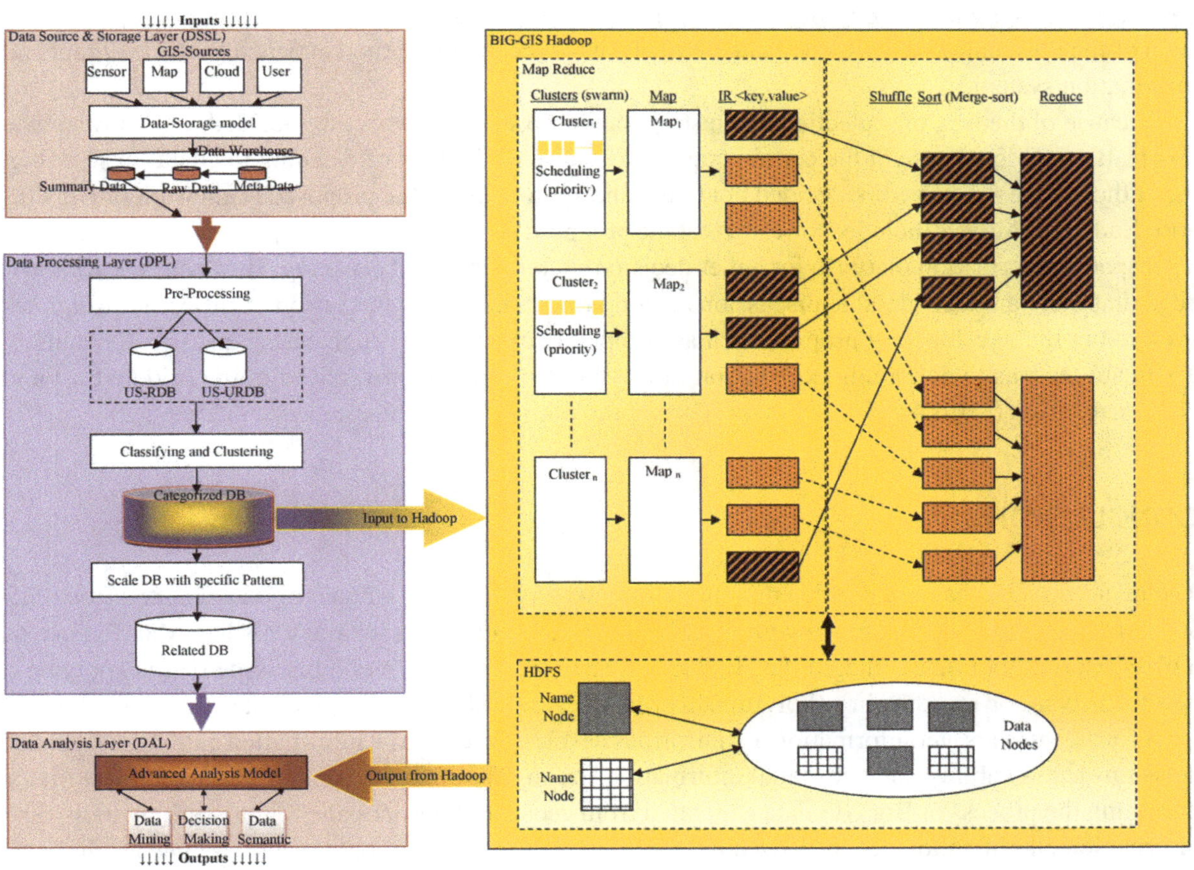

- The last part covers the Data Analysis Layer (DA Layer). If the user is not dealing with the Hadoop framework, in some special cases, then this layer may be used to save time and efforts. The layer deals with a special kind of analysis with the help of data mining tools like Neural Networks to help in decision support systems.

Part two represents Big GIS Map Reduce. It takes its classified inputs from the previous part, to be organized into multiple clusters, where each cluster may contain a queue of multiple blocks, in this work priority scheduling algorithm takes place to solve this issue and selects the suitable block to be forwarded to the map part. After forwarding all the blocks, the results are shuffled-sorted that depend on specific segment <key,value>. For better results in the sorting process, the merge-sort is used. The resulted blocks are reduced according to their similarities in features, related to the segment <key,value>, and output has resulted. Moreover, HDFS manages the whole work of part two as it is in the Hadoop philosophy.

DISCUSSION AND A FUTURE DIRECTION

It performs well, giving good results in using the proposed framework a research tool in the urban planning system. This delivers significantly better results due to mixing two or more discipline such as computer science with the city building. Furthermore, the results demonstrated in this chapter match state of the art method. In this, the example is given for the list of method attached to the geographical information system.

Because of the lack of collecting and testing the proposed framework, we decided to not investigate this matter in a local case in the certain context. Because of the lack of reaching the data. In so doing, the author of the present work decided not to test the application of the proposed framework of Big GIS with Hadoop. This was included to verify that the proposed framework of Hadoop can help in developing a geospatial information platform for better decision-making in the city making. However, we acknowledge that there are considerable discussions among researchers as to the fusion of different disciplines involved in the development projects such as computer science, civil engineering and urban planning. There are reasons to doubt this explanation of missing curricula in universities of the Middle East and Northern Africa (MENA) of the platform that let the discipline to integrate.

CONCLUSION

The main problem that leads to this research was how to manage big amount of data resulted from GIS tools. The proposed M2BG framework used Hadoop's fundamental idea and tools like Map Reduce as a classification tool that manages the Big GIS data. However, there was a big issue that is applying a new framework on real data in a short time in order because it should be implemented for testing results.

As in recent years, geoinformation organizations developed geospatial information platforms for better decision-making ability and geospatial information sharing. This work proposed M2BG framework for managing the process of Big GIS data generated from geospatial services and remote sensing tools. The information classification becomes even more critical. For making big-data secure, we need techniques for security enhancement, i.e. Encryption, Logging, and Privacy issues related to protecting some critical data types that may affect future investments in some lost regions on the map; this is an issue for future research to be explored. At the same time, the proposed framework should be tested in the simulated environment to check its weakness and strength point and optimize it before online applying, which is also an interesting topic for future work.

REFERENCES

Aji, A., Wang, F., Vo, H., Lee, R., Liu, Q., Zhang, X., & Saltz, J. (2013). Hadoop gis: a high performance spatial data warehousing system over map reduce. *Proceedings of the VLDB Endowment*, 1009-1020.

Cho, W., & Choi, E. (2015). A GPS trajectory map-matching mechanism with DTG big data on the HBase system. *Proceedings of the 2015 International Conference on Big Data Applications and Services*, 22-29. 10.1145/2837060.2837062

Elshater, A. (2015). Urban Design Redux: Redefining a Professional Practice of Specialization. *Ain Shams Engineering Journal*, *6*(1), 25–39. doi:10.1016/j.asej.2014.08.004

Elshater, A., & Abusaada, H. G. (2016). What is Urban Design? Teaching and Learning. London: Penguin Reader.

Gao, P. C., Liu, Z., Xie, M. H., & Tian, K. (2016). Low-cost cloud computing solution for geo-information processing. *Journal of Central South University*, *23*(12), 3217–3224. doi:10.100711771-016-3387-3

Giachetta, R. (2015). A framework for processing large scale geospatial and remote sensing data in Map Reduce environment. *Computers & Graphics*, *49*, 37–46. doi:10.1016/j.cag.2015.03.003

Helmi, A. M., Farhan, M. S., & Nasr, M. M. (2018). A framework for integrating geospatial information systems and hybrid cloud computing. *Computers & Electrical Engineering*, *67*, 145–158. doi:10.1016/j.compeleceng.2018.03.027

Klein, L. J., Marianno, F. J., Albrecht, C. M., Freitag, M., Lu, S., Hinds, N., & Hamann, H. F. (2015). A scalable geo-spatial data analytics platform. In *International Conference In Big Data (Big Data)* (pp. 1290-1298). IEEE. 10.1109/BigData.2015.7363884

Pasha, M. R. (2016). *Data Warehousing and the Unstructured Data.* Academic Press.

Pons, X., & Masó, J. (2016). A comprehensive open package format for preservation and distribution of geospatial data and metadata. *Computers & Geosciences*, *97*, 89–97. doi:10.1016/j.cageo.2016.09.001

Sebaa, A., Chick, F., Nouicer, A., & Tari, A. (2017). Research in big data warehousing using Hadoop. *Journal of Information Systems Engineering & Management*, *2*(2). doi:10.20897/jisem.201710

Sheshikala, M., Rao, D. R., & Prakash, R. V. (2016). Parallel Approach for Finding Co-location Pattern–A Map Reduce Framework. *Procedia Computer Science*, *89*, 341–348. doi:10.1016/j.procs.2016.06.081

Singh, H., & Bawa, S. (2016). Spatial data analysis with ArcGIS and Map Reduce. *International Conference on Computing, Communication and Automation (ICCCA)*, 45-49. 10.1109/CCAA.2016.7813687

Yeh, A. G. (1999). Urban Planning and GIS. *Geographical Information Systems, 2*(1), 877-888.

Yue, P., Baumann, P., Bugbee, K., & Jiang, L. (2015). Towards intelligent GIS Services. *Earth Science Informatics*, *8*(3), 463–481. doi:10.100712145-015-0229-z

Yue, P., & Jiang, L. (2014). BigGIS: How big data can shape next-generation GIS. In *Third International Conference on In Agro-geoinformatics (Agro-geoinformatics 2014)*. IEEE. 10.1109/Agro-Geoinformatics.2014.6910649

Zhao, L., Chen, L., Ranjan, R., Choo, K. K., & He, J. (2016). Geographical information system parallelization for spatial big data processing: A review. *Cluster Computing*, *19*(1), 139–152. doi:10.100710586-015-0512-2

Chapter 3
Digital Simulation for the Outdoor Thermal Comfort Assessment

Alshimaa Aboelmakarem Farag
https://orcid.org/0000-0003-1539-6928
Zagazig University, Egypt & Effat University, Saudi Arabia

Sama Badawi
Mansoura University, Egypt

Rahma M. Doheim
Assiut University, Egypt

ABSTRACT

Creating a successful public outdoor space for the city people is significantly affected by the sensation of thermal comfort, which is influenced by the local microclimate. With the increase in the complexity of the urban environment and the excessive change in the climate, there is a great need to quantitatively assess the thermal comfort in outdoor spaces to help designers and decision makers to establish the optimum scenario from the first stages of design. The traditional methods of assessment have been criticized for dependency on people's subjective perceptions and responses to different thermal environments. This chapter aims to explore the potentials and limitations of the "Digital Simulation" as one of the methods that allow objective assessment. This chapter tackles several related issues and discusses researchers' recent attempts and contributions, which would help the readers to understand the pros and cons of using the digital simulation method in outdoor thermal comfort assessment and proposes possible solutions for several challenges.

DOI: 10.4018/978-1-5225-9238-9.ch003

INTRODUCTION

Since the middle of the 20th century, creating a socially successful public outdoor space for city people has been considered a significant goal in urban design. One of the most important factors that significantly affect the outdoor spaces is outdoor microclimate, especially for pedestrians. The pedestrians using the outdoor spaces are exposed directly to all of the environmental elements; such as sun and shade, humidity and dry climate, changes in wind speed, etc. In fact, the people's decision of whether to use the outdoor space or not depends much on their sensation of thermal comfort, which is affected by the local microclimate (Chen & Ng, 2012).

In the global context with the aggravation of environmentally harmful practices that have led to the phenomenon of global warming, ensuring the outdoor thermal comfort is essential to improve the urban quality and livability. The outdoor space design should contribute to improving the local microclimate, especially in hot weather countries, which are suffering because of the continuous temperature rise. The cooling capacity of greenery, trees, and multi-building strategies are used to reduce air temperature. Therefore, enhancing the thermal comfort sensation and controls electrical energy consumption is also needed. Thus, designing appropriate urban open spaces is essential to enhance the people sensation of thermal comfort, to reduce the urban energy consumption and to develop the quality of the outdoor environment (Tariq & Arch, 2014).

Providing people with the sensation of outdoor thermal comfort is a complicated issue with multi-layers of concern that have to be studied in depth. There is a great need to assess the thermal comfort in outdoor spaces to help designers and decision makers to establish the optimum scenario from the first stages of design (Chen & Ng, 2012).

THE ASSESSMENT OF OUTDOOR THERMAL COMFORT

The assessment of thermal comfort is a multi-dimensional process. In order to understand the nature of the assessment process of the outdoor thermal comfort (OTC) concerning the thermoregulation aspect, two main methods of assessments are explained; Steady-state and Non-steady-state assessment methods.

First, *the Steady-state assessment method* is based on the assumption that the biometeorological data are reliable indices to describe human thermal comfort. The researchers who have adopted this assumption develop numerical solutions governing the thermoregulation with the human level of comfort.

Predicted Mean Vote Index (PMV), Predicted Percentage Dissatisfied Index (PPD), and the Physiological Equivalent Temperature (PET), are the most widely used steady-state assessment methods. Those methods are the quantitative prediction and analytical tools of people's thermal response to the local thermal environment that can be easily understood and interpreted. However, the main problem of the steady-state methods is that those methods are not flexible with the dynamic aspects of the course of human thermal adaptation due to the fact that the steady-state methods are based on thermoregulation, which is constructed for a passive state.

The Non-steady-state assessment methods can provide detailed investigations of the human thermal adaptation dynamics, but the *Non-steady-state assessment methods* do not have internationally accepted indices as in the steady-state methods. Those assessment methods depend on other indicators such as human skin temperature, sweating rate, and skin blood flow. Those Non-Steady-state indicators are hard to be monitored especially in the outdoor environment. In addition, these indicators require the main knowledge of Biometeorology and Physiology and are not informative enough to provide useful implications for design practice. Instead, the non-steady state assessment methods might be much more appropriate for assessing the design after implementation and occupation.

Concerning the social aspects, the assessment of people's thermal comfort in different studies shows a subjective nature since it is based on people response and their previous experiences. The fact is that people sensation of thermal comfort is mainly affected by the progressive human adaptation to the ambient thermal condition. In addition, the thermal comfort condition of people is not always consistent with the objective climatic or biometeorological condition. The people might feel thermal comfort regardless of the thermoregulation when they are enjoying and being engaged in social activities with friends (Chen & Ng, 2012). In addition, the social function of outdoor space has a noteworthy effect on people's thermal comfort (Knez & Thorsson, 2006; Lin, 2009). Like this, the feel of outdoor thermal comfort is determined not only by the ''state of body'' but also by the ''state of mind."

Nikolopoulou, et al., (2001) have conducted a study in Cambridge to assess the impact of the outdoor thermal comfort on people's behavior. They have interviewed a big number of people asking them to assess their thermal comfort in the outdoor space. The key finding of this study was that there was a significant contradiction between the subjective actual thermal comfort sensation of the interviewees and the objective predicted thermal comfort condition PMV, with only 35% of the interviewees' evaluation aligned with the PMV. This contradiction in results was linked to the psychological adaptation factors such as people past experience, perceived control, time of exposure, and their expectations of the weather. Similarly, Thorsson et al. (2004) have studied the influence of thermal bioclimatic conditions on the behavioral patterns of people in an urban park in Gothenburg. The most critical finding in this study is that the steady-state models such as PMV may not be appropriate for the assessment of short-run outdoor thermal comfort, because PMV may not consider either the transit exposure of people for sunny areas or the psychological adaptation. In another similar study in Germany, the research concluded that people's behavior is influenced by the outdoor thermal condition along with their individual expectations (Katzschner, 2006). It was noted that the people who are moving from air-conditioned buildings to the outdoor space are seeking sunshine even when the objective PET indicator exceeds neutral conditions. Another study explained the inconsistency caused by cultural and climatic differences, in which people's preferences of thermal comfort in terms of sun and shade are totally different between Japanese and Swedish people (Knez & Thorsson, 2006). Lin (2009) conducted her study in a hot and humid subtropical climate in Taichung City, Taiwan to uncover an interesting finding that the acceptable thermal range in Taichung City was 21.3–28.5 C PET, which was significantly higher than the European scale of 18–23 C PET. This indicates that people living in different climates have different thermal preferences.

Indead, the OTC assessment process has always been used during the previous years with very good attempts to assess the users' thermal comfort in the outdoor space. In order to understand the ongoing development in the assessment process, it is essential to go through the diligent previous attempts in this area of research.

TRADITIONAL METHODS OF OUTDOOR THERMAL COMFORT ASSESSMENT

The outdoor thermal comfort is a significant issue in designing open spaces, many pieces of research tackled this issue with different approaches. Previously, some studies have been carried out to investigate the effect of thermal comfort on people's behavior. The used methods were considered as simple in terms of procedures and results.

The people's behavior in the outdoor spaces was used for the first time as a mean to assess the thermal comfort in 1970, in which the impact of the microclimate on the outdoor activities was studied by counting people sitting on sunny and shady benches. The study concluded that the local sunny or shady conditions significantly impact the desire of people to either stay or leave (Gehl, 2011). The researchers since then started to study the link between the local microclimate and the usage level through multiple regression analyses and ANOVA tests (Zacharias, et al., 2001). In this area of knowledge, the microclimatic variables were demonstrated, in which the preponderant temperature and sun count for about 12% of the presence variance in studied open spaces, place counts for 38% and time of day counts for 7%. There was also a strong linear relationship between sitting behavior and air temperature. Notably, the authors anticipated the usefulness of microclimatic considerations in predicting the presence of people in site planning processes. However, they also noted that presence of people did not necessarily imply their satisfaction and suggested that the perception of comfort, as complementary to the levels of presence and activity type, should be considered if refinements of space design standards are achieved. In this regard, the major drawback of their study is the lack of a physiological analysis of the sensational thermal responses of the subjects.

Applying interviews with users to examine thermal comfort in outdoor spaces was a common method that was widely used. A study investigated a diurnal usage pattern of outdoor spaces in a neighborhood square in Athens through observing and interviewing 1503 individuals using a statistical approach (Nikolopoulou & Lykoudis, 2007). The authors of the present work notably considered the socioeconomic characteristics of ''popular locations'' in interpreting usage variations (e.g., distinctions among kiosk, coffee shop, and playground). The use of space was presented as a function of various meteorological parameters; air temperature and solar radiation were found to be the dominant factors affecting the use of space. However, the interesting part of this method is the distinguishable pattern of the influence of microclimate on sites with different functions. The example is given to the presence of people in relation to the sun).

Based on the previous discussion, the thermal behavior is assessed through different dimensions. One of these dimensions is the thermoregulation oriented, and the other one is the socially based dimension. It is clear that the traditional method serves the social dimension of the assessment process. All of these traditional research methods represented serious attempts to assess the thermal performance of the outdoor space, which follows the design and implementation phases, in this spot the designer can only improve the current situation as much as possible. Moreover, the traditional methods have been criticized for dependency on people's subjective perceptions and responses to the thermal environment that is almost various and not yet well understood.

However, by the increasing complexity of the urban environment and the excessive change in the climate, the previous traditional methods used for studying outdoor thermal comfort are not enough or effective anymore. There is a growing need for comprehensive design tools that can be used by developers and architects to assess thermal comfort and assist in the design process. The rapid advance in technology and software in the fields of urban climatology and biometeorology that happened in the past decade enabled performing detailed microclimatic analysis and thermal comfort assessments through a vast number of research projects in various climates around the world (Chen & Ng, 2012).

URBAN RECENT CHALLENGES

Due to the accelerated global growth of cities in the last century, the urban challenges are increasing especially with regard to the environment and the impact of urbanization on it, which certainly affect the human thermal comfort. The accelerated urban development, the reduction of green spaces, the great dependency on fossil fuel in many industries, electricity generation and the increase in heat discharge from vehicles and buildings have contributed to urban sickness (Pastore, et al., 2017). The degradation of the urban thermal environment is one of the urban sickness symptoms and has become a serious problem that needs global attention.

This led to worsening the urban thermal environment and rose the air temperature up to 2oC over the last century (Chen, et al., 2009). The Urban Heat Island (UHI) phenomena is another urban sickness symptom that is characterized by the rise of temperature in the built areas than the surrounding rural areas by 1oC–3oC. It negatively affects the outdoor thermal comfort by using paving materials that retain heat. The heat generated by energy usage is another contributor (Yang, 2013). In addition to affecting the outdoor thermal comfort, it affects the energy cost and the urban life quality (Pastore, et al., 2017) significantly. It is expected for UHI phenomena to increase in intensity as a result of global warming and the growth of urbanization (Guattari, et al., 2018).

It has become inevitable to reduce the effects of UHI in order to improve the outdoor thermal environment (Lin, et al., 2008). The urban designers can contribute to the mitigation of such phenomena through considering the thermal environment in their architectural and urban designs and implementing different strategies such as increasing the number of greening areas and using appropriate plants and trees that have a cooling effect on the urban environments. In order to reduce the effects of UHI and improve the outdoor thermal environment, it is necessary to adopt the environmental variables that affect the thermal behavior at the early stages in the design process through the use of quantifying measuring tools (Srivanit & Hokao, 2013; Guattari, et al., 2018).

DIGITAL SIMULATION

Simulation is defined theoretically as "the representation of the behavior or characteristics of one system by using computer programs designed for this purpose". The simulation research tools allow replicating real-world objects, forms, and settings. It allows to create a copy of reality and to test the impact of different variables without undergoing through the danger or risk of the real experiment (Groat & Wang, 2013). Using the digital simulation in urban studies field provides a quantitative and qualitative understanding of the environment thermal behavior (Chen & Ng, 2012). It can be utilized during the design process to generate different design alternatives that help to predict the impact of different variables on the thermal behavior of a space (Sousa, 2012).

For more than two decades, there has been a continuous and significant technological development, which helped and facilitated the expansion of research in the field of assessing thermal behavior (Zak, 2015). The digital simulation technology helped research to simulate the thermal comfort at different environments to provide certain references for future urban planning and spatial arrangement (Torrisi-Steele, et al., 2015). Digital Simulation has been highly recommended as outdoor design assistance. It contributes to multi-aspects of thermal comfort issues and helps in measuring the impact of changing design details on the outdoor thermal comfort as well as the impact of climate changes on the thermal comfort of persons staying outdoor.

The digital simulation method provides a complementary platform of integrated knowledge that enables designers to compare various design scenarios for thermal behavior and to establish the best thermal retrofitting measures to adopt in the urban space under analysis (Sousa, 2012). Digital simulation is a fast, effective research method, short research cycle, and clear visual display of the results. It saves manpower and material resources and can also fill the subjective assessment gap in the use of traditional methods (Yang, 2013).

Simulation Tools

The impact of urban microclimate significantly affects the thermal behavior of outdoor space. To assess the outdoor thermal comfort in urban spaces, several tools and software have been used to simulate a microclimate model (Fabbri, et al.,2015). Studying and analyzing the thermal behavior of space helps designers to take better decisions, increases users' satisfaction and improves their thermal comfort. The simulation tools can be used separately or combined to maximize their advantages. Understanding the capabilities of each tool and software helps in integrating them in the outdoor thermal performance analysis. Some examples of simulation tools are:

- **CityEngine,** this software is used to generate a detailed three-dimensional urban scene based on the geographic information system (GIS) databases and CityCAD software, to provide data reporting for urban master planning.
- **Umi** is integrated urban modeling that provides integrated operational energy, daylighting, outdoor comfort and walkability analysis of neighborhoods (Reinhart, et al., A., 2013). Umi software also includes a sustainable transportation module that concentrates on neighborhood walkability Walkscore. The Walkscore is the popular North American walkability metric. It is worth mentioning that the walkability is directly related to the outdoor thermal comfort, which can open new opportunities to integrate Umi in thermal behavior studies (Rakha & Reinhart, 2013).
- **Galapagos in Grasshopper** is combined with Walkscore to enhance land-use allocation in a neighborhood (Rakha, et al. 2013).
- **UrbanSim** is another tool used for modeling large-scale urban performance measures such as land use combined with transportation (Reinhart, et al., A., 2013). This tool and the Galapagos tool can provide a platform for other software to enhance the thermal comfort by allocating the sustainable land-use that conserves the natural resources and enhance the thermal behavior in terms of air temperature, wind fluency and speed, and humidity.
- **Energy-Plus,** is a simulation software that is mainly used in many studies for assessing the indoor thermal comfort. It can also be implemented in outdoor assessments. One of its outdoor applications was establishing a green roof module that allowed the Energy-Plus user to specify the outer layer of a rooftop construction. Accordingly, the user then can either accept default values or specify various parameters of the green roof construction (Sailor, 2008).
- **ENVI-met** Simulation Software is a Computational fluid dynamics numerical model that is capable of simulating complete built environment; surface-air-plant, thermal interactions and calculate air quality in urban structure and open spaces (Koerniawan, 2015), based on the principles of fluid dynamics, thermodynamics (Tariq & Arch, 2014). It can be used alone to simulate thermal com-

fort in open spaces or used with other programs. For example, to quantitatively assess the microclimate effects of different neighborhood and building retrofit scenarios, the simulation workflow moves from a micro-urban to a building scale assessment by linking the ENVI-met software to the building energy simulation program (Energy- Plus) (Pastore, et al., 2017).

- **Multi-Agent Systems (MAS)** software allows investigating the impact of the microclimate conditions on human thermal comfort on the level of the individual pedestrian. By way of pedestrians move through the urban environment, they are exposed to different microclimate conditions. With MAS it is possible to analyze the impact of such conditions on the thermal state and continuously monitor them using a simple transient 2-node model of the human thermoregulatory system. The simulation system integrates the biometeorological thermal comfort model ITCM, and the pedestrian traffic simulation model PedWalk. The required microscale meteorological data are taken from ENVI-met simulations (Bruse, 2009).

However, there is still a need to develop new modeling tools to expand the application of simulation programs from simulating the performance of a single building to model multiple buildings. This will guide urban planners and municipalities in their decisions at the planning level (He, et al., 2009). These tools should be developed to overcome the problems that are facing the process of modeling buildings, such as the problem of creating accurate models for simulating microclimate effects (Reinhart, et al., 2013). It would help as well to overcome the technical difficulties to enable users who lack the knowledge of heat transfer calculations to use the support tools easily (He, et al., 2009).

Simulation Process

The simulation process has steps to follow:

- The first step is the current state analysis, it starts by obtaining the Meteorological Data in the local micro- urban scale (e.g., temperature, wind speed, direct solar radiation) (Pastore, et al., 2017). This can be obtained from local microclimate stations, that are located in several sites within and out of the city. For achieving reliability, it is recommended to obtain those data from the closest station to assure that the data used matches the intended area climatic conditions (Guattari, et al., 2018).

Next, it is required to create an urban form through getting the input data that will reproduce the 3D urban forms including the coordinates through exporting AutoCAD or Google Sketchup files (Sousa, 2012) or through using the remote airborne mapping system (Huang, et al. 2005). The input data includes dimensions of the organizational structure, geometry and the materials used (Sousa, 2012).

The second step is the simulation is conducted on an urban area, in order to feed in the simulation conditions including all the variables that differ from scenario to another (Sousa, 2012). The designer proposes several places setting scenarios to explore and select the optimum scenario for implementation that would behave better in respect to thermal comfort.

The third step comes after running the software to check if there is any mismatch introduced in the variables, which in some cases could be automatically produced by the simulation software tool (Sousa, 2012).

THERMAL BEHAVIOR SIMULATION: APPLICATIONS AND CHALLENGES

In regard to the advantages of digital simulation, the researchers of the current work are eagerly desirous to its applications. However, there are some general areas of concern. These concerns reveal the limitations and challenges in addition to the unlimited potentials of simulation research. The followings describe these limitations.

The Accuracy of Simulation

Simulating the urban environment and weather condition requires high accuracy. Groat and Wang (2013) have introduced the simulation as a "projection" case for real life and not a "prediction". The projection case allows simulating an advanced replica of a specific sitting scenario in real life without the exposure to real risks or ethical barriers, which is not a prediction. When it comes to the people behavior and their physical adaptation, or the unexpected change in the weather condition, the pattern of possible behaviors could be projected in a realistically prepared replica of the actual circumstances, which increases the range of predictive outcomes.

Another challenge is the variation in thermal properties from point to another in the urban space, in which the thermal measures in a specific scenario are made at one point at street level in a specific time and location dealing with the urban space as a whole without considering the spatial, geographical, or climatic variations (Ali-Toudert & Mayer, 2006). The changes in any real setting affect the whole measurements, such as the wind speed, air temperature or the planting pattern. Therefore, to apply any change from real life to the simulated model, the whole situation should be re-evaluated (Lin, et al. (2008).

In terms of accuracy, the simulation process is going through a trial of imitating the realistic world, and this could not be completed unless an infinite number of variables is implemented. Despite the computational power, researchers are still struggling with the challenges of data availability and accuracy when assessing thermal behavior. The challenge that researchers face is if the input data they use to run a simulation is complete or covering all variables to fulfill the objectives of the simulation. In order to prove its certainty, the designer has to conduct multiple enactments. In addition, the data from the simulation can be triangulated with data yielded by other means for more robust results (Groat & Wang, 2013).

The Complexity of Variables

In the context of the outdoor environment, the huge amount of urban variables that affected the outdoor thermal comfort, and are involved in the assessment process, has become a real challenge for research studies in this area. Thus, the use of digital simulation provides a significant advantage over the regular field measurements and traditional methods in terms of the possibility of considering manifold variables and processes. In respect to the research development in the area of urban climatology, the digital simulation proves to be a perfectly suited methodology to deal with the complexities and non-linearities of urban climate systems (Ali-Toudert & Mayer, 2006).

Limitations of Tools

There are some challenges that face the use of simulation tools. Such tools are still not linked to different elements of the environment, such as vegetation, plant characteristics and types of soil. However, these elements have an important influence on the environmental conditions. The difficulty of simulating green roofs and living walls is another shortage of simulation tools (Pastore, et al., 2017). Recently, numerous researches in the field are trying to overcome these problems by integrating different simulation tools to have a comprehensive simulation for the outdoor environment (Lin, et al., 2008; Pastore, et al., 2017).

A case of tool limitation is ENVI-met simulation software that is capable of simulating a complete built environment based on the principles of fluid dynamics and thermodynamics (Koerniawan, 2015; Tariq & Arch, 2014). The results of ENVI-met are visualized as a 2D colored plan or section (Fabbri, et al., 2015). This case represents a shortage regarding creating a good graphic visualization of the results. In order to solve this problem a new tool called ENVI-BUG is used to obtain 3D output results. In addition, ENVI-BUG can accelerate the calculations of the software (Fabbri, et al., 2015). Another limitation is the cost limitation, it is a common challenge for applying simulation research, its equipment costs, professional actors, the spent time to track down numerous databases, in addition to the permissions required to access them (Groat & Wang, 2013).

Reliability

Reliability is a very important issue in research. It reflects the degree to which an assessment tool produces stable and consistent results. In simulation research, it is important to consider the reliable input data that is obtained from specific stations to assure that the results are truthful. The climate measures and their effects on human thermal comfort and air quality represent a critical issue. The inaccurate climatic measures might cause variations in thermal simulations results. Therefore, the climatic measures for simulating outdoor spaces and its thermal conditions have to be derived by means of a reliable database of meteorological information. Consequently, a Typical Meteorological Years (TMY), the measured climatic conditions for a specific location over a long time period (e.g., 10- 20- or 30-years) are used in the simulation research.

Reviewing the recent research, Guattari, et al., (2018) aimed to assess the effect of Urban Heat Island phenomena on the building in a dense neighborhood in Rome, and highlighted an important concern in the reliability aspect. The Meteorological Data that is used for simulation measurements is obtained mainly from airports which are usually far away from the built environment. In this study in Rome, the most populated Italian city, the researchers employed meteorological data derived from three different geographical locations for two years, with regard to temperature, wind velocity, and relative humidity. Significant differences have been noted between the data obtained from the two ear city airports and the station that is located in a dense neighborhood in Rome city. There were great differences among the locations especially for the wind speeds that were higher in Ciampino (near-airport) than the one in Rome.

The lower values of wind speed in Rome is due to the complex urban terrain and tall buildings that reduce the wind speed and flow. Regarding the Relative Humidity, the airport locations that are located on the Latium coastline close to the sea has relatively higher humidity levels during summer.

This points out the importance of the reliability of the weather-data files used in the simulation. The meteorological data obtained out of the urban context cannot be considered reliable for such locations. Using data from outside the urban context may lead to inaccurate simulation method. Instead, frequent updates are advised to avoid inaccuracy.

Validity

The use of validity is common in quantitative research; it proves truthful of research measurements and results (Golafshani, 2003). In the research of digital simulations, the researchers are facing the problem of lacking validity (Ali-Toudert & Mayer, 2006). However, Simulation software such as ENVI-met results is often validated by comparing with field measurement. By this means, improvements in the model could be reached (Elnabawi, et al., 2013).

Huang, et al. (2005) studied the actual status of an urban thermal environment in a complex urban area and covered a large district heating and cooling (DHC) system in Tokyo. Authors explained the procedures carried on to investigate the results validity, using both field measurements, and a numerical simulation program as below:

- The researchers investigated the actual status of the urban thermal environment, (e.g., Wind velocity, temperature, and humidity) within the built environment by means of field measurements;
- They develop a digital simulation for the urban thermal environment adapted to an unstructured computational grid which is proper for complex urban areas.
- The Spatial distributions of wind velocity, air temperature, and humidity are analyzed by the numerical simulation program.
- A comparison between the results of digital simulation and field measurements was carried out. The comparison confirmed the accuracy and validity of the simulation program.

In another study conducted by Sailor (2008) to assess the effect of roof garden on the indoor thermal comfort, the researcher started by implementing a simulation module in the building energy simulation program. In the next step to prove the method validity, the module was compared to the data obtained from the field study. Moreover, additional sensitivity experiments were conducted to explore the model response to variations in key input parameters such as exploring the effect of a thicker soil layer of the roof garden on energy saving and the effect of the increase of vegetation density on electricity usage, etc.

In addition, it has been discussed earlier in this chapter that the steady-state assessment methods as PET might produce overestimated values compared to people responses on their thermal comfort level and it is still difficult to interpret the meaning of one PET value for the thermal comfort of people accurately, especially in the outdoor spaces. Thus, in order to validate the simulation methods against subjective responses, it requires further sociological studies. The subjective interpretation of people may be influenced by acclimatization, expectation, or thermal history, which makes the issue of outdoor thermal comfort much more complex than described by simulation models (Ali-Toudert & Mayer, 2006).

Similarly, in a study about the cooling effect of greening on outdoor thermal comfort, the researcher used the on-site measurements from mobile and fixed weather stations, and they used the Envi-met model to verify the accuracy of the digital simulation. In addition, a paired t-test was used to determine if a significant difference occurred between the values measured on-site and the simulated values. No significant differences were found except that the solar irradiance was higher in the simulation module than the on-site measured results. The reason that the cloud-free sky condition was selected for field measurements as well as model simulation, as the spatial and temporal differences cannot be observed under existing cloud cover during the field survey periods. To overcome this challenge, the selection of a cloudless day was found to be more appropriate for simulating the maximum incoming solar irradiance in order to prove validation and accurately represented the real environments (Srivanit & Hokao, 2013).

CONCLUSIVE REMARK AND FUTURE DIRECTIONS

This book chapter aims to explore the potentials and limitations of the "Digital Simulation" as one of the methods that allow objective assessment. This chapter tackles several related issues and discusses researchers' recent attempts and contributions, which would help the readers to understand the pros and cons of using the digital simulation method in outdoor thermal comfort assessment and proposes possible solutions for several challenges.

The high potential of simulation has been explored, along with the limitations and challenges of simulation research as follows:

- The accuracy of the simulation is the main concern. Simulating the urban environment and weather condition gives an advance image of what may be the case in a specific sitting in real life that enables researchers to select the optimum design scenario. However, the replicated model might not consider the social behavior of people within the outdoor space and their natural abilities of physical adaptation. The researchers have to consider the social dimension of assessment by conducting interviews and user observation tactics. Moreover, simulation tools need more development in order to involve the pattern of possible user behaviors in thermal comfort assessment. In addition, testing different scenarios are essential to increase the range of predictions for the unexpected change in the weather condition or planting pattern.

Despite the computational power, the data availability and accuracy are still challenging the researchers' efforts to assess the thermal behavior. In order to prove its certainty, the designer has to conduct multiple enactments. In addition, the data from the simulation can be triangulated with data yielded by other means for more robust results.

- Regarding the complexity of variables, there is a need for a huge amount of urban variables that will be involved in the assessment process, which affect the outdoor thermal comfort. This is a real challenge for research in this area. Thus, the use of digital simulation provides a significant advantage over the regular field measurements and traditional methods in terms of the possibility of considering manifold variables and processes.
- Designers are challenged with the fact that the simulation tools are not linked to different elements of the environment (e.g., vegetation, plant characteristics and types of soil). To overcome this problem, the designers could integrate different simulation tools to have a comprehensive simulation for the environment.

Another software challenge is the shortage regarding creating a good graphic visualization of the simulation results. In order to solve this problem, new tools can be added to the original software to obtain a 3-D output result.

The high cost of software is another limitation due to the novelty of the subject. However, the initial cost of using this method will be cost effective in the long run when avoiding the post-retrofit to enhance the urban thermal environment after implementation and occupation.

- In order to ensure the reliability of input data, the data used must be obtained within the urban context or close to it. In addition, frequent updates are advised to avoid inaccuracy. Inaccurate climatic measurements result from data obtained out of the studied urban context, which result in misleading variations in thermal simulations results.

- The researchers are facing the problem of lacking validity in the simulation research. Accordingly, the simulation software results are often validated by being compared with field measurement. Moreover, additional sensitivity experiments can be conducted to investigate the model response to variations in key input parameters. By this means, improvements in the simulation model could be reached.

Finally, the steady-state assessment methods as PET might produce overestimated values compared to people responses on their thermal comfort level, and it is still difficult to interpret the meaning of one PET value for the thermal comfort of people especially in the outdoor spaces. The subjective interpretation of people may be influenced by acclimatization, expectation, or thermal history, which makes the issue of outdoor thermal comfort much more complex than described by simulation models.

As discussed in this chapter, the outdoor thermal behavior is assessed through two dimensions: the thermoregulation dimension (objective), and the socially based dimension (subjective). Although there is an urgent need to develop new modeling tools to expand the application of simulation programs in the area of outdoor thermal comfort to cope with the increase in the complexity of the urban environment and the expected excessive change in the climate in the coming years; however, further sociological traditional studies are highly needed to balance the objective measurements with the subjective responses to achieve a balanced assessment of thermal comfort for outdoor space design.

This can guide urban designers and municipalities in their decisions during the early stages of the design process by using the simulated model and examine the thermal comfort in the outdoor space using the traditional methods in the post-occupied stage, to provide socially successful open spaces in livable cities.

REFERENCES

Ali-Toudert, F., & Mayer, H. (2006). Numerical study on the effects of aspect ratio and orientation of an urban street canyon on outdoor thermal comfort in hot and dry climate. *Building and Environment*, *41*(2), 94–108. doi:10.1016/j.buildenv.2005.01.013

Asawa, T., Hoyano, A., & Nakaohkubo, K. (2004, September). Thermal design tool for outdoor space based on numerical simulation system using 3D-CAD. In *Proc. 21th Int. Conf. on PLEA* (Vol. 2, pp. 1013-1018). Academic Press.

Bruse, M. (2009, June). Analysing human outdoor thermal comfort and open space usage with the Multi-Agent System BOTworld. In *Seventh International Conference on Urban Climate ICUC-7* (*Vol. 29*). Academic Press.

Chen, H., Ooka, R., Huang, H., & Tsuchiya, T. (2009). Study on mitigation measures for outdoor thermal environment on present urban blocks in Tokyo using coupled simulation. *Building and Environment*, *44*(11), 2290–2299. doi:10.1016/j.buildenv.2009.03.012

Chen, L., & Ng, E. (2012). Outdoor thermal comfort and outdoor activities: A review of research in the past decade. *Cities (London, England)*, *29*(2), 118–125. doi:10.1016/j.cities.2011.08.006

Elnabawi, M. H., Hamza, N., & Dudek, S. (2013, August). *Use and evaluation of the ENVI-met model for two different urban forms in Cairo, Egypt: measurements and model simulations*. In *13th Conference of International Building Performance Simulation Association*, Chambéry, France.

Fabbri, K., Di Nunzio, A., Antonini, E., & Boeri, A. (2015). *Outdoor Comfort: the ENVI-BUG tool to evaluate PMV values point by point*. Building Simulation Applications BSA.

Gehl, J. (2011). *Life between buildings: Using public space*. Island Press.

Golafshani, N. (2003). Understanding reliability and validity in qualitative research. *Qualitative Report*, *8*(4), 597–606.

Groat, L. N., & Wang, D. (2013). *Architectural research methods*. John Wiley & Sons.

Guattari, C., Evangelisti, L., & Balaras, C. A. (2018). On the assessment of urban heat island phenomenon and its effects on building energy performance: A case study of Rome (Italy). *Energy and Building*, *158*, 605–615. doi:10.1016/j.enbuild.2017.10.050

He, J., Hoyano, A., & Asawa, T. (2009). A numerical simulation tool for predicting the impact of outdoor thermal environment on building energy performance. *Applied Energy*, *86*(9), 1596–1605. doi:10.1016/j.apenergy.2008.12.034

Huang, H., Ooka, R., & Kato, S. (2005). Urban thermal environment measurements and numerical simulation for an actual complex urban area covering a large district heating and cooling system in summer. *Atmospheric Environment*, *39*(34), 6362–6375. doi:10.1016/j.atmosenv.2005.07.018

Katzschner, L., Bosch, U., & Röttgen, M. (2006, September). Behaviour of people in open spaces in dependence of thermal comfort conditions. *Proceedings of 23rd Conference on Passive and Low Energy Architecture*.

Knez, I., & Thorsson, S. (2006). Influences of culture and environmental attitude on thermal, emotional and perceptual evaluations of a public square. *International Journal of Biometeorology*, *50*(5), 258–268. doi:10.100700484-006-0024-0 PMID:16541241

Koerniawan, M. (2015). *The Simulation Study of Thermal Comfort in Urban Open Spaces of Commercial Area Using EnviMet Software*. Academic Press.

Lin, B., Li, X., Zhu, Y., & Qin, Y. (2008). Numerical simulation studies of the different vegetation patterns' effects on outdoor pedestrian thermal comfort. *Journal of Wind Engineering and Industrial Aerodynamics*, *96*(10-11), 1707–1718. doi:10.1016/j.jweia.2008.02.006

Lin, T. P. (2009). Thermal perception, adaptation, and attendance in a public square in hot and humid regions. *Building and Environment*, *44*(10), 2017–2026. doi:10.1016/j.buildenv.2009.02.004

Nikolopoulou, M., & Lykoudis, S. (2007). Use of outdoor spaces and microclimate in a Mediterranean urban area. *Building and Environment*, *42*(10), 3691–3707. doi:10.1016/j.buildenv.2006.09.008

Nikolopoulou, M.-H., Baker, N., & Steemers, K. (2001). Thermal Comfort in Outdoor Urban Spaces: Understanding the Human Parameter. *Solar Energy*, *70*(3), 227–235. doi:10.1016/S0038-092X(00)00093-1

Park, M., Hagishima, A., Tanimoto, J., & Narita, K. I. (2012). Effect of urban vegetation on outdoor thermal environment: Field measurement at a scale model site. *Building and Environment, 56*, 38–46. doi:10.1016/j.buildenv.2012.02.015

Pastore, L., Corrao, R., & Heiselberg, P. K. (2017). The effects of vegetation on indoor thermal comfort: The application of a multi-scale simulation methodology on a residential neighborhood renovation case study. *Energy and Building, 146*, 1–11. doi:10.1016/j.enbuild.2017.04.022

Reinhart, C. F., Dogan, T., Jakubiec, J. A., Rakha, T., & Sang, A. (2013, August). *Umi-an urban simulation environment for building energy use, daylighting and walkability*. In *13th Conference of International Building Performance Simulation Association*, Chambery, France.

Sailor, D. J. (2008). A green roof model for building energy simulation programs. *Energy and Buildings, 40*(8), 1466-1478.

Sousa, J. (2012, September). *Energy simulation software for buildings: review and comparison*. In International Workshop on Information Technology for Energy Applicatons-IT4Energy, Lisabo.

Srivanit, M., & Hokao, K. (2013). Evaluating the cooling effects of greening for improving the outdoor thermal environment at an institutional campus in the summer. *Building and Environment, 66*, 158–172. doi:10.1016/j.buildenv.2013.04.012

Tariq, T., & Arch, M. (2014). *An ENVI-met Simulation Study on Urban Open Spaces of Dhaka*. Academic Press.

Thorsson, S., Lindqvist, M., & Lindqvist, S. (2004). Thermal bioclimatic conditions and patterns of behaviour in an urban park in Göteborg, Sweden. *International Journal of Biometeorology, 48*(3), 149–156. doi:10.100700484-003-0189-8 PMID:12955614

Torrisi-Steele, G., Wang, V. C., Sedivy-Benton, A., & Boden-McGill, C. J. (2015). Research Goes Digital: Some Methods, Frameworks, and Issues. *The Reference Librarian, 56*(4), 239–258. doi:10.1080/02763877.2015.1057467

Yang, L. (2013). Research of urban thermal environment based on digital technologies. *Nature Environment and Pollution Technology, 12*(4), 645.

Zacharias, J., Stathopoulos, T., & Wu, H. (2001). Microclimate and downtown open space activity. *Environment and Behavior, 33*(2), 296–315. doi:10.1177/0013916501332008

Žák, Š. (2015). The identification of innovative research methods and techniques utilized in marketing research in the digital era. *Studia commercialia Bratislavensia, 8*(29), 139-152.

Chapter 4
Internet of Things Interpreting a Myth:
Internally Displaced Persons Transform Cities of Hardship Into Chaos Cities

Hisham Abusaada
https://orcid.org/0000-0001-6530-7714
Housing and Building National Research Center (HBRC), Egypt

Abeer Elshater
https://orcid.org/0000-0002-5061-6861
Ain Shams University, Egypt

ABSTRACT

The livability standard still has not considered the chaos city that may stem from or lead to cities of hardship. This chapter rectifies this by making the phenomena of chaos and hardship the centerpiece of the analysis. It depends on the internally displaced persons (IDPs) to display the characteristics of liability and the hardship of living and be the indicators of chaos city. This chapter addresses the non-perceptible processes of the IDPs from outside and inside Cairo in Egypt. This internal displacement supposes the lead-in to chaotic changes in the lifestyles of the cities; it can even be said that they become cities of hardship. The theoretical reading depends on conventional and digital methods (content analysis and the internet of things) to follow these changes, which occur not only due to migrations but also due to ignoring decentralization. The outcomes provide an action plan to create cities free from hardship, displacement, and chaos.

DOI: 10.4018/978-1-5225-9238-9.ch004

INTRODUCTION

Over the last four decades of the twentieth century, Western literature discussed the issues of the "liveable cities" (Lash, 1976; Lennard & Lennard, 1995; Casselati, 1997; Hahlweg, 1997; Salzano, 1997; Vuchic, 1999) and the "quality of life" (Pacione, 1990; Ley & Newton, 2010; Okulicz-Kozaryn, 2013; Kraftl, 2014; European Environmental Agency (EEA), 2009; Mohit & Iyanda, 2016). At the beginning of the 2000s, the terms "material hardship" and "chaos city" have referred to the opposing nature of liveability and quality of life. The first focuses on inequality in the provision of minimum goods to all citizens to achieve a decent living; it leads to extreme poverty (Allen & Browne, 2010; Pilkauska, Currie, & Garfinkel, 2012; Nelson, 2011). The second indicates to city-life systems (Robinson J., 2000; Pulsellic, Ratti, & Tiezzi, 2011).

Urban chaos and visual or spatial chaos are supposed to cause cities of hardship, which seem difficult to live in (Mumford, 1938; Lynch, 1960; Jacobs, 1961). They have been characterised by the deterioration of the societal situation economically and socially and of the politics of many traditional cities in the developing world. The hardship of living remains one of the main reasons that people are disadvantaged economically and lack the equality needed to flee from their cities, which do not provide a good standard of living, particularly in economic and socio-cultural terms, so they try to settle in other cities located in the same country. In addition, the hardship makes them move from one place to another within the same city to provide a better standard of living. Those people, known as the IDPs, not only become part of the cities' poverty belts (Carrillo, 2009, p. 546) but may also lead to chaos city, which is increasing the hardship of living. Now, the questions are as follow: Does the hardship of living cause chaos city or vice-versa? Is there a real role for the IDPs in influencing all these phenomena? Is there a relationship between hardship, chaos and the concepts of Liveability?

The fundamental problem that faces us is the chaos displaced persons cause in the city; few studies claim that the displaced persons in the host communities cause chaos (Fan, 1999; Fielden, 2008; Carrillo, 2009). Some scholars believe that urban chaos occurs for two reasons (Abusaada, 2016; Abusaada & Elshater, 2017; Abusaada & Elshater, 2019). The first reason is the policies of governmental centralisation that chooses projects that generate substantial economic returns in the capital and large cities, depriving the smaller towns of these projects[1]. These policies produced cities of hardship, and consequently, forced some citizens to be displaced from their primary cities (Downing, 2002; World Economic Forum, 2017, pp. 16-17). The second reason is the non-enforcement of laws in large cities, mainly in traffic congestion[2], particularly *non-adherence to traffic regulations* and *traffic mismanagement* (Kiunsi, 2013, p. 95; Frances Agyapong, 2018, p. 87), as well as the street children, street vendors and marginalised people (Sande-Friedman, 2013; Abusaada & Elshater, 2019). To the best of our knowledge, no study has yielded actual results linking the three causes of the three elements of the problem (cities of hardship, the IDPs, chaos city). To illuminate this undiscovered area, we investigate this issue in the Egyptian context by using content analysis and the Internet of Things (IoT) to interpret the big data analysis on the internet.

Thus, this chapter is designed to explore the mutual link between cities of hardship and cities of chaos as it relates to understanding the phenomenon of internal displacement in the Egyptian context during the 30-year 1981–2011 period. It includes the period that promoted the idea of economic liberalism and the establishment of major cost-effective projects (Weiss & Wurzel, 1998; Dobronogov & Iqbal, 2005). This chapter aims at extending the debate on situations of chaos regarding the concepts of liveability and hardship. The ultimate goal of this chapter is to propose an action plan which involves a reduction of the chaos in a city. This action plan advocates decentralisation[3], supports the empowerment of people and sees solutions in many phases of the levels of regional, national, urban planning and neighbourhood unit development.

Since1980, due to *stagnant economies and inefficient central bureaucracies*, decentralisation has been one of the best solutions to our problems in developing countries (Schneider, 2003, p. 34). Decentralisation also improves opportunities for talented people to solve problems, provides time for higher level executive authorities, provides solutions that are in line with actual issues and helps in expanding democratic institutions (The World Bank, 2013; Swartz, 2012). In addition, decentralisation facilitates *matching resources with citizen needs more precisely and cost-effectively* (Faguet, Fox, & Poeschl, 2014, p. 16). It encourages poor and vulnerable people to participate in work and in decision-making; *participation and empowerment of civil society is a core element of good governance* (United Nations, 2005, p. 21; Faguet, Fox, & Poeschl, 2014). Notably, the empowered groups include *historically excluded groups such as women, poor people, and racial, ethnic, sexual and religious minorities* (IDRC, CDDI, UN-HABITAT, 2008, p. 3).

The current chapter adds to the literature that has yielded mixed outcomes for the impact of the IDPs on the host communities. Not only do we offer an attempt to end the myth that the displaced are the only cause of the chaos of the city, but we also provide an action plan to create cities free from hardship, displacement, and chaos based on identifying the priorities of urbanisation policy related to these three intricacies.

This chapter is structured as illustrated in Figure 1 as follows: after the introduction and background description, the first section highlights the definition of a chaotic city and explores internal displacement as a chaotic phenomenon. This section identifies internally displaced persons (IDPs) through a literature review along two axes. The first is to identify the causes of forced displacement, natural and human-made, and the places where displaced people are found: in the city centre, city fringe, or new settlements. The second is to determine IDPs' experiences of risk and examine the host community's related political, cultural and social attitudes. The second section describes the materials and methods, which includes identifying the reality of the relationship between IDPs and chaos and hardship in Cairo, Egypt, and the effect of the link between a city in chaos and internal displacement. The third and fourth sections present results, discussion and findings. The Fifth section highlights the characteristics of liveability versus hardship living as they are related to the indicators of a chaotic city. The sixth section addresses the public policy implications for resolving the phenomena of internal displacement, chaotic cities and cities of hardship. The concluding chapter consists of outcomes.

BACKGROUND

William Hogarth's illustrated document, *The Four Times of Day*, seems to be the first document to describe social chaos and human misery in the streets of London in 1730, just before the Industrial Revolution (Sande-Friedman, 2013). Chaos arose in the city, from Hogarth's view as an artist, from overcrowding, pollution, unhealthy places, vegetable and sugar vendors, drunken and slothful people, drug traffickers across roads and prostitutes in homes behind cafés. Immorality leads to poverty, misery and disease, which are visible in his portrayals of children eating remnants of food that have fallen from the windows as well as the distinction between the wealthy elite and the working class of artisans and laborers. Mumford used *solidification of chaos* in 1938 in his book *The Culture of City*, with urban

Figure 1. The structure of this chapter

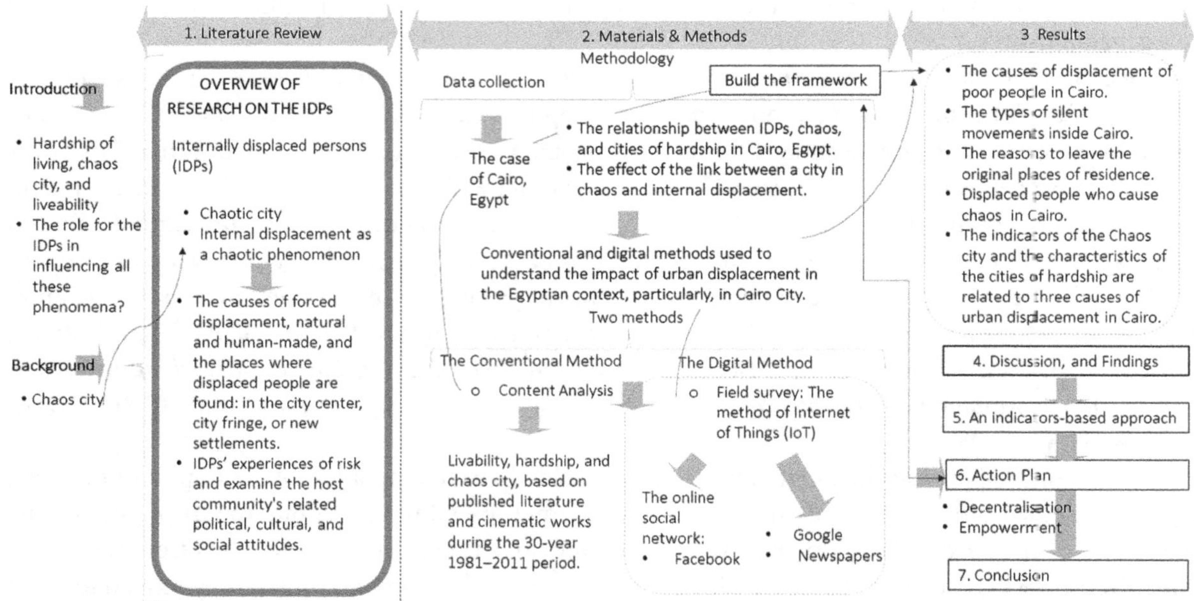

theorists in the 'Decentrist school' to describe the chaos in the American urban cities because of the *cultural rag-picking, ...every avenue was designed in accordance with the most inexorable of laws-the law of chance* (Mumford, 1938, p. 201). In the wartime, in the Greater London Plan 1944, prints were used consistently to *evidence urban chaos and suburban sprawl, and to provide a visual mapping of spatial planning* (Wakeman, 2016, p. 40).

In 1960, Kevin Lynch also used the terms urban chaos and visual or spatial chaos of the modern city to describe the perceptual form of the city (pp. 5, 42). In 1961, in Jacobs' book that entitled *The Death and Life of Great American Cities*, she pointed out that Lewis Mumford described the lined streets in New York's midtown is *solidified chaos* (p. 21). In addition, based on Catherine Bauer, Jacobs wrote: *the centres of cities amounted to 'a foreground of noise, dirt, beggars, souvenirs and shrill competitive advertising' (p. 21)*. Jacobs used 'chaos' and 'order' in contrary to describe the diversity of uses in the city, when she wrote *the intricate minglings of different uses in cities are not a form of chaos*, but it is *a complex and highly developed form of order* (p. 222). Jacobs also indicates that monotony and scenes of thoroughgoing sameness carry a deep disorder of conveying no direction. They are deeply confusing, and she wrote *this is a kind of chaos (p. 224)*. She added, chaos of cities *lies in the misconception about the relationship of cities with the rest of nature* (p. 243). Thus, in the field of urban planning and design, the chaos of the streets happens for two reasons; the first is the visual chaos and the second, which represents the subject of this article, is the behavioural chaos.

Chaos in architecture is *the absence of order, shape, or structure* (Johnson, 1994, p. 242), *the opposite of geometric order* (Rubinowicz, 2000, p. 197), and chaos and geometric order *exist in architecture together* and interfere one in the other (2000, pp. 197, 180), and it is *created by the natural transformations* (p. 202). In addition, both chaos and order exist in *the urban structure* (p. 205), discussed in the study of urban planning and design. Raymond Williams (1975) argued that the nineteenth century bequeathed the social inequalities and physical chaos in the city, and at the turn of the century the city represented the physical embodiment of a decisive these problems, such as housing and overcrowding (Beaufoy, 2010, p. 138). Chaos is referring to the imbalance of the performance of tasks and responsibilities: *a state of utter confusion* (Merriam Webster, 2018) and *the absence of order* in management (Çakar & Alakavuklar, 2011, p. 392).

In the social sciences, commentators use implicitly or explicitly the concept of chaos as an indicator to a political, social and economic phenomenon in the city (Weick, 1979; Gleick, 1987; Lorenz, 1993; Robinson J., 2000; Farazmand, 2003; Eriksson, Johnson, & Estep, 2004; Pulsellic, Ratti, & Tiezzi, 2011; Beauregard, 2013). According to Ali Farazmand (2003), chaos and transformation theories uses in systems analysis and futuristic studies, *social phenomena develop or evolve on strictly linear paths so that we can predict crises that would need to be avoided or controlled* (p. 349). This phenomenon is known as the "butterfly effect": *small changes in a system may produce consequences far beyond expected or planned outcomes* (p. 353). The concepts of *self-organization* and *attractors* is fundamental characteristics of chaos theory (pp. 353-354). Moreover, in social science, the concept *transformation* is used as *a preferred term to subsume chaos* (p. 355). In addition, today cities *have become the sites of structural transformation* (United Nations Human Settlements Programme, 2016, p. 35).

In some developing countries, mainly in Africa and Asia, many reports and journals indicate that the behaviour of people in their daily life in city streets are becoming terrible, which seems a state of organised chaos has been dominated (United Nations Human Settlements Programme, 2008; United Nations Human Settlements Programme, 2016; Tafirenyika, 2016). The causes of this chaos seem to have enough order to achieve limited goals. This chaotic situation not only can be limited in city-life systems, regarding the change of the style of architecture, which builds for accommodations, shopping, and working, but also appeared in traffic congestion, air pollution, noise, street children and street vendors. Notably, Jakarta and Bangkok and Cairo have *become famous for its traffic congestion, noise, and pollution* (United Nations Human Settlements Programme, 2008, p. 56). *Mega-cities such as Cairo suffer particularly from high levels of noise and air pollution* (p. 57).

Many literatures indicate that the chaotic changes are due to people displaced from rural to urban areas (Guthrie, 2014; Abusaada, 2016; Abusaada & Elshater, 2017). The problem of this chapter seeks to search about the reality of this statement and prove that the chaos of behaviour in the streets of Cairo is due to two factors. The displacement represents the first factor, and the second is the absence of discipline and the presence of thugs; hence, *urban legislation must be enforced, not just enacted* (United Nations Human Settlements Programme, 2016, p. 25).

Chaos has become an accepted concept in the field of humanities *to explore the implications of chaos theory for their disciplines* (Mosko & Damon, 2005, p. 7). It is the natural order of things; minor changes happen lead to massive changes over time (Lorenz, 1993). These gives definitive clues to what could happen in the future, including a glimpse into the underlying systems at work. This is different from randomness, wherein it is impossible to predict it's the next step. Often, chaos seems to be a random response, but it is not. Further, it does not adjust the models, which are considering the chance to reach completeness. Consequently, this phenomenon could be including street vendors, marginalised people, urban displacement, spreading of informal settlements, unruly traffic, and building disturbances each one different from any other (Abusaada & Elshater, 2019). In this vein, the internal displacement could be classified as a chaotic movement having the following:

- Characterised by the repetition, the ability to predict the future and to be counted by statistics.
- Carried the complicated features that lead to increasing the diversity in the irregular shapes, evolving with time and its future is related precisely to the past.

Now, the relationship between IDPs and chaos depends on the answers of three questions; why the people flee from their homeland? Which places in the city are they choosing to be their new place of living? What are the risks of displacement about the IDPs themselves and the host communities?

OVERVIEW OF RESEARCH ON THE IDPs

The number of IDPs in the year 2017 estimated about 30.6 million displaced people, *the equivalent of more than 80,000 people displaced each day; and Africa accounted for 5.5 million* (Internal Displacement Monitoring Centre, 2018), most of them live in the urban areas. Notably, *one in every 122 humans is now either a refugee, internally displaced, or seeking asylum* (United Nations High Commissioner for Refugees (UNHCR), 2015). Urban displacement includes refugees and IDPs; the first are fleeing between the nations due to natural and human-made disasters, keeping of the personal rights and international stability. The second describes people who are moving within the borders of their country and dependent the legal protection of their governments. Many instantiations defined the IDPs (The Commission on Human Rights and the General Assembly, 1998; The UN Refugee Agency, 2007; United Nations Human Settlements Programme, 2016).

According to the guiding principles of the United Nations (1998), Alexandra Fielden (2008) and Donald Brown; Camillo Boano; Janani Vivekananda; Julian Walker (2015) IDPs are the persons or groups of persons forced or obliged to flee their homes or places of habitual residence (be it rural or urban, at home or abroad), remain in their country under the legal protection of their government. Fielden (2008) indicated the ignorance of UNHCR to the *natural disasters or development activities.* She sees *those persons who subject to discrimination and human rights violations* [during] *their displacement (p.2).* In 2007, UNHCR classified IDPs as *probably the largest group of vulnerable people in the world* (The UN Refugee Agency, 2007, p. 4).

The IDPs phenomenon is focused on the slum areas, the predominance of an informal economy, lack of access to a decent living, females who find themselves in lives without breadwinners, homelessness, and marginalization (health risks and disruption of formal educational activities) (Downing, 2002, p. 11). In addition, the IDPs are the negative aspect of the process of gentrification (Byrne, 2003; Beauregard, 2013), which is known as a *chaotic process* (Benediktsson, Lambert, & Larsen, 2015; Bunio-Mroczek, 2017), a spatial and social practice, and a tool for measuring the physical declining of urban neighbourhood . People are called IDPs under two conditions: compulsion or moving within the borders of their state. The IDPs circumscribed six group of reason, as shown in Figure 2; to avoid natural and human-made disasters, the consequences of the political, social, and the developmental of rapid urbanization. Besides, the sever living conditions, such as extreme poverty and the lack of services and facilities. The last two reasons are the scope of this work, which needs to search about solutions for the IDPs problems depends on decentralisation and empowerment[4].

The IDPs due to the fleeing from the severe living conditions and material hardship are at the key. Inequality and the economic downturns lead to lost income and the growth of extreme poverty as a global phenomenon; it leads to material hardship (Castells, 2010; Nelson, 2011; Pilkauska, Currie, &

Figure 2. Six reasons for IDPs
(The Commission on Human Rights and the General Assembly, 1998; Robinson W. C., 2003; Mosel & Jackson, 2013; Beauregard, 2013; Brown, Boano, Vivekananda, & Walker, 2015; Benediktsson, Lambert, & Larsen, 2015; World Economic Forum, 2017)

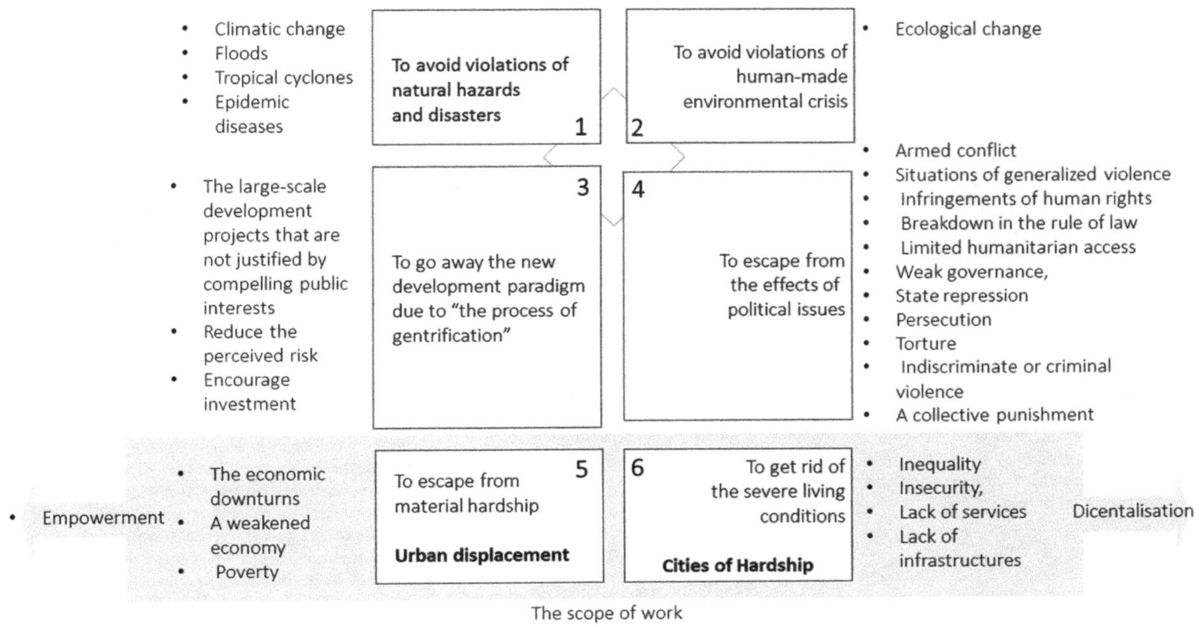

Garfinkel, 2012), which is defined as *a consumption-based indicator of economic well-being;* mainly *the foregone consumption that threaten health and well-being, such as going without food, housing, or needed medical care* (p. 403).

Understandably, it's hard to classify poverty as a humanitarian disaster in itself—because it is fatalistic—but human decisions influence it—which leads to vulnerability when dealing with any disasters. Poverty does not be the only and decisive factor in the urban displacement processes. Due to aggravating the manifestation and notices of inequality and social exclusion, poverty pushes quite a few people to leave their communities in the environment that has had less than the level of urbanity with the worst of life's hardships. These people not only financially incapable but also aspire to a better life. They tend to live in the major cities to get the best opportunities (Abusaada, 2016).

Poverty drives informal urban settlers *to live in unsuitable locations,* such as unhealthy, prone to natural hazards, flimsy makeshift homes, little access to basic services and sanitation). In addition, informal urban settlers are *noticeificant proportion of those affected and displaced* (McCallin, Scherer, & Duyne, 2015, p. 12). According to the United Nations Human Settlements Programme (2016), cities that are not ready for urbanization face challenges, such as slums, poverty, and the predominance of the informal economy, besides the difficult conditions for youth, women, and child labour exploitation *(p.144).* The displacement occurred because of severe state repression, deprivation of basic needs (Boswell, 2002, p. 7). Displaced families from the poor cities *seek to settle in cities with greater institutional capacity for getting aide and service* (Carrillo, 2009, p. 3).

The reasons that lead most poor people to be displaced are due to their geographic proximity, possible familiarity with the location, cultural ties, or familial and personal relationships. Besides, opportunity

Figure 3. Types of IDPs
(Fielden, 2008; Carrillo, 2009; Abusaada, 2016; World Economic Forum, 2017, pp. 15-16)

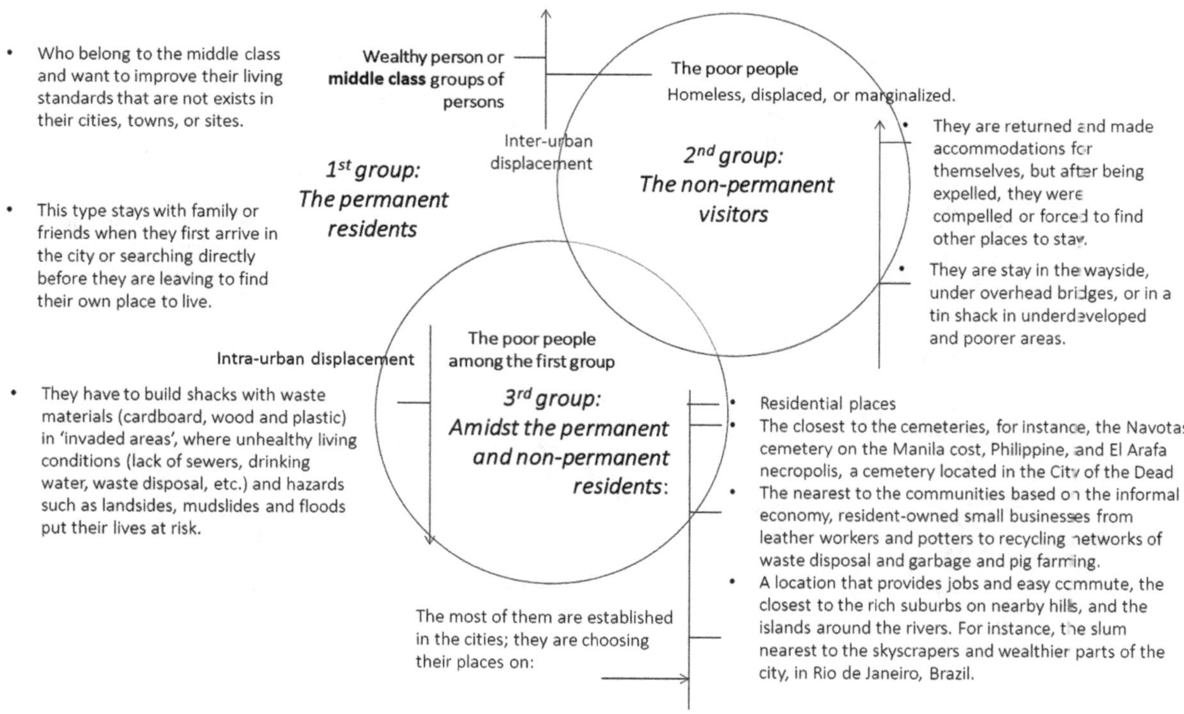

for work, income generation and better education (Humanitarian Fact Sheet, 2016). Based on 'severe living conditions' and 'material hardship', Figure 3 shows two types of IDP residencies: (a) inter-urban displacement, which involves the permanent residents who move between the cities and stay for extended times; and (b) intra-urban displacement, which includes the permanent and non-permanent residents, whom they are running continuously inside the city, particularly between the sites.

In this vein, the risk of the IDPs is their focus on the temporary or permanent residents who move to get rid of the severe living conditions, the poor, the uneducated and the illiterate who are living in the informal settlements and on the wayside. The movement of the IDPs dismantles living patterns, social continuity, and existing modes of production; this change has profound humanitarian, and economic and social consequences emerge in landlessness, joblessness, homelessness, marginalization, food insecurity, increased morbidity and mortality, and decay in income and livelihood levels (Muggah, 2000; Downing, 2002; Carrillo, 2009; Ruiz & Vargas-Silva, 2014). These risks have two-sides.

First: the problems of the IDPs themselves are:

- *The political attitudes*: violation of civil or human rights, and the risk of communal violence (Robinson W. C., 2003, p. 11).
- *The economic attitudes:* the people lose human-made capital, the ability to use earlier acquired skills in the new locations, the excellent production systems, the local labour markets due to that trade linkages between producers and their consumer base crashed, and to access to public and community services. Moreover, the informal economy emerged which are caused in the illegal

substance, weapons trafficking, prostitution, gambling, production, traffic, and distribution of a synthetic drug, manufacture of fake goods, in addition to income insecurity.

- *The social attitudes*: the IDPs lose the shelter, a quintessential person of the family, and the confidence in society. In addition, the social consequences related to the emotional impact destroy the social fabric which is cutting the physical and psychological ties with the past and cutting the roots of cultural identity, including six issues (Robinson W. C., 2003, pp. 11-15; Carrillo, 2009, pp. 540-541):
    - The loss of the importance,
    - the role of formal and informal associations,
    - the self-organised services,
    - leaders of traditional management systems; the family structure,
    - gender roles and identities; the coerced abandonment of symbolic markers (ancestral shrines and graves), and finally spatial contexts (mountains and rivers considered holy, or sacred trails).
    - Besides, the absence of social mobility, protection of labour rights, and the legal contracts.

Second: the problems of the IDPs toward the host community are (López, Arredondo, & Salcedo, 2011; Davies, 2012; Asplet, 2013)[5]:

- *The encounters and aspirations:* IDPs deal with the city according to their traditions, without regard to the rights and duties imposed by the nature of the host community of the major cities. Based on the intentions of adhering to the traditions, IDPs arrive at the major cities support limited implicit knowledge about the new urban areas, as well as varying degrees of experience about urban residents. Moreover, due to the low level of education, they are jumping with their intellectual beliefs which linked with their former lifestyles. Besides, IDPs refuse to relinquish their habits and thus behave unpleasantly. They believe that adhering to their traditions is a matter related to their social prestige. Therefore, they try to push the city to yield to their desires.
- *The expansion of urbanization:* without following clear instructions and determinations.
- *The non-discrimination*: the IDPs and the people of host community should have the same right to the liberty of movement and freedom to choose his or her residence and prefer to live in host families have the equal rights to protection, security, care, and equality (Davies, 2012, p. 10).
- *The risks of creating "new poverty"* (Downing, 2002, pp. 8-12): Landlessness, Joblessness, Homelessness, in addition, emerging the impacts on reduced the employment rates of natives, the greatest impact on the informal sector, decrease in real wages, and on females working in the informal sector (Ruiz & Vargas-Silva, 2014, p. 780).

MATERIALS AND METHODS

Arriving at the process of ending the hardships of living concerns as a prerequisite to preventing chaos city necessitates the need to extend the scope of conventional and digital research methods. The methodology applied, in this case, was primarily structured into two phases:

1. The first phase is to collect the data of the Egyptian context during the 30-year 1981–2011 period, which uses the conventional and digital methods to understand the impact of urban displacement and the real causes of urban chaos in Cairo City. The data collection methods are content analysis and field survey which depend on the method of Internet of Things (IoT).
2. The second phase is to build the theoretical framework uses the previous data to develop an action plan for the definition of local government authorities, how to solve the problems of IDPs, the hardships of living, and chaos city in traditional settlements. It is also aiming to investigate the chaos city in the current context and present the guidelines to solve the behavioural problems in the cities with hardship.

The first conventional method is content analysis, which focuses on the interrelationships between the IDPs and chaos in the cities of hardship.

- *The first review* is started with three Egyptian films for Youssef Chahine, "Al-Qahera Munawara be-Ahlaha [Cairo is enlightened with its people) or Cairo as Told]" (1981), "Hena Maysara [In Better Times]" (2006), and "Heya fawda…? [It is a mess...?] (2007)".
- *The second review* includes an article entitled "Critical Commentary, Cairo's Downtown Imagined: Dubaization or Nostalgia?" for Mona Abaza (2011), and a report published by the UN-Habitat (2015), prepared by Yahia Shawkat, Omnia Khalil, and Ahmed Zaazaa.
- *Finally, the last review* for a work which is developed by Hisham Abusaada and Abeer Elshater (2017) in the book entitled *Back to the Order: The Chaos in the Cities of Hardship*: which are focused on three questions: 1) Is there chaos in the city where you live? 2) Are IDPs the reason for the chaos city? 3) Does the city or site where you live in it feel you have the hardship or live-ability? In this book, the sample was varied from street children to homeless residents and persons residing in huts made of tin. There were also taxi drivers, passengers of mass transit, janitors, or porters, as well as faculty members at universities and research centres. It focused on ordinary people who attend scientific symposia and wedding parties, as well as individuals who live in places where wealthy people meet, such as the closed residential complexes (compounds, resorts, beaches, and sports and social clubs). An impression is concerning a significant part of Cairo city started with a displaced person from rural to urban areas.

This study adopts the Internet of Things (IoT) as a practical digital method depends on four steps: data acquisition, information extraction, knowledge extraction, and action-taking (L'Heureux, Grolinger, Higashino, & Capretz, 2017, p. 74). *The IoT is a novel paradigm in the scenario of the modern wireless telecommunications* (Atzori, Iera, & Morabito, 2010, p. 2787); enables connection and communication among real-world objects, environments, software, and people. It is related to *the pervasive presence around us of a variety of things or objects: Radio-Frequency Identification (RFID) tags, sensors, actuators, and mobile phones* (p. 2787). Through a unique addressing scheme, the IoT can make interact of one of those things or objects with each other and cooperate with their neighbours to reach common goals (Atzori, Iera, & Morabito, 2010; Giusto, Iera, Morabito, & Atzori, 2010). This network of things is capturing and exchanging enormous amounts of data and fuelling the Big Data movement.

In this article, the method of IoT focuses on two axes:

- *The first axis used the IoT based on Google Search Engine*; it searched for the three keywords "IDPs", "chaos", and "hardship" on Google, and newspapers from Egypt were browsed for the Journalists Salama A Salama (2006), Rasha Geddah, (2017), and Hany Ghoraba (2018), published in the Egyptian Journals Al-Ahram Weekly.
- *The second digital pragmatic axis is used the IoT*; the majority post in *Facebook* refer to chaos city and cities of hardship. This axis depends on the online social network in Facebook[6]. Different answers can be received by asking the users questions. Significant knowledge can be collected by tracking and reading what everyone writes on his or her post, through which a view can be made on a specific issue and a range of the discrete problems. Presently, people are continuously writing about the hardships of living and chaos in the city on social networks. In the last six months, between January and July 2018, several pieces of data about these issues have been collected.

RESULTS

This section includes two groups of results related to understanding and documenting the real role for the IDPs in influencing the phenomena of hardship living and chaos in Cairo city.

The first group focuses on the conventional method for exploring the characteristics of urban displacements, cities of hardships, and the indicators of chaos city related to Cairo, Egypt are as follows:

- According to "New & Improved Website, Coming Soon" (2018) online report, an abstract of the film "Cairo as Told" showed the emerging Cairo scene as a chaotic mix of poverty, overcrowding, luxury, and increased religious extremism. This film addresses the issues of unemployment, Nile pollution, poor services, and human misery. The film "Hena Maysara" discusses the issue of homelessness among slum dwellers, particularly socially and economically marginalised people such as street children and homosexuals who suffer from the characteristics of urban hardship: harassment, rape, drug-use, bullying, living in rubbish dumps and working in nightclubs for a living. Finally, "It is a mess…?" is a film that discusses the types of political corruption embodied in direct repression, bribery, nepotism, rigging of elections, brute control of power, and sexual repression.
- Abaza (2011) used the word chaos to describe the fantastic disharmony which *the city's public space has been famed* in Cairo downtown. And, she has criticised the attempts to paint *the façades of buildings* to *restore the centre primarily (p. 1078).*
    - Many of new migrants from rural areas to the cities resorted to living in unplanned areas in informal housing that has established around the cities. In these areas, all houses not only built without licenses but also lacked proper infrastructure and sanitation.
    - The other type is the forcible displacement due to the violent removal of the of livelihoods. For instance, the slums of 'ishash al-turguman and around the Maspero area.
- The relationship between the IDPs, chaos city, and ccities of hardship are as follows:
    - The rich people have taken their resident places in the suburbs and new communities, where is there is no chaos. Their places in the large cities replaced by the inhabitants whose coming from the lowest civilised and least developed places, and they are bringing what brought of the native culture, then the chaos happened.

- Often, in the large cities the problem inherent in the management of the city due to the absence of laws and legislation governing and unethical conduct.
- The problem in the city is in the difficulty of how to earn money for living, so IDPs are moving continuously between locations for one purpose is to subsistence.
- IDPs in slum areas not only suffer from chaos but also nobody can describe or explain what happens in these places; it is only positioned to spend their lives.
- The strangers IDPs are the ones who cause the confusion that takes place. For females, the impacts of chaos city happen; whether in traditional or modern, historical, or contemporary places, and high or lower level of education, social class, and income due to the eccentric behaviour with them as chaos city under the name of harassment or bullying. It occurs from the people whose are not classified as permanent residents. Working females' feelings of the chaos is more than men because of their inability to withstand excessive congestion on the one hand, and their prior knowledge about the nature of the behaviour of some young men in open spaces, such as streets, squares, and public markets on the other hand. Females with small children, teenagers, and older women undergo the feeling of chaos more than other.

The second group of results *based on analysis the data in* the *Facebook* and *Google Search Engine* are as follows:

- The causes of chaos city are due to traffic congestion, people overcrowding, waste on the roads, street children, street vendors, the owners of mass-transit cars that are medium (microbuses) and small sized (tuk-tuks).
- The chaos of roads and accidents because of entering the vehicle and driving in opposite of the normal direction; twenty hours after the publication of the Post, 377 comments were received. Many comments include discussion and examples about street chaos (Abada, 2018).
- The forced removal of buildings of historical value on 26 July Street in the centre of the country occupies a significant proportion of the comments provided by specialists of architecture and urbanization and many of the ordinary people (sokan26july, 2018; Abusaada, 2018).
- In Upper Egypt and some governorates of the Delta, Poverty rates is high, many households are not connected to sewage and may not have easy access to drinking water. Accordingly, families move from rural to urban areas in search of a better living for them and their children.
- Egyptian newspapers use the word "migration" and do not use the term "human displacement" regarding the person or the group of persons who are displaced from the rural to urban areas.
- Some of these aspects relate to chaos because of what can know the characteristics of cities of hardships due to poor planning, mismanagement, and inadequate municipal financial resources, including tax collection and the inability of cities to provide a good life for expatriates.
- Chaos in urban areas represented in homelessness children, traffic congestion, a lack of parking lots, street vendors, advertising signs, garbage, the dumping of waste, sculptures, other artwork, the random growth of cities, informal sectors, the interplay between residential and commercial use, and bullying is prevalent.

FINDINGS AND DISCUSSION

From the review above, key findings appear the characteristics and indicators related to the IDPS In Cairo. Material hardship and chaos city are not only closely related to the phenomenon of IDPs but because of the people who live in the city and do not respect its rules. In Cairo, the displacement of poor people appeared among two cases. The first is the migration from the rural areas bearing of the characteristics of the hardship of living for settle in the big cities. The second seems in the continuous silent movements inside the city for finding a better place for living.

These results go beyond previous reports (Fielden, 2008; Carrillo, 2009; World Economic Forum, 2017, pp. 15-16), This chapter reveals that the chaos one of the reason to migration; emerging that there are two types of silent movements occur the permanent residents and the non-permanent visitors. As the poor and disadvantaged escape to get money and better living, the wealthy classes leave their original places to live in the city for two purposes; (a) if they have deep cultural roots, searching of the quality of life and chaos city will be the most reasons for leaving. (b) If they do not have deep cultural roots, but they have the money, the reasons to leave their home will be to increase the money and political and social greatness. It means quality, chaos, money, and greatness could be significant and relatively reasons to leave the original places of residence.

- The first category of poor people settles in independent housing, the informal settlements', and in "Al-Ashwa'yat [slums]". Predominately, it lacks comforts and satisfactory life, to be a real example of the two terms cities of hardship and chaos city. However, the IDPs settle in these settlements because they are becoming nearest to their work and make much profit than from if they are still living in their origin areas, which are lacking the opportunity to work or live.
- The second category of poor people seems to have taken the movement from city to city and from site to site mainly to make a better living based on three types of displaced:
    - Who are arriving with each daylight from the cities located very closest to the major cities? Those people are working in various activities of the city such as teaching, healthcare, advertisement, entertainment, and manufacturing. Moreover, they are returning to their cities at the end of the day.
    - Who are coming to work in complementary work that do not need specific skills such as security of residential, commercial, or religious buildings, factories, or companies? They are staying for several weeks or months and return on official holidays to their origin places.
    - Who are coming in non-permanent displacement to make a living by having no legal cover, such as hawkers and beggars, who are often staying in their workplaces on sidewalks, bridges, train stations and river shores?

Thus, the second category, particularly the third type of displacement is what is causing predominantly chaos city, which occurs primarily due to the behavioural changes. Based on the content analysis and the IoT, the indicators of the chaos city and the characteristics of the cities of hardship are related to three causes of urban displacement in Cairo.

The first cause is the attempt to get rid of the harsh living conditions, such as lack of public services, infrastructure, and transportation choices. Four main characteristics linked with these harsh living conditions are the loss of access to community services, lack of resources, lack of the expertise to finance and maintain infrastructure and the limited means of transport. In this vein, the indicators of chaos that appear in five features are the randomness of buildings, the dire situation of drainage networks, traffic congestion, overcrowded streets, and garbage everywhere. This chaos needs three efforts at three levels: regional planning, national and urban planning. Since this chaos occurs on a large scale, it requires three governmental efforts at three levels of regional, national, and urban planning.

The second cause is the attempt to escape from material hardship due to economic downturns, a weakened economy and extreme poverty. Three characteristics of cities of material hardship have concentrated on the inability to a) provide solutions to poverty, b) work with non-governmental organisations and c) work with the government to find real solutions to the problem. In this case, the indicators are homelessness, street children, racketeers, drug dealers and child gangs, vandalism, street fighting and female sexual harassment. To solve all the above-mentioned problems, it is necessary to combine the policies of governmental and non-governmental organisations. For instance, non-governmental organisations organised anti-sexual harassment teams in Egypt in 2008 (Abdelmonem & Galán, 2017). Moreover, a set of laws were promulgated in 2013 (Guarcello & Koseleci, 2009; Ghandour, 2017). This chaos needs to solve the problems of urban governance and municipal corporation. In addition, it needs to provide solutions on the level of neighbourhood units regarding the behavioural settings in the public realm.

The third cause is the renovation deteriorated urban neighbourhoods, known as the gentrification process, leading to the process of displacement movement. The issue of gentrification emerged in 1986 in the book titled *Gentrification of the City*, which was edited by Neil Smith and Peter Williams (Smith & Williams, 2013). In this book, Robert Beauregard, the Professor of Urban Planning at Columbia GSAPP, presented his chapter entitled 'The chaos and complexity of gentrification' (Beauregard, 2013). The characteristics of this 'chaotic concept' are inequalities and injustices created by capitalist policies, the change of urban lifestyle amenities, the patterns of consumption, the abandonment of assets and customs and the predominance of chain stores, restaurants, and service providers. In addition, the indicators of the chaos in the city are seen in vandalism, aggressive behaviour and the vibrant sidewalk life diminished substantially. Moreover, gentrification causes the displacement of low-income groups and leads to socio-spatial segregation (Paulina, 2017, p. 72). We suggest two aspects in the face of the process of displacement:

The first follows urban planning and design solutions, which improve the deteriorated urban neighbourhoods and raises the living standard of people simultaneously.

The second depends on the idea of preventing the causes through activating the principles of city management and urban control. This solution aims to stop the causes which lead to deterioration to distinguish urban areas in the traditional cities. All the findings above may raise concerns about urban chaos in cities of hardship which the concepts of decentralisation, empowerment and liveability can address.

AN INDICATORS-BASED APPROACH

Grasp the outgrowths of causes, characteristics, and indicators of the IDPs, city of chaos, and the cities of hardship needs an integrated approach to consider these three synergies collectively. This approach offers a proposed action plan to promote three steps of eradicating the chaos city related to the IDPs.

- The first discusses the transformation of hardship to become liveable.
- The second is how dealing with the IDPs within a reasonable framework that helps them to stay in their cities based on the concept of decentralisation.

Figure 4. Timeline of the concept of liveability between 1976-2016

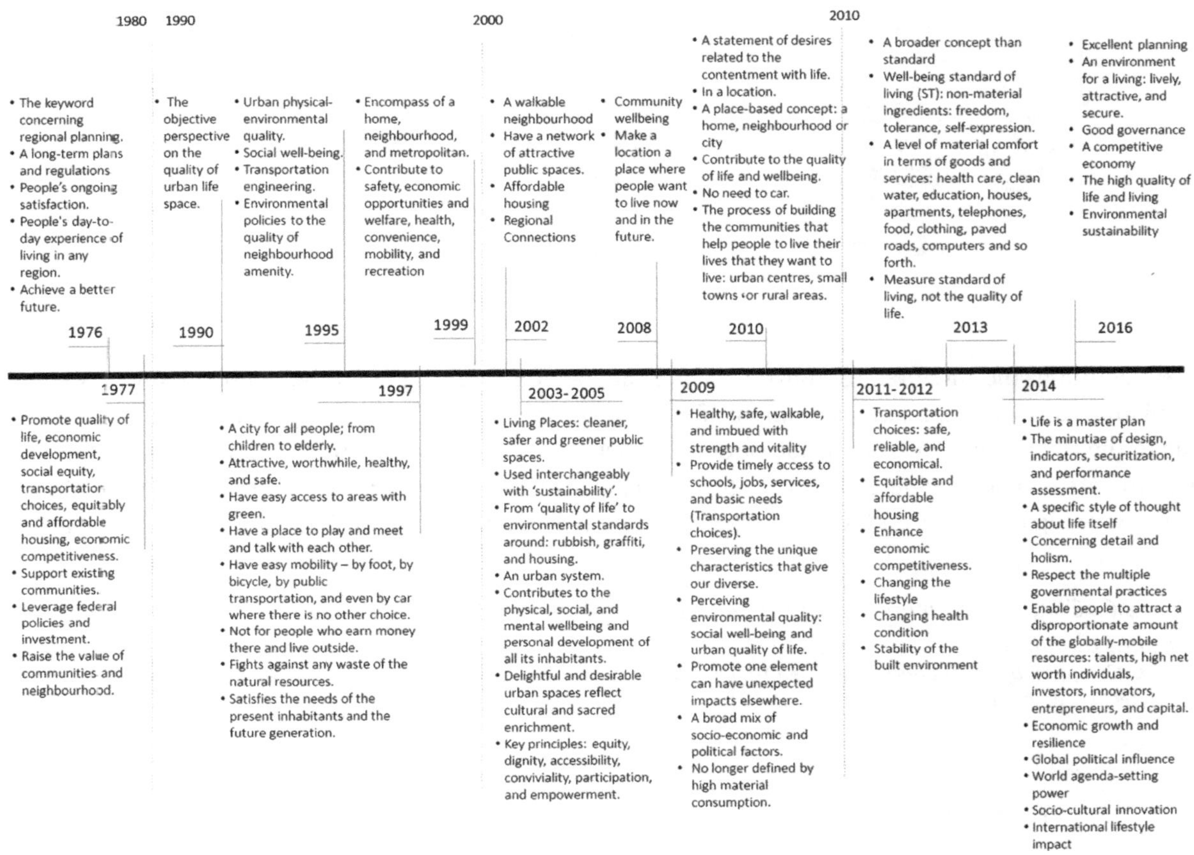

- The third step is according to how to prevent the chaos city before going it in one side, and what is the process to improve the cities which are already suffering from the chaos due to both behavioural of IDPs and the resident themselves in the other side.

Distinctly, this study focuses on the urban displacement due to the fleeing from the severe living conditions and material hardship. Since the mid-seventies, around the world, numerous institutions and analysts have dealt with liveability as a concept appeared to help us understand the right to live as well as we can. We present a theoretical framework that focuses on the concept of liveability (Figure 4 and Appendix 1), which links between two aspects: standard of living and quality of life. The first focuses on the physical components that can be measured through rates and constants. The second aspect deals with nonmaterial elements such as satisfaction and consequences of the impact of all physical features.

Figure 5 shows the three phases of the theoretical framework based on that the elements of this study encompass of a home, neighbourhood, and metropolitan area that contribute to safety, economic opportunities and welfare, health, convenience, mobility, and recreation (Vuchic, 1999, p. 7).

The current Figures 6, 7, and 8 build the framework based on the indicators of chaos city, which regards the debate between the characteristics of liveability and the hardships of the city [7]

Figure 5. three phases of the theoretical framework in three levels

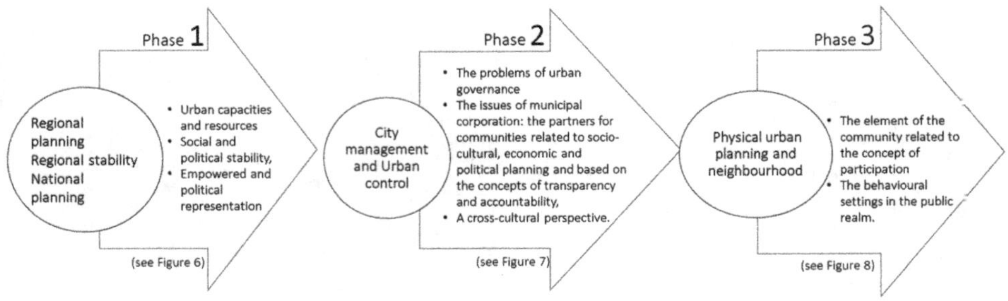

In addition, the three tables above measure satisfaction resulting from the application of laws and regulations and the implications of all this for what call "chaos city". Chaos city not only struck the city due to the IDPs, whether from inside or outside, but also appeared in the informal areas and slums and near the areas that provide economic returns such as heritage, commercial, and entertainment places. Chaos investigates the attitudes of people in the public realm of the city; its indicators imply the changes

Figure 6. Chaos city amidst liveability and hardship of living related to the level of the regional and national planning.
Source: Compiled by the authors based on (Young & Hermanson, 2012; Metcalf, 2002; Young & Hermanson, 2012; Lash, 1976; deChazal, 2010; Young & Hermanson, 2012).

Phase 1		Characteristics of Liveability		Indicators of Chaos City
Two levels of city planning	Urban capacities and resources	• Natural environment • Natural resources	• Climate change • Availability of food and daily items • Satisfies needs of the present inhabitants and the future generations	• The physical and mental stress • Black market. • Speculation
Regional planning	Promoting social and political stability	Non-material ingredients	• Social justice and equity • Cities for all; attractive, worthwhile, safe for children, and older people • Freedom, tolerance, self-expression, availability of free media, human values, freedom, tolerance, and self-expression • Life itself is a master plan	• Injustice, vileness, disrespect, selfishness, individuality, rejection. • Immorality: drunken and lazy people, drug traffickers across roads, and prostitutes in homes behind cafés. • The kitsch distorts the walls and misshapen sculptures. • Terrorism and extremism. • Small petty and violent crimes. • Insurgency and civil unrest.
		Political Environment	• local democracy • political stability of law enforcement • prevent crime	
National planning	Empowered and political representation	• The planning efforts	• Administrative practices in time • Proper development plans • A detailed and holistic view • A specific style of thought • Emergence solutions • Living places now and in the future	• Circumvention of the law: • Corruption • Access from the back doors (brokerage and bribery). • A blame game for the others • Apathetic responses of the executive

Figure 7. Chaos city amidst liveability and hardship of living related to city management.
Source: Compiled by the authors based on (Betsill, 2001; Lowe, et al., 2013; Salzano, 1997; Hahlweg, 1997; Young & Hermanson, 2012; Okulicz-Kozaryn, 2013; Kraftl, 2014; Lowe, et al., 2013).

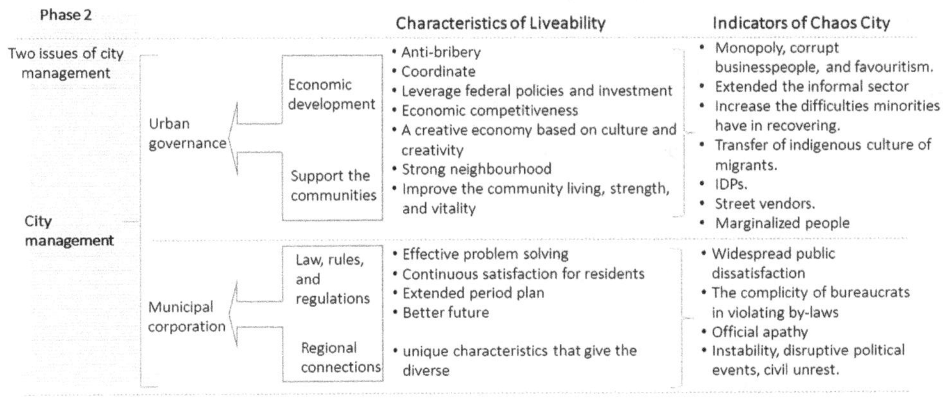

in the urban lifestyle, including homelessness, street children, street vendors, aggressive behaviours, vandalism, harassment, and traffic congestion. Chaos happens due to not applying the laws, but it has increased because of the lack of strong law enforcement. Now, the question is, if chaos city occurred due to one or more reasons above, then how can the practitioner deal with the city to accommodate these circumstances, whatever the actions and desires of people?

ACTION PLAN OF URBANISATION POLICIES

The agreement between the authorities of the city and the citizens should include an implicit belief required to prove that all situations are constant to protect the citizens from the hardship of living.

Many urban policies have focused on the problems of urban displacement, chaos city, and the hardship of living separately (Downing, 2002; Ley & Newton, 2010; Nelson, 2011; Young & Hermanson, 2012; Brown, Boano, Vivekananda, & Walker, 2015; United Nations Human Settlements Programme, 2016; Internal Displacement Monitoring Centre, 2018). Therefore, it should be combining the three terms to identify the relationships with each other and offer the priorities for solving the problems. Whichever solutions in the first will necessarily help address all the issues in the long run. Precisely, decentralisation seems an approach to achieve the effect on the relationships between those three terms. Action plan is starting from decentralisation and improvement of the smallest cities around and nearby the greater ones. Figure 9 shows that the action plan depends on the ordering three priorities based on two dimensions of city-systems framework are as follows:

First: the politicians, decision-makers, should be aware of what is going on in the local communities so that the pressing needs of these communities can be met. The way to do this is to enhance group efforts to bring about the most successful political goals that encompass competitiveness on the one hand, and liveability and quality of life on the other.

Figure 8. Chaos city amidst liveability and hardship of living related to the level of urban planning and neighbourhood units.
Source: Compiled by the authors based on (Hahlweg, 1997; Casselati, 1997; Metcalf, 2002; Timmer & Seymoar, 2005; ODPM, 2003; Young & Hermanson, 2012; Young & Hermanson, 2012; Shamsuddin, Hassanb, & Bilyaminb, 2012; Kraftl, 2014; Young & Hermanson, 2012).

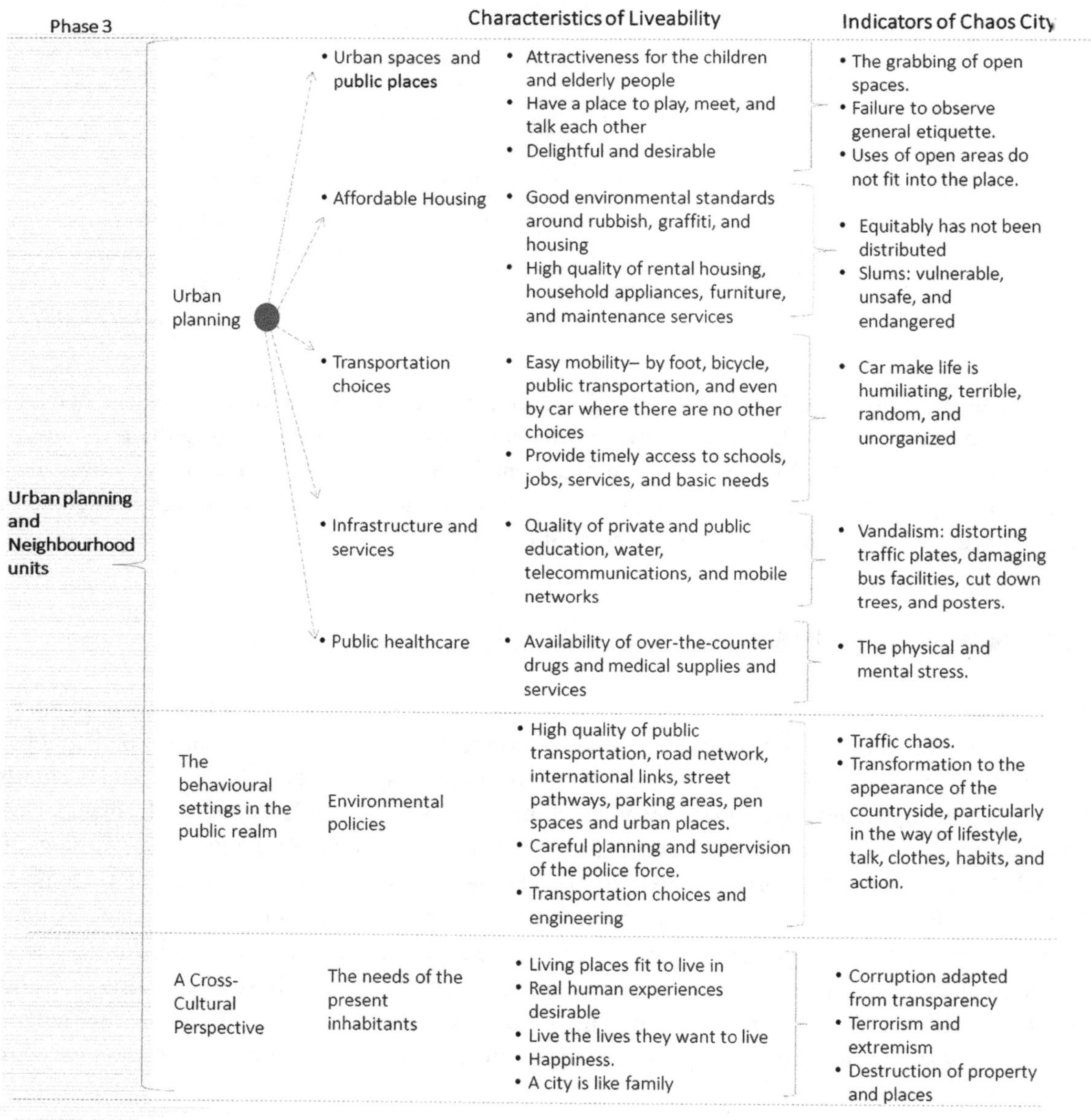

- High authorities should acknowledge that many urban cities have become cities of hardship on two sides, the first for individuals who have become poor people, and the second related to the physical configuration that lacks services, infrastructures, and facilities.

Figure 9. There priorities and two dimensions of city-systems framework

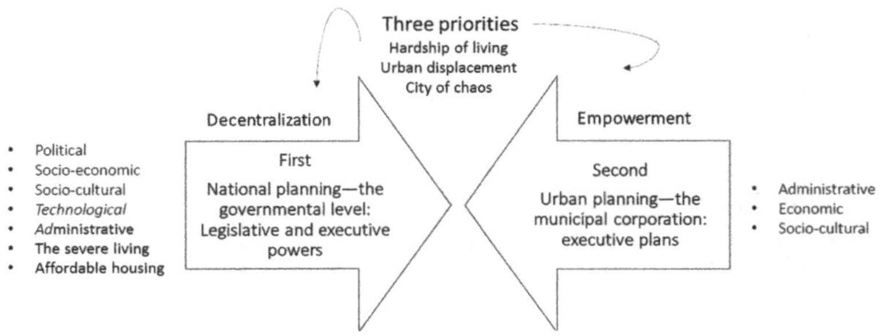

- The new national development agendas should focus on the socioeconomic stability of both IDPs and poor people at the same time.
- Raising the incentives of cities that have a more significant share of displacement, such as: expanding mega-development projects, improving salaries, and reducing taxes more than other cities, decreasing the cost of healthcare and entertainment while improving service levels.

Second: the development and prevention actors should activate the combined efforts to prevent hardship, chaos, and displacement either concerning the inter-urban displacement and intra-urban displacement. It happens by integration and overlap between the stakeholders with the liability of the persons who caused in the three types of degeneration equally be attributed.

- There is no link in most of the indexes of liveability between hardship, chaos, and displacement.
- Displacement began due to hardship and chaos; hardship implies by displacement and chaos; chaos has made hardship and displacement. These three interrelationships need to three-dimensional characterization in one level and a comprehensive approach in another level.

Figure 10 and Figure 11 displays the prioritising of urbanisation policies of urbanisation policies to solve the problems due to three priorities based on three steps of sequencing through address a hardship, displacement, and chaos crises collectively. Notably, these policies respect the ideas of decentralisation and empowerment.

- The first step focuses on the national level and improving the cities that are suffering from the hardship of living, which are located around and nearby large cities. Accordingly, the displacement from those cities to the large cities would be reduced.
- The second step, in large cities, is based on the municipality's corporation by executive plans. Herein lies the chaos caused by hardship and displacement, and it should be discussed by proposing the urban planning policies.
- The third step deals with chaos city arising from specific causes, which have not occurred from chaos and displacement but are latent in the typological and morphological perspective of urban areas. Urban planning and design policies apply proper solutions, aside from the aspects that deal with local communities and individual persons.

Figure 10. Prioritizing of urbanization policies in the level of national and urban planning (decentralisation).
Source: Compiled by the authors based on (Economic and Social Affairs, 2001; Schneider, 2003; Beard, Miraftab, & Silver, 2008)

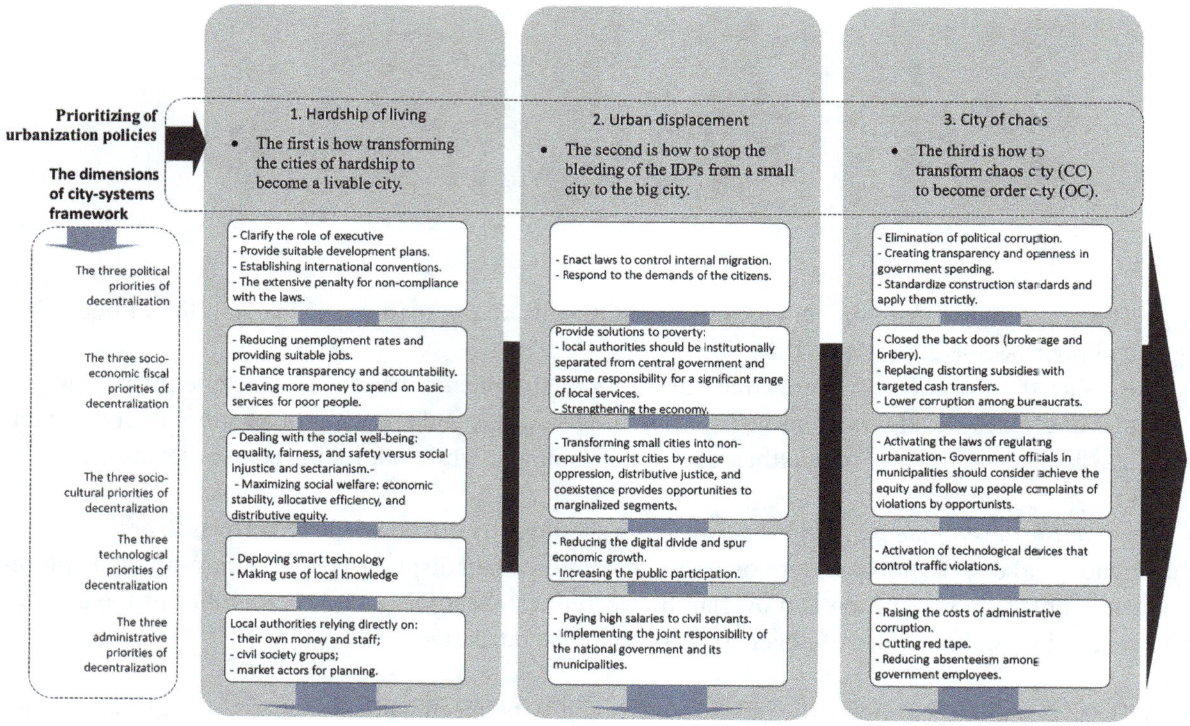

CONCLUSION

In summary, this chapter argued that the IDPs phenomenon cause city chaos; consequently, this chaos creates cities of hardship. We used two conventional and digital research methods to test the hypothesis above. These methods are the content analysis for a literature review of three Egyptian films and three published works (an article, a report, and a book). The second method is the IoT to investigate the aspects of the problems by using Google engine and Facebook.

The main outcome is Cairo has the characteristics of some indicators of cities of chaos, which not only arise because of displaced persons who are coming to the city to stay continuously or intermittently, but it also arises due to a lack of application of the rules and regulations. This result may end the myth that chaos is caused just by the IDPs, especially those who came from the nearby cities for work and live in various places inside the city. However, some IDPs remain a cause of chaos that is transforming traditional cities into cities of hardship due to the fact that their culture differs from the culture of the citizens of the host community. In this vein, the solution lies in dealing with the three elements of the problem (the IDPs, chaos, and in apply the rules) within a general framework governed by the ideas of decentralisation and empowerment. This solution includes the following two points:

Figure 11. Prioritizing of urbanization policies in the level of urban planning (empowerment) — executive plans for the municipal corporation
Source: Compiled by the authors based on (Musgrave, 1958; Mawhood & Davey, 1980; Rose-Ackerman, 1999).

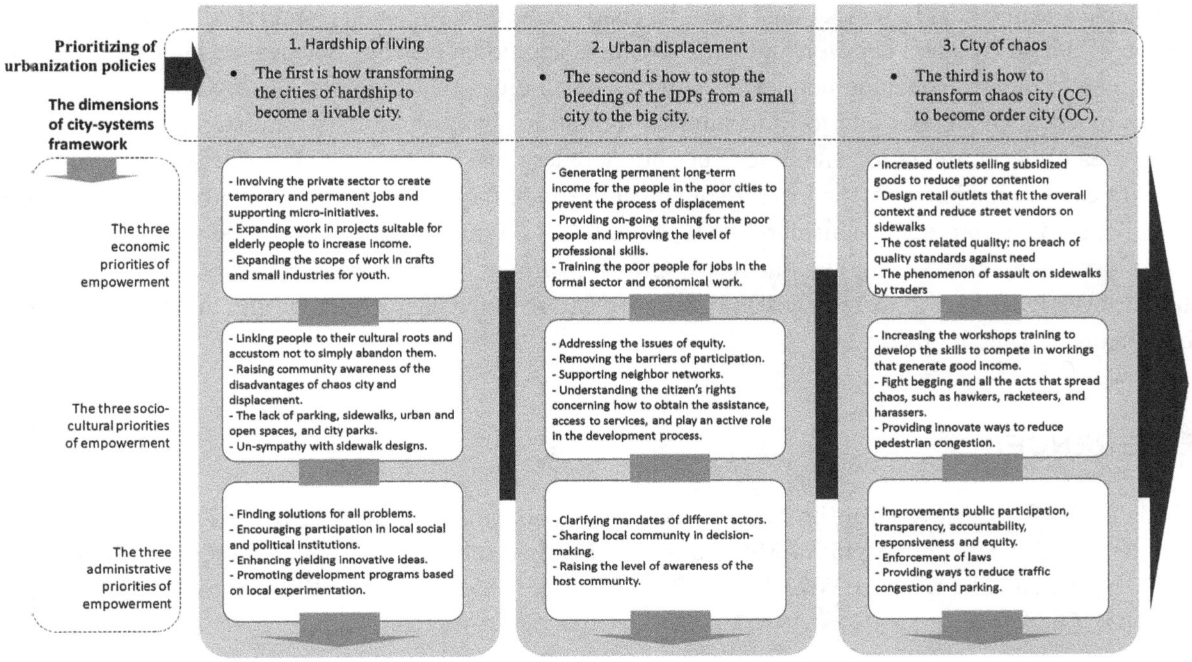

- The first is how to deal with cities of hardship surrounding the big cities, which reflect a high percentage of internal displacement. This should occur through the expansion of large-scale development projects that offer job opportunities and raise the income of poor people. Ending the causes of poverty and empowering people to work in their cities will help directly to reduce the manifestations of people's frustration and, consequently, reduce the rates of displacement from cities of hardship to large cities.
- The second side begins by recognising that many areas in some developing countries have harsh living conditions, which makes people move internally to improve their standards of living. This internal movement should be organised to prevent chaos in the city, whose indicators appear as several phenomena, such as street vendors, street children, harassment, rape, drug use, bullying, living in rubbish dumps, and traffic congestion. Moreover, the areas that experience hardship should be improved.

This chapter proposes that the solution lies in improving the level of community services, providing affordable housing, and the stricter enforcement of the laws. Broadly translated, our findings indicate that there are three priorities in the urbanization policies. There has been a significant outcome in the understanding of the causes of urban chaos and the hardships of cities related to the raising of the IDP phenomenon. These priorities depend on each level of planning in the three phases. Phase One includes regional and national planning, Phase Two focuses on city management and urban control, and Phase Three deals with physical urban planning and neighbourhood units.

Future research should consider more carefully the potential effects of the chaos within a city; for example, the impact of overcrowding on user satisfaction and causing urban chaos in the new residential areas. Research should focus on the interaction between traffic congestion, liveable cities and the hardships within cities. Laws that deal with the chaos caused by traffic, the spread of slums, as well as the mess created by street children and street vendors, should be enacted. The researchers should think about how to increase development projects in the small cities that surround the big cities, in order to absorb much of the labour and to limit internal migration. They should also investigate the relevance of the idea of decentralization and empowerment in emerging societies, which suffer not only from the difficulties of living but from the low rate of education and high rate of illiteracy. Further studies should investigate how to provide a considerable database based on the exemplary use of new digital methods and information, such as Big Data analysis, Crowd-sourcing and the IoT. Furthermore, these studies should investigate how to use these data to identify and map the characteristics and indicators of cities that are undergoing hardships.

REFERENCES

AbadaG. (2018). Retrieved from https://www.facebook.com/galal.abada/posts/10156270326245283?__xts__[0]=68.ARAVZlYtnWxJ5Ww3hK4sqpgrzDRFFCbA-4GGR4njWL9OTzcyuSGX02GlFd-i71-fW2hJH_JHauqV2kqmBcvSP3WOUYwF_BPy_oSCD46RG5J1m8DJNwrKnQcXJ6889exAwanRsI-40J41p&__tn__=-R

Abaza, M. (2011). Critical Commentary. Cairo's Downtown Imagined: Dubaisation or Nostalgia? *Urban Studies (Edinburgh, Scotland)*, *48*(6), 1075–1087. doi:10.1177/0042098011399598

Abdelmonem, A., & Galán, S. (2017). Action-Oriented Responses to Sexual Harassment in Egypt: The Cases of HarassMap and WenDo. *Journal of Middle East Women's Studies*, *13*(1), 154–167. doi:10.1215/15525864-3728767

Abusaada, H. (2016). Internal Displacement as a Silent Crisis: Managing Urban Chaos and Hardship—Back to the Order. In *RSA Annual Conference: New Pressures on Cities and Regions*. London, UK: Regional Studies Association.

Abusaada, H. (2018). *Sympathizing Withthe the Footpaths (in Arabic)*. Retrieved from www.facebook.com/sympathizingwiththefootpaths: https://www.facebook.com/sympathizingwiththefootpaths/?modal=admin_todo_tour

Abusaada, H., & Elshater, A. (2017). *Back to the Order: The Chaos in the Cities of Hardship*. Saarbrücken, Germany: Lambert Academic Publishing.

Abusaada, H., & Elshater, A. (2019). From Chaos to Order: Articulating the Urban Policies for Cities of Hardship. In U. G. Benna (Ed.), *Industrial and Urban Growth Policies at the Sub-National, National, and Global Levels*. IGI Global. doi:10.4018/978-1-5225-7625-9.ch003

Allen, J., & Browne, M. (2010). Sustainability Strategies for City Logistics. In *Green Logistics: Improving the Environmental Sustainability of Logistics* (pp. 282–305). London: Kogan Page Limited.

Asplet, M. (2013). *Internal Displacement: Responsibility and Action.* Geneva: Inter-Parliamentary Union and UNHCR No 20.

Atzori, L., Iera, A., & Morabito, G. (2010). The Internet of Things: A survey. *Computer Networks, 54*(15), 2787–2805. doi:10.1016/j.comnet.2010.05.010

Bartlett, S., Dodman, D., Hardoy, J., Satterthwaite, D., & Tacoli, C. (2011). Social Aspects of Climate Change in Urban Areas in Low and Middle-Income Nations. In *Cities and Climate Change: Responding to an Urgent Agenda: Main report (English)* (pp. 670–726). Washington, DC: World Bank, Urban development series. Retrieved from http://documents.worldbank.org/curated/en/613201468149671438/Main-report

Basiago, A. (1999). Economic, social, and environmental sustainability in development theory and urban planning practice. *The Environmentalist, 19*(2), 145–161. doi:10.1023/A:1006697118620

Beard, V. A., Miraftab, F., & Silver, C. (2008). *Planning and Decentralization: Contested Spaces for Public Action in the Global South.* Routledge.

Beaufoy, H. (2010). Order out of chaos': The London Society and the planning of London 1912–1920. *Planning Perspectives, 12*(2), 135–164. doi:10.1080/026654397364708

Beauregard, R. (2013). The chaos and complexity of gentrification. In N. Smith & P. Williams (Eds.), Gentrification of the City (pp. 35-55). Routledge.

Benediktsson, M. O., Lambert, B., & Larsen, E. (2015). Taming a "chaotic concept": Gentrification and segmented consumption in Brooklyn, 2002–2012. *Urban Geography*, 1–21. doi:10.1080/02723638.2015.1096113

Betsill, M. (2001). Mitigating Climate Change in US Cities: Opportunities and obstacles. *Local Environment, 6*(4), 393–406. doi:10.1080/13549830120091699

Boswell, C. (2002). *New Issues in Refugee Research, Addressing the causes of migratory and refugee movements: the role of the European Union.* Institute for Peace Research and Security Policy University of Hamburg.

Brown, D., Boano, C., Vivekananda, J., & Walker, J. (2015). *Urban Crises and Humanitarian Responses: A Literature Review. The DFID Humanitarian Policy and Partnerships Group.* CHASE.

Bunio-Mroczek, P. (2017). Gentrification, revitalization and children raising. Family gentrifiers in a post-socialist city. *Przegląd Socjologii Jakościowej, 13*(4), 71-91. Doi:10.18778/1733-8069.13.4.05

Byrne, J. P. (2003). Two Cheers for Gentrification. *Georgetown Law Faculty Publications and Other Works*, 405-432. Retrieved from https://scholarship.law.georgetown.edu/facpub/930

Çakar, U., & Alakavuklar, O. N. (2011). Inherent Chaos in the Organizational Order: An Epistemological Approach. *International Journal of Business and Management Studies, 3*(1), 391–405.

Carrillo, A. C. (2009). Internal displacement in Colombia: Humanitarian, economic and social consequences in urban settings and current challenges. *International Review of the Red Cross, 91*(875), 527–546. doi:10.1017/S1816383109990427

Casselati, A. (1997). The Nature of Liveability. In *Making cities liveable. International Making Cities Liveable Conferences*. Gondolier Press.

Castells, M. (2010). End of Millennium: The Information Age: Economy, Society, and Culture, Volume III (2nd ed.). Wiley-Blackwell.

Davies, A. (2012). IDPs in Host Families and Host Communities: Assistance for hosting arrangements. United Nations High Commissioner for Refugees (UNHCR).

deChazal, J. (2010). A systems approach to livability and sustainability: Defining terms and mapping relationships to link desires with ecological opportunities and constraints. *Systems Research and Behavioral Science*, *27*(5), 585–597. doi:10.1002res.1058

Dobronogov, A., & Iqbal, F. I. (2005). Economic Growth in Egypt: Constraints and Determinants. *SSRN Electronic Journal*, 1-29. doi:10.2139srn.872328

Downing, T. E. (2002). *Avoiding New Poverty: Mining-Induced Displacement and Resettlement*. International Institute for Environment and Development.

Economic and Social Affairs. (2001). *Decentralization and Poverty Reduction: Africa and Asia Experience*. Dakar, Senegal: United Nations.

EEA. (2009). *Ensuring quality of life in Europe's cities and towns, Tackling the environmental challenges driven by European and global change*. Copenhagen: European Environmental Agency.

Eriksson, K., Johnson, C., & Estep, D. (2004). Lorenz and the Essence of Chaos. In K. E. Authors (Ed.), *Eriksson, Applied Mathematics: Body and Soul* (pp. 849–858). Berlin: Springer. doi:10.1007/978-3-662-05800-8_5

European Environmental Agency (EEA). (2009). *Ensuring quality of life in Europe's cities and towns, Tackling the environmental challenges driven by European and global change*. Copenhagen: European Environmental Agency.

Evans, P. (2002). *Livable Cities? Urban Struggles for Livelihood and Sustainability* (Vol. 4). Berkeley, CA: University of California Press. doi:10.1525/california/9780520230248.001.0001

Faguet, J.-P., Fox, A. M., & Poeschl, C. (2014). *Does decentralization strengthen or weaken the state? Authority and social learning in a supple state*. London, UK: Department of International Development, London School of Economics and Political Science.

Fan, C. (1999). Migration in a Socialist Transitional Economy: Heterogeneity, Socioeconomic and Spatial Characteristics of Migrants in China and Guangdong Province. *The International Migration Review*, *33*(4), 954–987. doi:10.1177/019791839903300406 PMID:12349707

Farazmand, A. (2003). Chaos and Transformation Theories: A Theoretical Analysis with Implications for Organization Theory and Public Management. *Public Organization Review: A Global Journal, 3*, 339–372.

Farquhar, M. (1995). Definitions of Quality of Life: A Taxonomy. *Journal of Advanced Nursing*, *22*(3), 502–509. doi:10.1046/j.1365-2648.1995.22030502.x PMID:7499618

Fielden, A. (2008). *Ignored Displaced Persons: the plight of IDPs in urban areas.* UNHCR, Policy Development and Evaluation Service, United Nations High Commissioner for Refugees.

Frances Agyapong, T. K. (2018). Managing traffic congestion in the Accra Central Market, Ghana. *Journal of Urban Management*, *7*(2), 85–96. doi:10.1016/j.jum.2018.04.002

Geddah, R. (2017, January 26). The costs of marriage. *Al-Ahram Weekly,* (1329).

Ghandour, S. E. (2017). Towards More pedestrian-friendly streets in Cairo. *PPAD Working Paper Series,* (3), 1-19. Retrieved from https://documents.aucegypt.edu/Docs/GAPP/PPAD-Working-Paper-Series-3.pdf

Ghoraba, H. (2018). Defusing the population bomb. *Al-Ahram Weekly,* (1404).

Giap, T. K., Thye, W. W., & Aw, G. (2014). A new approach to measuring the liveability of cities: The Global Liveable Cities Index. *World Review of Science, Technology and Sustainable Development*, *11*(2), 176–196. doi:10.1504/WRSTSD.2014.065677

Giusto, D., Iera, A., Morabito, G., & Atzori, L. (Eds.). (2010). *The Internet of Things.* Springer. doi:10.1007/978-1-4419-1674-7

Gleick, J. (1987). *Chaos: Making a New Science.* New York: Knopf.

Guarcello, L., & Koseleci, N. (2009). *A profile of Cairo street children.* ILO, World Bank, and UNICEF programme.

Guthrie, R. A. (2014). *The Causes and Consequences of Migration: How Displacement and Change Affect the Well-Being of Humli Tibetans in Katmandu.* Missoula, MT: Graduate Student Theses, Dissertations, & Professional Papers.

Hahlweg, D. (1997). The City as a Family. In S. H. Lennard, S. v. Ungern-Sternberg, & H. L. Lennard (Ed.), *Making cities liveable. International Making Cities Liveable Conferences.* Gondolier Press.

Humanitarian Fact Sheet. (2016). *Forced Displacement: Poverty and Financing, Seven facts you need to know.* Global Humanitarian Assistance and development initiatives. Retrieved from http://devinit.org/wp-content/uploads/2016/09/Forced_Displacement_Poverty_and_Financing_DI_Sept_2016.pdf

IDRC, CDDI, & UN-HABITAT. (2008). *Decentralization, local power, and women's rights.* Ottawa, ON, Canada: International Development Research Centre. Retrieved from https://www.idrc.ca/sites/default/files/sp/Documents%20EN/policy-recommendations-on-decentralization.pdf

Internal Displacement Monitoring Centre. (2018). *2018 Global Report on Internal Displacement (GRID 2018).* Retrieved from reliefweb: https://reliefweb.int/report/world/2018-global-report-internal-displacement-grid-2018

Jacobs, J. (1961). The Death and Life of Great American Cities. Random House Inc.

Johnson, P.-A. (1994). *The Theory of Architecture: Concepts Themes & Practices.* New York: John Wiley & Sons.

Kiunsi, R. (2013). A review of traffic congestion in Dar es Salaam city from the physical planning perspective. *Journal of Sustainable Development*, *6*(2), 94–103. doi:10.5539/jsd.v6n2p94

Kraftl, P. (2014). Liveability and urban architectures: Mol(ecul)ar biopower and the 'becoming lively' of sustainable communities. *Environment and Planning. D, Society & Space*, *32*(2), 274–292. doi:10.1068/d21012

L'Heureux, A., Grolinger, K., Higashino, W. A., & Capretz, M. A. (2017). A Gamification Framework for Sensor Data Analytics. In *IEEE International Congress on Internet of Things (ICIOT 2018)* (pp. 74-81). Honolulu, HI: IEEE. 10.1109/IEEE.ICIOT.2017.18

Lash, H. (1976). *Planning in a human way: personal reflections on the regional planning experience in Greater Vancouver*. Ottawa: Ministry of State for Urban Affairs.

Lennard, S. H., & Lennard, H. L. (1995). *Livable cities observed: a source book of images and ideas for city officials, community leaders, architects, planners and all other committed to making their cities livable*. Southampton, UK: Gondolier Press.

Ley, A., & Newton, P. (2010). Creating and sustaining liveable cities. In S. Kallidaikurichi & B. Yuen (Eds.), Developing Living Cities: From Analysis to Action (pp. 191-230). Singapore: World Scientific. doi:10.1142/9789814304504_0008

López, R. C., Arredondo, C. I., & Salcedo, J. (2011). *The Effects of Internal Displacement on Host Communities: A Case Study of Suba and Ciudad Bolívar Localities in Bogotá, Colomb*. Bogotá: Brookings Institution-London School of Economics, Project on Internal Displacement.

Lorenz, E. (1993). *The Essence of Chaos*. Seattle, WA: University of Washington Press. doi:10.4324/9780203214589

Lowe, M., Whitzman, C., Badland, H., Davern, M., Hes, D., Aye, L., . . . Giles-Corti, B. (2013). *What Are the Key Indicators for Melbourne Neighbourhoods?: Place, Health and Liveability Research Program.* Melbourne, Australia: Melbourne Sustainable Society Institute.

Lynch, K. (1960). *The Image of the City*. Massachusetts Institute of Technology.

Mawhood, P., & Davey, K. (1980). Anglophone Africa. In D. C. Rowat (Ed.), *Anglophone Africa International Handbook on Local Government Reorganization* (pp. 404–414). London: Aldwych Press.

McCallin, B., Scherer, I., & Duyne, J. (2015). Urban informal settlers displaced by disasters: challenges to housing responses. Geneva, Switzerland: The Internal Displacement Monitoring Centre (IDMC), Norwegian Refugee Council.

McCormick, B., & Wahba, J. (2003). *Migration and Mobility in the Egyptian Labour Market*. Cairo, Egypt: Economic Research Forum.

Merriam Webster. (2018, September 30). *Chaos*. Retrieved from www.merriam-webster.com: https://www.merriam-webster.com/dictionary/chaos

Merriam-webster. (2018). *livability*. Retrieved from www.merriam-webster.com: https://www.merriam-webster.com/dictionary/livability

Metcalf, G. (2002). *The path to a livable city.* San Francisco: Transportation for a Livable City.

Mohit, M. A., & Iyanda, S. A. (2016). Liveability and Low-income Housing in Nigeria. *Social and Behavioral Sciences,* (222), 863 – 871.

Montiel, L. M., Nathan, R. P., & Wright, D. J. (2007). *Divided they Fall: Hardship in America's Cities and Suburbs*. Albany, NY: The Nelson A. Rockefeller Institute of Government.

Mosel, I., & Jackson, A. (2013). *Sanctuary in the city? Urban displacement and Urban displacement and Pakistan*. Humanitarian Policy Group, Overseas Development Institute.

Mosko, M. S., & Damon, F. H. (2005). *On the Order of Chaos: Social Anthropology and the Science of Chaos*. Berghahn Books.

Muggah, R. (2000). Through the Developmentalist's Looking Glass: Conflict-Induced Displacement and Involuntary Resettlement in Colombia. *Journal of Refugee Studies*, *13*(2), 133–164. doi:10.1093/jrs/13.2.133

Mumford, L. (1938). *The Culture of Cities*. Orlando, FL: Harcourt Brace Jovanovich.

Musgrave, R. A. (1958). *Multi-Level Finance*. New York: McGraw-Hill.

Nelson, G. (2011). Measuring Poverty: The Official U.S. Measure and Material Hardship. *Poverty & Public Policy*, *3*(3), 1–35. doi:10.2202/1944-2858.1077

New & Improved Website. Coming Soon. (2018, August 15). *misrinternationalfilms.* Retrieved from Cairo as Told: http://www.misrinternationalfilms.com/ar/%D8%A7%D9%84%D9%82%D8%A7%D9%87%D8%B1%D8%A9-%D9%85%D9%86%D9%88%D8%B1%D8%A9-%D8%A8%D8%A3%D9%87%D9%84%D9%87%D8%A7

ODPM. (2003). *Sustainable Communities*. London: Building for the Future Office of the Deputy Prime Minister.

Okulicz-Kozaryn, A. (2013). City Life: Rankings (Livability) Versus Perceptions (Satisfaction). *Social Indicators Research. An International and Interdisciplinary Journal for Quality-of-Life Measurement*, *110*(2), 433–451.

Pacione, M. (1990). Urban liveability: A review. *Urban Geography*, *11*(1), 1–30. doi:10.2747/0272-3638.11.1.1

Paulina, B.-M. (2017). Gentrification, revitalization and children raising. Family gentrifiers in a post-socialist city. *Przegląd Socjologii Jakościowej*, *13*(4), 70–91. doi:10.18778/1733-8069.13.4.05

Pilkauska, N., Currie, J., & Garfinkel, I. (2012). The Great Recession, Public Transfers, and Material Hardship. *The Social Service Review*, *36*(3), 401–427. doi:10.1086/667993 PMID:24379487

Pulsellic, R., & Tiezzi, E. (2011). City out of Chaos: Social Patterns and Organization in Urban Systems. In C. Brebbia (Ed.), Ecodynamics: The Prigogine Legacy (pp. 193-202). WIT Press.

Robinson, J. (2000). Cities on the Move: Urban chaos and global change–East Asian art, architecture and film now. *Ecumene*, *7*(1), 105–109. doi:10.1177/096746080000700105

Robinson, W. C. (2003). *Risks and Rights: The Causes, Consequences, and Challenges of Development-Induced Displacement*. Washington, DC: The Brookings Institution-SAIS.

Rose-Ackerman, S. (1999). *Corruption and Government: Causes, Consequences, and Reform*. Cambridge, UK: Cambridge University Press. doi:10.1017/CBO9781139175098

Rubinowicz, P. (2000). Chaos and Geometric Order in Architecture and Design. *Journal for Geometry and Graphics*, *4*(2), 197–207.

Ruiz, I., & Vargas-Silva, C. (2014). The Economics of Forced Migration. *The Journal of Development Studies*, *49*(6), 772–784. doi:10.1080/00220388.2013.777707

Salama, S. A. (2006, December 14 - 20). Forgotten children. *Weekly Online*.

Salzano, E. (1997). Seven Aims for the Livable City. In *Making Cities Livable, International Making Cities Livable Conferences*. Gondolier Press.

Sande-Friedman, A. (2013). *The Four Times of Day by William Hogarth*. Retrieved from The Commononline: a modern sense of place: https://www.thecommononline.org/the-four-times-of-day-by-william-hogarth/

Schneider, A. (2003). Decentralization: Conceptualization and Measurement. *Studies in Comparative International Development*, *38*(3), 32–56. doi:10.1007/BF02686198

Sen, J. (2013). Sustainable Urban Planning. New Delhi: The Energy and Resources Institute (TERI).

Shamsuddin, S., Hassanb, N. R., & Bilyaminb, S. F. (2012). Walkable Environment in Increasing the Liveability of a City. *Social and Behavioral Sciences*, (50), 167 – 178.

Shawkat, Y., Khalil, O., & Zaazaa, A. (2015). *Parallel Urban Practice in Egypt 2015. Nairobi, Kenya: United Nations Human Settlements Programme*. UN-Habitat.

Smith, N., & Williams, P. (Eds.). (2013). *Gentrification of the City*. New York: Routledge. doi:10.4324/9781315889092

sokan26july. (2018). *Residents of July 26 - Cairo Khedive (in Arabic)*. Retrieved from www.facebook.com/sokan26july/: https://www.facebook.com/sokan26july/photos/pcb.207063756833259/207062963500005/?type=3&theater

Swartz, S. (2012). Decentralization, Local Development and Women's Empowerment: Innovating Knowledge and Practice. *Universitas Forum*, *3*(1).

Tafirenyika, M. (Ed.). (2016). Africa's cities of the future. *Africa Renewal*, *30*(1).

The Commission on Human Rights and the General Assembly. (1998). *Guiding Principles on Internal Displacement, E/CN.4/1998/53/Add.2*. New York: United Nations.

The National Association of Regional Councils (NARC). (n.d.). *Livability Literature Review: A Synthesis of Current Practice*. Washington, DC: The National Association of Regional Councils (NARC).

The UN Refugee Agency. (2007). *Internally Displaced Persons: Questions and Answers.* Geneva, Switzerland: United Nations High Commissioner for Refugees (UNHCR). Retrieved from http://www.unhcr.org/basics/BASICS/405ef8c64

The World Bank. (2013, June 6). *Decentralization.* Retrieved from www.worldbank.org: http://www.worldbank.org/en/topic/communitydrivendevelopment/brief/Decentralization

Timmer, V., & Seymoar, N.-K. (2005). *The livable city.* Vancouver: The World Urban Forum 2006, Working Group, International Centre for Sustainable Cities.

United Nations. (2005). *Decentralisation: Poverty Production, Reduction, Empowerment and Participation.* New York: United Nations. Retrieved from https://publicadministration.un.org/publications/content/PDFs/E-Library%20Archives/2005%20Decentralization_Poverty%20Reduction,%20Empowerment%20and%20Participation.pdf

United Nations High Commissioner for Refugees (UNHCR). (2015, June 18). *Worldwide displacement hits all-time high as war and persecution increase.* Retrieved from http://www.unhcr.org: http://www.unhcr.org/news/latest/2015/6/558193896/worldwide-displacement-hits-all-time-high-war-persecution-increase.html

United Nations Human Settlements Programme. (2008). *The State of African City.* Nairobi, Kenya: UN-HABITAT.

United Nations Human Settlements Programme. (2016). *Urbanization and Development: Emerging Futures.* Nairobi, Kenya: UN-Habitat.

United Nations Human Settlements Programme (UN-HABITAT). (2009). *International Guidelines on Decentralisation and Access to Basic Services for all.* Nairobi, Kenya: UN-HABITAT.

Victorian Competition and Efficiency Commission. (2008). *A state of liveability: An enquiry into enhancing Victoria's liveability.* Melbourne, Victoria: Victorian Competition and Efficiency Commission.

Vuchic, V. R. (1999). *Transportation for Livable Cities.* Center for Urban Policy Research.

Wakeman, R. (2016). *Practicing Utopia: An Intellectual History of the New Town Movement.* Chicago: University of Chicago Press. doi:10.7208/chicago/9780226346175.001.0001

Weick, K. (1979). *The Social Psychology of Organizing.* McGraw-Hill Humanities/Social Sciences/Languages.

Weiss, D., & Wurzel, U. (1998). *The Economics and Politics of Transition to an Open Market Economy: Egypt.* Paris, France: OECD. doi:10.1787/9789264163607-en

Williams, R. (1975). *The Country and the City.* Oxford University Press.

World Economic Forum. (2017). *Migration and Its Impact.* Cologny: World Economic Forum. Retrieved from http://www3.weforum.org/docs/Migration_Impact_Cities_report_2017_low.pdf

Young, E., & Hermanson, V. (2012). Livability Literature Review: A Synthesis of Current Practice. Washington, DC: The National Association of Regional Councils (NARC).

ENDNOTES

1 The notion of 'urban bias' used to explain this phenomenon (McCormick & Wahba, 2003, p. 6).

2 Traffic congestion can be viewed as an indicator of the deterioration of urban life (Kiunsi, 2013, p. 95).

3 See also Figures 6, 7, and 8.

4 Serve living, poverty, and hardship needs to review the way of thinking in decentralisation and empowerment (Economic and Social Affairs, 2001; United Nations, 2005; IDRC, CDDI, UN-HABITAT, 2008; United Nations Human Settlements Programme (UN-HABITAT), 2009).

5 The impact of the IDPs on the receiving communities seems need to be identify more areas for future research, such as, the impacts of forced migration on the communities of origin economics analysis, on housing markets, and host children (Ruiz & Vargas-Silva, 2014, p. 783).

6 Which the author can deal with them directly and continuously.

7 See also Appendix 2 and 4.

APPENDIX 1

Table 1. Commentators and Institutions, Date, and Concepts of Liveability

Commentators and Institutions	Date	Concept of Liveability
(Lash, 1976, p. 33)	(1976)	Liveability became the keyword for the regional planning. Therefore, experts searched to find effective ways to deal with many problems. Not only to produce a plan and regulations but also through a deal with long-term future Liveability, but also with people's ongoing satisfaction, their day-to-day experience of living in any region. Moreover, tomorrow's liveability needed as much attention as the attainment to achieving a better future.
		Liveability is *The proof of the planning would be in the living.*
The Partners for Liveable Communities (PLC) (The National Association of Regional Councils (NARC), p. 7)	(1977)	Improve the liveability of any community being through *promoting quality of life, economic development, and social equity.*
		There are six principles of liveability that are *provide more transportation choices, promote equitable, affordable housing, enhance economic competitiveness, support existing communities, coordinate, and leverage federal policies and investment, and value communities and neighbourhood.*
(Pacione, 1990, p. 398)	(1990)	The developing approach which focuses on the concept of urban liveability paralleled the objective perspective on the quality of urban life space.
(Lennard & Lennard, 1995) (Farquhar, 1995, p. 503)	(1995)	Liveability *focuses on urban physical-environmental quality, social well-being, transportation engineering, environmental policies to the quality of neighbourhood amenity Cited by* (Sen, 2013, p. 44).
(Hahlweg, 1997)	(1997)	A city for all people; attractive, worthwhile, and healthy. Safe for our children and older people to have easy access to areas with green, where they have a place to play and meet and talk with each other. And provide an *easy mobility – by foot, by bicycle, by public transportation, and even by car where there is no other choice.* In addition, it is not for people who earn money there and then go and live outside in the suburbs and the surrounding communities.
(Timmer & Seymoar, 2005, p. 3)		Liveability *means that we are experiencing oneself as a real person in the city, cited from* (Casselati, 1997).
(Salzano, 1997)		*A city that fights against any waste of the natural resources.*
		A city that satisfies the needs of the present inhabitants without reducing the capacity of the future generation to satisfy their needs.
(Vuchic, 1999, p. 7)	(1999)	Its elements encompass of *a home, neighbourhood, and metropolitan area that contribute to safety, economic opportunities and welfare, health, convenience, mobility, and recreation*
(Metcalf, 2002, p. 4)	(2002)	There are five fundamental aspects of great, liveable cities: strong neighbourhood, walkability, a network of attractive public spaces, affordability, and regional connections.
(Evans, 2002, pp. 1-2)		Livelihood is one of the two faces of the coin of Liveability.
Building for the Future Office of the Deputy Prime Minister (ODPM), (Kraftl, 2014, pp. 279-280)	(2003)	*Creation of so-called 'Living Places': 'cleaner, safer and greener' public spaces* (ODPM, 2003, p. 1).
		Used interchangeably with 'sustainability' to encapsulate ideas from 'quality of life' to environmental standards around rubbish, graffiti, and housing (ODPM, 2003, p. 1).
(Timmer & Seymoar, 2005, pp. 2, 32, 32)	(2005)	*Liveability refers to an urban system that contributes to the physical, social, and mental wellbeing and personal development of all its inhabitants.*
		Delightful and desirable urban spaces that offer and reflect cultural and sacred enrichment.
		Key principles that give substance to this theme are equity, dignity, accessibility, conviviality, participation, and empowerment.
(Victorian Competition and Efficiency Commission, 2008, p. xxi)	(2008)	*Liveability reflects the wellbeing of a community.* It includes many *characteristics that make a location a place where people want to live now and in the future.*
The Chicago Metropolitan Agency of Planning (CMAP). (The National Association of Regional Councils (NARC), p. 10)	(2009)	*Liveable communities are healthy, safe, and walkable. It offers transportation choices that provide timely access to schools, jobs, services, and basic needs.* In addition, *they are imbued with strength and vitality, features which emerge from preserving the unique characteristics that give our diverse.*
European Environmental Agency (EEA, 2009, pp. 5, 8, 43)		Liveability *perceiving environmental quality as a fundamental building block of social well-being and urban quality of life. The complex interaction between the many determinants of quality of life means that efforts to promote one element can have unexpected impacts elsewhere.*
		Liveability defined by a broad mix *of socio-economic and political factors.*
		[n]o longer defined by high material consumption.

continued on following page

Table 1. Continued

Commentators and Institutions	Date	Concept of Liveability
(deChazal, 2010, p. 587)	(2010)	Liveability is a *statement of desires related to the contentment with life in a particular location.*
Ley & Newton, 2010, pp. 191-19)		*A place-based concept that refers to those elements of a home, neighbourhood or city that contribute to the quality of life and wellbeing.*
The United States Environmental Protection Agency & Smart Growth Network (The National Association of Regional Councils (NARC), p. 7)		Liveability *being able to take your kids to school, go to work, see a doctor, drop by the grocery or post office, go out to dinner and a movie, and play with your kids at the park, all without having to get into your car.*
		Liveability means the process of building the communities that help people to live their lives that they want to live: urban centres, small towns, or rural areas.
The North Central Texas Council of Governments (NCTCOG) (The National Association of Regional Councils (NARC), p. 9)	(2011)	Liveability *offer safe, reliable, and economical transportation choices; contain equitable and affordable housing; and enhance economic competitiveness which support the HUD-DOT-EPA Interagency Partnership Principles of Liveability.*
The Economist Intelligence Units' (EUI) (Shamsuddin, Hassanb, & Bilyaminb, 2012, p. 169)	(2012)	Represents one of the aspects which contribute to achieving *a high quality of living.*
		Changing the lifestyle of the citizens and their health condition. Besides, it presents the proof on *the stability of the built environment.*
(Okulicz-Kozaryn, 2013, p. 437)	(2013)	Liveability is *a broader concept than standard.*
		Overall well-being: well-being standard of living (ST)/ liveability includes non-material ingredients freedom, tolerance, self-expression. ST is a level of material comfort in terms of goods and services: health care, clean water, education, houses, apartments, telephones, food, clothing, paved roads, computers and so forth.
		Liveability rankings measure standard of living, not the quality of life.
(Kraftl, 2014, pp. 279-280)	(2014)	How life itself be a *master plan.*
		Liveability achieves through considering *the minutiae of design,* indicators, *securitization, and performance assessment.*
		A specific style of thought about life itself; take into consideration the *both detail and holism.*
		Respect *the multiple governmental practices* constitute a scale at *which intervention into life itself is possible.*
(Giap, Thye, & Aw, 2014, p. 177)		The characteristic of liveable cities is that enable people to *attract a disproportionate amount of the globally-mobile resources (such as talents, high net worth individuals, investors, innovators, entrepreneurs, and capital.* These resources are *recognised to make positive contributions to economic growth, economic resilience, global political influence, world agenda-setting power, socio-cultural innovation, and international lifestyle impact.*
The Centre for liveable Cities Singapore (Mohit & Iyanda, 2016, pp. 863-864)	(2016)	*The city with excellent planning, create a lively, attractive, and secure environment for people to live their life, work and play.*
		Liveability includes *good governance, a competitive economy, high quality of life and environmental sustainability.*
		The term *connotes the ability of a living place to support well-being or quality of life.*
		Concerning what ought to be *the reality of the situation of the inhabitants* for an *existing condition* of an area or a city.

APPENDIX 2

Table 2. Chaos City Amidst Liveability and Hardship of Living: Regional and National Planning

Systems of Framework	The Elements	Characteristics of Liveability	Indicators of Chaos City	Characteristics of Hardship of Living
Regional planning: urban capacities and resources	Natural environment	Mitigating climate change (Betsill, 2001, p. 394).	The physical and mental stress (Bartlett, Dodman, Hardoy, Satterthwaite, & Tacoli, 2011).	The Inability to control in climate change, floods, and ecological change.
	Natural resources	• Availability of food and daily items (Lowe, et al., 2013, p. 16). • Satisfies needs of the present inhabitants and the future generations (Salzano, 1997).	• Black market. • Speculation.	• The inability to stop the waste of the natural resource. • Do not meet the needs of the future generations.
Regional instability: promoting social and political stability	Include non-material ingredients	• Social justice and equity: cities for all; attractive, worthwhile, safe for children, and older people (Hahlweg, 1997; Young & Hermanson, 2012, p. 7). • freedom, tolerance, self-expression, availability of free media, human values, freedom, tolerance, and self-expression (Okulicz-Kozaryn, 2013, p. 437) • How life itself be a master plan (Kraftl, 2014, pp. 279-280	• The dominant of injustice, vileness, disrespect, selfishness, individuality, rejection. • The spread of immorality: drunken and lazy people, drug traffickers across roads, and prostitutes in homes behind cafés. • The proliferation of the kitsch distorts the walls and misshapen sculptures.	• It is a repellent city: unsafe, difficult for children and the elderly to live well. • The benefits are oriented toward a certain category have immunity and nepotism. • The absence of the values of social justice and fairness, dignity and coexistence, vulgarity, artistic decadence, and cultural knowledge, participation and empowerment. • The social misery is dominant. • A lack of ideas to get rid of trash.
	Political Environment	The ability to local democracy, political stability of law enforcement, and prevent crime (Lowe, et al., 2013, p. 16).	• Terrorism and extremism. • Small petty and violent crimes. • Insurgency and civil unrest.	• The spread of disruptive political events, threat of terror, military conflict, and civil unrest/conflict. • Authoritarian context.
National planning: empowered and political representation	The planning efforts on the various aspects of life (Kraftl, 2014, pp. 280-282)	The ability to provide the administrative practices in time, proper development plans, detailed and holistic view, presents a specific style of thought, emergency solutions in the planning, and living places now and in the future.	• Circumvention of the law. • Corruption, access from the back doors (brokerage and bribery). • A blame game for the others. • The apathetic responses of the executive.	• The inability to provide the governmental practices, solutions in time, proper planning policies, and holistic view of the problem-solving • Lack of clarity in practical assignments; mainly the multiplicity of the provider of urban service delivery.

APPENDIX 3

Table 3. Chaos City Amidst Liveability and Hardship of Living: City Management

Systems of Framework	The Elements	Characteristics of Liveability	Indicators of Chaos City	Characteristics of Hardship of Living
Urban governance	Economic development	• Promoting the economic development: anti-bribery, coordinate, leverage federal policies, and investment, as well as enhance economic competitiveness • (Young & Hermanson, 2012, p. 7). • Creating a creative economy based on culture and creativity.	• Monopoly, corrupt businesspeople, and favouritism. • Extended the informal sector • Increase the difficulties minorities have in recovering. • Transfer of indigenous culture of migrants.	• The economic volatility causes the material hardship, increasing the unemployment rates, the proportion of nonworking-age residents, poverty levels, low income per person, and recession. • Affecting economic development become complicated, confusing, inflexible, and are not workable.
	Support the communities	Strong neighbourhood: improve the community living, strength, and vitality (Metcalf, 2002, p. 4; Young & Hermanson, 2012, p. 7).	• IDPs. • Street vendors. • Marginalised people (Abusaada & Elshater, 2019).	• Ignoring the local communities and are neighbourliness units unrespect their value. • Decline standard of living
Municipal corporation	Law, rules, and regulations	• Find effective problem solving (Lash, 1976, p. 33). • Create continuous satisfaction for residents (deChazal, 2010, p. 587). • Extended period plan • Achieve better future.	• Widespread public dissatisfaction. • The menace of stray cattle. • The complicity of bureaucrats in violating by-laws. • Official apathy	• Obsolete laws, ill-conceived policies, and the low capacity of institutions. • The utter disregard for the law. • The apathetic responses of the executive. • Short term plans. • Future is unknown.
	Regional connections	Preserving the unique characteristics that give the diverse (Young & Hermanson, 2012, p. 10)	Instability, disruptive political events, civil unrest.	Increased the political and catastrophic challenges.

APPENDIX 4

Table 4. Chaos City Amidst Liveability and Hardship of Living: Urban Planning and Neighbourhood Units

Systems Framework	The Elements	Characteristics of Liveability	Indicators of Chaos City	Characteristics of Hardship of Living
Urban planning	Urban spaces and public places	• Easy access to areas with green and attractive public spaces for the children and elderly people, where they have a place to play, meet, and talk each other (Hahlweg, 1997; Metcalf, 2002, p. 4_ • Delightful and desirable urban spaces (Timmer & Seymoar, 2005, p. 33)	• The grabbing of open spaces. • Failure to observe general etiquette. • Uses of open areas do not fit into the place.	• The inability to find green spaces and if exists are unhealthy, cheerless, unwanted, unsafe, repellent, and not enjoyable. • Miserable and unacceptable urban spaces and places.
	Affordable Housing	• High environmental standards *around rubbish, graffiti, and housing* (ODPM, 2003, p. 13). • High quality of rental housing, household appliances, furniture, and maintenance services (Metcalf, 2002, p. 4; Young & Hermanson, 2012).	• Equitably has not been distributed • Slums: vulnerable, unsafe, and endangered.	• The extent of crowded housing. • Limited livelihood. • Fall of environmental standards around rubbish, graffiti, and unproper housing.
	Transportation choices	• Easy mobility– by foot, bicycle, public transportation, and even by car where there are no other choices. • Provide timely access to schools, jobs, services, and basic needs.	Used the car make life is humiliating, terrible, random, and unorganised.	• Limited means of transport. • Loss of access to community transportation.
	Infrastructure and services	Quality of private and public education, water, telecommunications, and mobile networks.	Vandalism: distorting traffic plates, damaging bus facilities, cut down trees, and posters.	• Inefficiency of the infrastructure networks, public services, and facilities. • The lack of provide public services.
	Public healthcare	Availability of over-the-counter drugs and medical supplies and services.	The physical and mental stress.	Infectious diseases, sewage, waste disposal, air and sewerage pollution.
The behavioural settings in the public realm	Environmental policies	• High quality of public transportation, road network, international links, street pathways, parking areas, pen spaces and urban places. • Careful planning and supervision of the police force. • Transportation choices and engineering (Young & Hermanson, 2012, p. 10).	• Traffic chaos. • Transformation to the appearance of the countryside, particularly in the way of lifestyle, talk, clothes, habits, and action.	• Traffic congestion, low public transport, lack of adequate parking spaces, and violation of the provisions of the law or its executive regulations • The adverse environmental influences: air pollution, noise, visual intrusion, physical intimidation (of pedestrians and cyclists), and accidents.
A Cross-Cultural Perspective	The needs of the present inhabitants (Salzano, 1997)	• Fit to live in (Shamsuddin, Hassanb, & Bilyaminb, 2012, p. 169) • Creates living places (Kraftl, 2014, pp. 279-280). • Real human experiences desirable (Casselati, 1997). • Live the lives they want to live (Young & Hermanson, 2012, p. 7). • Happiness. • A city is like family (Hahlweg, 1997).	• Corruption adapted from transparency. • Terrorism and extremism • Destruction of property and places.	• Low level of censorship • Social or religious restrictions. • Living places that are not clean. • Spreading of informal settlements • Increasing the IDPs • Marginalised people. • Loss of hope, miser, and frustration.

Chapter 5
GeoDesign and Urban Impact Analysis

Mohammed Ahmed Alfiky
University of Stuttgart, Germany

Hans-Georg Schwarz-v Raumer
Universität Stuttgart, Germany

ABSTRACT

During the past decades, geographic information systems (GIS) have developed massively to be used in various fields. In this regard, within the urban planning and geodesign fields, practitioners, urban planners, developers, designers, and scientists started using GIS as a tool to digitalize and analyze reality. Hence, in this research, besides an introduction to GIS-based modelling and geodesign, the authors analyze an example of the "Filder" area in Stuttgart city based on a set of pre-identified parameters. This case study demonstrates how GIS can be used to avoid the employment of complex methods and tools.

INTRODUCTION

Since computers help to analyze spatial aspects, urban environments are—besides nature conservation—the most visited field of interest. In the beginning mapping of urban land-use, urban zoning plans and conflict analyses by overlay technique were proudly presented as decision background or as a lay down of decision results. However, the capacity of Geographical Information System (GIS) to support spatial scenario storytelling was immediately used when raster GIS systems have been brought to the market. For example, studies on urban sprawl and forecast of future urban expansion got a common GIS-based exercise.

Since 2010, *geodesign* is a term which summarizes the help of GIS in regional and urban planning. Geodesign uses GIS not merely as a tool to have better-informed decisions, but also to address planning as an iterative GIS supported process in which suggestions, interventions, and visions can be tested against its impacts which result from the change of a region's geography (Schwarz-v. Raumer & Stokman, 2014; Steinitz, 2012). This idea leads to the question of how to estimate the environmental impacts of urban structures when being restricted to the tools GIS commonly provides. Most of those

DOI: 10.4018/978-1-5225-9238-9.ch005

impacts only can be quantified by models which are able to map complex physical processes based on not easy to get empirical data. Within this research, the authors show that it is possible to use the limited capacities of GIS to at least implement rough screening models which indicate the severity of impacts of existing urban infrastructures or the direction of changes when considering interventions or future developments. In many cases, this is enough, especially when being used in a communicational stage in urban planning. Besides an introduction into GIS-based impact modeling and geodesign, this contribution presents a summary of a case study which demonstrates how GIS can be used to avoid the employment of complex methods and tools.

BACKGROUND

Geoinformation is booming. People orientate via GPS, locate points of interest by smartphones, and thanks to satellites we can always and everywhere find out who or what is currently in our environment. Geodata surveying "by the way" is just as possible as the precise control of events "on the way." It seems natural to assume that this development influences techniques and methods in which our built and natural-oriented living environments are planned and designed. However, this assumption quickly gives way to skepticism when one compares the wishes and hopes of digital landscape planning with what has remained. The big promises of web-accelerated planning today are delivered only (but at least) by the offer that actors and interested persons have the chance to be quickly and comprehensively web-informed (Schwarz-v. Raumer & Stokman, 2013). To successfully transport geoinformation into the sphere of the design process, a field between technology and creativity must be opened. The attempt to engage geo-IT and design as "Geodesign" is burdened with some obstacles (Schwarz-v. Raumer & Stokman, 2012), but it works and is illustrated in this chapter.

Geodesign is not an emerging activity or practice. Nevertheless, it has been practiced for a long time ago. Li & Milburn (2016) and Miller (2012) argued that any action that incorporates dealing with geography, space design or allocation, hunting animals or selecting materials for shelter construction is considered as geodesign activity. Many designers and architects used the methodology of geodesign without mentioning the term though. For example, Ian McHarg (1920–2001), in his book *Design with Nature* (1969), introduced a geo-based methodology for landscape architecture and dealing with the surrounding environment. Not only MacHarg but also Frank Lloyd Wright (1867–1959) and Richard Neutra (1892–1970) gave full attention not merely to their clients but also the context of the design (Miller, 2012).

Nevertheless, the term *geodesign* did not appear until the year 1993 (Kunzmann, 1993); Kunzmann uses it to discuss scenarios communicating spatial structure ideas (Schwarz-v. Raumer & Stokman, 2011). Geodesign is made up of two syllables: *geo-* and *design*. According to the *Cambridge Dictionary* (n.d.), *geo-* is a prefix that is "of or relating to the earth." Hence, anything that is earth related like land, water, material, in or above the earth, 2D or 3D is considered as *geo*. While design could be defined either as a noun or verb, as a noun, it is "the way in which something is planned and made," while as a verb it is, "to make or draw plans for something that will be produced" (*Cambridge Dictionary*). Hence, geodesign could be defined as the process of planning or developing an entity within a geographical context. Alternatively and more scientifically, Flaxman (2010, p. 29) defined geodesign as "a design and planning method which tightly couples the creation of a design proposal with impact simulations informed by geographic context."

Li & Miburn (2016) developed four phases of historical development for geodesign for landscape architecture. However, those phases are valid for other geodesign activities, not landscape architecture inclusive. Firstly, from 1850 until 1950, they called it the *analogue era*, when the technology was not yet developed to pursue complicated analysis. At that time, the hand drawing and analogue representation were dominant. Additionally, designers neglected natural and social aspects during this era until Patrick Geddes[1] criticized that in 1917 (Li & Milburn, 2016; Meller, 1990). This was another start for geodesign, along with Warren H. Manning in 1912, using an overlaying mapping technique to develop a landscape map for the entire USA from 363 maps (Li & Milburn, 2016; Miller, 2012). Manning's technique fostered integrating information and analysis within the design process, along with providing a base for GIS development as a spatial decision-making tool (Li & Milburn, 2016).

The second stage, named the *poor data era*, dates back from 1950 until the mid-1970s. On the one hand, the awareness of environmental challenges and means of facing them increased, while on the other hand, there was a lack of accurate data. However, during this period, three developments worked as a cornerstone for geodesign: academic institutions focused more on environmental issues, ecological determinism, and the advent of digital data and GIS[2]. Additionally, the quantitative revolution of spatial statistics played a major role in spatial data analysis (Goodchild & Haining, 2004).

Third, the *small data era* lasted from the mid-1970s until 2000, when GIS developed rapidly as a tool for politicians and academics for decision making within the planning practice (Li & Milburn, 2016; Thakur & Sharma, 2009). This development is based on the development of the gestalt inventory, the Laboratory for Computer Graphics and Spatial Analysis and the Synagraphic Mapping System (SYMAP[2]) program developed by Howard Fisher (Haddad, 2015; Miller, 2012). Finally, the fourth phase, the *big data era* (2000 until now) is when the human mind can no longer handle the amount of increasing information, developed technologies and geospatial science. Since then, geodesign is focused more on design scenarios evaluation and impact assessment (Crooks, 2013; Li & Milburn, 2016; Schwarz-v. Raumer & Stokman, 2011).

Scenarios evaluation and impact assessment are keystones for geodesign process. In this regard, Steinitz (2012) developed a framework for geodesign. Within his framework, he managed to propose six questions that need to be answered and agreed upon with the participating stakeholders for any design task (Steinitz, 2010, 2012; see Figure 1). Each of the questions targets a different aspect, yet all of them together draw the whole image of the geodesign framework.

According to Steinitz (2012), the first question targets the representation models, while the second targets the process model. The third and the fourth questions target evaluation and change models respectively. Finally, the fifth and sixth model, they target the impact and decision models respectively.

Those questions go through three iterations to reach a coherent design. The first iteration is sequential, aiming to develop an understanding of the context and the scope of the study. In other words, *why* we are doing this study. On the contrary, the second iteration goes from the last question to the first. This is the *how* iteration. It focuses on the methodology. The last iteration returns to sequential order. It is targeted at the implementation phase. Here the focus is on the *what, where* and *when* questions. Within those three iterations, the evaluation and impact models are present and vibrant within the study. This is what makes geodesign discrete. It is not static design, however. It is a dynamic design process (Steinitz, 2016) with various loops of iterations, trying to develop a better understanding for the surrounding context, aiming at developing a comprehensive design.

Figure 1. The stakeholders, the geodesign team, and the framework for geodesign
Source: (Steinitz, 2012, p. 25)

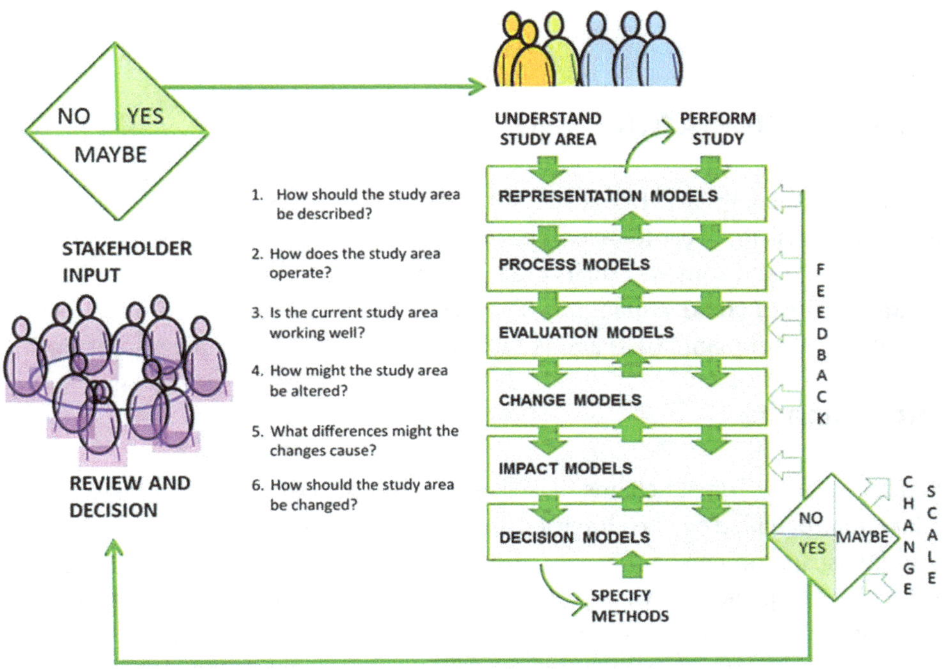

An understanding of the surrounding context is one of the factors that makes geodesign unique. The context is always vital for the design process. Hence, there are three aspects to study within the context: 1) geography; 2) scale; and 3) size (Crooks, 2013; Steinitz, 2012, 2016). Regarding *geography*, Steinitz (2012, 2016) argues that from place to place the design and the design process will be adjusted as per people culture. Additionally, he explained that identifying the *scale* is vital for the design process. Are we designing for the whole humanity (i.e., Global scale), are we designing for locals (i.e., Local scale), or is it something in between (i.e., Regional scale) that is being influenced by the other two scales? Finally, Steinitz clearly shows that the *size* of geodesign projects impacts a couple of other factors. As the project size increases, the risk and the scientific complexity are directly related. On the other side, public knowledge and public understanding are inversely related[4] (Steinitz, 2012, 2016).

It is worth mentioning that geodesign is just a buzzword, that ESRI (the GIS industry leading company) is trying to market within the past decade (Haddad, 2015). Yet, this does not mean that the word geodesign is misleading; on the contrary, it is setting a framework for integrated spatial planning. GIS as a tool for geodesign helped in engaging and integrating communities within the decision-making process (Thakur & Sharma, 2009). Along with developing new ways of storing and overlaying data. However, according to Thankur & Sharma (2009), GIS has a couple of challenges. Certainly, data comes on top of those challenges. Lack of data or lack of reliable data is such a nightmare for most of the development project. Additionally, accessing or collecting accurate data is often costly (Stewart Fotheringham & Rogerson, 1993; Thakur & Sharma, 2009). Additionally, GIS and its applications need well-developed infrastructures that are expensive. Moreover, incorporating non-spatial datasets with spatial datasets is not always easy or sometimes does not represent the real environment. The issue of scale that was mentioned by Steinitz pops up as a real challenge within the representation of the data in GIS (Thakur & Sharma, 2009). Finally, it is notable that GIS is a tool for digitalizing geodesign and screening scenarios and not for detailed analysis.

A SET OF IMPACT INDICATORS AND TOOLS

This chapter illustrates the help of GIS in estimating ecological impacts of urban land change (Table 1). Firstly, the authors discuss main ecological impacts, afterward, introduce a GIS-model to estimate them and finally, present a tool and demonstrate its usage in a design process. This process is assumed as *iterative*, a core characteristic of creative planning. The use of iteration loops is a core concept of geodesign[5] (Schwarz-v. Raumer & Stokman, 2013) and is illustrated in a case study on urban extension planning.

Land Change Impacts

The main impacts or better impact chains that planners have to expect when transferring cultivated or uncultivated land to urban land can be classified as changes in land-cover, changes in landscape patterns and changes in landscape processes (Forman, 1995). At the level of land-cover change impact can be described as direct consequences of what planners often address with "land consumption." The rareness of a specific landscape type and reversibility of land cover change must be considered when evaluating this type of impacts. When considering landscape patterns spatial relations like a neighborhood, distance or connectivity should be taken into consideration (see Figure 2). Finally, planners must address the specific processes related to the change in cover and patterns. These changes are related to both human activities as well as to movements and dispersal of other organisms and fluxes in mater and materials.

To respect land change effects in strategic urban planning on a scientific or engineering basis huge effort is necessary. Thus, in an explorative planning phase, the authors must estimate impacts on a low level of evidence using a rough indicator model. The indicator calculations cannot be considered as the truth but as a hint, a direction. The indicators play the role of a design control variable rather than of a forecast.

Soil Related Effects

Soil protection is one of the most challenging issues in landscape related urban planning. Worldwide the development of settlements is closely related to the presence of fertile soils and water availability. In the case of soils, this initial advantages of the location under modern societal and economic conditions are turned into conflicting claims for land when urban land-use is due to expand. The importance of soils for nutrition security and as a bio-economic resource for energetic or material purposes has led to a widespread and detailed application of methods for soil survey and evaluation. Hence, the conflict between competing for land-use types can be solved using a soil map and the related description of soil properties including an evaluation of suitability for specific land use types.

The State Institute for the Environment, Baden-Württemberg (LUBW) provides a classification of the importance of soil regarding to crop cultivation (2010). Nutrient availability (cation exchange capacity) and water supply (based on soil texture dependent field capacity) are factors under consideration and

Table 1. A set of simple impact indicators (Change parameter = Parameter necessary for new land-use type)

Affected Landscape Sphere	Effects	Change Indicator	Data	GIS-Method	Measurement Unit	Short Name
Soil	• Loss of fertile soils	• Area of land suitable for crop cultivation	• LUBW (2010) soil evaluation data	• Overlay and summary statistics	• Number of Raster cells indicating high soil quality	• Fertile soil
	• Reduced risk of soil loss by erosion from precipitation	• Soil loss by erosion from precipitation	• DHM • C-factors • K-factor • LS-Factor	• raster calculation	• Number of Raster cells indicating high soil erosion	• Erosion
	• Reduced risk of eutrophication of creeks	• Nutrient input from agricultural fields	• Location of arable land, creeks and surface-watersheds	• neighborhood analysis	• Number of Raster cells indicating an input of sediment into creeks	• Eutrophication
Water	• provision of infiltration • retention capacity • flood risk	• Area of land with high importance to water balance	• LUBW (2010) soil evaluation data and percentage of sealed area	• Overlay • summary statistics	• Number of Raster cells indicating high importance for water retention	• Water balance
Biota	• Risk of species extinction by a decline of provided habitat • habitat quality • connectivity	• Area of land inside biotope network	• LUBW (2012) polygons for wet, dry and grassland network.	• Overlay • summary statistics	• Number of Raster cells indicating high habitat quality	• Habitat quality
Air, noise and local climate	• Heat island effect	• Change in surface temperature	• Land cover map	• Calculation from regression model	• Sum of temperature change per raster cells	• Heat island
	• Ventilation	• Loss in cold air production and flow relevant land	• Map on cold air production and flow	• Overlay • summary statistics	• Map on cold air production and flow	• Ventilation
Visual quality and recreation	• Visual quality decline	• Scenic beauty	• Calculations from regression model	• Overlay • summary statistics	• Amount of raster cells above the mean value of visual quality	• Visual quality

Source: the authors

lead to a delineation of areas of "high" and "very high" suitability for crop cultivation. The decline of land categorized as such is taken as an indicator for the urban land change.

A second rather contrasting impact of the land change is the reduction of negative effects from land cultivation practice when soils are used for agriculture. Two severe fields of such effects exist: (a) eutrophication of natural biotopes and surface water and (b) groundwater contamination through the lack of filtering and buffering capacity or through nitrate leaching caused by unadjusted nitrogen fertilization. A GIS analysis calculates soil erosion according to the Universal Soil Loss Equation (Wischmeier &

Figure 2. Effects of land change
Source: the authors.

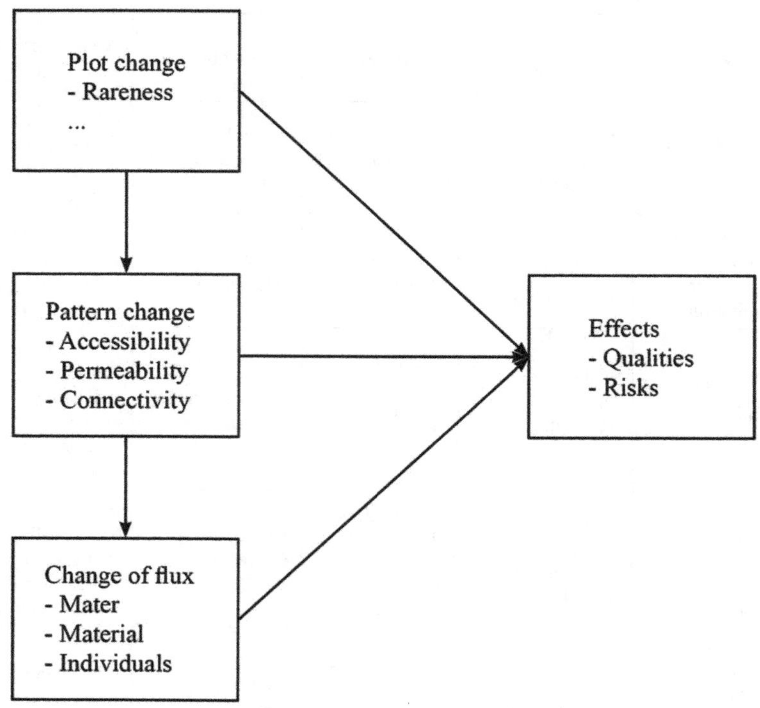

Smith, 1978) and reports the sum area assigned to a specific soil erosion risk class. The sum of cells which have direct input of eroded sediments into biotopes or creeks, the authors assume as an indicator for eutrophication.

Water Balance Related Effects

Any change in land-use covers the change in the land cover which leads to different hydrological characteristics of a plot. In this regard, the share of evapotranspiration, surface runoff, and infiltration from precipitation as well as water storage and percolation will alter as a consequence of land-use type transfer. Flood management takes this into account and plot-wise accounts the total flood generating surface runoff in a catchment area to estimate flood risk. When considering different types of urban plots, the percentage of a sealed surface is a key indicator for the contribution to flood generation, assuming that infiltration, percolation and retention capacity of unremoved soils and unsealed underground leads to significantly less fast surface transport of water towards the creeks and rivers. LUBW (2010) suggests an assessment of the importance of soil for the water balance. This indicator is based on soil infiltration and storage capacity. When assessing land change, the reduction of the area with assignments "high" and "very high" is used as a control indicator. Land categorized as such helps to prevent flood and to recharge ground-water reservoirs.

Biota Related Effects

The loss of habitat and the decline of permeability of open space are severe effects of urbanization. The problem is, that, when urban land overtakes agricultural land this is the second "big killer" (Maxwell, Fuller, Brooks, & Watson, 2016) which removes the scattered relicts of habitat necessary for the remaining species. A metapopulation theory (Levins, 1969) approach and treatment of landscape in planning (Fahrig, 2003; Hanski, 1999) when considering biota related effects is important, but it is neither well developed nor established. To respect nature protection area in planning is not enough. As a guardrail for urban planners, it is necessary to establish both a target species concept that includes the delineation of habitat structures as well as a comprehensive setting of biotope network structures which are reserved for species daily, seasonal or strolling movement, migration, shift in species home range (e.g., due to climate change or disturbance by human activities) and population dispersal.

LUBW (2012) suggests three networks of biotopes and connecting areas as a guideline for planning. The concept considers species assemblages which use either wet, dry or grassland habitat and results in a delineation of three accordant polygon datasets which include the area connecting core patches within a distance of 500m. As an indicator, the authors calculate the area which is covered and affected by urban developments.

Local Climate, Noise and Air Quality Effects

Urban atmosphere as human environment mainly suffers from two main causes: overheating materials and the presence of fuel combustion of vehicles, and there is an aggravation effect by reduced air ventilation which has the capacity to mitigate both concentrations of harmful substances and thermal stress in the urban canyons (Steel 2001, Erell et al. 2011). It is broadly accepted, that green and blue infrastructures are mitigating these effects and so can be considered as the backbones of livable cities. Likewise, fostering appropriate developments leads to a regain of urban environmental outdoor comfort. To indicate air and climate-related components of urban comfort it is necessary to have a comprehensive monitoring system or to apply complex physical models, and better to have both.

Simple and rough change indicators should follow the three main factors of atmospheric environmental stress occurring in urban areas: surface material, ventilation and amount of traffic. Some authors estimate surface temperature assigned to a location depending on land cover in a neighborhood using a regression model. Zhang (2012) suggests a model that is derivate for the Stuttgart Region (see Table 2). Therefore, to indicate effects with respect to ventilation, it is important to know the location of cold air production area and cold air flow channels. The sum of area related to an intervention which is inside such locations is a rough indicator of ventilation related effects. Hence, to estimate the effects of additional traffic it is not possible to suggest a simple indicator. Neither current daily vehicle load on streets and roads is easy to estimate, nor it is easy to make rough assumptions on the directions of traffic generation by new urban development.

Visual and Recreational Quality Effects

There is a need to preserve and to provide natural landscapes due to its visual quality (Bishop,1994; Roth et al. 2018). Planning conventions and targets fixed in planning regulations and nature conservation laws, as well as broad findings from social science, prove that human beings are evaluating landscape quality

Table 2. Regression model for heat island. The dependent variable is land surface temperature on 29 August in 2005 (14:04–18:45).

	Coefficients	
	Unstandardized B	**Standardized Beta**
Constant	33.784	
Share Forest	-0.063	-0.766
Share Green Garden	-0.03	-0.213
Share commercial and industrial	0.029	0.084
Share Grassland	-0.024	-0.088
Share Open spaces in industrial area	0.019	0.070
Share Railway station	0.019	0.060
Share Housing area	1.937	0.062
Share Open spaces in residential area	-0.012	-0.081
Share Green area	-0.009	-0.057
Share Lake, reservoir, pond	-0.013	-0.035
Share Mixed use area	0.017	0.027
Share Other kinds of buildings	0.013	0.014

Source: (Zhang, 2012, pp. 31, 34)

from an immaterial point of view and relate aspects of identity, sensation, esthetics or mystics to visually perceived landscapes. In addition, many recreational effects of landscape as a human environment are generated by its visual appearance. The authors have also observed that given more associations with nature and fewer associations with recent anthropogenic activity, visual quality is typically assessed as high.

Urban and landscape planning have developed a series of methods to estimate the visual quality of a given landscape. In general, the authors can distinguish between local methods, which consider a single scenic view, and regional methods, which try to assign a score of visual landscape quality to a spatial extent at a regional scale. In this project, the authors use data which have been developed by Roser (2011). They result from individual scores given to a set of pictures in an interview campaign transferred to the whole state of Baden-Württemberg. The transfer is done by a regression model Table 3 for visual quality depending on the density of different landscape elements. The recreational effects are closely related to visual landscape quality; higher is better. However, recreation depends on the activity specific attractiveness of infrastructures provided at and the disturbances joint with a location. Due to its complexity, detailed recreational aspects cannot be included in a simple indicator model.

AN EXAMPLE OF APPLICATION IN STRATEGIC URBAN PLANNING

The case study region is located at the urban-rural fringe of Stuttgart City. Here, rural and urban landscapes are mixed up, and rural land use is intensively affected by urban activities and urbanization processes and vice versa. Complex interrelationships lead to specific conflicts and synergies between land traditionally called "rural" and "urban." Daniels (1999) broadly describes this collision of rural and urban interests

Table 3. Regression model for visual quality. The dependent variable is the mean subjective grade given by people evaluating an image.

Landscape Element	Max. Appearing Distance [m]	Standardized Beta
Crop land	250	-.125
Transmission lines	250	-.174
Grassland	250	.112
Industry and commerce	250	-.143
Industry and commerce	1000	-.166
Land cover diversity	250	.152
Plantations	250	-.028
Topography	250	.164
Cultural points of interest	1000	.088
Main roads	250	-.178
Extensively used orchards	250	.102
Scattered settlements	250	-0.81
Forest	250	.152
Surface water	250	.140
Wind turbines	250	-0.067
Vineyards	250	.050
Other built-up area	250	-.082

Source: (Roser, 2014, p. 17)

of an area which is called *Zwischenstadt* by (Sieverts, 1997). Schwarz-v. Raumer, Jörg, & Alfiky (2016) introduced the region to students, to have some exercises in using geo-design tools. This contribution now shows the ability to use the tools described above to be applied in a strategic planning process.

Short Description of the Case Study Area

The so-called "Filder" (which means "fields") region is located on a fertile plateau of a slightly undulated topography. In this region, some of the best soils for agricultural production in the South of Germany can be found. So, the region is a hot spot in intensive agriculture, with a high production rate of cabbage as a basis for "Sauerkraut," Germany is famous for. Filder is located 15 km far from Stuttgart City center and represents an important suburban residential area hosting more than 300.000 dwellers Figure 3. Situated in a flat, open plain and located in an intersection of the national road network, Filder suffers from much land-consuming infrastructure projects like Highway A8, Stuttgart Airport or the Stuttgart Trade Fair Appendix 1. Railway project "Stuttgart 21" includes a high-speed rail line from Stuttgart City center to Filder. This will be the driver for further housing and industrial development. As a consequence, the existing lack of green infrastructure will be aggravated, and the pressure for additional land for construction will lead to a loss of primary agriculture land.

Figure 3. Land-use map
Source: The authors

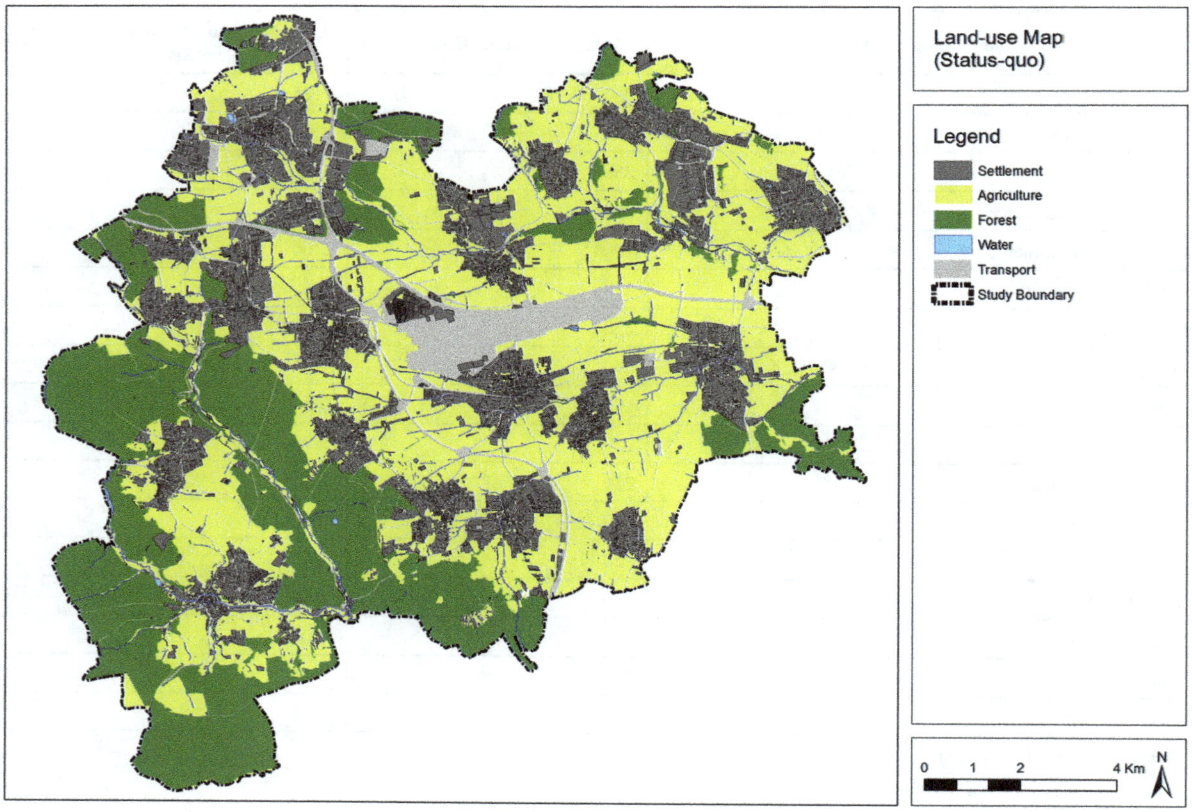

Methodology: A Series of Planning Iterations

One of the significant challenges in geodesign is to realize fast iterations which evaluate design decisions and provoke alterations of the proposed interventions in the next step (Schwarz-v. Raumer & Stokman, 2013). To demonstrate this, a planning scenario for the case study region is described. As explained above, Filder is supposed to go through extraordinary infrastructure development. To develop an ecologically optimized suggestion to control such development strategically, the authors follow a planning procedure which starts with a suggestion that follows past tendencies of settlement growth (initial step). Having edited this suggestion in a geo-dataset, it can be assessed by geodesign tools which report on key indicators. The analysis of the assessment leads to the identification of hotspots of conflicts to be avoided by a first alteration of the initial design (iteration step). Several iteration steps follow, maybe going back to the proposal some steps before. A protocol of the assessment indicators then shows their development during the planning process.

The methodology of proposing any change in land-use cover (LUC) is following five main steps: (a) *Reasoning*, why and where this change is taking place; (b) *Editing geometry*, where the power of GIS is used to change the geometry of a selected area from one form to another; (c) *Assign new LUC*, another advantage of the GIS tool, which the authors used to update the land-use of the selected area for development; (d) *Specify model parameters*, in which the authors defined a set of impact indicators to calculate the impact change for different LUC; and finally, (e) *Recalculate*, to recalculate the new impact indicators based on the different iterations. It is worth mentioning that the methodology the authors adopted in this example is contributing only to phases 4 and phase 5[6] (i.e., Change models and Impact models) proposed in the Steinitz model (see Figure 1).

Discussion

Urban development is booming all around Stuttgart Region as a result of the Stuttgart 21 project. Since the airport is an attraction pole for economic development, the authors suggested that the first step of development (i.e., Iteration 1; see Appendix 3) in the Fildern area will take place around the airport (see Figure 4). Hence, a new settlement will be allocated. Afterward, the impact indicators are calculated to compare the impact of such development on the overall area of the Fildern. Although most of the indicators are negatively affected by such development (see Table 4), there is an improvement in both erosion and eutrophication indicators. Subsequently, the next step is trying to alter the negative impacts and to maintain the improvements. Thus, the authors propose two alternatives for further development plans (see Appendix 2), either maximizing economic benefits or preserving agricultural land. Hence, alternatives 2a or 2b were introduced.

On the one hand, alternative 2a suggests that development will take place over forests instead of agricultural land[8], breaking the long-lasting planning tradition to spare forests. On the other hand, alternative 2b proceeds spatially the development near the airport suggested in the first iteration. Afterward, the authors run the iteration model to compare the impact indicators of the two alternatives. Despite there are slight differences between the two alternatives, yet alternative 2a has a better result for most indicators (see Figure 5). Therefore, the decision is to make alterations on alternative 2a in order to

Figure 4. Iterations of GIS supporting planning process (authors)

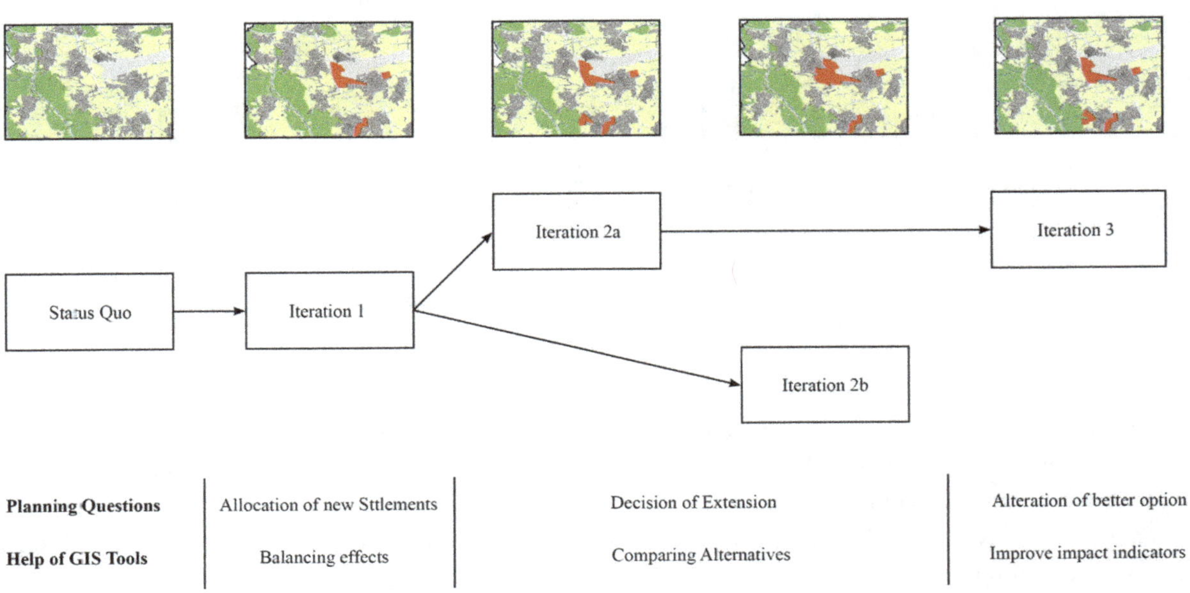

Table 4. Change of indicator values (= raster cells according to column (6) in Table 1)

Indicator Short Name	Status Quo	Iteration			
		1	2a	2b	3
Fertile soil	53728	49992	49990	49450	47053
Erosion	12099	11861	11861	11833	11861
Eutrophication	1575	1567	1567	1532	1567
Water balance	85104	80699	80292	80508	80355
Habitat loss	74547	72621	72621	71806	72556
Heat island	-186.108086	-141.838117	-137.230168	-136.736459	-137.50445
Ventilation	29576	29110	29110	28612	29045
Visual quality	______[7]	33	58	55	56

Source: The authors

Figure 5. Comparing change indicators
Source: The authors

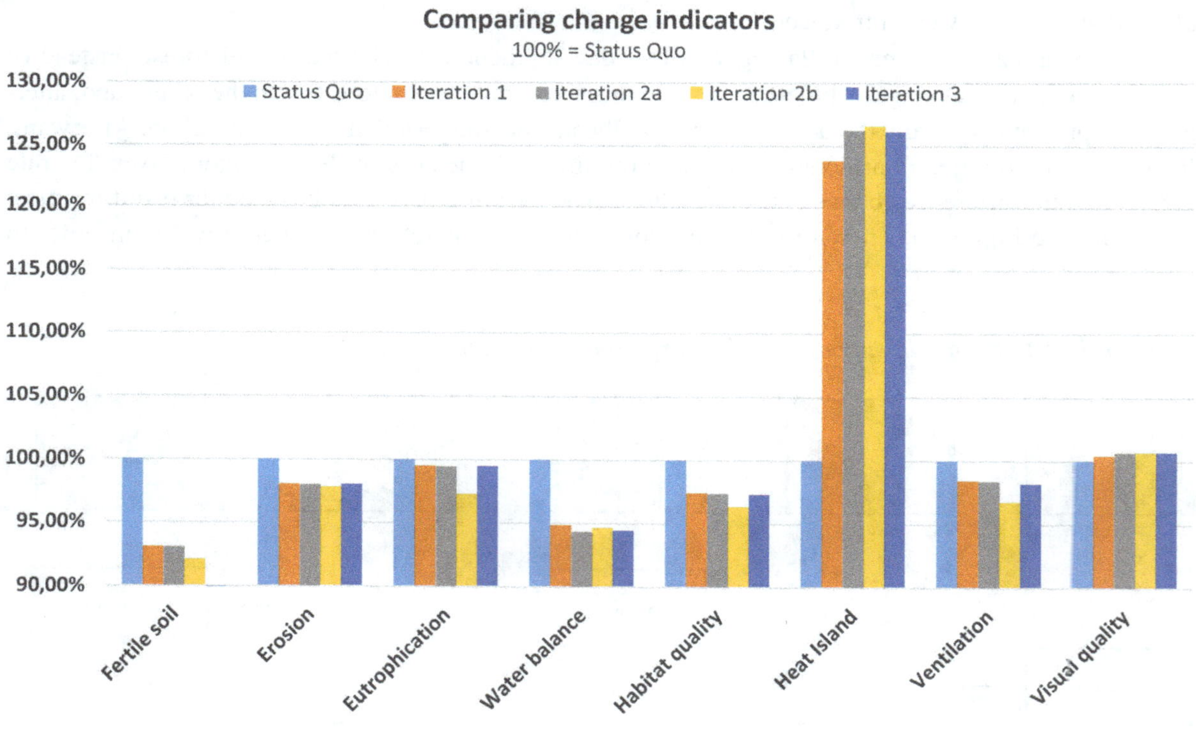

enhance the impact indicators. Consequently, alternative 3 is introduced. In this regard, the authors try to compromise between further development over forests and agricultural land to maintain a minimal negative effect over impact indicators.

This case study shows the main principle of how GIS-informed decision is used in a geodesign process. According to Steinitz (2010), we have shown a small example of a sequential design strategy.

Additionally, this example shows the power of geodesign methodology in identifying change models and impact models. As well as, highlighting the power of GIS as a tool in running iterations and comparing the results in order to choose the suitable strategy for development. The GIS methods applied in the case study use common techniques to identify changes in land qualities caused by development. Those techniques are an approximation which can be done by geographical representation coincidence and distance analysis. This is enough for geodesign in a communicative environment. For detailed impact analysis more, sophisticated models must be applied.

FUTURE RESEARCH DIRECTIONS

The ideal concept of geodesign communicated so far is iterative-cyclical based on the idea of working with GIS and models in a short-circuiting instead of a usually linear input-processing-output procedure. The role of using geodata analysis, models, and visualization is no longer limited to a one-off production process. In ideal cases, fast repetitions are possible that allow for creative development through trial and assessment. The design workflows follow the paradigm that spatial information (also in the refined form of complex spatial analysis) and the anticipation of the effects of spatial interventions are indispensable ingredients for a successful design process. Related challenges can be summarized as following (Schwarz-v. Raumer & Stokman, 2013):

- **Technical Challenges:** Geodesign claims to be able to introduce complex spatial relationships into creative decision-making contexts on the fly. High demands on input devices can now increasingly be satisfied. Concerning user interfaces, however that, there is a need for development. Even after the legendary invention of ArcSketch by ESRI (a tool which tries to offer better facilities to improve mapping language), the existing GIS systems do not offer any convenient functionality to transform an intuitive sketch into an operable geo-dataset. Another challenge is the computational capacity that is available (commonly) for geodesign projects. To put it briefly: the speed of geo-processing in ArcGIS is not sufficient to deliver the result of complex valuation models within a response time corresponding to a workflow which prevents idle time and breaks down of creativity. The use of GIS geoprocessing functionalities is not an option when it comes to "high-speed" calculations of complex process models. As stated above, GIS is only an option if practitioners and researchers can accept simple evaluation models.
- **The Problem of Collaboration:** The described ideal of a geodesign includes in which planners/designers, IT professionals, possibly scientific specialists, and other stakeholders work together in a network. There are hardly any systematically worked-up experiences about basic mechanisms and formats for how to optimally organize this collaboration. Steiniz (2012) develops planning case studies and documents examples of collective planning. *GeoDesign in the Curriculum* discusses educational geodesign formats, and Schwarz-v. Raumer & Stokman (2014) presents a suggestion on how to structure collaboration. Basically, the geodesign approach here offers a wide and not simple experimental field.
- **Claim and Reality:** A cross-sectional analysis of what is currently known under the label "geodesign" revealed that the framework concepts of geodesign are only partially realized in the documented projects, yet they are communicated with much euphoria as geodesign and proclaimed as methodological-technical innovations. On the other hand, Artz's (2010) claim that, "We've

been doing GeoDesign for years" was a statement commonly overheard at the first Geodesign Summit held in January 2010. The question about being an innovation or not can hardly be decided. However, we now benefit from the fact that in using the label "geodesign," we have a term for everything geographers and planners have been doing for a long time. So, it is no problem to embrace a lot of GIS applications (including small ones) with the term *geodesign* including geodata-based design, anticipative design, collective-participatory landscaping, or even spatio-temporal scenario vision and GIS-based multi-criteria suitability assessment.

CONCLUSION

The analysis of the Filder area using the proposed model, helps in understanding the consequences of various urban development decisions. The authors give a short example of how to apply a logical urban planning technique using geodesign methodology along with GIS tools. In this regard, the four proposed development options take place around the airport—a very attractive development pole—and over the forest, trying to preserve the agricultural land. Each of those proposed iterations shows the impact of development over the pre-defined indicators. It is crucial to mention that not all the indicators are always negatively impacted by development. Yet, stakeholders have then to compromise their needs according to the estimated results. For example, on one side, Iteration 1, proposing development on the periphery of the airport has a great financial value for developers. On the other side, the city of Stuttgart is losing a very good agricultural land? And here arise a couple of questions: Are we ready to make such a trade-off? And what are the impacts of losing such land not only on the food production but also on the environment?

To conclude, GIS addresses urban planners (rather than architects) and for them, GIS is well known as a convenient tool to produce maps for better-informed planning. The authors' contribution shows that there is a big potential of GIS beyond mapping. Geodesign acts as an interactive assistant in decision making, not easy to establish, but more and more a must in a world of increasing complexity.

REFERENCES

Artz, M. (2010). GeoDesign. Changing Geography by Design. *GeoDesign. Changing Geography by Design. Selected Readings in GeoDesign.* ESRI Press. Retrieved from www.esri.com/library/ebooks/GeoDesign.pdf

Cambridge Dictionary. (n.d.). *Cambridge Dictionary | English Dictionary, Translations & Thesaurus.* Retrieved September 4, 2018, from https://dictionary.cambridge.org/

Crooks, A. (2013). Reviews: A Framework for geodesign: changing geography by design. *Environment and Planning. B, Planning & Design*, *40*, 1122–1123. doi:10.1068/b4006rvw

Fahrig, L. L. (2003). Effects of Habitat Fragmentation on Biodiversity. *Annual Review of Ecology Evolution and Systematics*, *34*(1), 487–515. doi:10.1146/annurev.ecolsys.34.011802.132419

Flaxman, M. (2010). Fundamentals of Geodesign. In *Digital Landscape Architecture 2010* (pp. 28–41). Dessau: Wichmann. Retrieved from http://www.kolleg.loel.hs-anhalt.de/landschaftsinformatik/fileadmin/user_upload/_temp_/2010/Proceedings/Buhmann_28-41.pdf

Forman, R. T. T. (1995). Land Mosaics. *Networks*.

Goodchild, M. F., & Haining, R. P. (2004). GIS and spatial data analysis: Converging perspectives. *Papers in Regional Science*, *83*(1), 363–385. doi:10.100710110-003-0190-y

Haddad, M. A. (2015). Review: Steinitz, Carl. 2012. A Framework for Geodesign: Changing Geography by Design. *Journal of Planning Education and Research*, *35*(2), 228–230. doi:10.1177/0739456X15581606

Hanski, I. (1999). Habitat Connectivity, Habitat Continuity, and Metapopulations in Dynamic Landscapes. *Oikos*, *87*(2), 209. doi:10.2307/3546736

Kunzmann, K. R. (1993). Geodesign: Chance oder Gefahr? In Bundesforschungsanstalt für Landeskunde und Raumornung (Ed.), Planungskartographie und Geodesign (pp. 389–396). Informationen zur Raumentwicklung.

Levins, R. (1969). Some demographic and genetic consequences of environmental heterogeneity for biological control. *Bulletin of the Entomological Society of America*, *15*, 237–240. doi:10.1093/besa/15.3.237

Li, W., & Milburn, L. A. (2016). The evolution of geodesign as a design and planning tool. *Landscape and Urban Planning*, *156*, 5–8. doi:10.1016/j.landurbplan.2016.09.009

LUBW (Landesanstalst für Umwelt Baden-Württemberg). (2010). Bewertung von Böden nach ihrer Leistungsfähigkeit. Leitfaden für Planungen und Gestattungsverfahren. *Reihe Bodenschutz, 23*.

LUBW. (2012). *Landesanstalst für Umwelt Baden-Württemberg*. Fachplan Landesweiter Biotopverbund. Arbeitshilfe.

Maxwell, S. L., Fuller, R. A., Brooks, T. M., & Watson, J. E. M. (2016). Biodiversity: The ravages of guns, nets and bulldozers. *Nature*, *536*(7615), 143–145. doi:10.1038/536143a PMID:27510207

McHarg, I. L. (1969). *Design with nature*. New York: Natural History Press.

Meller, H. (1990). *Patrick Geddes: Social Evolutionist and City Planner* (N. Wrigley, Ed.). London: Taylor & Francis e-Library.

Miller, W. R. (2012). *Introducing Geodesign : The Concept Director of GeoDesign Services*. Redlands: Esri Press. Retrieved from http://www.esri.com/~/media/Files/Pdfs/technology-topics/introducing-geodesign.pdf

Roser, F. (2011). *Entwicklung einer methode zur großflächigen rechnergestützten Analyse des landschaftsästhetischen Potenzials*. Berlin: Weißensee.

Schwarz-v. Raumer, H.-G., & Stokman, A. (2011). GeoDesign-Approximations of a catchphrase. *Proceedings of Digital Landscape Architecture*, 106–115.

Schwarz-v. Raumer, H.-G., & Stokman, A. (2012). *GeoDesign : Herausforderungen an einen verständigen Umgang mit GIS Einleitung Wunsch und Wirklichkeit*. Academic Press.

Schwarz-v. Raumer, H.-G., & Stokman, A. (2013). GeoDesign : Herausforderungen an einen verständigen Umgang mit GIS Einleitung Wunsch und Wirklichkeit. In 18. Münchner Fortbildungsseminar Geoinformationssysteme (pp. 297–306). Academic Press.

Schwarz-v. Raumer, H.-G., & Stokman, A. (2014). Integrating Technology, Science and Creativity-A Challenge for Collaborative Settings in Geodesign. *Digital Landscape Architecture*, 15–26. Retrieved from http://dla2014.ethz.ch/talk_pdfs/DLA_2014_Keynote_Schwarz_v_Raumer.pdf

Schwarz-v. Raumer, H., Jörg, J., & Alfiky, M. (2016). Respecting the Role of Agriculture for an Untegrated Landscape Development at the Urban-rural Fringe Using Geodesign Tools. *Journal of Digital Landscape Architecture*, (1–201), 327–336. doi:10.14627/537612037.This

Sieverts, T. (1997). *Zwischenstadt zwischen Ort und Welt, Raum und Zeit, Stadt und Land.* Bauwelt Fundamente.

Steinitz, C. (2010). Landscape Architecture into the 21st Century-Methods for Digital Techniques. In *Digital Landscape Architecture 2010* (pp. 2–26). Burner: Wichmann. Retrieved from http://www.kolleg.loel.hs-anhalt.de/landschaftsinformatik/fileadmin/user_upload/_temp_/2010/Proceedings/Buhmann_2-26.pdf

Steinitz, C. (2012). *A framework for geodesign : changing geography by design.* Esri. Retrieved from https://esripress.esri.com/display/index.cfm?fuseaction=display&websiteID=218&moduleID=0

Steinitz, C. (2013). Beginnings of Geodesign: A Personal Historical Perspective. In *Geodesign: Past, Present, and Future* (pp. 4–14). Redlands: ESRI Press. Retrieved from http://www.esri.com/~/media/Files/Pdfs/library/ebooks/geodesign-past-present-future.pdf

Steinitz, C. (2016). *GeoDesign Dynamics - Carl Steinitz.* Coastal Regional Commision- Planning Department. Retrieved from https://www.youtube.com/watch?v=Jf_R4rB7MIQ&t=2021s

Stewart Fotheringham, A., & Rogerson, P. A. (1993). GIS and spatial analytical problems. *International Journal of Geographical Information Systems*, *7*(1), 3–19. doi:10.1080/02693799308901936

Thakur, R., & Sharma, M. (2009). Local Government Use of GIS in Comprehensive Planning. In J. D. Gatrell & R. R. Jensen (Eds.), Planning and Socioeconomic Applications (pp. 125–137). Springer. 13 doi:10.1007/978-1-4020-9642-6

Wischmeier, W. H., & Smith, D. D. (1978). Predicting rainfall erosion losses - A Guide to Conservation Planning. *Agriculture handbook no. 537.* doi:10.1029/TR039i002p00285

Zhang, J. (2012). *Environmental quality assessment of outdoor public space in urban area: development of a Geoprocessing method.* Universität Stuttgart.

ADDITIONAL READING

Ervin, S. (2011). A system for geodesign. In Buhmann, Ervin, Tomlin, & Pietsch (Eds.), *Teaching landscape architecture* (pp. 145–154). Dessau. Retrieved from http://193.25.34.143/landschaftsinformatik/fileadmin/user_upload/_temp_/2011/Proceedings/305_ERVIN_2011May10.pdf

Goodchild, M. F., Anselin, L., Appelbaum, R. P., & Herr Harthorn, B. (2000). Toward spatially integrated social science. *International Regional Science Review*, *23*(2), 139–159. doi:10.1177/016001760002300201

Lee, D. J., Dias, E., & Scholten, H. J. (Eds.). (2014). *Geodesign by Integrating Design and Geospatial Sciences* (Vol. 111). Switzerland: Springer, Cham. doi:10.1007/978-3-319-08299-8

Steele, C. (2001). A critical review of some traffic noise prediction models. *Applied Acoustics*, *62*(3), 271–287. doi:10.1016/S0003-682X(00)00030-X

Steinitz, C. (2013). Beginnings of geodesign: A personal historical perspective. In *Geodesign: Past, present, and future* (pp. 4–14). Redlands, California: ESRI Press. Retrieved from http://www.esri.com/~/media/Files/Pdfs/library/ebooks/geodesign-past-present-future.pdf

ENDNOTES

1 Sir Patrick Geddes was a Scottish biologist, sociologist, geographer, philanthropist and pioneering town planner known for his innovative thinking in the fields of urban planning and sociology.

2 For more information regarding the three developments please see (Li & Milburn, 2016).

2 Invented by Howard Fisher in 1963 and based on the line printer technology. It is the first automated computer mapping system that included spatial analytic capabilities applied to spatially distributed data (Steinitz, 2013, p. 4).

4 For more information regarding the three aspects please refer to (Steinitz, 2012: pp. 19-22).

5 Iterative means, that a quick and close planning loop is established between sketching an intervention, followed by GIS-based impact evaluation of the intervention which is the basis of alterations in a next sketch.

6 The authors focused only on those two phases, since their main concern is how the study area is expected to develop and how this development will impact it either positively or negatively and to compare the results to the status quo. But not to evaluate the status quo itself and how it should be enhanced.

7 Since it is the amount of change of raster cells above mean value of visual quality, therefore the status-quo has no number.

8 This is following a general agreement between government, administration and planning authorities to preserve the agricultural land.

APPENDIX 1

Status-Quo Maps

Figure 6. Land-use map: Status-quo (authors).

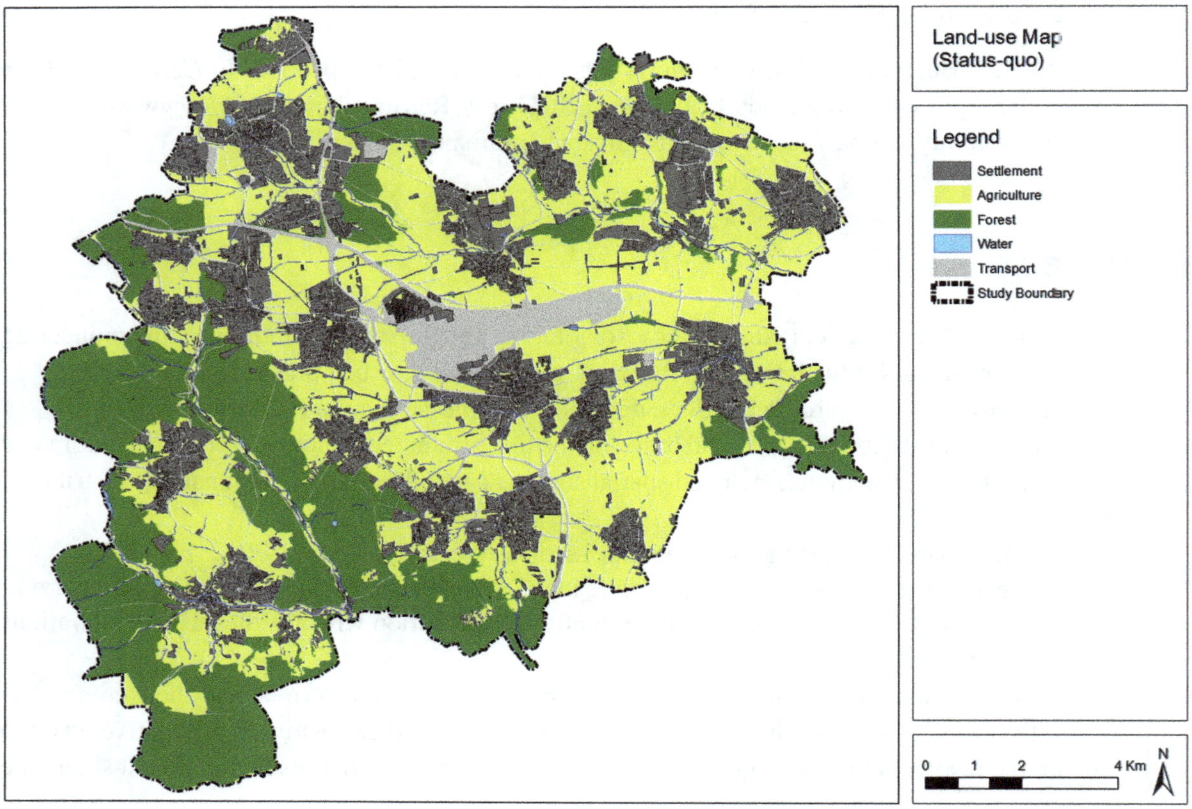

**For a more accurate representation see the electronic version.*

Figure 7. Status quo maps. Top left: soil quality; bottom left: habitat quality; top right: cold air production; bottom right: erosion risk (authors).

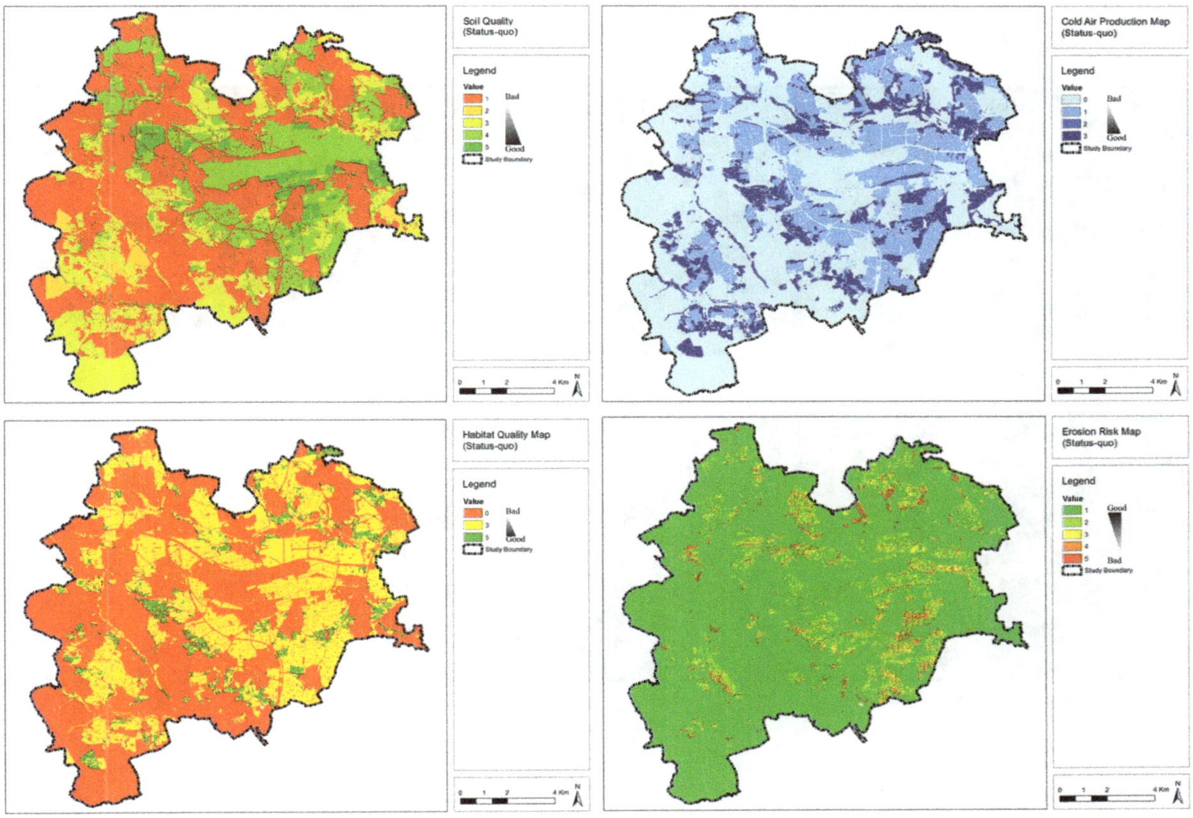

**For a more accurate representation see the electronic version.*

APPENDIX 2

Figure 8. Change land-use maps (authors).

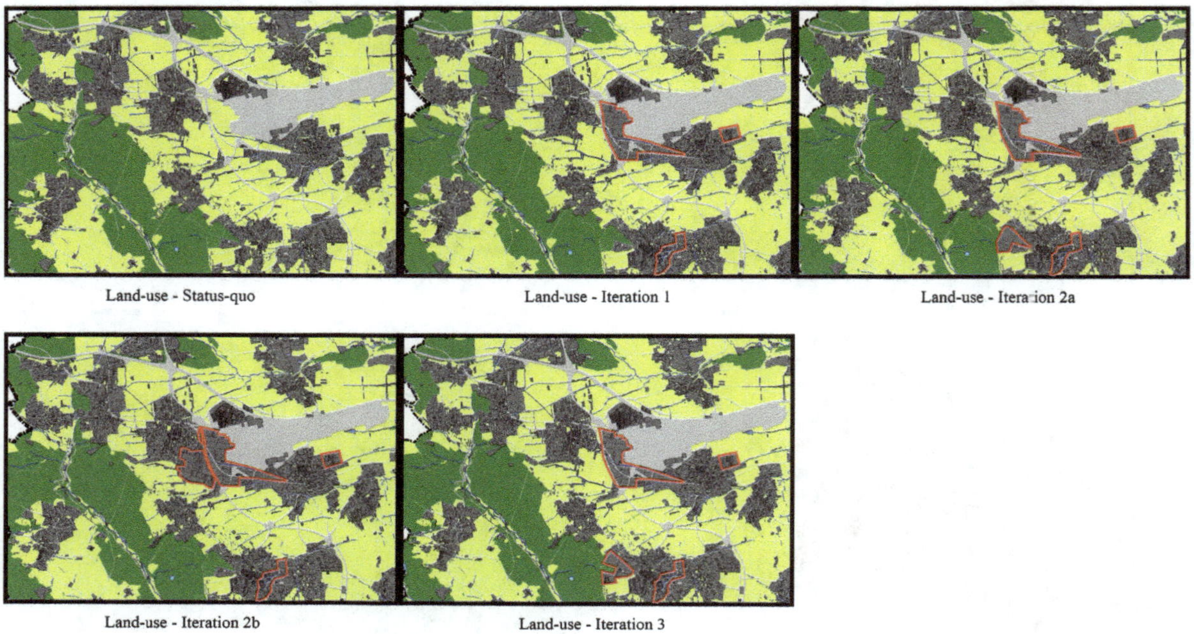

APPENDIX 3

Change Example

Figure 9. Land-use map: Iteration 1 (authors).

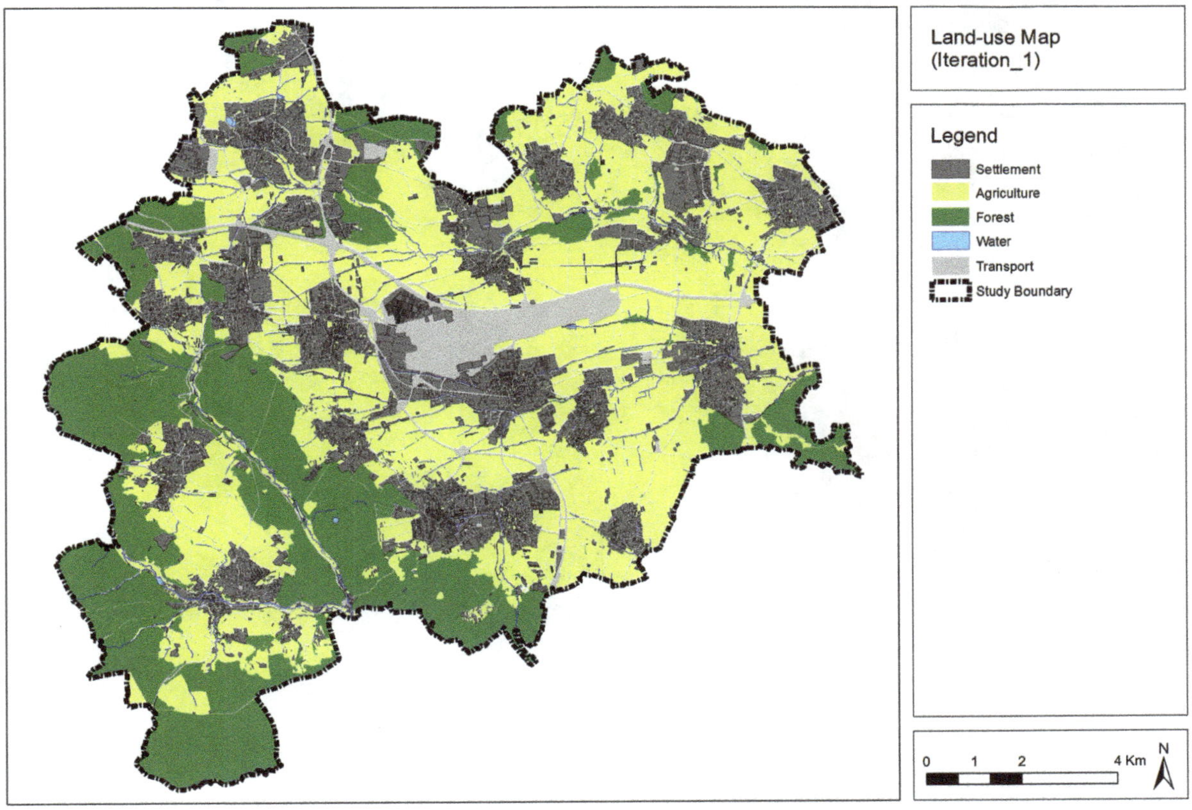

Figure 10. Iteration 1 maps. Top left: soil quality; bottom left: habitat quality: top right: cold air production; bottom right: erosion risk (authors).

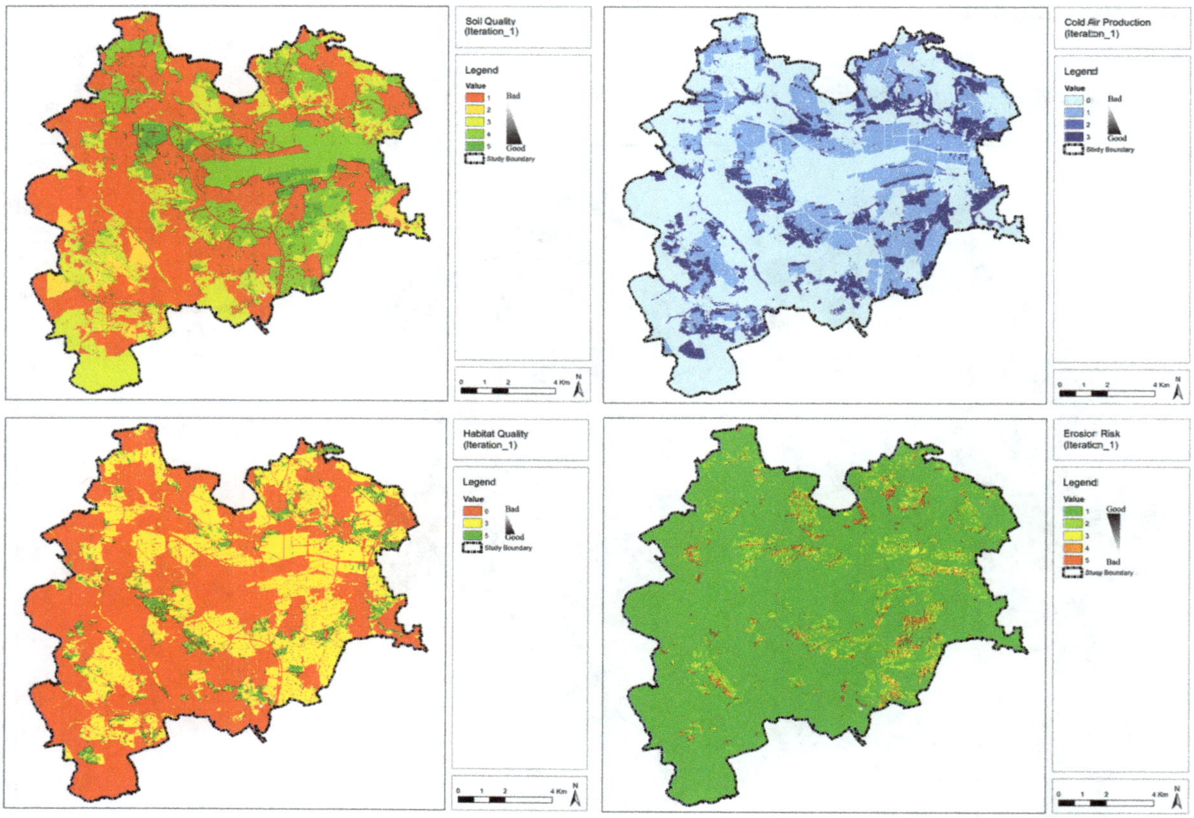

Chapter 6 Towards City Information Modeling: A Multidisciplinary Platform for Urban Planners

Abdalrahman T. Y. Alashi
Istanbul Technical University, Turkey

Turgay Kerem Koramaz
Istanbul Technical University, Turkey

ABSTRACT

In the past two decades, building information modeling (BIM) has been widely adopted by architectural, engineering, and construction firms. Its technology is based on the integration and coordination of different disciplines and professions. It involves creating three-dimensional models containing data that can be organized and manipulated to serve design, construction, and operational phases. This raises the debate over how the existence of a platform that enables urban planners and decision makers of different disciplines is so crucial. Several research studies have recently been done to integrate BIM with geographic information systems (GIS) for numerous purposes, as illustrated by several case studies. This chapter comparatively analyzes different cases, given general acknowledgment of each. Problems and potentials of the existence of such integration will be defined, in order to estimate the need for such a platform. Finally, a model of integration between different disciplines was illustrated as a motivation for further studies in the future concerning this topic.

INTRODUCTION

Different firms are experiencing dynamic changes due to technological developments that introduce a frequently high amount of software, making wider controllability and manageability for different disciplinary (Fosu, 2015). As the industry rises, there is more need to stabilize the relationship between people and process (Isikdag & Zlatanova, 2009a). Lately, new technology was introduced to the archi-

DOI: 10.4018/978-1-5225-9238-9.ch006

tecture, engineer and construction industry, named Building Information Modeling BIM. The reason behind its widespread is the need for a platform or software that is specifically developed to coordinate and integrate the different electrical, mechanical, plumbing, and structural designs with the architectural model, giving a chance for the coordinator to track fatal faults that are hard to appear in 2D construction documents. This rapid expansion was a reason that it was yet possible to see a complete integrated model that has all construction details, drawings, and tables, including the work of all people in various disciplines, decreasing the risks due to lack of coordination and its related costs.

Geographic Information System (GIS) is software that is widely used by urban planners, due to its ability to present, analyze, and model spatial data. The software's main, reliable features of interest to GIS developers is its integration of planning models, visualization, and the Internet (Fosu, 2015). GIS obtains detailed spatial data indicated for a certain defined location or area defined by a border. Recently, there have been many attempts to integrate BIM software (and the amount of building information that it carries) with GIS software (and spatial characteristics that can provide environmental spatial information).

Ever since urban planning as a discipline has generally agreed on the need to propose courses of action based on massive data stores collected from different disciplines and professions. Such actions mainly aim to fulfill public interests throughout evidence-based analysis. These sets of actions are responsible for reshaping and reimaging communities' lives in a spatial scale and are immediately reflected in terms of their built environment (Altshuler, 1965). In this chapter, we introduce different approaches illustrated with case studies, each done for its own purpose and using its own method of integration. A comparative analysis of such approaches is achieved; and a new model is extracted from all previous models. This could be the motivation for developers to develop such a coordinated and integrated model of City Information Modeling Technology for the use of urban planners, urban designers, regional developers, and other stakeholders.

BACKGROUND

Building Information Modeling (BIM)

BIM is a multidisciplinary technology introduced to the world through a range of computer software options. Its aim is that of providing an integrated and coordinated 3D visual model of a building, its site, and geographical information. It facilitates the manageability of project phasing from a project's conceptualization through to its implementation and future development (Azhar, Khalfan, & Maqsood, 2012). It provides the ability to provide detailed construction documentation, coordination plans, and a phasing management plan (Arayici, 2008). BIM models are generally considered to be very rich with model information which it carries at the building scale, as it includes a number of elements (windows, doors, beams, columns, etc.), and their materials (area, volume, etc.). Such data can be used in cost estimation, in construction sequences, in forensic analysis (Azhar, 2011), and even (in the case of highly detailed models) in some analytical techniques relative to the assessment of energy consumption.

The concept of BIM arises firstly from its ability to save time for the architects using it. Unlike 2D software, BIM has the ability to conduct 2D construction drawings (elevations, sections, and plans; etc.) from a 3D model; and unlike similar 3D software, it provides detailed drawings. When architects are dealing with BIM software, the elements are identified regarding their discipline, where commands are defined at the very beginning by the developers as architectural elements. Walls, windows, and doors, etc., are the defined commands for BIM Revit Architecture. Beams, columns, and slabs, etc., are the defined elements for BIM Revit structure. Pipes, ducts, mechanical and electrical equipment, etc., are the defined elements for BIM Revit MEP. All are then combined and coordinated to form one model of a building—a model rich with data. The model shows faults, conflicts, collisions, and errors defined by developers, according to architectural bases. These complicating factors are then managed by technicians, to avoid ongoing faults in the various implementation phases of the project. All elements of each discipline are combined in a tool with an easy interface which can easily be used by normal software users (Vandezande & Krygiel, 2015), as shown in Figure 1.

Accordingly, over the last few years, the world's architectural firms have experienced widespread adoption of BIM software. Its speed, effectiveness, integrity, coordination, and ease of usage for normal software users are the main reasons for this. International firms are beginning to shift their 2D and 3D software accreditation efforts towards BIM technology. They are providing paid courses for their technical and design teams, stimulating BIM training by offering higher salaries for those so trained, and giving a broader platform for new graduates who have knowledge of BIM software. It is of particular use in dealing with complicated architectural buildings like hospitals and hotels, etc., that are congested with fire systems, electrical equipment, structural elements, plumbing system, etc., and which are more exposed to the risk of faults and collision.

Geographic Information Systems (GIS)

Geographic Information Systems (GIS) were developed in 1960 and have been widely used by planning departments. It is considered to be one of the platforms that are capable of the integration of data with different formats, illustrations, and analyses of data for making reasonable decisions in urban planning (Han, Kim, 1989). Its principal operational functions are overlaying maps, connectivity measurements,

Figure 1. Tools preview in BIM platform, including Architecture – Structure – MEP panels
Source: (Vandezande & Krygiel, 2015)

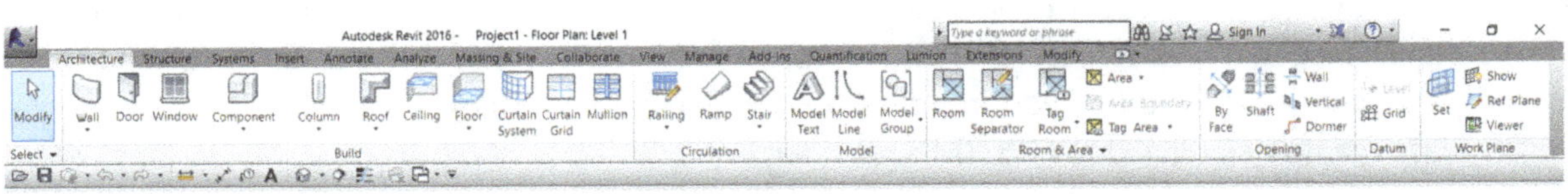

Revit 2016 Architecture Panel

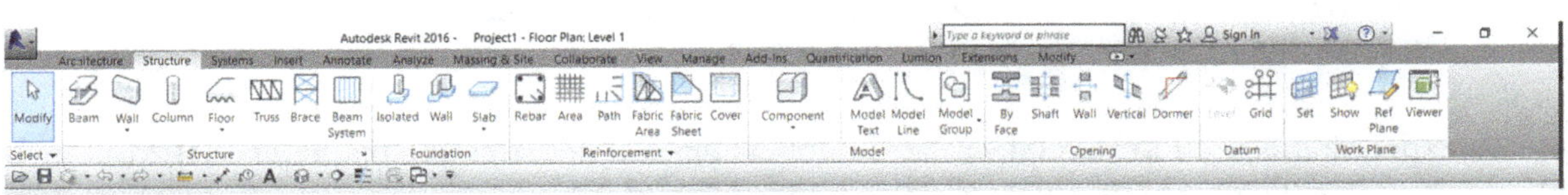

Revit 2016 Structure Panel

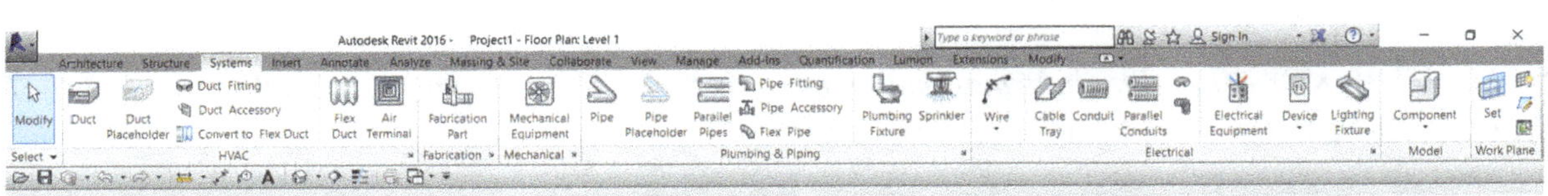

Revit 2016 MEP Panel

and buffering. These are considered to be effective GIS tools—ones from which urban planners effectively derive such decisions (Berry 1987). Land suitability analysis, for example, is another important tool for urban planners, can be easily performed. GIS programs also have the ability to manage land use maps, socio-economic maps, and environmental spatial maps, as well as the ability to visualize and overlap the data in spatial dimensions.

Urban planning is a discipline that can work on many scales. The scale of the planning area can vary from the whole city scale, to the neighborhood scale, or even the street block scale. Also, the plan passes through different stages; and from the development of objectives through to the monitoring and control stages, GIS is used throughout most of these stages. Different scales and stages of the plan mostly require different types of data and different analytical techniques. Whereas in strategic development plans, high detailed data are considered to be less effective in the decision-making process. GIS is more open to end-users' data manipulation in their operations. Although GIS provides functions of validity, accuracy and diagnosis, BIM is more robust, while giving volumetric measures and quantities. Aside from the ability to analyze data, GIS has the ability to simulate future projections according to the given data—including economic, environmental, and population data estimations (Longley, Higgs, Martin, 1994).

BIM and GIS Integration

BIM is widely known for the rich information that it carries and for the manageability of such data from the design phase and throughout the construction phases. However, GIS offers spatial-dimension geographic visualization, which is widely used for decision making (Song, 2017). The literature contains a significant amount and variety of studies on GIS and BIM. Besides, the need for a platform that can provide integration and coordination of these two different systems has been a crucial topic throughout the last decade. Several research studies and conference papers have been held in an attempt to integrate Building Information Modeling (BIM) and Geospatial Information Systems (GIS). The total number of these journal-published papers (indexed in the SCOPUS academic bibliographic database) is 96. Fosu (2015) and Song (2017), in separate literature review studies, presented the distribution of studies which integrate these two systems. In order to understand the emergence of such integration inquiries on GIS and BIM, searching GIS- and BIM-related publications individually would provide more explicit evidence for the importance and significance of the motivations behind them. For this chapter, GIS- and BIM-related publications are investigated through SCOPUS database searches, separately. When searching the database for studies related to GIS, the search term

"(TITLE-ABS-KEY ("geographic information system") AND NOT TITLE-ABS-KEY ("building information modeling" or "building information modeling"))"

The expression is used to find GIS-related publications but not stressing the topic of BIM in their title, abstract, and keywords. For BIM-related publications, the expression sequence had been shifted vice versa:

"(TITLE-ABS-KEY ("building information modeling or "building information modelling") AND NOT TITLE-ABS-KEY ("geographic information system"))"

Then, the number of publications related to only the GIS topic was found to be 69,076. The number related to only the BIM topic was found to be 4,067 (in the SCOPUS database search, as of November 2018).

From the mid-1970s, scholars in related professions started to work on GIS; but especially in the last two decades, the number of research studies and publications has enormously increased. Before 2000, there were 8,047 publications concerning GIS; but this increased more than triple in the period of 2000–2009 (for this decade: 25,460). This decade was also the period of emergence of the BIM concept in academic publications. There were 256 publications which used only the BIM concept in their title, abstract, and keywords. In the more recent decade that followed, the number of publications about GIS has reached the figure of 35,569 publications. The increase in literature discussions of the BIM concept was more than the increase of GIS publications after 2000 until 2018. The number of publications about BIM has reached to 3,811 in the recent decade. as shown in Figure 2.

The publications, which integrated GIS and BIM concepts in SCOPUS-indexed sources, indicated that the emergence of such an integration idea dates back to 2007, but that—especially after 2012—the number of GIS BIM integration related publications increased significantly. There were 22 publications between 2012 and 2014; and for the next three years cumulatively, the number increased more than twice—reaching 55 publications (2015–2017). Among 96 entire publications in the SCOPUS database—publications having bilateral use of both GIS and BIM—there are 53 conference papers, 37 articles, 5 review notes and short papers, and 1 book chapter. There are 158 total publications, from the same periods of publication, in the SCOPUS database. Distinctively, however, between 2015 and 2017, the total number of publications related to both BIM and GIS—according to Web of Science database—has increased to 93 publications—nearly twice the cumulative of those in the SCOPUS database for the same years. as shown in Figure 3.

Figure 2. The annual list of publications related to GIS, BIM and GIS and BIM both in Scopus database
Source: The authors of the present study

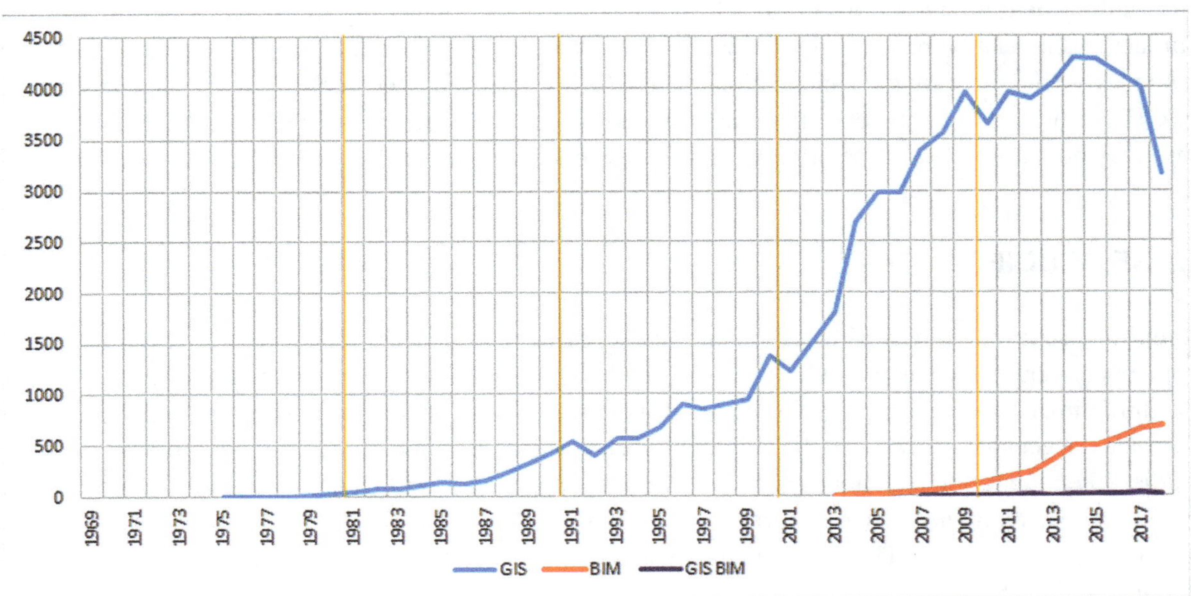

Figure 3. The annual list of publications related to GIS and BIM both in WoS and Scopus databases
Source: The authors of the present study.

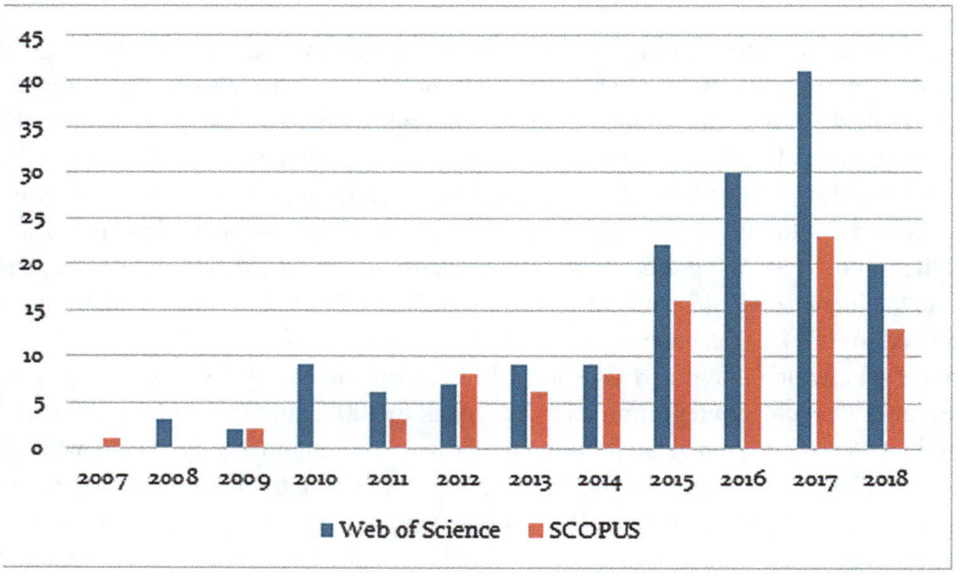

Among 96 publications related to both GIS and BIM, subject areas mostly concentrated on (1) engineering, (2) computer science, and (3) social science—as similar as the subject areas of publications related to GIS and BIM separately. The publications, which cited these 96 publications related to both GIS and BIM, also have similar concentrations as to the aforementioned three subject areas. The fourth major subject area is Earth and Planetary Sciences for GIS-BIM publications, as with GIS publications. By contrast, publications related to BIM individually have a different type of major subject area as Business, Management, and Accounting. This finding indicates that the Business, Management, and Accounting field of study in BIM has a lesser majority while integrating its applications with GIS. A similar finding can be reflected in the Environmental Science field of study in GIS which has a lesser tendency while integrating its applications with BIM. Additionally, the publications related to the uses of GIS in Environmental Science (such as publications related to Business, Management, and Accounting) may hold a significant majority, while citing publications related to both GIS and BIM (Table 1).

CASE STUDIES

Rapid urban constructions worldwide, the city development in 3D, and the need for developed infrastructure which is integrated between building and city scale—in addition to the need for analysis techniques for the third dimension of the city, including wind and energy—are some of the reasons for such integration between the geospatial scale and the building information scale. However, throughout different research studies, the integration techniques were found to be complicated; and from this is derived the question of whether we really need such integration (Xu, Ding, Luo, Ma, 2014). Three case studies were chosen and are illustrated below. Then, a comparative analysis between them—along with some problems and potentials of such integration—is discussed.

Table 1. Subject areas of publications related to GIS, BIM and GIS and BIM both in Scopus database

Subject Area	GIS Related Publication	BIM Related Publication	GIS BIM Integrated Publications	Citing Publications
Engineering	18514	3106	56	429
Computer Science	19670	1336	40	189
Social Sciences	11519	338	25	122
Earth and Planetary Sciences	16953	133	11	75
Business, Management and Accounting	1689	472	7	91
Mathematics	6630	320	7	44
Environmental Science	18760	220	6	94
Energy	2713	145	5	45
Physics and Astronomy	3636	70	4	31
Agricultural and Biological Sciences	8435	24	2	7
Multidisciplinary	575	19	2	4
Materials Science	3054	158	1	28
Arts and Humanities	774	147	1	18
Medicine	6576	38	1	19
Chemical Engineering	526	34	1	7
Biochemistry, Genetics, Molecular Biology	1698	19	1	9
Decision Sciences	1131	141		17
Chemistry	558	62		14
Economics, Econometrics and Finance	400	30		6
Pharmacology, Toxicology and Pharmaceutics	454	3		
Health Professions	602	2		1
Psychology	263	2		1
Immunology and Microbiology	824	1		
Nursing	369			
Veterinary	348			
Neuroscience	131			
Dentistry	31			

Source: The authors of the current study

The targeted choice of the case studies was aimed to represent various worldwide perspectives of BIM-GIS integration into the context. Thus, different publications included in the Web of Science database and SCOPUS were selected. Cases reflect various types of major subject areas (see Table 1), these cases are Social Sciences, Computer Science, Engineering, and Mathematics (see Table 2). Diverse methodologies of BIM-GIS integration were considered, along with authors' different backgrounds. These three cases were chosen to be considered as being the original, most adequate studies, reflecting the major subject areas of BIM-GIS integration.

Table 2. Case studies comparison according to their methodologies, application, country, year and study's scope

Methodology	Application	Vanguards	Country	Case Study	Year	Scopus
Integration of both BIM & GIS	Traffic Planning	Wang, J., Hou, L., Chong, H. Y., Liu, X., Wang, X., and Guo, J.	China & Australia	Traffic Planning FOR A High-Rise Building	2014	Mathematics and Computer Science
Autocad 3d, BIM and GIS integration	Create a 3D model for the city	Mijic, N.; Sestic, M.; Koljancic	Bosnia & Herzegovina	Banja Luka City Center	2017	Civil Engineering, Mechanical Engineering and Computer Science Artificial Intelligence
integration of 3D Building Information Modelling and the 3D Geographic Information System (GIS)	restoration of historical monument	A. Baik a, R. Yaagoubi b, J. Boehm	Saudi Arabia & United Kingdom	the Nasif Historical House, Jeddah historical monument	2015	Social Science and Computer Science

Source: The authors of the current study

Traffic Planning for a High-Rise Building

Traffic problems as a result of cities' developments derived the author through the integration of both BIM & GIS to create an optimization model of traffic planning for a high-rise building, aiming to solve the congestion problem that citizens face in cities (Wang, 2014). Many inputs were taken into consideration; static data of traffic using GIS, and dynamic data of departure and arrival, were also included. Other economic, geological, and constructional factors took part in this study. Combining and coordinating these inputs provided an allocation analysis of parking lots, streets, entrances, and exits. Traffic-peripheral data were taken from "Beijing City Planning Technology Centre". The data were then added to BIM simulation software (VISSIM). A 3D model was drawn, showing entrances, exits, streets, and pick-up zones (see Figure 4).

The hourly traffic flow was represented using ARCGIS software, in the form of spatial analysis, in order to identify high-traffic routes, as seen in the above figure. Vehicles types were then classified according to speed, stop time, and paths, and were then distributed spatially according to entries and exits. The findings from the analysis showed that there was no congestion in the traffic flow, where the distribution of entries and exists was fairly distributed and avoidant of any jam. The study showed that the design of multiple entries and exits was successful in avoiding traffic jams around the building area. The cooperative system analysis, according to the author, proved that the success was achieved by the design team of the building.

According to the author, Further data regarding the "… main evacuation [route] of the building, [and] precise information [regarding] the traffic flow of the north main artery" needed to be included in the study (Wang, Hou, Chong, Liu, Wang, Guo, 2014, p.149). Although the ArcGIS data format was

Figure 4. Three Dimension Model of the building and surrounding on left, spatial analysis of high traffic routes on right
Source: (Wang, 2014)

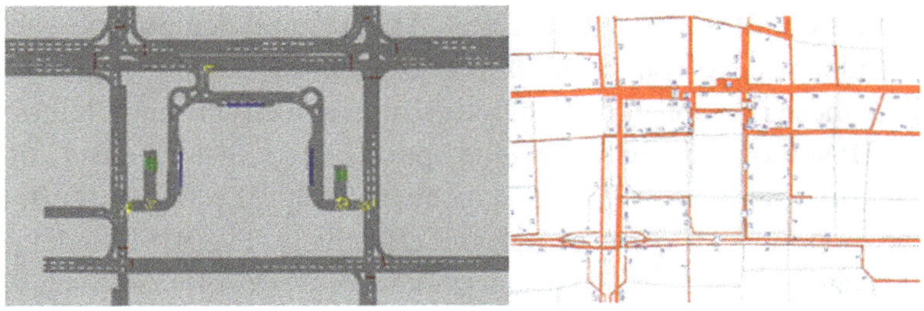

easily integrated into the input data of the VISSIM simulation model, the study shows only a localized scale analysis. Also, the building itself was excluded from its physical environment, whereas the model showed the surrounding streets—an inclusion which, in this case, was successful in serving the specific building scale objective of the study.

Case Study of Banja Luka City Center

The experimental research done by the author aims to clarify how multiple platforms can be integrated in order to convert 2D plans and documentation into a 3D model for the city of Banja Luka and perform 3D analysis (Mijic, Sestic, & Koljancic, 2017). 2D plans were digitalized. Then, GIS building height data and other platform usage were employed in order to automatically generate a 3D model of the city without the need of manually creating it. Other platforms were used to provide materials, shading, walkthroughs, and vegetation. Lastly, real-time wind simulation analysis was performed.

The study shows how (for of the same developing company) different data sets, using multiple platforms, can integrate easily. This is despite the massive amounts of time and cost consumed in such studies—where each platform, carrying a certain tool, was used in one step of the study. The integration of different data formats into various platforms seemed slight and smooth (due to the unifying factor of the developing company), where companies take into consideration the smoothness of file transfer through its various platforms.

The Case Study of Jeddah Historical Monument Restoration

The study aims at creating a model that is rich with information besides modeling geological attributes of a historical building and its surrounding physical environment in a short amount of time (Baik, Yaagoubi, & Boehm, 2015). BIM platforms provide the ability to create detailed 3D architectural models with a high level of detail, but the modeling phase in this high level of detail for buildings takes time (Details are needed in this case for conservation and restoration purposes.) For solving the time issue, when modeling several buildings that have common parametric elements, historical BIM was used, where these parametric elements can be drawn once and then be used in the modeling of more than one building. HBIM also provides more detailed data about the parametric elements and their relationship with other BIM defined components. A library, including parametric elements was developed, using LIDAR point cloud images and a book which included these patterns.

Yet, by using BIM, it is easier to create a detailed 3D model of the building and represent all detailed views automatically (Section, Elevations, Plans, etc.). Architectural components such as windows were drawn in the finest details, where it is now easier to document the materials, context, and techniques of construction for each of these components. Also, Mechanical, Electrical, and Plumbing (MEP) information were added. Still, BIM models were lacking the option of adding attributes to the non-architectural parametric models. This is where the 3D GIS role came into play. 3D GIS deals effectively with simple geometries; but in the modeling, BIM performs more effectively. The BIM model (including architectural and MEP detailed data) was integrated to Autodesk Infra Work software (including GIS data; Road network, landmarks in Vector format, Ground imaging in raster format, and building historical data in descriptive format). The result of such integration is a BIM model which is rich with information on both an architectural and an urban scale. Such data can be useful for restoration, documentation, and visualization, besides its having geographical attributes of both the building and its surroundings.

DISCUSSION

The case studies were done by several researchers—different people in various disciplines, and with a different method of integration. These research studies were done for different reasons, as mentioned in Table 2. Common observations were illustrated among them all, where BIM-GIS integration was performed. Also, the publications are mostly dated to no more than 10 years prior. The integration technique was the most concentrated material, and sometimes was even accorded greater importance than the purpose of such integration. It is also observed that such integrations were done by developers and professionals, and that such integration can hardly be performed by end users with access to standard software, where it can lead—if performed wrongly—to the loss of data during a change in the file's format. Nevertheless, the results and the outcome of the studies and the creation of such urban models—which contain suitable amounts of data, according to the scale needed—are found to be successful. Such an integration, which creates a 3D spatial model with a specific amount of data needed for analysis, can lead to the derivation of reasonable decisions, coordinate between different disciplines, improve efficiency, and perform manageability tasks during construction phases (Song, 2017).

In the first case study, the method was that of creating the model, then integrating the simulation data of traffic. Peak points and the areas of congestion around certain buildings were estimated given the entrances and exits. The scale was observed to be the building block scale. The main purpose was to estimate whether or not the allocation of cars entrances and exits of the building were wisely located in the design. Simulation was performed using GIS, and BIM was the tool for creating the building and its surrounding streets.

In the second case study, multiple software programs from the same developer company were used to create a model of the city. Wind analysis of BIM software was then performed using this model, as were the steps of model creation from PDF file format of the plans to the 3D model. Then, the GIS environment was added. Finally, performance of the analysis was illustrated. It is also observed that the main concentration was on technical integration techniques, which was due to the simplicity and the small amount of data needed, was found to be applicable. In this case, the BIM was used as an analytical tool, and the buildings did not need to carry information except data regarding the physical environment, which was created using 3D software.

In the third case study, and unlike the second case study, the model was created using BIM software, the scale is found to be the building scale and its surroundings; and the aim, as mentioned; was the documentation of visual characters of a historical building, located in a historical site. The model included fine architectural detailing, where the components' model created was a 3D image of the real ones. The building had different detailed views. Then, geospatial information was integrated in order to document the location of the building, the distances from landmarks, and the relation to the surrounding physical environment. Unlike both the first and second cases, the usage of BIM-GIS integration is for documentation purposes, which is not common among different research studies done with the same techniques. The main aim for such documentation is that the model can be used for restoration, reconstruction, and education purposes—and that it is considered to be a reference for heritage professionals and decision makers.

After illustrating the differences, there is a need to estimate the problems and potentials of such integration and to estimate whether the BIM-GIS integration enables us to derive a model of a platform for modeling of the cities? Table 3 gives some of the problems and potentials of the BIM-GIS integration. This table illustrates some of the problems and potentials of such integration, and also raises the debate regarding the need for a platform that is designed for, and can be used mainly for, urban and regional scale decision making.

FUTURE RESEARCH DIRECTIONS

Throughout the course of all previous research studies, models of integration were illustrated, where GIS contained geological, environmental, social, and economic spatial data. Meanwhile, BIM research studies contained building data—including energy consumption analysis; and material, mechanical, electrical,

Table 3. BIM GIS integration problems and potentials

Problems	Potentials
BIM contains a huge amount of information that can be less effective on bigger scale planning decisions	BIM provides geometrical representation for the geospatial environmental maps
BIM architecture contains indoor visualization which makes a load of file huge, while there is no need for such views	BIM provides the third dimension as a complimentary for the 2D geospatial data
Data are not controllable, where the useful data can't be filtered from the others	BIM provides more Collaboratory environment to the model, where different disciplines collaborate easier
The need for professionals or developers to perform the integration, which is hard to be done by end-user software	BIM provides more hierarchy of elements
The hardness of integration of different file formats, and the need in some cases for third software	Using BIM-GIS, it is easy to perform optimum allocation of facilities or housing according to given input data
The loss of some data or inconvertibility of formats, which may affect the results and decisions	In specific scale, different analysis and estimations can be performed, like wind analysis, energy analysis
	The manageability of different phases from design throughout construction and towards monitoring phases

Source: the author based on (Isikdag & Zlatanova, 2009a)

and plumbing data. In most cases, GIS analysis is considered to be for Outdoor elements, whereas BIM analysis is considered to be for Indoor elements. By integration, a full image of the various elements of the city are formed. Clifton, Ewing, Knaap, & Song (2008), classified the quantitative approaches of urban form analysis into five perspectives, Landscape Ecology concerned with regional and undeveloped areas, their reliable source of data is Aerial photographs and remote sensing in producing land cover, which is then categorized using GIS into 9 categories (Anderson, 1976), where environmental landscape maps are created and are ready for analysis. Economic Structure is concerned with spatial economic activities, statistical data concerning economic structure in most the economic research are represented in metropolitan or country scale (Clifton, Ewing, Knaap, Song, 2008).

Transportation Planning perspective focuses on sub-metropolitan and sub-regional scale, statistical data used in this perspective are spatial geographic data. "The community design perspective is the focus of land-use planners"(Clifton, Ewing, Knaap, Song, 2008), where GIS data are analyzed at different scales and resolutions, the focused scales in this perspective researches are urban and neighborhood scales (Elshater, 2017; Ericson & Lloyd-Jones, 2001). Urban design perspective is considered the highest resolution and lowest scale; where it focuses on street network design, land use intensity, land use pattern, and urban design characteristics (Elshater, 2018; Elshater & Abusaada, 2016).

Accordingly, City Information Model's multidisciplinary platform followed Clifton's (2008) categorization. Economic, Social, Environmental data along with physical environment is integrated, coordinated and analyzed according to variable researches' approaches in three dimensions (see Figure 5). The CIM platform provides end-users with access to standard software the ability to integrate different file formats, as long as data is provided in 3D coordination system. The 3D coordination of cities' infrastructure, physical structure, transportation system, water bodies, etc. are provided with a graphical warning system, which aims to illustrate the faults, conflicts, errors, and collision—ones that can hardly be observed in 2D. CIM platform have also the ability to display spatial data in various graphical ways, across a different range of scales, with the ability to manage the type and range of data according to the research perspective. 3D drawing tools, analysis tools, platform's interface is disciplinary or respectively categorized, to avoid the conflicts between disciplines, 3D models will then be combined and coordinated—this is where the conflicts are illustrated and can be resolved.

CONCLUSION

The chapter illustrates the emergence of technological software that is used by different disciplines for more convenient results and analysis for the benefit of users who are working under limited time constraints. Firstly, the chapter discusses BIM technology, its uses, and its huge influence on architectural firms from the design procedure phase to the construction phases. It discusses how BIM technology makes coordination between different disciplines concerned with building construction more applicable and easier. Secondly, GIS offers huge benefits in providing environmental, economic, social data, etc. and presenting such data in spatial dimensions. Also, it is widely used among urban and regional planning firms, as well as solving transportation problems. Thirdly, the tracking of different trials and research studies is done in order to integrate these two kinds of software in an attempt to generate a city model that relays these amounts and kinds of information and which, through wise analysis, can drive us towards more reasonable decisions.

Figure 5. City Information Modelling platform, showing data integration & multidisciplinary coordination (Clifton, Ewing, Knaap, Song, 2008)
Source: The authors of the present work

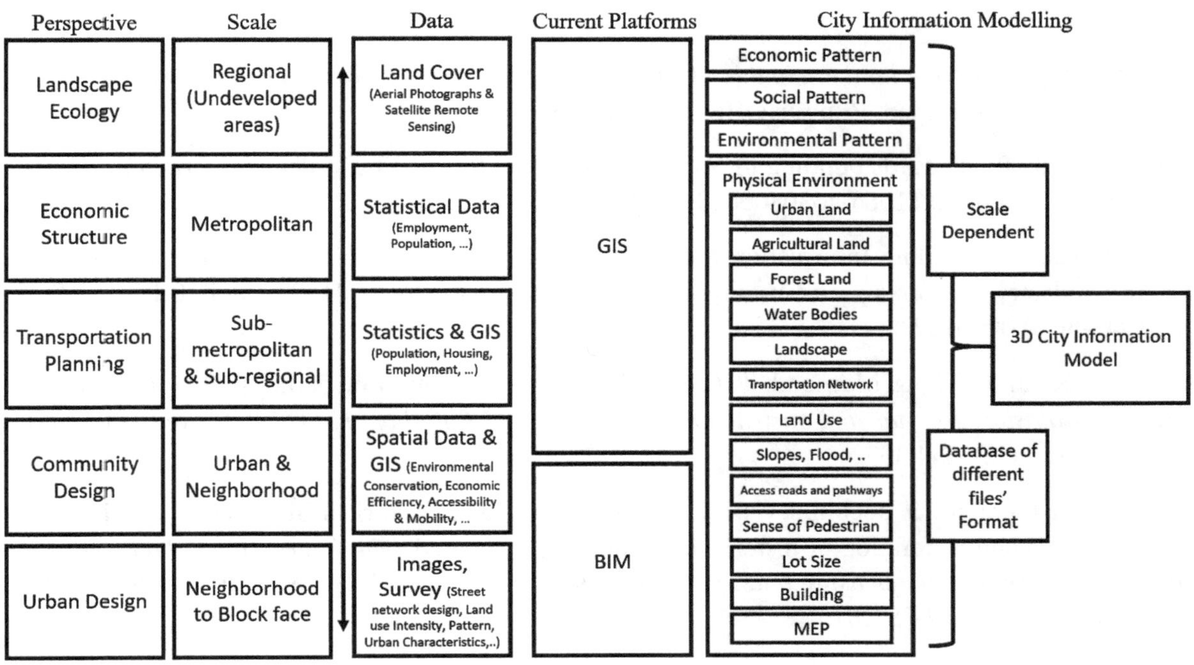

In the SCOPUS research, Business, Management and Accounting field of study in BIM and Environmental Science field of study in GIS were observed to have significant majority when compared with both BIM- and GIS-cited publications in relevant fields. As illustrated in the case studies, integration was used at many scales and with different purposes. However, the most obvious among the cases was that the application of such integrations needed one or more kinds of software, and that most of them were done by developers or software professionals. Also, the cases were not applicable to software end-users.

Yet, it is time to introduce such a platform—one which is capable of modeling a neighborhood, city, or region, and which can carry such information in both 2D and 3D. Merging the benefits of both GIS and BIM, such a platform would provide more coordination between different people in various disciplines. Each profession would work on a certain interface, with tools that are defined to his profession. Then, models are to be combined and integrated into one model—one that carries these huge amounts of data. Then, with certain tools and analytical techniques, this can help the professional coordinator towards deriving wise and reasonable decisions.

REFERENCES

Altshuler, A. A. (1965). *The City Planning Process: A Political Analysis*. Ithaca, NY: Cornell University Press.

Anderson, J. R. (1976). *A land use and land cover classification system for use with remote sensor data* (Vol. 964). US Government Printing Office. doi:10.3133/pp964

Arayici, Y. (2008). Towards building information modelling for existing structures. *Structural Survey*, *26*(3), 210–222. doi:10.1108/02630800810887108

Azhar, S. (2011). Building Information Modelling (BIM): Trends, benefits, risks and challenges for the AEC industry. *Leadership and Management in Engineering*, *11*(3), 241–252. doi:10.1061/(ASCE)LM.1943-5630.0000127

Azhar, S., Khalfan, M., & Maqsood, T. (2012). Building information modeling (BIM): Now and beyond. *Australasian Journal of Construction Economics and Building, 12*, 15. Retrieved from http://doi:10.5130/ajceb.v12i4.3032

Baik, A., Yaagoubi, R., & Boehm, J. (2015). Integration of Jeddah historical BIM and 3D GIS [data] for documentation and restoration of historical monument [structures]. *The International Archives of the Photogrammetry, Remote Sensing and Spatial Information Sciences*, *XL-5*(W7), 29–34. doi:10.5194/isprsarchives-XL-5-W7-29-2015

Berry, J. K. (1987). Fundamental operations in computer assisted map analysis. *International Journal of Geographical Information Systems*, *1*(2), 119–136. doi:10.1080/02693798708927799

Clifton, K., Ewing, R., Knaap, G., & Song, Y. (2008). Quantitative analysis of urban form: A multidisciplinary review. *Journal of Urbanism*, *1*(1), 17–45. doi:10.1080/17549170801903496

Elshater, A. (2017). Widen the Scale of Urban Design to the Level of City Planning: Argument Beyond a Case of Two Cities. UPLanD – Journal of Urban Planning. *Landscape & Environmental Design*, *2*(2), 207–221.

Elshater, A. (2018). What can the urban designer do for children? Normative principles of child–friendly communities for Responsive third places. *Journal of Urban Design*, *23*(3), 432–455. doi:10.1080/13574809.2017.1343086

Elshater, A., & Abusaada, H. G. (2016). What is Urban Design? Teaching and Learning. London: Penguin Reader.

Ericson, B., & Lloyd-Jones, T. (2001). Design Problem. In M. Roberts, & G. Harlow (Eds.), Appreciating Urban Design: The Design Process. Longman.

Fosu, R., Suprabhas, K., Rathore, Z., & Cory, C. (2015). Integration of Building Information Modeling (BIM) and Geographic Information Systems (GIS) – A literature review and future needs. *Proceedings of the 32nd CIB W78 Conference*, 196–204.

Han, S. Y., & Kim, T. J. (1989). Can expert systems help with planning? *Journal of the American Planning Association*, *55*(3), 296–308. doi:10.1080/01944368908975417

Isikdag, U., & Zlatanova, S. (2009a). A SWOT analysis on the implementation of BIM within [the] geospatial environment. In Urban and Regional Data Management, UDMS Annuals 2009. CRC Press.

Longley, P., Higgs, G., & Martin, D. (1994). The predictive use of GIS to model property valuations. *International Journal of Geographical Information Systems*, *8*(2), 217–235. doi:10.1080/02693799408901995

Mijic, N., Sestic, M., & Koljancic, M. (2017). CAD—GIS BIM integration—Case study of Banja Luka city center. In M. Hadžikadić & S. Avdaković (Eds.), *Advanced technologies, systems, and applications, Lecture Notes in Networks and Systems, 3*. Cham: Springer. doi:10.1007/978-3-319-47295-9_22

Song, Y., Wang, X., Tan, Y., Wu, P., Sutrisna, M., Cheng, J., & Hampson, K. (2017). Trends and opportunities of BIM-GIS integration in the architecture, engineering and construction industry: A review from a spatio-temporal statistical perspective. *International Journal of Geo-Information, 6*, 397. doi:10.3390/ijgi6120397

Vandezande, J., & Krygiel, E. (2015). Mastering Autodesk Revit Architecture 2016: Autodesk Official Press 1st. Alameda, CA: SYBEX.

Wang, J., Hou, L., Chong, H. Y., Liu, X., Wang, X., & Guo, J. (2014). *A cooperative system of GIS and BIM for traffic planning: A high-rise building case study. In Cooperative design, visualization, and engineering* (pp. 143–150). Springer.

Xu, X., Ding, L., Luo, H., & Ma, L. (2014). From building information modeling to city information modeling. *Journal of Information Technology in Construction, 19*, 292–307.

ADDITIONAL READING

Isikdag, U. (2006). *Towards the implementation of building information models in geospatial context* (Doctoral dissertation). Salford: University of Salford.

Kim, J. B., & Clayton, M. J. (2010). Support form-based codes with building information modeling: The parametric urban model case study. In *Proceedings of the 30th Association for Computer[-]Aided Design in Architecture Conference (ACADIA)*, New York, 133–138.

Thompson, E. M., Greenhalgh, P., Muldoon-Smith, K., Charlton, J., & Dolník, M. (2016). Planners in the future city: Using city information modeling to support planners as market actors. *Urban Planning*, *1*(1), 79. doi:10.17645/up.v1i1.556

APPENDIX

Abbreviation

BIM: Building Information Modelling
GIS: Geographic Information System
AEC: Architecture, Engineering and Construction
MEP: Mechanical, Electrical and plumbing
SCOPUS: Database powered by ELSEVIER
VISSIM: Traffic flow simulation software powered by PTV
ARCGIS: Geographic Information System Software powered by Esri company
LIDAR: method for measuring distance using laser light and sensors
HBIM: Historic Building Information Modelling
3D GIS: Three-dimensional geographic information system software

Chapter 7
Q Sensor for the City on the Threshold of Stress:
Degenerative Dimension of the Sound Atmosphere

Sana Layeb
University of Carthage, Tunisia

Mohsen Ben Hadj Salem
University of Carthage, Tunisia

ABSTRACT

The urban atmosphere evokes several sensory registers that participate in our perception of singular tonalities, of our daily situations. Tunis is, in this chapter, the space-time that would serve as a framework for our hearing. The experimental protocol is threefold. The authors quantify users' feelings through the commented walk method and especially by objective measures of electrodermal activity. The authors conducted in situ metrological work on the sound signal. These measurements were taken using a device "Q sensor." This device quantifies emotional arousal by measuring electrodermal activity (EDA). The data collected were compared and crossed to identify the links between the architectural configurations of the public space, the sound signals, and the ways in which the feeling of stress appears. The results indicate that urban stress situation seems complex and enjoyable to explore using a multidisciplinary approach. A future direction was presented to the urban settings through the draw on a variety of disciplines, including urban planning, architecture, and psychology.

INTRODUCTION

The contemporary city tends to be more and more inhospitable (Paquot, 2015)[1]. Designers often forget that human beings are in search of a place to live, to flourish and to withdraw. Today, this quality of the city is in danger, or even on its way to extinction. Current research on the city and architecture can no longer consider public space as a simple container. It is a modular and combined universe of physi-

DOI: 10.4018/978-1-5225-9238-9.ch007

cal and experienced properties. Indeed, it is through our different senses that it is heard, seen, felt and perceived. Sound phenomena are one of the major phenomena that influence man's relationship with his daily urban environment. They have an important role in our experience, our perception and our representations of the built space.

This chapter aims to reveal a particularity of this sensitive dimension, that of its responsibility in the appearance of the feeling of stress. In the absence of deep knowledge of its involvement in the perception of urban sound atmospheres, architects and urban planners have difficulties in integrating it into their decisions, which can sometimes have serious consequences on our urban experiences. We are looking, in this chapter, for suggestions of reflections and actions based on this sensory component. Cities could and should re-invent themselves, freeing them from these degenerative trends.

Observations made on different dense urban fabrics show that urban intersections condense a part of the city's emotionality. With the incessant movements of the automobile and pedestrians and the difficult sharing of public space, several perceptual phenomena are intertwined to give meaning to the urban character. Streets and boulevards present the structure that ensures a dual role of driving, distributing and organizing urban spaces. They can be wide and narrow in a regular pattern, such as that of colonial cities, or twisted arborescent like medieval or Arab cities. Aldo Rossi defines the street as *a rectangular scene, it is the place of meetings, conversations, games, disputes, demonstrations, pride, seduction companies* (Rossi, 1966, p.105). The street is the support for various human and social activities.

Hearing is no longer *the sense that gives us the company of the street* (Chelkoff, 1996, p.13), as Marcel Proust described it, but rather a real discomfort and a real sound stress, as shown by the statistical data of the CSTB (Scientific and Technical Center for Building)[2], since the late 1980s. These data concern France, but we can affirm that this observation remains true in the countries of the southern Mediterranean, as attested by the AFP (France Press Agency) report, describing Cairo. *In addition to record air pollution, Cairo has a permanent cacophony that makes it one of the most unbearably noisy capitals in the world, according to scientific studies. Horns honking, loud music, calls for prayer from the speakers and constant traffic, all these noises turn into cacophony in an overcrowded megalopolis of seventeen million people [...]* (AFP, 2008). The first condition for rigorous analysis, of the occurrence of the noise nuisance, is a careful exploration of the complexity of the sound situation, in its different aspects: morphological, sensory and physical. The urban atmosphere brings together several sensory registers that contribute to our singular perception of our daily situations. Sight, hearing, smell, etc., are involved in the sketches of our urban apprehensions. More specifically, the sound dimension is a way to capture the world in a collective and shared experience. A process that can be disruptive and stressful.

Tunis is here the space-time that would serve as a framework for our hearing experience. The experimental protocol is composed of three parts. We will simultaneously quantify the users' feelings through objective measurements of their electrodermal activity (EDA) and capture the present sound signal, in addition to the different perceptual investigation techniques.

BACKGROUND

Sound Atmosphere to Apprehend Urban Emotions

The authors of the present work should remind that our problem concerns the links between the urban and architectural characteristics of the built environment, the urban sound atmosphere, and more precisely the ways in which feelings of sound stress are expressed in the contemporary urban fabric. We will use the notion of atmosphere as a way of reading and analyzing architectural and urban forms centered on the sensitive and perceptive aspect.

Before developing the notion of sound atmosphere, we should first consider the general theory of ambience[3]. In the context of research on architectural and urban environments, Jean Pierre Péneau combines the perception with the presence of the five senses and the movement of the human body in space. It is a transition from the exclusive privilege given to visibility in architecture, (Augoyard & Torgue, 1995) to a sensitive approach that he calls *cénesthésie* [4] (Ricœur, 1949, p. 141), that combines the different senses of the human body (hearing, tactility, olfaction, …) in movement. The theory of architectural and urban environments aims to make the sensitive intelligible. The research on Ambiences is mainly based on the analysis of the different signals present: light, sound, heat, etc. This analysis is conducted by crossing quantitative and qualitative methods, i.e. physical data, the designed space, the fundamental sciences, the perceptual modalities, and the perceived space. For architectural and urban environments, they are generated, according to the philosopher Jean-François Augoyard, by two parameters. A physical device and a reaction expressed by a perceptual impression composed of cognitive and sensitive elements.

To define the sound atmosphere, we must first start by identifying the sound and its components. As Jean François Augoyard and Henry Torgue present, *the sound is immediately linked to a circumstantial cause, i.e. linked to the characteristics of the built environment and the physical conditions of hearing and listening* (Augoyard & Torgue, 1995, p.9). It can be deduced so that the treatment of the sound environment depends on the sound propagation factors and the perception conditions. They can be summarized in physical sound signals, social and cultural representation, the differentiated perception of users, the interaction between their actions and perceptions and the relationship between form, space and time. We can then affirm that the analysis of the sound environment is multidisciplinary because it mobilizes physical and quantitative techniques, psychophysiology, musicology, the sociology of interactions and the practices of architectural and urban design.

Because of the privilege of the visual dimension that strongly marks the attitude of urban designers, and like other atmospheres, sound atmospheres are often neglected or completely excluded during the conceptual process. Due to a lack of knowledge and tools for space designers, sound ambience is often the result of an essentially normative approach based on physical acoustics as an objective response to the noise problem. A clarification is needed to clarify the differences between sound and noise. In the West, the distinction between noise and sound was a prelude to the analysis of musical pleasure[5] in the 18th century. Noise would ultimately be a less complex sound and less pleasant to the musical ear[6]. The distinction would be made because of a strong sensory difference and cultural appreciation. We will say that music that we don't know well is noise. Listening to noises now makes hearing one of the forms of sensitivity. All sounds then become mediators of a culture, a way of being in and through the city.

Stress in Urban Environment

Until the 1960s, psychological research on urban noise had focused on its auditory effects. It was only after the 1970s, with Gabriel Moser (1980) and Claude Lévy-Leboyer (1983)[7], that we began to talk about the effects of non-auditory noise on individuals and micro-societies. Among these effects, the notion of "urban stress", which is often used in our daily modern life. But what is stress? And what are the elements of the urban environment that contribute to its emergence? Georges Vigarello (2014) calls it as *discomforts guiding everyday life (p.15)*, but in the scientific literature, the definitions vary with the authors, but they can be classified into two categories: stress as an urban condition (a logical consequence of urbanity), and stress as a reaction to these conditions.

According to Gabriel Moser[8], two approaches to stress research are identifiable. *The physiological approach (stress as a particular situation or as a specific response of the individual) and the psychological approach that considers stress as a relational variable between the individual and the situation* (Moser, 1997, p.14). These two methods can reflect three main categories of stress: stress as a source of nuisance, stress as a physiological reaction, and stress as a way of asserting oneself in an environment. The third class, which considers stress as a particular relationship to the environment, is a psychological approach that depends on the subject's reaction to environmental factors and his or her ability to cope with them. Gabriel Moser then distinguished two conceptual models[9]: interactive and transactional (Moser, 2009).

The interactive model defines stress as the dynamic interaction between the individual (the subject), the environment (environmental conditions) and the perceptual resources of each subject. According to him, everyone evokes specific emotions to his/her experience, to his/her life experience. In this sense, Gabriel Moser adopts the definition of Andrew Baum, Jerome Singer and Carlene Baum[10], who state that *Stress is the result of an individual's understanding of the situation and reaction to his environment.* (Moser, 1997, p.15)

Following several psycho-environmental studies conducted on the different types of stress, we have been able to distinguish three categories: stress due to catastrophic events (earthquake, fire, flood, etc.), stress related to one's personal life, this is what Sheldon Cohen[11] (1977) calls Background stress (related to memory, or urban situation experienced negatively), and environmental stress related to the physical sources that surround us. According to Hans Selye, this type of stress is stable and repetitive. It is the most common because environmental factors are omnipresent in our daily lives. They include density, intensity, heat, pollution, and noise. Regardless of the nature of the stress or its effects on potential subjects, these multiple sources can always be identified. We focus particularly on the sources of environmental stress that take the urban noise environment as a supporter.

Urban Space and the Sound Component

Noise has long been an underestimated object of study, despite Michel Chion's subtle work[12]. Noise is an elusive, ephemeral, intrusive and often painful phenomenon. According to our tolerance gradient, we react differently in its presence. It should be noted that we are also noise-makers, noise is not only the other(s). Researchers have shown that long-term exposure to noise can have very serious consequences on human health. These impacts fall into two categories: auditory and non-auditory effects. Non-auditory effects include stress, neurological effects, and safety risks. Auditory effects include alteration[13] and degradation of hearing ability.

This classification is taken over by Gabriel Moser. He considers noise as an undesirable sensation that generates several subjective and objective effects on social behavior. It combines annoyance with the measurable volume of noise, its predictability and the possibility of its control. From a threshold of 90 dB(A)[14], noise can damage the ear up to complete deafness by irreversibly destroying the sensory cells of the inner ear.

Indeed, sound discomfort is not only due to the intensity and volume of the sound. As Claude Lévy-Leboyer states, sound nuisance depends rather on other factors that are perceived differently from one person to another, depending on the social significance of the noise. Indeed, noise is only accepted if its source has a noble reason to exist (a fountain is accepted while a truck is not tolerated). Noise is a danger to human physical and mental health, a feeling accentuated by urban density. Let us now take a closer look at the ways in which the feeling of stress related to exposure to urban noise appears. What methods should be used?

FROM SPATIAL CONTAINER TO AUDIBLE CONTENT: MATERIALS AND METHODS

The experimental protocol combines urban reading and a sensory characterization. For the spatial analysis of the chosen urban itinerary, it is based on reading out the plan and the elevations of the path, identification of the reference buildings, the realization of the urban sections and urban profiles, etc. It is followed by an investigation phase using the commented itinerary method (Grosjean & Thibaud, 2001). This experimentation implies a major shift: a radically new way of questioning oneself, of appreciating one's own existence in a hasty manner, of circumscribing it from physical presence, from one's immediate, embodied limits. To do so, new and finer methods are then mobilized. Each respondent[15] must wear an on-board physiological biosensor[16] before the experiment begins. At the same time, a third person accompanies us equipped with a sound level meter to measure the sound level before, during and after we pass through each urban crossroads.

The Walk as a Way of Capturing Urban Atmospheres

Our choice was based on the commented walk method developed by the sociologist Jean Paul Thibaud. This approach combines, simultaneously, space, time and movement. (Grosjean &Thibaud, 2001) The term "commented" announces the word, description and narrative. It is about walking, perceiving and describing, at the same time. The description, therefore, reveals the instantaneous perception during the action (perception in motion). This method involves several disciplines. There are physical measurements, architectural analysis and social science approaches. In parallel to this method, which is based on classical ethnological approaches, based on persons as sensors (via their perceptions), we use, also, the latest emotion capture technology, through a wearable sensory sensor, called the Q sensor.

Sensory Capture

As previously mentioned, stress can be identified using physiological indicators (Boucsein, 2012). The most significant physiological indicator is the variation in the heart rate reinstituted through the conductance of the skin (Picard, 2009). These measurements were made using a device called a Q sensor. The basic technology was developed at Affectiva[17] Laboratory, founded in 2009 by the Massachusetts Institute of Technology (MIT). It is a wireless connection through a wearable biosensor[18]. This device quantifies emotional arousal by measuring electrodermal activity (EDA)[19] (electro cutaneous activity), also known as Galvanic Skin Response (GSR). It also quantifies the movement and temperature of the user. This new technique is used to evaluate frustration, stress, and mood.

The result is illustrated by a curve that shows the stress value measured in micro siemens (μs: X-axis) as a function of time (Y-axis). The data collected by these two methods: "the commented walk" and by sensory capture, will be compared in order to identify significant stress situations. To objectify the results, we must complete our experimental protocol with an analysis of the sound environment.

COLLECTING THE SOUND DATA

The Seizure of the Sound Level

This quantitative approach allows us to measure sound levels. These measurements are made by a sound level meter. The unit generally used is the A-weighted decibel, noted as dB (A). We are generally interested in Leq (equivalent sound levels), which informs us about the average energy content of a sound during the measurement. Noise in the environment is rarely stable (variable intensity). It should be recalled that in this research, we are examining the links between the physical characterization of the urban sound phenomenon, the user's stress related to this signal and the morphological specificity of the built environment. We will take measurements before, during and after each urban intersection crossed, to highlight the variation in sound intensities during the passage of the urban crossroads. Frequency analysis of quantitative values must be completed by qualitative surveys to identify sound sources.

The Sound Recordings

The audio recording has audio characteristics such as sampling frequency. It is expressed in hertz and determines the number of samples used per second. The best compatibility is obtained with 44100 Hz. This qualitative phase aims to characterize the sound sources in urban crossroads. The resulting frequency composition will be treated by the Audacity software, which allows us to display the sound level according to the recording time, called "waveform". The sound recordings represented in the form of frequency spectral curves inform us about the frequency dynamics through the different frequency bands along the chosen path.

At the end of this empirical phase, we will cross the results of each phase of the experimental protocol: observation and analysis of the spatial environment, acoustic measurements, sound recordings and the comments of the respondent (collected from the commented walk method). This confrontation of all the data aims to objectify our results to confirm or refute our questioning of the departure.

But before that, let's first define our field of study.

Audition Walk in Tunis

The city center of Tunis is known for its colonial buildings. Its urban fabric is composed of the main avenue and perpendicular secondary streets. People walk alone, in pairs, or in groups. Their trajectories depend on the occupation of the sidewalks. The sound atmosphere is very mixed, composed of horns, the squeaks of the subway, ... etc. After several strolls in the center of Tunis, our choice was the neighborhood of Little Sicily in Tunis. This district now has a privileged situation in the Tunisian agglomeration. It is delimited to the north by *Habib Bourguiba* Avenue, to the east by Republic Avenue, to the west by Carthage Avenue and to the south by *Jallez* Hill. Today, this district represents the articulation of the capital's urban growth towards the south and north. Since its genesis, this quarter has known a great diversity of urban activities and types of social interaction (bourgeois, middle-class, workers, pedestrians, motorists, etc.), and today it has very rich potential in terms of sound atmospheres. Our field of study begins at the intersection of Yugoslavia street and Greece Street and ends at the crossroad of United Nations Avenue, Yugoslavia street and *Fares El Khoury* street, as we see in Figure 1.

Figure 1. Trajectory of the chosen path in the city of Tunis
Source: (Layeb, 2012)

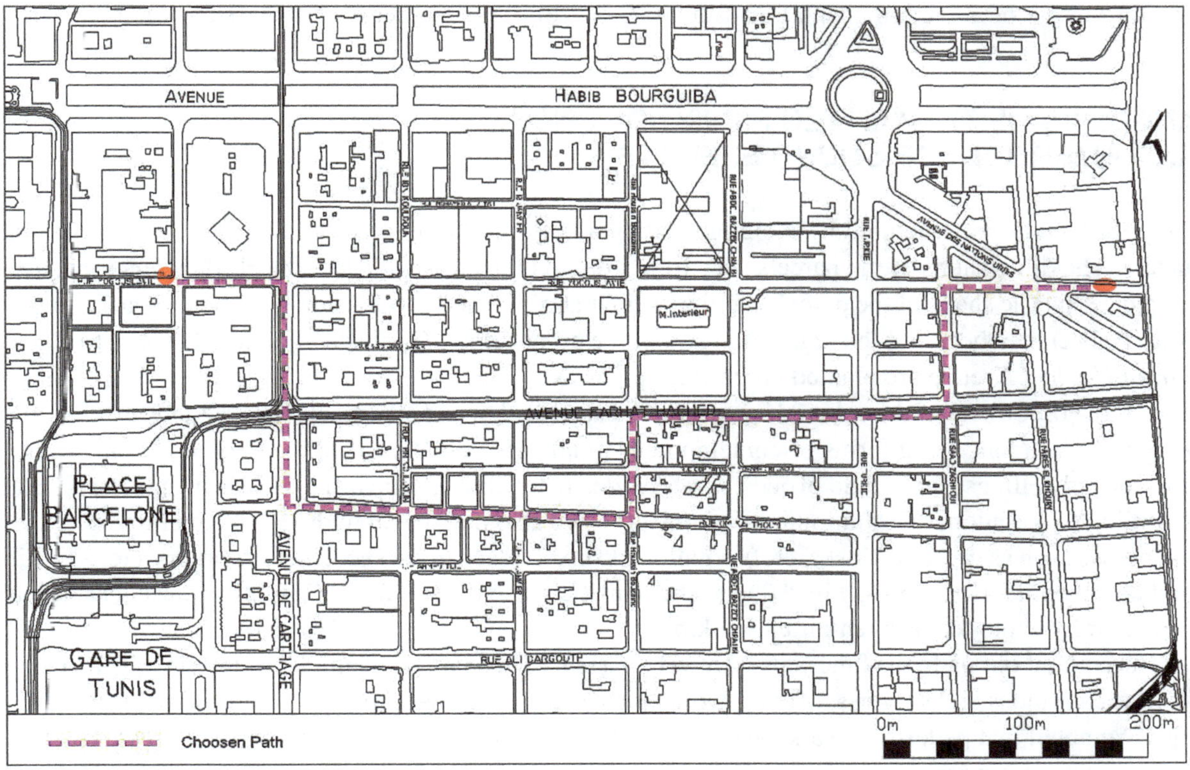

This itinerary cross several strategic vehicular routes, which connect *Habib Bourguiba* Avenue to the multi-modal station Place Barcelona. We first pass by the Carthage Avenue, a green street, crossed by a metro line. We stop at the crossroads of *Om Kalthoum* Street, which leads to Barcelona Square and its bus shelters. Then we take the pedestrian street *Houssine Bouzaine*, then the avenue *Farhat Hached*, which is a very crowded, planted avenue crossed by the subway, to end up on the United Nations Avenue and *Fares El Khoury* street through *Saad Zagloul* street. This walk can be divided into three main sequences. The first sequence is very frequented and animated. This is due to the types of shops (street vendors, the Palmarium shopping center), and its proximity to the multimodal station Place Barcelona. The second sequence is characterized by the dominance of pedestrian crossings and the multiplicity of one-way traffic flows. In the third and last sequence, we notice the decrease of pedestrians and small businesses. The street becomes wider, and the traffic more intensive, as it connects the southern part of *Moncef Bey* and *Bab Alioua* bus station to *Habib Bourguiba* Avenue.

We opted for this route because of the variety of commercial and administrative functions and consequently the multiplicity of the sound source types. The small number of residential units that exist is confined to the upper floors to make way for the shops on the lower levels. The majority of buildings have mixed functions combining commerce, services and sometimes habitation. We also note the richness of the building materials and architectural styles present: Italian, neoclassical or even the so-called "modern" style with refined facades and banded windows. Among the specificities of this site, we mention the variety of angle ornamentation as loggias, balconies, and other more or less decorated devices.

ANALYSIS OF THE "LITTLE SICILY" DSTRICT AND EMERGING OF SOUND EFFECTS

The urban fabric is structured by a simple orthogonal grid, whose street widths vary from 8m to 12 meters. The dimensions of the parcels vary from 40m to 60 meters wide by 80 to 100 meters long. The simple geometric shape of the parcels facilitates their addition and subdivision on a regular basis (in two or four). Islet then becomes the organizing unit of the urban fabric. Carthage Avenues, Turkey Street and *Farhat Hached* Avenue are planted (vegetated) pathways.

Today, some traces of the colonial epoch have disappeared, while others are slowly fading away to give place to new architectural forms of contemporary trends. (Layeb, 2008) Our field of study has several functions: dwellings, shops, and offices, ... etc. Indeed, dwellings, shops, and offices invade the district according to a random distribution in the same plot. We also note the abundance of small trades (small businesses such as grocers, kiosks, etc.), small workshops (mechanics, sheet metal workers, etc.). Virgin land has become increasingly rare[20] due to the density of the urban fabric. Currently, these vacant lots are generally exploited as open-air car parking.

Throughout our path, the functions were very various. There are residential buildings, hangars, many small urban trades and services, offices, workshops, small businesses, in old or newly renovated buildings. We noted that pedestrian walkways are abundant in addition to traffic and parking problems. This district is undergoing major changes, depots and hangars are disappearing and scrap metal and mechanical workshops are closing due to the rarity of vacant sites in the city center. Old buildings (in a dilapidated condition) are demolished and new[21] office and commercial buildings are constructed.

The repartition of building heights shows that three and four-floor buildings (higher than 12 meters) are scattered throughout the sector. Most of these buildings correspond to tertiary and mixed-use buildings. They are located mainly at the beginning of our chosen pathway. As we move eastward (towards the port), the height of the buildings decreases. The dominant heights are then 7 meters, which corresponds to buildings of one floor or a high first floor, used as stocking areas.

We have noted in several commented walks that the configuration of the built environment influences the user's feelings and emotions. Indeed, the urban configuration of streets (height of buildings, the width of streets, etc.) affects respondents' feelings. This is what we find with respondent n°4, during his passage through two different urban configurations, as we see in Figure 2. In the first narrow sequence, the respondent states *It's narrow, the sounds are amplified*, or when crossing the next, wider sequence, *it's an open area, there's a lot of noise issuing from all sides*, the sound cause is then clearly expressed (Layeb, 2012). The correlation of the physiological, verbal and acoustic data collected has allowed us to identify different stress situations related to different factors: we find sound, social, cultural, climatic and even political factors.

Figure 2. The variation of the EDA curve of respondent 4 in a stressful situation due to different urban configurations
Source: (Layeb, 2012)

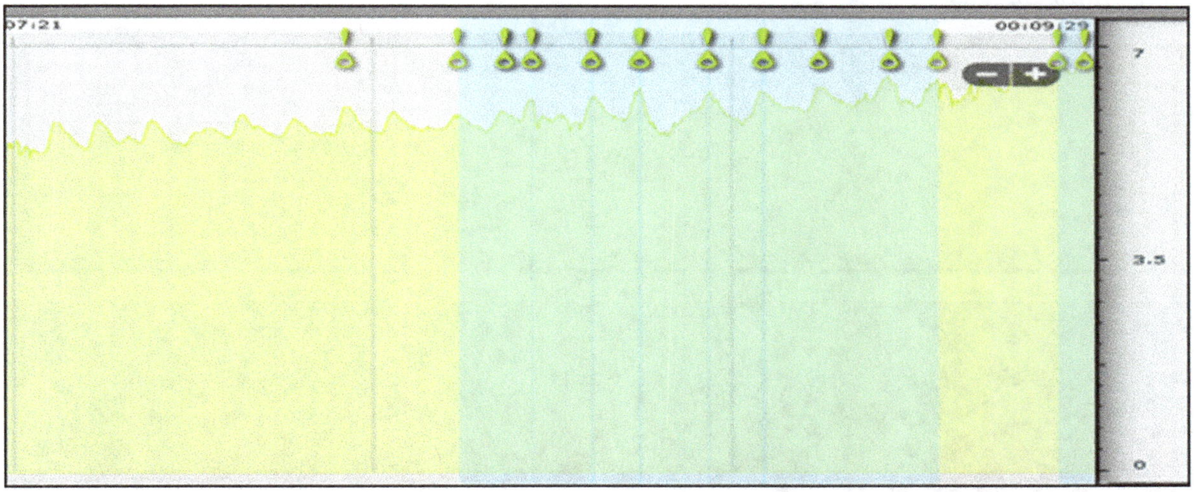

Sound as a Vector of Urban Stress

The correlation of the various data showed that the high noise level does not necessarily present a source of stress for the urban users. Some expressed their depreciation of the sound atmosphere judged as too calm and slow. Indeed, an agitated sound environment presents, for them, a reason for assurance and integration. This finding was confirmed by physiological data and verbal surveys, as we see the two situations in Figure 3. This is the case of respondent n°8, with a sound level of 69dB: *[...] But once you cross it's a bit monotonous. It's boring. I want to sleep (laughs). I feel the break-up. I don't feel motivated to stay long. I want to move to a more dynamic area. It's a monotonous calm. It's boring* (Subject 4 in Layeb, 2012, p.96).

In other contrary cases, the high sound level and the multiplicity of sound sources are expressed as the cause of sound stress situations. This is the example of a respondent who expressed his discomfort with the sound atmosphere at the intersection of Turkey Street and *Farhat Hached* Avenue. The noise level measured at this junction is 78 dB. *The hissing is increasing more and more. All sounds mix here. The subway is coming. It makes a special noise. It's getting stressful. The light turns green. All drivers are accelerating. The rhythm of the policeman's whistle makes me nervous. It's stressful* (Subject 2 in Layeb, 2012, p. 89).

Urban Stress Through the Other Senses

In addition to sound stress situations, we also identify different stress situations related to different sensory phenomena. Several respondents mentioned in their speeches the sensations and emotions linked to climatic: *What a heat! There is a breeze here. I can even hear the wind blowing*, says respondent 8, or *(...) what I'm hearing right now is the wind whistling. I think it's a little loud and intense. And it annoys me a little*, with respondent n°6 (Layeb, 2012). These perceptions were physiologically confirmed by

Figure 3. The confrontation of physiological data (on the top) and sound data (on the bottom): on the left, first stress situation due to a low sound level, on the right, second relaxation situation due to a high sound level
Source: (Layeb, 2012)

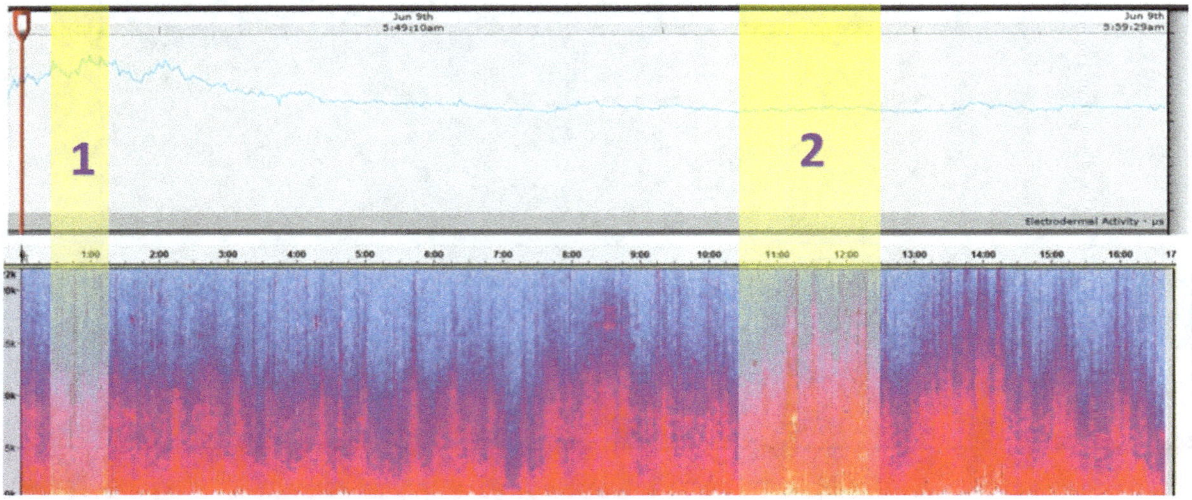

variations in the two respondents' electrodermal curve (EDA), as we see in Figure 4. The background sound produced by the draught is identified as a stress generator.

Other commented walks revealed the implication of the sociopolitical and security situation in the manifestation of the stress feeling. This is what we noted with respondent 10: *You know after the revolution... even if everything is fine, we are still stressed and panicked.* This speech was announced during our passage near the Ministry of the Interior. Following the various events of violence, looting and robbery, which appeared after 14 January 2011 revolution, and during the conduct of this investigation, soldiers of the National Army were still present on the streets. Barbed wire surrounds some buildings (Layeb 2012). We believe that this unusual landscape among Tunisian citizens, after the 2011 revolution, influences the apprehension of the urban landscape and may even be at the origin of this urban stress (Zairi, 2016).

In the same way, the particularity of Tunisian society is revealed in some speeches. This is the case with respondent 6: *But once you turn left* (she means with *Houssine Bouzaine* street) *it becomes very*

Figure 4. On the right, the plot of the EDA curve of respondent 8, on the left, the plot of the EDA curve of respondent 6 showing stress situations due to weather
Source: (Layeb 2012)

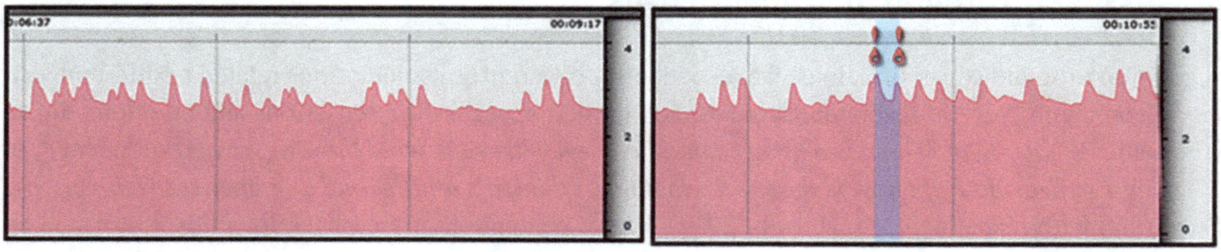

quiet, except for a few particular sources of noise. What bothers me is not the noise, but rather the mechanics. I am always very suspicious of this kind of activity. [...]. *If we have to turn right, I'd rather not pass in front of the café terrace. I prefer to go around with the newsstand. I avoid men sitting on the terrace. I'm moving fast. I don't like men's assemblies* (Subject 8 in Layeb, 2012, p. 115).

This respondent (female) expressed her discomfort and mistrust and even her feeling of insecurity about the gender of (male) traffic on this street. *Figure 5*. She also pointed out that this discomfort has nothing to do with the intensity of the noise or the types of sound sources, it is the dominance of mechanics' activity that makes this street very frequented by men. This stress due to the gender occupation (domination) and also detected by another respondent *Now I feel stressed. It's because of these (men) mechanic workers. I prefer to avoid them* (Subject 5 in Layeb, 2012, p. 82).

We end with the influence of personal memory and habituation on the manifestation of stress: *At this crossing, I am always very careful. It was because of a bad memory, I slipped while crossing quickly. I had a serious knee injury. Since then, I have always tried to be careful* (Slama, 2007). These comments were confirmed by the physiological data as shown in the following figure, as we see in Figure 6.

Figure 5. The variation of the EDA level of respondent 6 in a stress situation due to social factors
Source: (Layeb, 2012)

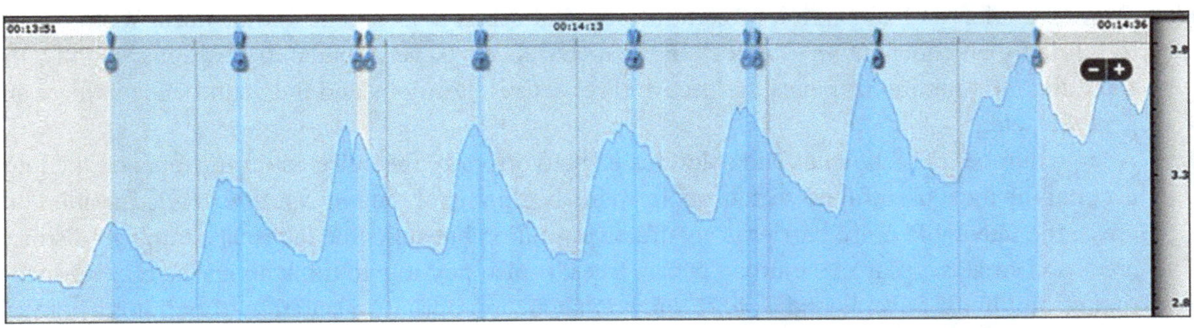

Figure 6. The variation of the EDA level of the respondent n°5 in a situation of stress due to a bad experience
Source: (Layeb, 2012)

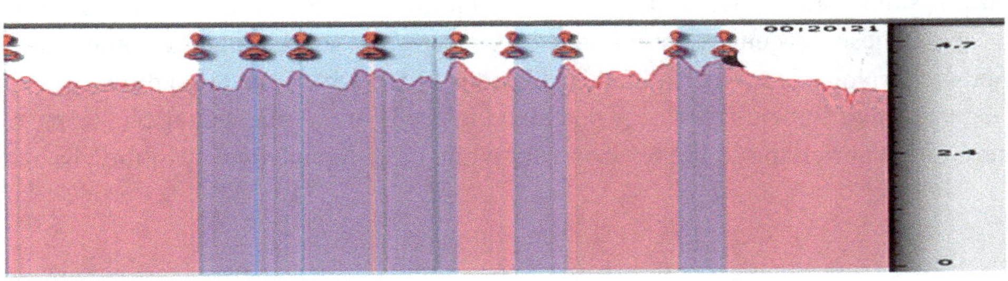

DISCUSSION

We have chosen to add to the perspectives of this work that have been formulated to emphasize the interactions between the issue raised by the article noise stress in urban areas and urban planning. We have tried to make adjustments to this section by updating the possible operational objectives of this research.

The development of cities has been followed by a profusion of varied sound productions, perceived by city dwellers as the result, or even the fatality, of the most urban activities. This situation has often led to degradation (nuisance, reduced speech intelligibility, etc.) of ordinary sound perception. The study of situations of auditory stress requires looking at the city from two points of view: ordinary users and urban planning (urban models, urban density, etc.).

Seen here from the user's point of view, urban sonic situations such as waiting, walking, crossing or even exchanging situations require more intelligibility of the sound conditions. In this context, sound reception is not a passive action, but an active process. It is then necessary to remedy the situations of discomfort that are increasing more and more, especially in urban areas. These disturbing experiences can be observed in several cases such as the dangerous situations by using urban transport, or in the cohabitation of pedestrians and vehicles. To limit these situations of discomfort, stress, and danger, three possible solutions can be identified at three levels of intervention: the transmitting source, the propagation environment and the sound receiver (the urban user).

In the end, and with the aim of improving listening conditions in the city, the operational objectives of this research are as follows:

- Transformation and adaptation of architectural and urban dispositions and devices. We propose to future urban contributors to reflect on the materials envisaged, spatial dispositions, sound barriers, the introduction of plants and to avoid reflective elevations and the sound effects they can generate, etc.
- A large number of techniques and solutions exist to optimize the noise environment and take into account the most harmful noise sources from the beginning of the project. In extended residential areas, the inevitable noise barrier is an effective solution but splits the landscape into two distinct areas and creates a scar in the urban fabric. But it is also possible to think about the relative location of buildings. Several studies have shown that an appropriate orientation of buildings in relation to each other and a good functional distribution (commercial, residential, etc.) often limits the transmission of noise in areas intended as quiet.
- Adjustment and modification of the diffusion of sound signals in urban spaces (warning, information, alert, etc.).
- A quiet area in the city contributes to well-being with the notion of resourcing while promoting the social cohesion that is conducive to welcoming and highly attractive. It is relevant to develop a culture of sound atmospheres in order to be able to recreate and enhance them through relaxation spaces. These new environments can present original sounds and the sound identities of the designed areas. Noise is a material that is shaped in space with a temporal dimension. The feeling of an acoustic environment is not only perceived in terms of volume, but also in terms of harmonious sources that enrich the sound environment while ensuring the quality of the visual ambiences.

FUTURE RESEARCH DIRECTIONS

These investigations open the way for new modalities of perception, urban sensing and human sensory in public space. This investigation was based on new methods about the impacts of contemporary digital technology (Wearable Technologies) associated with anthropological approach (commented walk) in the field of urban planning. This experimental protocol can be applied in design disciplines, especially in architecture.

Indeed, in recent residential complexes, new social problems emerged, the interior acoustic insulation of apartments is enough to guarantee comfort, but that attention must also be paid to all those interstitial spaces that serve as a basis for collective exchanges: streets, passages, intersections, squares, etc. The urban stress situation seems complex and interesting to explore using a multidisciplinary approach. This is what makes this work so relevant, innovative and original that it draws on a variety of disciplines, including urban planning, architecture, and psychology. Various other specific urban and architectural situations require in-depth research (autistic (Layeb, 2016), elderly, presbycusis (Hussein, 2012), as well as other spatial situations that involve study, further and deeper investigations.

CONCLUSION

Much more than sight, hearing is the sense by which we feel the corporal immersion in an ambient environment. Sound submerges us, stresses us or soothes us, always reactivating our consciousness of our bodies. It is the body that produces and receives a sound that places us in the urban environment: we can cope with a situation through our acoustic control or, on the contrary, be exasperated by our submersion in sound flows. In conclusion, we must remember that the results collected after the analysis and correlation of physiological, verbal and acoustic data confirm the role of emotions and senses in our perception of the urban space.

The atmosphere and more precisely the sound atmosphere, the main object of this approach, have a significant role in the sensitive qualification of urban space. The sound atmosphere, through the co-existence and dynamics of its sound sources, structures the perception of the city, giving it a sensory dimension. This perception is not always positive. The multidisciplinary approach used here examines these negative and stressful situations to identify them. We simultaneously analyzed the frequency content of the recorded soundtracks, the variation of the electrodermal activity (EDA) and the real-time description of the urban user through the commented path method. We have shown that the perception of stress in public spaces is linked to certain physical characteristics of the sound signal, and more precisely, certain frequency bands of the signal specifically of the contemporary urban environment (horns, vehicle acceleration, … etc.).

ACKNOWLEDGMENT

This research received no specific grant from any funding agency in the public, commercial, or not-for-profit sectors. The authors are, particularly, grateful to Professor Jean Pierre PENEAU for framing this work. Sincere gratefulness goes to the unit U2S (Unité de recherche système et signaux) of the National

School of Engineering of Tunis (ENIT), for help in analyzing the EDA data and frequency data. The authors would like to thank Mrs. Lilia Ezzedine for her precious proofreading.

REFERENCES

Augoyard, J.-F., & Torgue, H. (1995). *A l'écoute de l'environnement. Répertoire des effets sonores.* Marseille: Parenthèses.

Ben Slama, H. (2007). *Parcours urbains quotidiens, L'habitude dans la perception des ambiances.* Grenoble: PhD, ENSAG, CRESSON.

Boucsein, W. (2012). *Electrodermal Activity* (Vol. 2). Allemagne: Université de Wupperta.

Ghozi, R. (2015). Occupancy-based Analysis and Interpretation of Soundscape Auditory Complexity: Case of a Campus Restaurant. *J. Audio Eng. Sco*, *1*(1).

Grosjean, M., & Thibaud, J. P. (2001). *L'espace urbain en méthode*. Marseille: Parenthèses Editions.

Houssein, F. (2012). *Construction et expérimentation d'un protocole pour le traitementdes ambiances sonores urbaines adapté aux personnes âgées presbyacousiques.* Nantes: thèse de doctorat en cotutelle, ENSA nantes.

Layeb, S. (2008). *Intervention urbaine dans un tissu urbain ancien: cas de la petite Sicile de Tunis* (Unpublished doctoral dissertation). Tunis: ENAU.

Layeb, S. (2012). *Parcours commenté, parcours segmenté: recherche exploratoire sur les liens entre configurations urbaines et stress sonore en milieu urbain* (Unpublished doctoral dissertation). Tunis: Academic Press.

Layeb, S. (2016). Caractérisation ambiantale des espaces d'accueil pour enfants autistes:Vers une projectuelles des ambiances. *Séminaire de Conception Architecturale Numérique, SCAN'16*, 163-172.

Lynch, K. (1976). *L'image de la cité*. Paris: Bordas.

Moser, G. (1997). *Les stress urbains*. Paris: Armand Colin.

Moser, G. (2009). *Psychologie environnementale - Les relations homme-environnement.* Bruxelles: De Boeck Editions.

Paquot, T. (2015). *Désastres urbains*. Paris, France: Éditions La Découverte.

Picard, R. (2009). Future affective technology for autism and emotion communication. *Philosophy Transactions of the Royal Society, Biological Sciences*, *364*(1535), 3575-3584.

Picard, R., Vyzas, E., & Healey, J. (2001). Toward machine emotional intelligence: Analysis of affective physiological state. *IEEE Transactions on Pattern Analysis and Machine Intelligence*, *23*(10), 1175–1191. doi:10.1109/34.954607

Rossi, A. (1981). *L'architecture de la ville*. Paris: L'Equerre Editions.

Sitte, C. (1996). *L'Art de bâtir la ville, l'urbanisme selon ses fondements artistiques*. Paris: Seuil Editions.

Torgue, H. (2009). Composer le son des villes. WildProject, Ondes du monde: territoires sonores de l'écologie.

Vigarello, G. (2014). *Le Sentiment de soi. Histoire de la perception du corps. (XVIe - XXe siècle)*. Paris: Seuil Editions.

Zairi, M. (2016). *L'ambiance comme enjeu politique dans l'espace public de Tunis lors du processus révolutionnaire* (Unpublished doctoral dissertation). Grenoble: PhD, ENSA, CRESSON.

ADDITIONAL READING

Chelkoff, G. (1992). *La mise en vue de l'espace public*. Grenoble: Cresson.

Chelkoff, G. (1996). *Imaginaire sonore et environnement urbain. Les cahiers de la recherche architecturales et urbaines, n°38-39*

Dutheil, F., Chambres, P., Hufnagel, C., Auxiette, C., Chausse, P., Ghozi, R., ... Bertrand, P. R. (2015). Do Well B.': Design of WELL Being monitoring systems. A study protocol for the application in autism. *BMJ Open*, *5*(2), e007716. doi:10.1136/bmjopen-2015-007716 PMID:25710916

Ghozi, R., & ... . (2013). Affective State Real-Time Monitoring through Electro-Dermal Activity: Towards Daily Stimulus and Arousal Level Analysis in Autistic Children. *International Conference: Autism, Support Methods*

Healey, J. A. (2000). *Detecting Stress During Real-World Tasks, Using Physiological Sensors*. Cambridge: MIT.

Layeb, S., & Ben Hadj Salem, M. (2014). Commented walk, segmented walk: An exploratory study on the relationship between urban space and sound stress. *Invisible Places Sounding Cities*, (http://invisibleplaces.org/IP2014.pdf). Viseu.

Nocua, R. (2010). *Conception et développement d'un système ambulatoire pour la mesure de l'activité du Système Nerveux Autonome pour la surveillance de personnes âgées à domicile*. Grenoble: Université Joseph Fourier–Grenoble I.

Piolat, A. (2008). Emotions et affects: Contribution de la psychologie cognitive. In *D. P. Nagy, Le sujet des émotions au Moyen Age* (pp. 53–84). Paris: Beauchesne Editeur.

Sano, A., Picard, R. W., & Stickgold, R. (2014). Quantitative analysis of wrist electrodermal activity during sleep. *International Journal of Psychophysiology*, *94*(3), 382–389. doi:10.1016/j.ijpsycho.2014.09.011 PMID:25286449

Terrin, J. J. (2011). *Le Piéton dans la ville. L'espace public partagé*. Marseille: Parenthèses Editions.

ENDNOTES

1 The philosopher Thierry Paquot has chosen to address in this book several "devices" that he considers problematic, each one leads to *a situation that increases the confinement and subjection of the individuals who undergo them* (Paquot, 2015, p. 25). The book is divided into five main chapters devoted to what the author identifies as disasters: the large complex, the shopping center, the skyscraper, the gated community and the major project.

2 The CSTB, Scientific and Technical Center for Building, was created in 1947. Four main activities are developed: applied research, expertise, evaluation, and dissemination of knowledge that enable it to meet the objectives of sustainable development for construction products, buildings and their integration in neighborhoods and cities.

3 This theory was developed by the UMR (Unité Mixte de Recherche) AAU-CNRS 1563 (Ambiances Architecturales et Urbaines), which combines the two Cresson research laboratories (Centre de recherche sur l'espace sonore et l'environnement, founded in 1979 in Grenoble) and the CERMA (Centre de recherche en modélisation architecturale, founded in 1971 in Nantes). Since its creation in 1998, the laboratory has developed in a highly interdisciplinary approach, through its composition (architects, sociologists, computer scientists, anthropologists, urban planners, geographers, historians, philosophers, ...), through the challenges it addresses (architecture, environment, city), and through the tools it designs and deploys (in situ survey methodologies, modelling and simulation of environmental phenomena, virtual reality, etc.).

4 Organic sensitivity, emanating from all internal sensations, which arouses for the human being the general feeling of his existence: *All these conflicts, even the most spiritual and refined, are finally included in the confused cenesthesia.* (Ricœur, 1949, p. 141)

5 This statement is the result of an analysis of Jean Jacques Rousseau's work: Jean-Jacques Rousseau, "Art. Génie", in Dictionnaire de musique, Bern, Peter Lang Editions, 2008[1768]. See also: Denis Diderot, "Salon de 1767", in Salons, Paris, Folio Gallimard Editions, 2008[1767].

6 On this subject, see the recent work of the laboratory L'Autre musique (associated with UMR 8218 Paris1/CNRS Programme: Arts Sonores), and more specifically: Hélène Cussac, "Anthropologie du bruit au siècle des Lumières", Bruits, n°4, March 2016, pp. 1-12.

7 Claude Lévy-Leboyer is a professor at the University of Paris Descartes. She was President of the International Association of Applied Psychology

8 Gabriel Moser is Professor of Environmental Psychology at the Institute of Psychology of the University of Paris-Descartes where he is also Director of the Laboratory of Environmental Psychology of the CNRS. He is the director of the "Regards psycho-sociaux" collection and of the Social Psychology sector of the "Cursus" collection at Armand Collin. He created the specialization "Environmental Psychology" (professionalized training authorized in 1992, in charge of the Master 2 Research "Individual, Social and Environment", training authorized jointly with the EHESS (School of Higher Studies in Social Sciences). He is also a member of the "PRIMEQUAL" Scientific Committee and the Scientific Council of the "Concertation, Décision et Environnement" programme at the Ministry of Ecology and Sustainable Development.

9 Developed in the author's book: Gabriel Moser, Aspects des relations homme-environnement, Éditions De Boeck, 2009

10 Andrew Baum, Jérome Singer and Carlene Baum are professors of psychology. They are interested in studies of the health effects of stress.

11 Sheldon Cohen is a Doctor of Psychology. He has worked, with his team (Cohen, 1996) and (Cohen, Tyrrell, & Smith, 1991), on the negative impacts of urban stresses on our daily life.

12 After working on electro-acoustic questions (1976), analyzing the manifestations and functions of voice and sound in cinema (1982, 1985), and studying symphonic poetry and program of music (1993), Michel Chion developed an "acoulogie" essay, rich in fine notations, under the evocative title of the Listening Walker. Essais d'acoulogie, Paris, Plume, 1993.

13 The research presented here is a continuation of a research project entitled "Altered perceptions of sound atmospheres in urban environments. Characterization and corrections: contribution of audio textures" (2010-2014). The idea of associating the theme of urban sound atmospheres with the field of audio textures was at the origin of bringing together researchers in architecture and urban planning with a young team (the ERA- Équipe de recherche sur les ambiances: Reaserch team on Ambiences) from the ENAU (École Nationale d'Architecture et d'Urbanisme de Tunis: National school of architecture and urbanism), research engineers from the TIC Department (Information and Communication Technologies) of the National School of Engineers of Tunis, the U2S research unit and the team of researchers from UMR 1563 of the CNRS "Architectural and urban environments".

14 The decibel "dB(A)", it commonly expresses the sound level. This value expresses the power ratio between the measured value and a reference value set by a standard.

15 Ten respondents, five women and five men, each on a different day but in the same time slot.

16 These measurements were made using a device called a Q sensor.

17 Affectiva is a company based in Waltham, Massachusetts, founded in 2009 and the result of research conducted in collaboration with MIT's Media Lab by Rosalind Picard and Rana El Kaliouby, two scientists who have joined the National Science Foundation (NSF) to develop technologies on the autism spectrum.

18 The Q sensor is in the form of a digital watch without a reader. A warning light indicates the tuning on or off of the device and the battery charge level. Two silver electrodes are located below to measure the conductivity of the skin. The Q sensor includes downloadable software for data visualization or exportation in suitable formats. The recorded values are saved in the memory card of the bracelet and then transmitted through the USB port to a computer.

19 Electrodermal activity is a biological electrical activity recorded on the surface of the skin, which reflects the activity of the sweat glands and the autonomic nervous system. It informs us about the perception of the individual, his involuntary behavior and his emotion (and not a response he wishes to give).

20 As the example of the terrain between *Farhat Hached* Avenue and *Houssine Bouzaine* Street or even the one between *Sâad Zaghloul* Street and Yugoslavia Street, currently exploited as an open-air car parking.

21 This is the example of the commercial building at the intersection of Yugoslavia Street and Turkey Street.

Chapter 8
Introducing a Conceptual Framework for Spatial Decision SIM:
A Tool Supports Decisions for Stakeholders

Ahmad M. A. Toimah
https://orcid.org/0000-0001-6537-7392
Ain Shams University, Egypt

Samy M. Z. Afifi
Ain Shams University, Egypt

ABSTRACT

Planning is a time-sensitive process with spatial characteristics as its core. It is effective to formulate spatially-related decisions on an informative background to maximize benefits and minimize risks. Not only decision makers who affect the space, but also users and owners interact with it, affect the related decisions. Thus, it is healthful to widen participation. This chapter introduces a conceptual framework for the Spatial Decision Simulator "SD-SIM." This work aims to reach a platform that supports spatial decisions made by various stakeholders to provide a capability for integrated modeling of socio-economic, man-made, and natural environmental impacts. It contains four components as a logical target for expressing the evolution of spatial issues and reflecting them into a simulator. These four components are Districts Sub-System, Property Price and Living Cost Simulator, Interventions Sub-System, and Development Scenarios Sub-System. The SD-SIM depends on free-access data sources. Through its sub-systems, the platform integrates different analytical methods and tools.

DOI: 10.4018/978-1-5225-9238-9.ch008

INTRODUCTION

Urban Planning is a profession that is concerned with forming a future living environment (Frenchman, 2000). All individuals, groups, and institutions practice planning for their future. As Worral (1991) states, "It is estimated that 80% of the data used by managers and decision makers is geographically related". Therefore, it is desirable and effective to formulate spatially-related decisions on an informative background to maximize the benefits of these decisions and minimize associated risks.

It is not only managers and empowered decision-makers who affect the space, but users and owners also interact with the spaces in to affect decisions related to them. Accordingly, it is healthy to widen the range of participants in the decision-making process to include all potential stakeholders in the spaces. However, not all stakeholders have the same background, and they may not perceive the impacts of these decisions. As Khaldun (1377) explains, "It should be known that history, in a matter of fact, is information about the human social organization". Therefore, it is logical to inform stakeholders about the previous spatial decisions to understand specific conditions and their consequences. History repeats itself, and it is no wonder that patterns have evolved from an information system methodology, as Cellular Automata Model showed, over time, a degree of self-similarity (Torrens, 2000).

This chapter introduces a conceptual framework for Spatial Decision Simulator (SD-SIM). This research intended to provide a platform that supports spatial decisions taken by different stakeholders based on time series spatial and non-spatial data that are analyzed, correlated, and clustered logically. This framework is designed to provide the capacity for integrated modeling of socio-economic, human-made, and natural environmental impacts. The framework contains four components as a logical target focal point for expressing the evolution of a spatial problem and reflecting that evolution into the simulator.

The stakeholders targeted for this platform involve individuals including employees at services providers, mediators as contractors, banks and real estate developers, planners, and central and local governments. The spatial decisions taken by these groups based on the information provided by this platform have social, economic, environmental, and institutional impacts.

BACKGROUND

It has become increasingly evident that decision-making is becoming a highly complex process. Centralized "traditional comprehensive" planning has many difficulties related to its complexities. In this model, only planners and decision-makers have the right to decide what, where, and how to take actions. Nowadays, slums, urban poverty, urban heat islands, traffic jams, environmental degradation, and inflation, stagflation are among the common phenomena that our earth is suffering from. These phenomena negatively affect the environment. One of the leading causes of these phenomena is the monopolization of resources, information, responsibilities, and decision-making processes.

From another point of view, and assuming that people have the right to decide. A safe, attractive and healthy living environment, with adequate reachable services and employment is the average resident's desire (Koomen, 2008). Residents' desires are different, changeable by time, and perhaps inapplicable, especially regarding available resources. Decision processes must consider a wide range of often-conflicting objectives at different spatial and temporal scales, and there is a need to synthesize a viable alternative for communication between a wide range of stakeholders (Afifi, 2014). Acknowledgement and participation are considered a step towards gaining public satisfaction. In this way, residents should understand the details about decisions related to their future. They also need to be informed about the system of allocating different types of projects and their consequences, so that they can objectively make decisions..

Globally, in recent decades, many changes tackled spatial decisions either on the theoretical level or in the mechanisms and methodologies used for their implementation. Spatial decisions become more sophisticated to set out a strategic framework to guide future development and policy interventions, instead of being limited to technical and legal regulation. This approach is developed using a negotiated form of governance, rather than a government-led, top-down model (Albrerchts, 2010). However, to come to a strategic decision, those who are responsible for this decision face many challenges. They have to handle large amounts of information and big data with various accuracies, types and time bounds (Afifi, 2014). Moreover, the issues in the strategic decision-making processes are heavily dependent upon subjective assessments. Problems are unstructured and usually unique, uncertainty and risk are high, and alternatives cover a long time-period (Steiner, 1969). Therefore, it is necessary to support stakeholders and decision-makers by providing them with better understanding, methods, and tools to tackle spatial problems (Sudhira, 2008).

Spatial complexity can be easily seen, for example, as an underdevelopment land that has many interfaces. Spatial complexity has characteristics, such as value, neighbors, topography, soil, and other environmental conditions. It is located within a certain zoning, and it is subjected to specific planning and building regulations. Its owner and developer acquire maximum economic benefit, as they construct new buildings or redevelop existing ones, impact fees, collect real estate taxes to preserve the environment and maintain infrastructure. Residents have their own work, either clustered or dispersed, and need a set of different services. Services providers target these residents, and transportation agencies serve them to access works and service locations.

In this way, spatial decisions are complex systems. They interact within and across a wide range of stakeholders, addressing different patterns of socio-economic and/or environmental problems (Koresawa & Konvitz, 2001). Therefore, spatial decisions can be seen transformative and integrative processes through which visions are collaboratively articulated. In spatial decisions, coherent actions and means for implementation are produced, which form and frame the present and future of a place (Albrerchts, 2010).

SPATIAL DECISION-MAKING CHARACTERISTICS AND REQUIREMENTS

As spatial decisions are complex systems, their sub-elements are linked by non-linear interactions. These elements act independently in a parallel way to form a global order of local interactions, and make the system self-organized one (Cheng, Masser, & Ottens, 2003). The parameters controlling the system are self-organized systems themselves (Koresawa & Konvitz, 2001). From the non-linear interactions between elements in these complex systems emerge many phenomena. Dynamics behind these phenomena cannot be understood by decomposing this system to its elements (Bertolini, 2006). This complexity of the spatial decision-making process requires sufficient data, analysis tools, and methods, planning approaches, and a robust framework that can deal with a vast number of attributes, variables, indicators, and scenarios.

Data Supporting Decision Making

Data can be classified into different categories depending on their nature, behavior, and types of applicable analysis. Classically, data can be categorized as primary and secondary data. With the emergence of the digital era, the need to gather primary data is losing its importance. Primary data that have been collected can now be obtained from secondary data sources (Peersman, 2014). The scale of development affects the dependability of each of the two types of data. Small-scale interventions depend mainly on primary data, while in large-scale interventions, development of secondary data has great importance.

According to the type of data source, data can be classified into on-site, on-Office, and online (Hox & Boeije, 2005). Nowadays, the online data source is more common, and the importance of on-site and on-office is decreasing. Online data is associated with the growing process of data-mining and thus, the development of new methods and analysis tools is necessary. Such tools should be efficient enough to extract knowledge from raw data through seven steps: data cleaning, data integration, data selection, data transformation, data mining, pattern evaluation, and knowledge presentation (Han, Kamber, & Pei, 2012).

Another classification for data is quantitative and qualitative data. Although quantitative data represents the solid core for a spatial planning processes, the importance of qualitative data is growing with the emergence of participatory planning approaches. Qualitative data helps planners and decision-makers in observing a situation, generating issues, and developing a hypothesis, while quantitative data is reliable in terms of testing feasibility, monitoring consequences, and designing actions (Hox & Boeije, 2005). The usage of digital press, social media, and web photography by decision-makers and community groups as a source of qualitative data are becoming more noticeable.

Based on these classifications, we can classify data into conventional and non-conventional collection methods. In conventional collection methods, the subject includes primary data conducted from surveys, questioners, and field measurements, and secondary data gathered from governmental agencies. The non-conventional method includes a wide range of quantitative and qualitative open-access data. In this method, examples include imagery available from virtual globe software and services, such as Google Earth, SOS. They also include maps and traffic data from web mapping such as Google Maps. Furthermore, the non-conventional method includes data from ridesharing services, such as Uber. The non-conventional method also includes air quality data from apps and services based on phone and vehicle sensors, such as AirVisual, Ambiciti. Finally, the non-conventional method also includes market data from e-commerce sites, such as Amazon and OLX. Figure 1 illustrates the data flow and classification into conventional and non-conventional methods.

These different types of data must be gathered and uploaded onto a database to be analyzed using spatial data analysis software and tools (e.g., ArcGIS) and non-spatial tools (e.g., SPSS). Alternatively, researchers can develop a tailored analysis tools that combines the two groups of software and tools. The results of this process can directly support the spatial decision-making process or represent inputs to Spatial Decision Simulator that can, in turn, enhance the efficiency of this process.

Community and Planning

Managing spatial decisions is a complex process that is influenced by defining inputs for this process. Traditionally, a spatial planner collects and uses inputs to tackle issues raised by the society. Inputs are data, which are considered objective, unambiguous and mostly quantitative. In the traditional model, society is seen as pure and simple in its structure, controllable in its behaviors, and unaffected by ex-

Figure 1. Data flow and classification into conventional and non-conventional method
Source: The authors

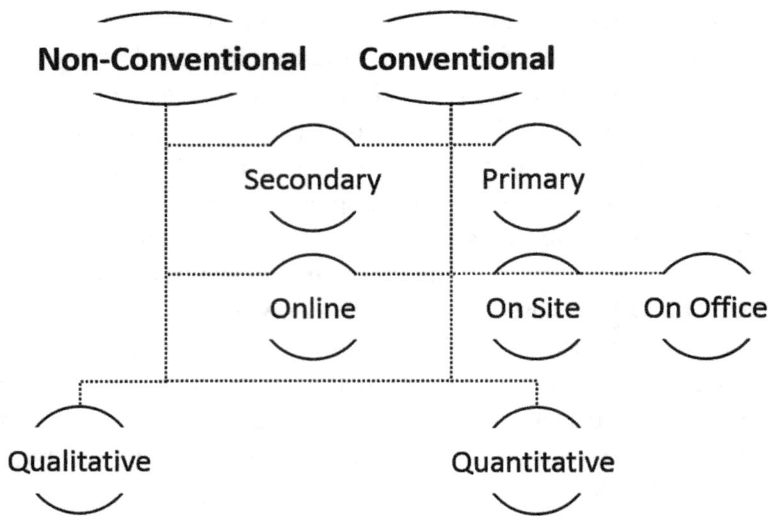

ternal factors. On the other hand, in the collaborative model, society is seen in a more realistically way. In this model, society is seen as complicated, critical, and uncontrollable. Knowledge, rather than data, is inter-subjective and ambiguous, and it is constructed by society rather than collected by the planner as in the traditional model. In this way, the planner facilitates societal interrelationships. The spatial planning has a paradigm shift from a "one-way direction"' planning for communities to new, complex planning with the communities' model (Forester, 1999) via (Pettit, Keysers, Bishop, & Klosterman, 2008, Bertolini, 2006).

In this new, complex model, all stakeholders should work together to analyze, compare, contrast, and prioritize different proposed development alternatives. Planners – as facilitators – are responsible for gathering and evaluating available "conventional" data, extracting and understanding the "potential hidden" knowledge, and objectively presenting future consequences of different interventions and development scenarios to all stakeholders. Furthermore, planners have the duty to understand the history of the community, to respond accurately to the forces of growth, and to predict the future of the social, economic, environmental, and cultural status of a community (Wang & vom Hofe, 2007). Considering the complex behavior of individuals and organizations, this approach may result in undesirable land use for stakeholders unless they are able to collaborate effectively (Sugumaran & DeGroote, 2011). Spatial decisions should be processed in a way that ensures effective participation of the people concerned (Council of Europe, 2009). Each person concerned should participate in a so-called "shared" vision to support decisions, rather than creating little more than a "wish list" that includes every project and policy intended by all stakeholders (Koresawa & Konvitz, 2001).

Intervention and Development Time Response

Spatial phenomena are much more involved in their pattern-temporal dynamics than described in static relations. Interventions have different response behaviors. Some interventions have a few weeks or months between a stimulus and the first sign of response (e.g. firm relocating or residential mobility). Other interventions require years for the first sign of response to become evident (e.g. changing in land uses and transport construction). As an extreme example, some interventions may need decades for their responses to work their way through urban stock (e.g. residential and industrial construction). Funding and political wishes can also affect the rate of interventions' development. On one hand, even small interventions may take years of development in communities that lack sources of funding. On the other hand, political wish can drive rapid development of mega interventions, especially in communities with small degrees of accountability. Interventions may also differ in their rates of change associated with the process in relation to the magnitude of the affected stock, and in the degree to which the process may reverse its direction (Wegener, Simmonds, & Waddell, 2011). The speed of change of any intervention is determined according to its descriptive dimensions. A proposed intervention may be influenced by a previous intervention. Spaces are not static, and they should not be represented as such (Eid, Attia, El-Refaei, & Afifi, 2014).

Spatial Decision Support (SDS)

According to Simon (1960), any decision-making process can be structured into three major phases. The first phase is the intelligence phase, which scans environmental conditions that need a decision through exploring data and information. This phase supports decisions by situational analysis by answering the questions, "Is there a problem? If so, is there an opportunity to change it?" The next phase is the design phase, which is responsible for developing and analyzing a set of possible alternate decisions for the problem identified in the previous phase using suitable modeling techniques. This phase answers the question, "What are the available decision alternatives?" Next, the choice phase ranks, evaluates, and analyses generated alternatives in the design phase. Decisions made in this phase are ranked according to evaluation criteria. The importance of decisions depends upon the decision-makers' preferences. The question that should be answered by the end of this phase is, "Which alternative is best?" (Malczewski & Keller, 1997).

Decision Support Systems (DSS) are computer-based information systems that are developed to support decision-makers in addressing ill-structured problems by allowing them to access and use data and analytic models (Engelen, White, & Uljee, 2005). Sprague and Carleson (1982) define DSS as "a class of information system that draws on transaction processing systems and interacts with the other parts of the overall information system to support the decision-making activities of managers and other knowledge workers in organizations" (Power, 2008).

The term "spatial" brings into focus the "where of things" - whether static or dynamic, the characteristics of certain locations and sites, the inter-relations among different activities and networks, and significant intersections and nodes in areas that are physically co-located (Albrerchts, 2010). The decisions related to spatial considerations are complex, multidisciplinary, characterized by uncertainty, and usually involve numerous concerned stakeholders. Thus, the spatial decision-making process cannot be preceded linearly. Instead, it is likely to follow an iterative process with various interactions between groups with adequate knowledge-base to participate in the decision-making process. This knowledge-base is either hard information, which is derived from reported facts, quantitative estimates, or systematic opinion, surveys, or soft information that is based on opinions, priorities, or preferences of decision makers or based on ad hoc surveys or comments (Sugumaran & DeGroote, 2011).

Stakeholders involved in spatial decisions processes should interact in a way that allows them the capacity to understand differences between stimuli. Alongside the planners, stakeholders may include experts in other disciplines or decision-makers, who approve the plan, or the community with whom the decisions are being made. The decision process should be interactive and participatory because a variety of information must be incorporated and a variety of stakeholders must participate in the process (Ayeni, 2005) (Sugumaran & DeGroote, 2011). In addition, the decision-making process is often iterative because alternative actions are analyzed, and information gained is used as knowledge to guide further analyses. Thus, it is important to justify the decision that is made. To achieve this requirement, related information concerning the issue must be acquired and organized to support problem analysis.

Although the Geographic Information System (GIS) can deal with spatial data and has spatial analysis capabilities, it is generally insufficient and not directly applicable to unstructured spatial decision problems. GIS usually does not have sufficient flexibility for interactive and recursive problem-solving skills. The development of SDS has been associated with a need to expand GIS capabilities for tackling complex, ill-defined, spatial decision problems (Densham, 1991). The development of SDS integrates GIS with a computer-based spatial analysis module to determine, analyze, and adopt development scenarios according to the planning standards used by planners and decision-makers or the preferences of the community.

GIS and SPSS can play a vital role in the intelligence phase by scanning spatial data and non-spatial data, respectively. They can help in coordinating decisions in situational analysis through integrating and exploring data and information from a wide range of sources. Moreover, GIS and SPSS can effectively present information in a comprehensive form to decision-makers. However, most available GIS and SPSS platforms lack the kinds of user-friendly analysis and modeling required by stakeholders, which is essential for the design phase or selection phase. They also lake the tools to complement the advanced spatial and non-spatial analysis and modeling, as they can only meet in primary levels of analysis and then each platform goes in a different way and uses different techniques of analysis, modeling, and simulation.

Successful application of SDS relies on the involvement of a wide array of potential actors. Each alternative proposed in the process is evaluated based on Multiple Criteria Analysis. Some of these criteria may be qualitative, while others may be quantitative. The relative importance of evaluation criteria and decision consequences are suggested by many different decision makers and interest groups involved in the decision-making process (Malczewski & Keller, 1997). In applications of SDS, besides decision makers and end users related to a certain decision, stakeholders also include developer and analyst. The developer collects all requirements and designs and programs the functionality of the system. The analyst is responsible for selecting models, carrying out simulations, analyzing data, and producing outputs to aid the decision makers. The analyst may be a team composed of multiple analysts from different disciplines (Sugumaran & DeGroote, 2011). The end-user team composition is based on the size and the nature of the spatial decision process, and the user's level of expertise.

AN ARCHITECTURAL FRAMEWORK

The conceptual framework for SD-SIM, as illustrated in Figure 2, can be developed as a platform that consists of four main interconnected sub-systems:

1. **Districts Sub-System:** this sub-system has data connected to districts at different time series to facilitate detection of change rate, pattern, and behavior. It supports the whole platform with the probabilities of allocating land uses over time series under a given scenario.
2. **Property Price and Living Cost Simulator:** this sub-system simulates property prices and living costs for basic daily needs and mobility schemas, including the characteristics of locations change over time. This sub-system also analyzes the effects of short-term variations in vacancy rates on overall property prices. This simulator provides other sub-system with the economic dynamics of the entire urban system.
3. **Interventions Sub-System:** this sub-system relies on agent-based modeling, as it contains the different behaviors that the interventions perform on the objects of the environment. These behaviors are set either by the decision-makers or as an outcome from other sub-systems. This sub-system feeds other sub-systems with the consequences of individual decisions.
4. **Development Scenarios Sub-System:** this sub-system produces semi-automatically development scenarios with its potential strategic consequences. It targets mainly governmental agencies, local administrations, and planners, as it facilitates developing spatial plans and actions without neglecting other stakeholders.

Districts Sub-System

This system is considered the central knowledge base for the framework. The main task of this sub-system is to perform the required in intelligence phase of Malczewski's Decision Support System (Malczewski, 1997). It scans environmental conditions that need a decision by exploring data and information so that this can help in decision situation analysis. The main question for this sub-system is, "Is there a problem?" The district-level is used in this sub-system for two main reasons: Firstly, in many communities, it is the level that has direct contact with people and their everyday life. Secondly, this level usually has an administrative governing body. The district's sub-system feeds the other sub-systems with the environmental conditions at the district level, and the probabilities of allocating different land use over time series under a given scenario. The elements for this sub-system are described in the following subsections.

Non-Spatial Database

A set of quantitative and qualitative data should be conducted through an analytical and collaborative process to be used as inputs for different sub-systems of the platform. These data sets include variables such as population and their demographic and social characteristics, crowdedness index, and types of housing for households. (UN, 2017). These variables must be checked for repetition, validity, and creditability. Then, computed percentile variables are deduced to be used as input to develop sets of indicators, which can be used as input for the different function held on the platform.

The determination of the relevant variables controls the decision-making process related to spatial development and conducts correlation analyses between the datasets and the main confirmatory variables. These confirmatory variables must be selected through a collaborative process that engages different stakeholders. The variables with significant correlation are the governing variables for spatial development. These variables are clustered in groups to facilitate calculating the related indicators. The resulting

Figure 2. Conceptual framework architecture for SD-SIM
Source: The authors

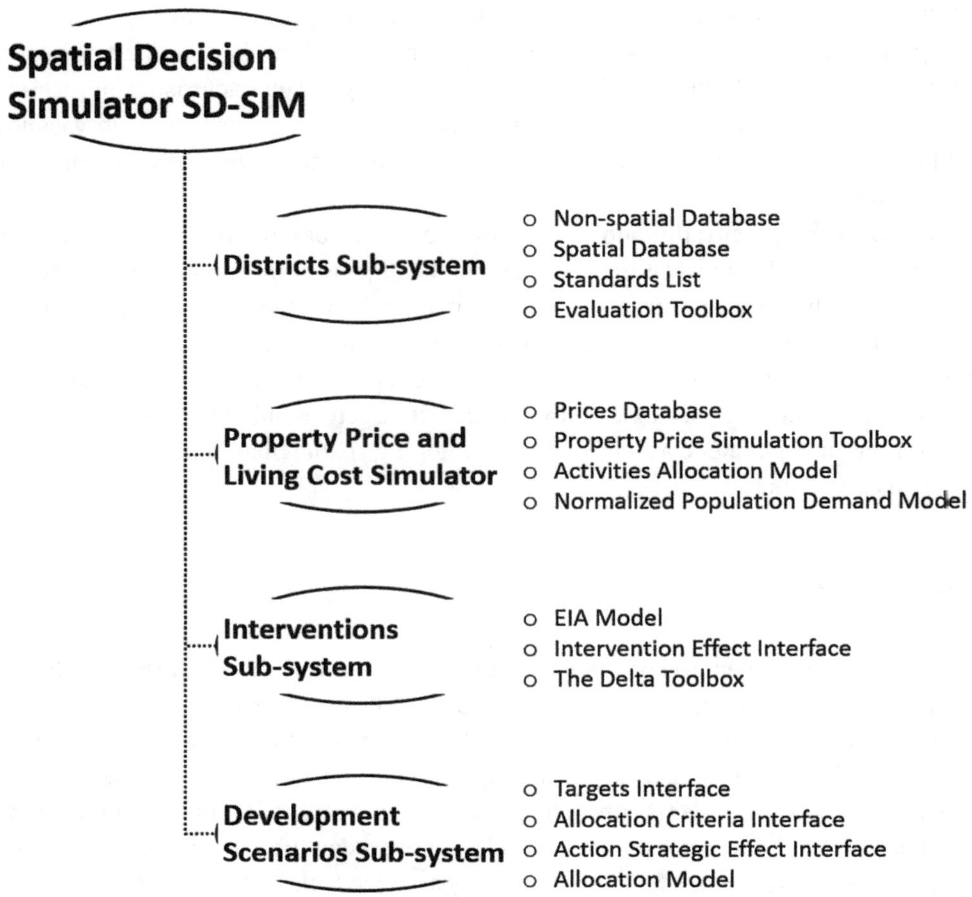

indicators cover different development sectors: demographic, social, economic, and spatial indicator can be used as inputs for different sub-systems and simulators.

Having identified the major variables that are highly correlated with spatial development, a modified version of an arithmetic method that was previously developed and used to calculate Balance Indicators for Egyptian Urban-Rural Continuum (Eid, Salheen, & Toimah, 2012) can be used to calculate the value of these indicators. This value is computed with the overall value of the indicator on the district level (V_{indD}). Each indicator consists of a group of homogenous variables; hence this calculation should consider each variable (the summation of the combined indicators), and the weight of variables according to their significance (relation to the higher administrative level) and impact (the correlation coefficient between each variable and the population variable, r). Hence, the following equation is concluded:

$$V_{indD} = \left[\frac{1}{n} \sum_{i=1}^{n} \left(r_{\mathrm{var}} * \frac{V_{\mathrm{var}\,D}}{V_{\mathrm{var}\,G}} \right) \right] * 100$$

Where:

V_{indG} = The indicators value on the district level
r_{var} = Correlation value for variables composing this indicator with the population variable
V_{varG} = The variable value on the district level
V_{varN} = The variable value on governorate level

n = The number of variables constituting the indicator

1. **Demographic Indicators:** are indicators resulting from the summation of variables highly correlated with population growth. These indicators can include:
   a. Basic collective population characteristics variables that include population, number of families, and age groups.
   b. A Migration and Population Movement Indicator composed of variables recording migration patterns.
2. **Social Indicators:** are indicators to be deduced from variables highly correlated with change on population characteristics. These indicators can include:
   a. Education Indicator as a collective indicator of variables that record the education level, education type, number of students, and school condition.
   b. Health Indicator as a collective indicator of variables that record the level of health services, number of beds per 1000 population, clinics for 1000 population, and number of members of health service providers.
   c. Occupational Classification Indicator, which includes variables that monitor the main population occupations and distribution.
   d. Marital Status Indicator, which is composed of variables that monitor the population in the marriage age and the impact on family formation.
   e. Living Condition Indicator as a collective indicator of variables that monitor family distribution, number of rooms, average family size, overcrowding rates. and housing type.
   f. Household Connected to Public Utilities Indicator, which consists of variables that monitor household distribution with reference to the water resources, sewage connection electricity, gas, and communication.
3. **Economic Indicators Group:** This group includes key economic indicators that represent variable that have oblivious correlations with market changes. These indicators can include:
   a. Employment Status Indicator, which consists of variables that monitor the employment status, unemployment, dependency rates, and participation of women in economic activities.
   b. Income Indicator, which consists of variables that monitor inequalities in income, poverty rate, and per capita GDP.
   c. Work Sectors Indicator, which consists of variables that monitor the key sectors of business, the percentage of employees, public sector employees, informal sector workers, and the distribution of public and private sector investment.
   d. Economic Activities Indicator, which consists of variables that monitor population distribution according to the sections of economic activities.

e. Economic Establishments Indicator, which consists of variables that monitor the status of these establishments in terms of employment, location description, legal entity, sectors of economic establishments, employment rates, classification of workers and their distribution on the types of facilities.

Spatial Database

The spatial database consists of spatial, natural, and human-made features within the district, such as roads, lots, buildings, ownership, infrastructure networks, services, waterbodies, and cliffs. Each of these features has a specific list of attributes. The data are gathered from different sources, including governmental agencies, local administration, and the community. By using an open platform, each contributor and participant can share data. Next, these data are verified, using a co-auditing process based on satellite imagery processing and users' reviews. Users must accept editing rules that prevent any user from editing after certain numbers of fake data entry or incorrect auditing. The database also contains the governing regulations and zoning for each sector and/or lot in the district. Spatial indicators, which are generated from the analysis of spatial database variables, can be used as inputs for the other modules of the framework and may include:

1. Land Indicators consist of variables that monitor the distribution of different land covers, land uses, land development trends, and building densities.
2. Building Indicators consist of variables that monitor building typology, building uses, building heights, and building components.
3. Tenure Indicators consist of variables that monitor different types of properties for land, buildings, and patterns of ownership.
4. Public Utility Indicators consist of variables that monitor the connection of building to different types of utilities (water, sewage, electricity, gas and communication) and their quality based on users' satisfaction.
5. Mobility Indicators consist of variables that monitor the mobility scheme, its adequacy, availability of public transportation, and their quality based on users' satisfaction.
6. Public and Open Space Indicators consist of variables that monitor the availability of public and open spaces, their areas, types, distribution, and quality.

Standards List

The standards list is a collective list of national, regional, and international quantitative standards and thresholds that concern the quality of life and sustainability of development (e.g., water supply, power consumption, emissions, and green areas per capita). Firstly, the items on the list are identified, defined, and quantitated by experts based on national regulations, international agreements, and experimental and research findings. However, each of these items provide a forum to discuss and enhance the methodology of its measurement and validity of each environment and community. These forums should be composed of community representatives, experts, and local administrations. The standards list should be periodically modified and justified with new circumstances.

Evaluation Toolbox

The evaluation toolbox has four toolsets: The first toolset calculates statistics for the district per each time series, and computes rates, patterns and behavior of changes. It also measures the space-mate per spaces and district sectors. The second toolset evaluates and aggregates statistics and measures with neighboring districts, across the city and the standards list. The third toolbox makes regression analyses between 'later' changes that take place, such as changes in density, land cover/use, prices with 'previous' status-quo of the district, and its neighboring ones. The last toolset predicts the zero-development scenario based on the outcomes from the other three toolsets. The aim of these toolsets and overall the toolbox is to create a better understanding of the change across the different variables and dynamics between them. This knowledge helps in developing future scenarios and the estimation for its impact on different variables and indicators.

Property Price and Living Cost Simulator

The Property Price and Living Cost Simulator simulates property prices and living costs for basic daily needs and mobility schemas as the characteristics of locations change over time. This tool incorporates the effects of short-term variations in local and regional vacancy rates on overall property prices. This simulator is based on the Lowry gravity model, which assumes that change in areas for urban settlements is a function of the change of size for the economic sector "non-local demands". This change in turn influences the employment of two other sectors, retail and residential (Rodrigue, 2013). However, the simulator considers population growth as a key factor. The Property Price and Living Cost Simulator has four sub-components:

Prices Database

The price database component includes time series data for prices "either selling or renting" for each land and building use. The data are a combination of data extracted from the spatial database in the district sub-system and data gathered from real estate agencies and e-marketing portals. The wide spread of different platforms and the application of e-marketing provides a huge set of time series data that can be freely accessed or obtained through win-win agreements between local governments and the developers of these applications, who can benefit from spatial data provided by local governments.

Property Price Simulation Toolbox

The Property Price Simulation Toolbox simulates prices for lands and properties within the district and different parts of the city over time. This toolbox is based on the urban economic axiom, which states that the prices adjust to achieve locational equilibrium (O'Sullivan, 2011). This toolbox is calibrated from the prices database and using a regression model that calculates the effect of site characteristics, neighborhood, accessibility, and policy effects on land prices. The toolbox interpolates the unknown prices for each activity type based on its spatial and economic characteristics, in term of known prices for similar activities. It has a margin to incorporate the effects of short-term fluctuations in prices.

Activities Allocation Model

This component allocates increment of employees to districts within the spatial database, based on the non-spatial database in the districts sub-system, the prices database and as per Lowery model. Firstly, the model filters acceptable locations for development and creates a list of possible allocation and transition alternatives for each development type. The probability of each chosen alternative is calculated in a multinomial logit model, as proposed by UrbanSim sub-models (Waddell & J. Evans, 2002). The output of the model is estimated by a spatial allocation for different activities within different alternatives and a priority list for locations based on their feasibility.

Normalized Population Demand Model

The Lowery model approach that is used in this sub-model predicts the flow of travel between locations as a function between: the sizes of origin and destination and the ease of travel between them from one hand, and prediction of trip distribution patterns in transportation modelling based on spatial-interaction approach from the other hand (Borning, Waddell, & Förster, 2008). Accordingly, after allocating increment of employees to districts in Activities Allocation Model, this model calculates modified increasing in residents and population working in the different activities. This, in turn, is used as an input in the Activities Allocation Model during allocating increment of services employment.

Interventions Sub-System

The interventions sub-system is an Agent-Based Model (ABM), which is used for simulating different behaviors that the interventions do on the objects of the environment. ABM is a computer-based simulation approach that uses heterogeneous "agents" to represent decision makers, possessing characteristics of autonomy, communication, reactivity and proactivity (Wooldridge & Jennings, 1995; Johnson & Sieber, 2009). This approach allows planners and decision makers to simulate the individual actions of diverse agents and to measure the resulting behaviors over time (Crooks & Heppenstall, 2012). These behaviors are generated through agent interaction or communication with other objects and their environment(s); therefore, they can be seen as properties of objects and/or environments. Behaviors are set either by the decision makers or as an outcome from Evaluation Toolbox in the district sub-system. The interventions of sub-systems are composed of three components:

EIA Model

The EIA Model models the environmental impact assessment for the intervention, as reported in the intervention documents, regarding soil, water, atmosphere, and physical agents. This set of data is considered the zero comparable data with output data from another sub model. Although detailed EIA is not a mandatory component for all interventions, the model can generate a basic EIA based on a set of variables provided by other sub-systems of the framework and a list of environmental norms that can be developed by experts in collaboration with different stakeholders. The EIA Model needs field data inputs to normalize the value of impact on various ecological components. These data must be updated periodically, depending on adequate measurement tools.

Intervention Effect Interface

The Intervention Effect Interface sets the list of objects that are prospected to be affected by the intervention. These objects are retrieved from the spatial database. Next, the model identifies the behavior of change per each subtype for the objects. This behavior is a function of the neighboring schema, area and time of effect generated from the intervention, and the pattern of change. The default initial values are imported from the Evaluation Toolbox in the districts sub-model. The outcome of this interface helps in the process of alternative selection and prioritization of different interventions.

The Delta Toolbox

The Delta Toolbox calculates the value of changes in objects in relation to different interventions. This toolbox generates comparative maps and reports. It differentiates between the EIA and the intervention effect simulation outcoming from the previous components. The Delta Toolbox can be used to evaluate the status quo in the Districts sub-system, before intervention implementation from one hand, and after the intervention life cycle from the other hand. The Delta Toolbox can also be seen as a monitoring tool for the implementation of interventions and could help the process of resources and projects management.

Development Scenarios Sub-System

The Development Scenarios Sub-System allows stakeholders to propose development scenarios and evaluate their strategic consequences using Discrete Choice Analysis, which illustrates understanding how actors select one option out of a number of mutually exclusive choices that are available to them (Weber, 2004). This tool seeks synergies, as it looks for combined solutions and mutually beneficial interactions. This sub-system mainly targets governments and planners, as it facilitates developing spatial plans and actions. However, it collects the community's opinions as a step towards a collaborative vision. This sub-system has four components:

Targets Interface

The Targets Interface ensures a participatory approach, while selecting a development scenario. This interface set values for short- to medium-term indicators and long-term targets, based on polls. These indicators are categorized as environmental, social, and economic. They are mostly quantitative, and thus, the model runs its iterations to propose the alternative that achieves it. Yet they may also be qualitative, in which case the community opinions are gathered and categorized through surveys, questionnaires, discussion forums, and observation protocols.

Allocation Criteria Interface

This component allows users to input values that are used as parameters for criteria applied to an entire development action. After setting the criteria that are used while allocating the development action, two steps are required. The first step is to compute rate for each criterion that influences the allocation for the development. The next step is to set the weight for each category under the same criterion. These two steps must be based on a wide community and professional participation.

Action Strategic Effect Interface

This interface is similar to Intervention Effect Interface in interventions sub-system. However, it targets the effects on the strategic level, either spatially or non-spatially. Firstly, the model sets the list of objects that are prospected to be affected by the intervention. These objects are retrieved from the spatial database and non-spatial database. Then, the model identifies the rate and percentage of change per unit area for each subtype of the objects. The model also sets the development constraints, if any. The default – initial - values are imported from the Evaluation Toolbox in the Districts sub-model.

Allocation Model

The Allocation Model allocates development actions composing a development scenario, based on the outcomes from the targets and assumptions interfaces. In iterative cyclical revisiting, each plot is tested to see if it is suitable for each of the development actions, then many alternatives for allocating development actions are generated. Next, based on the Action Strategic Effect Interface, the model projects the future land use for each development scenario and the spatial database with future land use for no-action – zero-development- scenario. The model generates a report for non-spatial effects and compares the outputs with targets and indicators for each scenario.

CONCLUSION AND FUTURE DEVELOPMENTS

In this chapter, a framework of Spatial Decision Simulator (SD-SIM) has been introduced. SD-SIM is an iterative, participatory, GIS-based, scenario-based, multi-criteria decision support platform, with a range of sub-systems that are inspired by a commonly-accepted conceptual framework of spatial (temporal) behavior. The suggested simulator is capable of supporting every stage of the decision-making process, from the identification of a problem, an issue, and the definition of (multiple) objective(s) to allowing stakeholders to produce spatial analyses and assessments required to generate alternatives, and to the evaluation/assessment of other option and proposing the probable future spatial profile.

Involving the community in decision-making process increases the degree of reality of decisions and their probability to be implemented. In this platform, community is included in all sub-systems through many methods and tools for collaborative spatial decision making (e.g., data gathering, polls and discussion forums that discuss targets and evaluation criteria). Using this platform makes it easier for communities to produce a shared vision for the future of the community and to coordinate interventions to produce outcomes that are consistent with this vision.

The framework considers several aspects while making spatial decisions. It allows all relevant stakeholders to tackle spatial issues with a more comprehensive specific view. The framework includes all aspects which must be considered, disassembling a spatial issue problem into more manageable components. The stakeholders may consist of legislature, inhabitants, and employees at service providers, mediators as contractors, banks and real estate developers, planners, central and local governments, media, and discipline specialists.

However, there are several issues that need to be discussed during the development of the platform and before publication. The first of these issues is tackling fake opinions and incorrect data. Although spatial data can be reviewed and verified automatically or semi-automatically, non-spatial entries need a high level and a high cost of verification. Analyzing of opinions and discussions – as individual-based – can be affected by web robots. This issue can be minimized using verified login. However, in many cases, individuals prefer not to share personal data or opinions, as they do not want to be observed. The second issue is the effects of extra knowledge in the social behavior, future expectations, and consumer preferences. These effects can profoundly influence demand structure in term of quantity, types and quality. This issue is related to the level of participation, according to the level of spatial decision. Individuals share issues related to their everyday and near future decisions. However, the participation probability is lower when discussing a more global issue, as visions and goals. This could result in a gap between goals and required interventions.

REFERENCES

Afifi, S. (2014). *Spatial Planning Support Systems in Egypt, PhD*. Ain Shams University.

Albrerchts, L. (2010). How to Enhance Creativity, Diversity, and Sustainability in Spatial Planning: Strategic Planning Revisited. In G. C. M. Cerreta (Eds.), Making Strategies in Spatial Planning, Knowledge and Values (pp. 3-26). Springer.

Ayeni, B. (2005). The design of spatial decision support systems in urban and regional planning. In H. Timmermans (Ed.), *Decision Support Systems in Urban Planning* (pp. 2–15). E & FN SPON.

Bertolini, L. (2006). *Why and how can complexity theories help spatial planners?* Amsterdam: De U niversiteit van Amsterdam.

Borning, A., Waddell, P., & Förster, R. (2008). Urbanism: Using Simulation to Inform Public Deliberation and Decision-Making. In Digital Government: E-Government Research, Case Studies, and Implementation (pp. 439-464). Academic Press. doi:10.1007/978-0-387-71611-4_22

Cheng, J., Masser, I., & Ottens, H. (2003). Understanding Urban Growth System: Theories and Methods. *Sendai: 8th International Conference on Computer Techniques for Urban Planning and Management.*

Council of Europe. (2009). *Resolution No. 2 on The European regional/spatial planning charter (Torremolinos Charter)*. Retrieved 11 30, 2012, from European Conference of Ministers Responsible for Regional/Spatial Planning: http://www.coe.int/t/dg4/cultureheritage/heritage/cemat/confminist1-15/6eresolution2_EN.asp

Crooks, A., & Heppenstall, A. (2012). Introduction to Agent-Based Modelling. In Agent-Based Models of Geographical Systems (pp. 85-105). Academic Press. doi:10.1007/978-90-481-8927-4_5

Densham, P. (1991). Spatial decision support systems. In M. S. J. Maguire (Ed.), *In Geographical information systems: Principles and applications* (pp. 403–412). London: Longman Publishing Group.

Eid, Y., Attia, A., El-Refaei, M., & Afifi, S. (2014). Spatial Planning and need to Spatial Planning Support Systems – UrbanSim as example. *Al Azhar University Engineering Sector, JAUES*, *9*(32), 1063–1074.

Eid, Y. Y., Salheen, M. A., & Toimah, A. M. (2012, July). Deriving Balance Indicators for Egyptian Urban-Rural Continuum (As a tool for Balanced Regional Development). *Al Azhar University Engineering Sector, 9*(24), 1673-1694.

Engelen, G., White, R., & Uljee, I. (2005). Integrating constrained cellular automata models, GIS and decision support tools for urban planning and policy-making. In H. Timmermans (Ed.), *Decision Support Systems in Urban Planning* (pp. 95–117). E & FN SPON.

Forester, J. (1999). *The Deliberative Practitioner: Encouraging Participatory Planning Processes.* Cambridge, MA: MIT Press.

Frenchman, D. (2000). Planning shapes urban growth and development. In L. B. Sanyal (Ed.), *The Profession of City Planning: Changes, Images and Challenges 1950-2000* (pp. 27–30). Center for Urban Policy Research, Rutgers, The State University of New Jersey.

Han, J., Kamber, M., & Pei, J. (2012). *Data mining: concepts and techniques* (3rd ed.). Waltham, MA: Morgan Kaufmann, Elsevier Inc.

Hox, J. J., & Boeije, H. R. (2005). Data Collection, Primary vs. Secondary. In *Encyclopedia of Social Measurement* (pp. 593–599). Elsevier. doi:10.1016/B0-12-369398-5/00041-4

Ibn Khaldun, A. b. (1377). *The Muqaddimah "Prolegomenon".*

Johnson, P., & Sieber, R. (2009). Agent-Based Modelling: A Dynamic Scenario Planning Approach to Tourism PSS. In *Planning Support Systems Best Practice and New Methods. The GeoJournal Library* (Vol. 95). Dordrecht: Springer. doi:10.1007/978-1-4020-8952-7_11

Koomen, E. (2008). *Spatial analysis in support of physical planning* (PhD thesis). Amsterdam: Vrije Universiteit Amsterdam.

Koresawa, A., & Konvitz, J. (2001). Towards a new role for spatial planning. In O. Secretariat (Ed.), Towards a New Role for Spatial Planning (pp. 11-32). Paris: Organisation for Economic Co-Operation.

Malczewski, J. (1997). *Spatial Decision Support Systems, NCGIA Core Curriculum in GIScience.* Retrieved 06 30, 2011, from http://www.ncgia.ucsb.edu/giscc/units/u127/u127.html

Malczewski, J., & Keller, C. P. (1997). *Unit 127 - Spatial Decision Support Systems.* Retrieved 11 20, 2018, from NCGIA Core Curriculum in Geographic Information Science: http://www.geo.upm.es/postgrado/CarlosLopez/materiales/cursos/www.ncgia.ucsb.edu/giscc/units/u127/index.html

Peersman, G. (2014). *Overview: Data Collection and Analysis Methods in Impact Evaluation, Methodological Briefs: Impact Evaluation 10.* Florence: UNICEF Office of Research. Retrieved from https://www.unicef-irc.org/publications/pdf/brief_10_data_collection_analysis_eng.pdf

Pettit, C., Keysers, J., Bishop, I., & Klosterman, R. (2008). Applying the What if? Planning Support System for Better Planning at the Urban Fringe. In C. C. Pettit (Ed.), *Landscape Analysis and Visualisation: Spatial Models for Natural Resource Management and Planning* (pp. 435–454). Berlin: Springer. doi:10.1007/978-3-540-69168-6_21

Power, D. (2008). Decision Support Systems: A Historical Overview. In F. B. Holsapple (Ed.), *Handbook on Decision Support Systems 1 - Basic Themes* (pp. 121–140). Springer. doi:10.1007/978-3-540-48713-5_7

Steiner, G. A. (1969). *Top Management Planning, Studies of the modern corporation.* New York: Macmillan.

Sudhira, H. S. (2008). *Studies on Urban Sprawl and Spatial Planning Support System for Bangalore, India, PhD*. Bangalore: Centre for Sustainable Technologies and Department of Management Studies, Indian Institute of Science.

Sugumaran, R., & DeGroote, J. (2011). *Spatial Decision Support Systems - Principles and Practices.* CRC Press, Taylor & Francis Group.

Torrens, M. P. (2000, November). *How Cellular Models of Urban Systems Work. (1. Theory).* Retrieved July 29, 2018, from http://discovery.ucl.ac.uk/1371/1/paper28.pdf

UN-Department of Economic and Social Affairs Statistics Division. (2017). *Principles and Recommendations for Population and Housing Censuses* (Vol. Revision 3). New York: United Nations. Retrieved from https://unstats.un.org/unsd/publication/seriesM/Series_M67Rev3en.pdf

Waddell, P., & Evans, J. (2002). UrbanSim: Modeling Urban Development for Land Use, Transportation and Environmental Planning. *Journal of the American Planning Association, 68*(3), 297–314. doi:10.1080/01944360208976274

Wang, X., & vom Hofe, R. (2007). *Research Methods in Urban and Regional Planning*. Tsinghua University Press, Beijing and Springer-Verlag GmbH Berlin Heidelberg.

Weber, B. (2004). Educating Multi-Disciplined Experts in GIS-Based Global Valuation Methods. *24th Annual Esri International User Conference*.

Wegener, M., Simmonds, D., & Waddell, P. (2011). *Equilibrium v. dynamics in urban modelling. In Applied Urban Modelling (AUM 2011) "Innovation in Urban Modelling*. Cambridge, UK: University of Cambridge.

Wooldridge, M., & Jennings, N. (1995). Intelligent agents: Theory and practice. *The Knowledge Engineering Review, 10*(2), 115–152. doi:10.1017/S0269888900008122

Worrall, L. (1991). *Spatial Analysis and Spatial Policy Using Geographic Information Systems*. Belhaven Press.

Chapter 9
Theory, Data, and Methods:
A Review of Models of Land-Use Change

Eda Ustaoglu
Gebze Technical University, Turkey

Arif Çagdaş Aydinoglu
Gebze Technical University, Turkey

ABSTRACT

Land-use change models are tools to support analyses, assessments, and policy decisions concerning the causes and consequences of land-use dynamics, by providing a framework for the analysis of land-use change processes and making projections for the future land-use/cover patterns. There is a variety of modelling approaches that were developed from different disciplinary backgrounds. Following the reviews in the literature, this chapter focuses on various modelling tools and practices that range from pattern-based methods such as machine learning and GIS (Geographic Information System)-based approaches, to process-based methods such as structural economic or agent-based models. For each of these methods, an overview is given for the advances that have been progressed by geographers, natural and economy scientists in developing these models of spatial land-use change. It is noted that further progress is needed in terms of model development, and integration of models operating at various scales that better address the multi-scale characteristics of the land-use system.

INTRODUCTION

Changing patterns of land-use development have significant implications for future changes in climate, land function, global food demand, biodiversity, and associated ecosystem services, and in return, these have implications on subsequent land-use change (de Groot et al., 2002; de Groot et al., 2010; FAO, 2016). During the past three centuries, almost 12 million km^2 of forest and woodlands were cleared, and 5.6 km^2 of grassland and pasture have been diminished on a global scale (Bryant et al., 1997; Ramankutty and Foley, 1999; Davidson, 2017). During the same period, cropland has increased by 12 million km^2 (Ramankutty and Foley, 1999; Davidson, 2017). These land-use changes have significant implications for future changes of global climate, and consequently, have implications for subsequent land-use change.

DOI: 10.4018/978-1-5225-9238-9.ch009

Rapid changes in land-use development patterns increase uncertainties related to possible future trends and the effectiveness of policies. This is of importance for the studies of land-change science (Turner et al., 2007) and land-system science (Rounsevell et al., 2012) considering that land is a scarce resource where competing claims interact. This is also important to governments, planners and policy-makers to understand the interactions between human activities and natural resources. The long-term impacts of policy decisions on land-use change may cover many diverse issues such as transportation, zoning, water supply, air quality, natural resources, recreation and ecosystem conservation. In this respect, there is a growing need for forward-looking information and scenario-based policy analysis that addresses these issues through ensuring consistency among past, present, and future assessments.

Considering that land-use change is the outcome of human decisions, land-use patterns reflect the decision-making processes of those who control scarce land resources (Verburg et al., 2006a). Land-use policies such as tax regulations and subsidies, urban growth restrictions, setting aside of land in return for monetary compensation, schemes to encourage urban regeneration and brownfield development, and land-zoning policies on different land uses have caused significant changes in urban landscapes (Randolph, 2004). As such policies and restrictions are influential on spatial development, it is important for policy-makers to develop appropriate policies and take necessary actions to keep pace with continuous spatial development. In this regard, land-use modeling approach can be considered as a tool to examine changes of land-use and pattern and to support planners and policy-makers in developing more informed policies and decisions (Verburg et al., 2004). Models are useful for resolving the complexities stemming from socio-economic and biophysical factors that influence the spatial pattern of land-use change (Figure 1), and for estimating the impacts of changes in land-use (Veldkamp and Fresco, 1996a; Verburg et al., 2004; Meyfroidt et al., 2013).

The initial studies of land-use modeling applications have mainly focused on biophysical factors (e.g. aspect, topography, soil quality etc.) given vast availability of such data. However, incorporation of data that is based on a wide variety of socio-economic and political drivers is more valued (Wilbanks and Kates, 1999; Veldkamp and Lambin, 2001). The urbanization process accompanied by population growth and in migration to the urban centers is one of the major factors resulting in land-use change (Geist and Lambin, 2002). Other than population change, the increasing volume of international trade of agricultural products and remote urban demand also result in land-use change (e.g. deforestation) (DeFries et al., 2010). In Lambin et al.'s (2001) explanation, in migration is triggered by government policies to open the non-urban land for development through settlements schemes, development projects and construction of industrial plants or commercial centers. Government decisions may include *the goal to secure territorial claims and national political support, to attract international capital, to facilitate market opportunities, or to promote the interests of specific groups through exploiting natural resources controlled by the state* (Lambin et al., 2001, p. 263). The programs and policies on environmental protection such as Kyoto protocol also have important role in driving land-use change (Veldkamp and Lambin, 2001).

Further to this, the influence of remote markets, diffusion of technologies and political forces were shown to influence the land-use system. Urban centers have significant impacts on land-use change through *changing consumption patterns associated with urban lifestyles and their wider networks of influence* (Meyfroidt et al., 2013, p. 440). These objectives, policies and technology-related and external factors, and the settlement-infrastructure development have subsequent impacts on land-use and cover; and their impacts vary by different regions (Lambin et al., 2001). Veldkamp and Lambin (2001) stated that the inclusion of these socio-economic, political and technology-related factors in land-use modeling applications is hampered by lack of spatially explicit data and methodological issues in linking social

Figure 1. Land-use change drivers
Source: Reproduced from Geist and Lambin (2002)

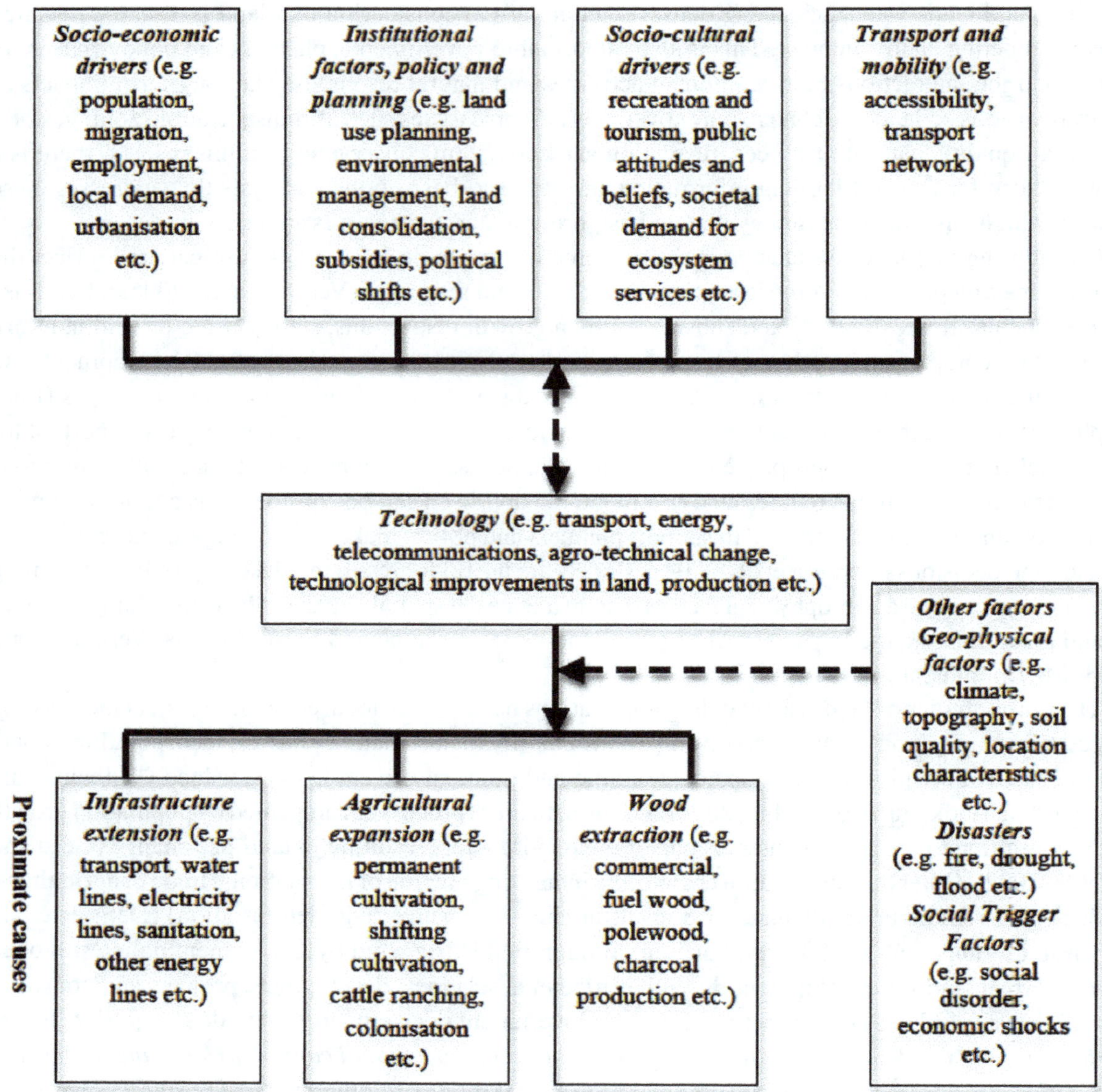

and spatial data. In their explanation, the use of proxy variables instead of driving factors might obscure causality; and further to this, *land-use change models need to incorporate the endogeneity of factors such as land management technologies, infrastructures and land-use policy* (Veldkamp and Lambin, 2001, p. 2). Despite the complexity of land-use systems and comprehensive data requirements, a variety of studies has focused on the development and application of land-use models in the literature. These include examples of land-use modeling ranging from regional/global applications (Verburg et al., 2006a; Verburg and Overmars, 2009; Seto et al., 2012) to applications of more local case study areas (de Nijs et al., 2004; Echeverria et al., 2008; Andersen et al., 2017).

The objective of this chapter is to review the currently available land-use models aiming at assessing the impacts of policy alternatives; and therefore, can be used as spatial decision support tools. The review focuses on descriptive models that aim to simulate the functioning of the land-use system and the spatially explicit simulation of near future land-use patterns (Verburg et al., 2004). Some examples of prescriptive models that aim at the computation of optimized land-use configurations under the constraints of a set of goals and objectives are also provided. More detailed reviews of prescriptive models can be found in Briassoulis (2000) and Van Keulen et al. (2000). Another difference between the groups of land-use models is based on the economic theory. Regarding this, deductive models are based on urban location or structure theories and the model simulations are compared to actual land-use changes to test the validity of the theory (Irwin and Geoghegan, 2001). By contrast, inductive models are based on observed patterns of land-use changes, but these are not based on the theoretical model. A further classification of the models is based on the difference between static and dynamic models. Static models explain the spatial distribution of the land-use pattern at one point in time and can be used to examine the driving factors of land-use change. By contrast, dynamic models incorporate temporal dynamics in explaining the land-use system and can be used for the projections of future land-use change.

There are a wide variety of models in the literature with differing modeling traditions and theoretical backgrounds. The reviews of land-use models are provided by Baker (1989) for landscape ecology, Lambin (1997) for deforestation, Lambin et al. (2000) for modeling agricultural intensification, Irwin and Geoghegan (2001) for economic modeling of land-use change, Wegener (2004), Cheong et al. (2012), and Acheampong and Silva (2015) for integrated modeling, and An (2012) for agent-based models. A general framework for the classification and evaluation of land-use change modeling is provided by Briassoulis (2000), Verburg et al. (2004), Heistermann et al. (2006), EEA (2008), Agarwal et al. (2010), Wicke et al. (2012), Brown et al. (2013), and Sohl and Claggett (2013).

In the current chapter, we have extended the review work of Brown et al. (2013) through broadly covering and updating all the modeling approaches and discuss the shortcomings and the advances that have been progressed with land-use change modeling. The study provides reviews and comparisons of methods and applications without providing a complete description of the models corresponding to each model category. Therefore, the readers are referred to the related publications to overview the modeling process, the technical specifications, and the case study applications.

Through this review of the literature, we can explore the main characteristics, data and methodological requirements and some details on practical applications of the covered models. This may contribute to researchers, practitioners and policy-makers in selecting the most appropriate model(s) for their study area given the purpose of the study, the scale of analysis, and availability of socio-economic, spatially explicit or other required data.

BACKGROUND

There are three recent developments that have increased interest in urban space and the means to study it (Clifton et al., 2008). First, rising awareness on the unwanted consequences of urban sprawl (e.g. traffic congestion and noise, air/water pollution, loss of green spaces, spatial segregation, increase in public service provision costs, distance traveled by private cars and traffic accidents etc.) has made urban spatial structure the main focus of studies in the land-use literature. Second, advances achieved with remote sensing and geographic information technologies have made it easier to analyze the spatial development and the consequences. Finally, the quality of spatial data has been enhanced in parallel to the developments in geographic information system technologies, which has increased the accessibility and availability of commercial and open source spatial data and computing tools not only for researchers, planners and policy-makers but also for public and stakeholders.

The use of computer-based models of land-use change and urban growth has considerably increased primarily due to the acceptance of these models in local collaborative decision making and in supporting urban planning, policy making and management (Sui, 1998; Herold et al., 2005). The increased data resources and the improved spatial datasets and tools are the other factors that are influential in the increased use of urban land-use models. In this respect, a wide variety of quantitative approaches have been created to quantify and analyze the urban spatial structure. A land-use modeling approach provides a framework for *the interactions between different non-linear systems (human and biophysical systems) that influence the dynamics of land-use/cover (Lavalle et al., 2011, p. 60).* Based on assumptions corresponding to different scenario conditions, land-use models can simulate the functioning of the land-use system in future; and therefore, are relevant elements for structuring discussion and assessing ex-ante consequences of different policy options (Verburg et al., 2004; Lavalle et al., 2011). Lee (1973) listed five key roles a model must play in a policy context. These include: (1) transparency, (2) robustness, (3) reasonable data needs, (4) appropriate spatio-temporal resolution, and (5) inclusion of key policy variables to allow for significant policy questions to be answered.

Given the complexity of the land-use system, there is no single model that captures all drivers of land-use change, nor is there a single theory that can explain the land-use/cover system (Verburg et al., 2008; Hersperger et al., 2011). For this reason, a variety of modeling methods have been developed to address the issue of complexity that is observed in the land system. Complexity in the land-use/cover system can be defined as *encompassing interactions within and among ecological systems, the physical systems on which they depend, and the human systems with which they interact* (Michener et al., 2001; Blumler et al., 2011). Here the main challenge for the modelers is to reduce the complexity of the system while representing the critical processes influencing system's behavior (Rindfuss et al., 2008).

MODELING APPROACHES

Recent reviews of land-use models indicate that there is a considerable amount of land-use models and applications. From review of the subject literature, some examples of operative land-use models are provided in Table 1. According to the Table, there are a variety of models operating at different scales, having different modeling goals and applications and are based on different principles in explaining the land-use pattern and change. There is no standard classification for the modeling approaches in the literature. Different classifications have been developed based on the focus of the study. Some specific examples are: Agarwal et al. (2000); Briassoulis (2000); Gertsev and Gertseva (2004); Verburg et al. (2004); Jorgensen (2008); and Brown et al. (2013). Agarwal et al. (2000), for instance, stated that space and time are the two components and provide a common setting in which *all biophysical and/or human processes operate in a temporal context, a spatial context, or both.* Models also incorporate human processes (Agarwal et al., 2000) which is an important component given that changes in land-use is determined by *the interaction in space and time of the biophysical factors and human factors like population, technology, economic conditions etc.* (Veldkamp and Fresco, 1996a). These three components are presented in Figure 2 with a particular focus on a few general types of models from Agarwal et al.

(2000). Agarwal et al. (2000) argued that temporal, spatial and human-decision making complexity of a model could be represented by an index where low values indicate simple components whereas high values indicate more complex behavior and interactions. A model of low temporal complexity may be a model with one- or few-time steps and a short duration. By contrast, a model with high temporal complexity is a model with large number of time steps, long duration and has the potential to deal with time lags or feedbacks among variables or different time steps for different sub-models (see Agarwal et al., 2000). Regarding spatial complexity, a model with a lower value is the one having little or no capacity to represent data spatially while a model with a high value may be a spatially interactive model in two or three dimensions. A model with a low value of human decision-making complexity indicates that the model does not include any human decision-making. By contrast, a model with a high value includes one or more types of decision makers or has the capacity of multiple decision makers interact across domains (see Agarwal et al., 2000).

In Figure 2, time-series statistical models have higher temporal complexity, GIS-based approaches have higher spatial complexity, and econometric and game theory models have higher decision-making complexity compared to their alternatives. Temporal, spatial and human decision-making components are all important for the land-use change models; and therefore, the models aim at incorporating all three components to support urban management and policy making.

An alternative approach for the classification of land-use models was developed by Brown et al. (2013, p. 452) which stated that *models provide the basis for encoding inferred or deduced relationships.* In this respect, models can be static or dynamic, spatial or non-spatial (explaining change of patterns or predicting the change), inductive or deductive (examines land-use change process based on statistical correlations or explores the process more explicitly), agent-based or pattern-based (Mas et al., 2014). For instance, models that are used to explain land-use changes differ from those that are describing *patterns*, and to those that expresses dynamics as *processes.* In Brown et al.'s (2013) explanation, models may describe *patterns* regardless of whether or not the reason for the patterns is based on land change theory. Alternatively, models may focus on processes that generate change in patterns based on observations in field and land change theory (Overmars et al., 2007; Brown et al., 2013). Brown et al. (2013) claim that models can be arranged according to their emphasis on projection versus explanation. Projection aims at *producing an estimate of the quantity and/or spatial allocation of land-use/cover at some point in the future, under specified set of conditions* (Brown et al., 2013, p. 453). Explanation, on the other hand, aims to identify *which land-use decision theory better explains the observed land-change patterns or the variable interactions producing non-linear dynamics or threshold effects* (Brown et al., 2013, p. 453). This chapter will follow Brown et al.'s (2013) classification for the identification of different land change models; and therefore, focus on five key types of modeling approaches.

MACHINE LEARNING APPROACHES

These models focus on projection patterns and rely on algorithms that envisage the relationship between land-use changes and the characteristics of locations where the changes are likely to occur. A group of models are based on remotely sensed data, and the spatial indicators are calculated using GIS such as distance measures (e.g. distance to roads and cities), biophysical factors (e.g. soil, slope, elevation), and socio-economic drivers such as population and economic output (Mertens and Lambin, 1997; An and Brown, 2008; Leh et al., 2013; Varga et al., 2016). Considerable progress has been achieved with the

Table 1. Examples of the characteristics of a number of land-use change models

Model Name	Classification	Application Area	Temporal Extent	Land-use Change	Explanation
General Ecosystem Model* (GEM) (Fitz et al., 1996)	Dynamic systems model	Florida Everglades/Big Cypress Area	Variable	Captures feedback among abiotic and biotic ecosystem components	Spatially dependent model with feedbacks between units and across time. Includes many sectors e.g. hydrology, algae, nutrients, fire, dead organic matter. The model can adapt resolution, extent and time step to match the process being modeled
Urban Growth Model** UGM/ SLEUTH (Clarke and Gaydos, 1998)	Cellular Automata model	San Francisco and Washington/Baltimore Region, Mid-Atlantic Region	1900-2100	Models urbanization process	Self-modifying cellular automata model. Four processes included: spontaneous growth (neighborhood, suitability); diffusive growth (slope determined); edge growth (neighborhood); road influenced growth. SELUTH enables spatial and temporal autocorrelation
Constrained CA Models: ** Ram Co, LOV, MODULUS, SIMLUCIA, MOLAND/ METRONOMICA (White and Engolden, 2000; Barredo et al., 2003; Engelen et al., 2007; Petrov et al., 2009)	Cellular Automata model and/ or integrated/ hybrid model	St. Lucia, Caribbean; Netherlands; City of St. Johns, Newfoundland, Canada; Ujung Pandang, Indonesia; Dublin, Ireland; Algarve, Portugal and others	±30 years	Multiple land use types	Demand for land is calculated in sectoral models at an aggregated spatial level and allocated using cellular automata. The cellular automata are quantified using expert knowledge
Common Agricultural Policy Regionalized Impact (CAPRI-Dynastat); Common Agricultural Policy Simulation Model (CAPSIM)*** (Britz, 2004; Witzke and Zintl, 2003)	Comparative static/partial equilibrium model	EU-27 plus Norway plus six Balkan countries	Variable (time series at regional level from 1985 onwards)	Agricultural land uses	Agricultural sector model developed to evaluate regional and aggregated impacts of CAP and trade policies on production, income, markets and environment. Scenario analysis covers areas cropped and herd sizes along with output and input coefficients and income indicators for each agricultural activity and each region. The model covers supply and demand positions at country level, environmental indicators at regional level.
Conversion of Land Use Change and its Effects (CLUE, CLUE-s, Dyna-CLUE)*** (Veldkamp and Fresco, 1996a; Verburg et al., 1999; Verburg et al., 2002)	Empirical-statistical model and/ or simulation model	Continental level: EU-27, Central America, China; National level: Ecuador, Central-American countries, Netherlands; Sub-national level: Philippines, Costa-Rica, Indonesia etc.	±20 years	Multiple land use types	Dynamic simulation model quantifying the most important bio-geophysical and human drivers based on either knowledge of the land-use system or empirical analysis of observed land-use patterns. The dynamic simulation model allocates changes in land-uses, of which the aggregate quantity is determined by trend analysis, scenario assumptions or macro-economic modeling.

continued on following page

Table 1. Continued

Model Name	Classification	Application Area	Temporal Extent	Land-use Change	Explanation
IIASA-LUC** (Fischer and Sun, 2001)	Integrated/ hybrid model and/or optimization model	China (integrated analysis for eight regions)	Projections to 2025	Multiple land uses	The model provides agro-ecological analysis for detailed pixels. It is a General Equilibrium Model based on multi-sector input-output tables
CORMAS** (Multi-Agent Simulation Software for Renewable Resource Management) (Barreteau and Bousquet, 2000; Rouchier et al., 2001; Bommel et al., 2016)	Agent-based model	Different case study areas at village level	Variable	Specific land use types relevant for application	Multi-agent simulations based on behavior of individuals towards other individuals and natural resources
PLM** (Patuxent Landscape Model) (Irwin and Geoghegan, 2001; Bockstael, 1996; Costanza et al., 2002)	Empirical/ statistical model and/ or simulation model	Patuxent watershed (several US counties) based on integrated ecological models for the Everglades.	Decades	Urbanization habitat change	Integrated economic/ecological model. Land-use change allocation is based on economic modeling (hedonic modeling combining distance and location operators). The neighborhood of the location is taken into account by quantifying its effect on land value
LUCAS* (Land Use Change Analysis System) (Berry et al., 1996)	Spatial stochastic model	Two watersheds, the Little Tennessee river basin in North Carolina, and the Hoh river watershed on the Olympic peninsula in Washington State	15 years (1975-1991)	Multiple landscapes, habitat change	Human choice is modeled via a probability function for land-cover change with basic socio-economic determinants with no feedback. It is modular and uses low-cost open source GIS software (GRASS)
Chomitz et al. (1996) *	Econometric (multinomial logit model)	Central and south Belize	Cross-sectional analysis for the data collected between 1989-1992	Agricultural landscapes	Predicts land-use aggregated in three classes: natural vegetation, semi-subsistence agriculture, commercial farming. It has a theoretical framework supporting Von Thunen's model. Human decision making is implicit in the inclusion of variables that impact rent (e.g. distance to market, soil quality)
CUF* (California Urban Futures) (Landis 1995)	Spatial simulation	Nine counties of the San Francisco Bay Area and five adjacent counties	Model takes 1990 base data and forecasts growth in 2010	Urbanization process	Explains land-use change in a metropolitan setting in terms of demand and supply of land. The model has not a feedback of mismatch between demand and supply on price of land.

continued on following page

Table 1. Continued

Model Name	Classification	Application Area	Temporal Extent	Land-use Change	Explanation
STARM (Spatio-temporal binary auto logistic model) (Zhu et al., 2008)	Spatial Markov/logistic model	Chilcotin Plateau in Western Canada	1972-1986	Forest landscapes	The probability of pine beetle outbreak is assumed to be conditional on current and past observations. A spatial weight matrix is used to control for spatial dependencies. A Markov chain assumption is made, where the distribution of the endogenous variable is dependent on the most recent time points, because historical tree conditions are considered to have an effect on current three conditions.
Mertens et al. (1997)*	Univariate spatial model	Southeast Cameroon	1973-1986	Deforestation process	Models deforestation by proposing a typology of deforestation patterns. Each model runs with four independent variables separately (e.g. Total study area, corridor pattern, island pattern, diffuse pattern). It does not model interaction between factors.
NELUP* (O'Callahan, 1995)	General systems model/ recursive linear planning model	River Tyne catchment in Northern England	1981-1988	Agricultural landscapes	Based on a regional economic agricultural model, hydrological model, and ecological model. Uses land cover to link market forces, hydrology and ecology in a biophysical model of land-use.
FASOM* (Forest and Agriculture Sector Optimization Model) (Adams, 1995; Adams et al., 1996)	Dynamic, non-linear, price endogenous, mathematical programming model	The entire US except Hawaii and Alaska	1990-2089 (policy analysis is limited to 1990-2039)	Agriculture and forest	Comprise three sub-models including: Forest sector model, agricultural sector that is optimized with the forest sector sub-model, carbon sector for terrestrial carbon. The model maximizes the discounted economic welfare of producers and consumers in agriculture and forest sectors over nine-decade time horizon.
CURBA* (California Urban and Biodiversity Analysis Model) (Landis et al., 1998)	Overlay of GIS with statistical urban growth projections	County level in California	1995-2010	Multiple land use types	Human decision making in the urbanization context implicitly a function of highway facilities, natural constraints, growth policies, and job and population growth. Calculates impacts at 1-hectare grid level, and includes the impacts of scale decision making (e.g. county, sub-county zoning level)

Source: Adapted from: *Agarwal et al. (2000) **Verburg et al. (2004) ***EEA (2008)

use of GIS for management and policy purposes (Webster, 1993; 1994). However, the literature suggests that GIS has not reached the potential to support and facilitate spatial modeling processes considering that the system does not provide optimization, iterative problem solving and simulation capabilities that is essential in urban planning especially at the strategic level (Jankowski and Richard, 1994; Wagner, 1997; Murray, 2010).

There are models in this group that are not based on economic theory but use statistical techniques to quantify the relationship between land-use change and the drivers of change. Relevant examples include:

Figure 2. Example of a framework for the evaluation of land-use change models
Source: Agarwal et al. (2000)

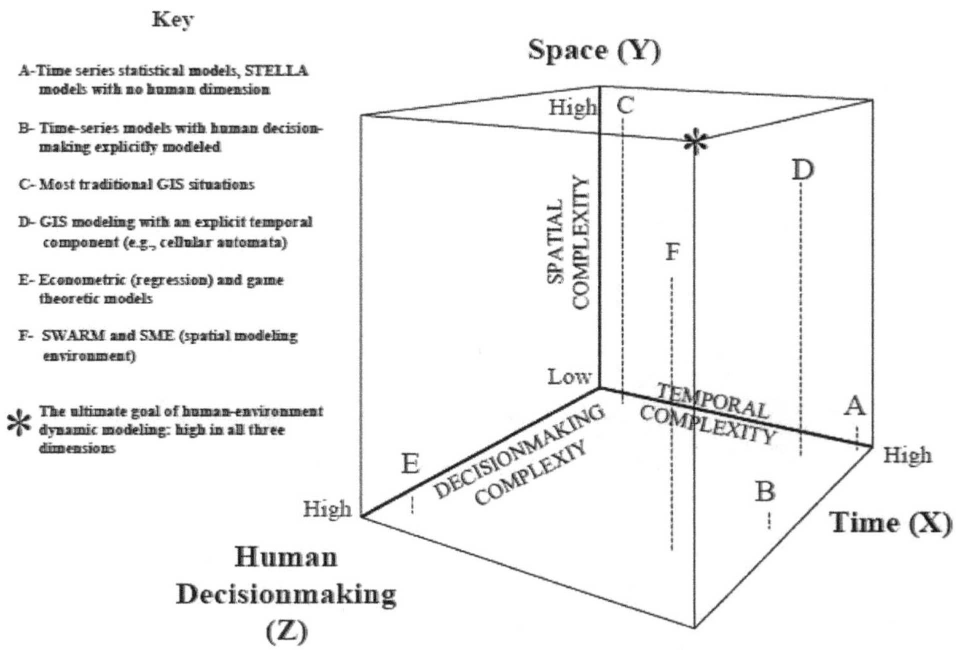

Turner et al. (1996), Wear and Bolstad (1998), Pontius et al. (2001), Basse et al. (2014), Sun and Robinson (2018) (the first three references are cited in: Verburg et al., 2004). These models mainly describe land-use conversions as a function of changes in driving forces and locational characteristics. Verburg et al. (2004, p. 315) noted that *the use of such statistical techniques results in a relatively low degree of explanation due to the relative short time-period of analysis, small sample size, and variations over the analyzed time-period* (see also Veldkamp and Fresco, 1997). In their explanation, *cross-sectional analysis of the actual land-use pattern, reflecting the outcome of long-term period of land-use changes, results in more stable explanations of the land-use pattern* (Verburg et al., 2004, p. 315). Cross-sectional analysis can be used to specify and quantify the drivers of land-use change; however, this approach does not take into account temporal feedback and path dependencies (Verburg et al., 2006a). By contrast, dynamic models describe the behavior of a land-use system in terms of processes within the time period examined, and the qualitative relationship between the parameters and the changes to the static model. Dynamic modeling, therefore, provides an improvement over cross-sectional analysis as it can more thoroughly represent the observed changes in the land-use system (Koomen and Stillwell, 2007). An example of the stages of land-use modeling using the machine learning approaches is presented in Figure 3.

For other group of models, the variable selection depends on theory and methods including artificial neural networks (Pijanowski et al., 2002; Rocha et al., 2007; Shooshtari and Gholamalifard, 2015), classification and regression trees (Etter et al., 2006; Shooshtari and Gholamalifard, 2015; Pazur and Bolliger, 2017), and logistic regression (Dendoncker et al., 2007; Li et al., 2012; Alsharif and Pradhan, 2014) (Figure 4). According to Irwin and Geoghagen (2001), the selection of economic parameters in these models is ad-hoc and is not based on structural models that aim to explain underlying economic processes explicitly. Another drawback of these models is that the models are less successful in explaining

Figure 3. Example of a land-use modeling framework using machine learning approaches
Source: Basse et al. (2014)

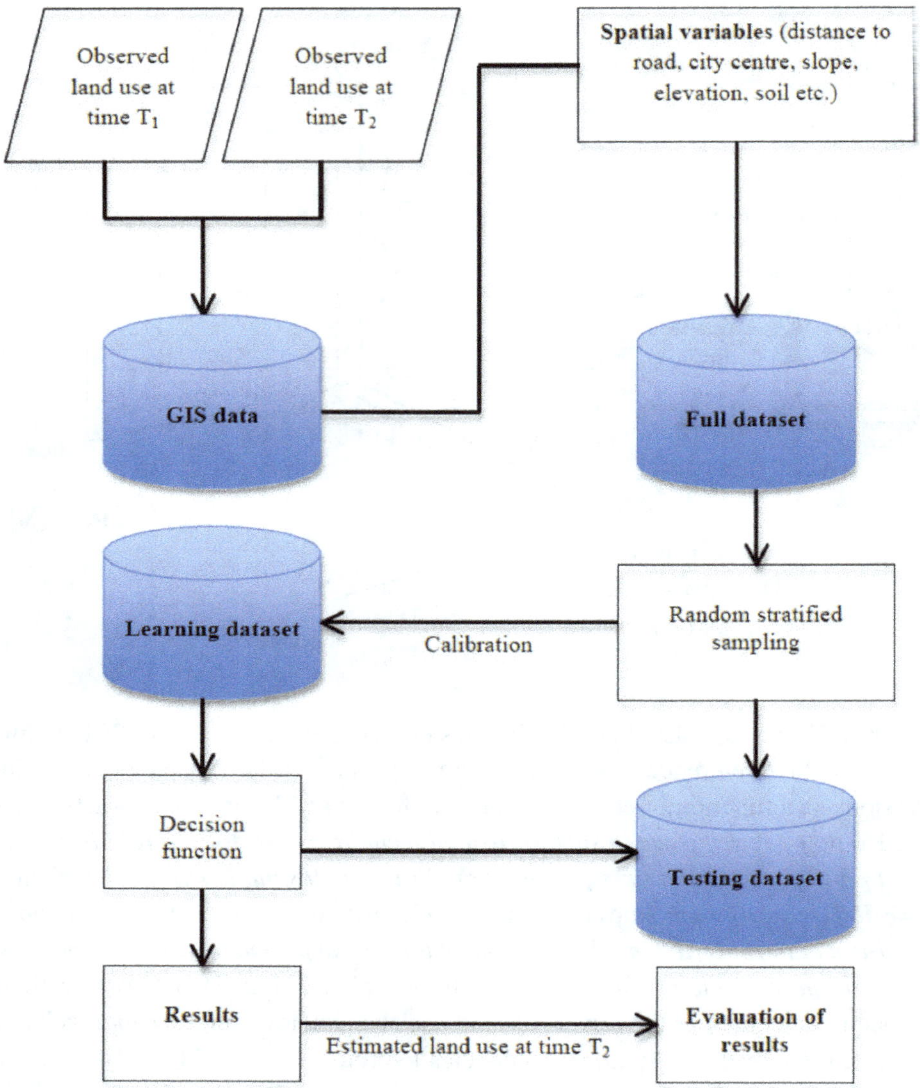

human behavior, which is the main focus of studies in social sciences where the individual behavior is linked with the spatial outcome of land-use change (Irwin and Geoghagen, 2001; Verburg et al., 2004). The researchers in social sciences study human behavior at the micro-level using either qualitative or quantitative methods of micro-economics and social psychology (Bilsborrow and Okoth Ogondo, 1992; Verburg et al., 2004). The group of models developed by geographers on the other hand has examined the land-use at the macro scale through using macro-scale characteristics of social organizations in order to explicate the relationship between social factors and the macro-scale patterns (Verburg et al., 2004).

Figure 4. Machine learning techniques
Source: https://www.mathworks.com/discovery/machine-learning.html

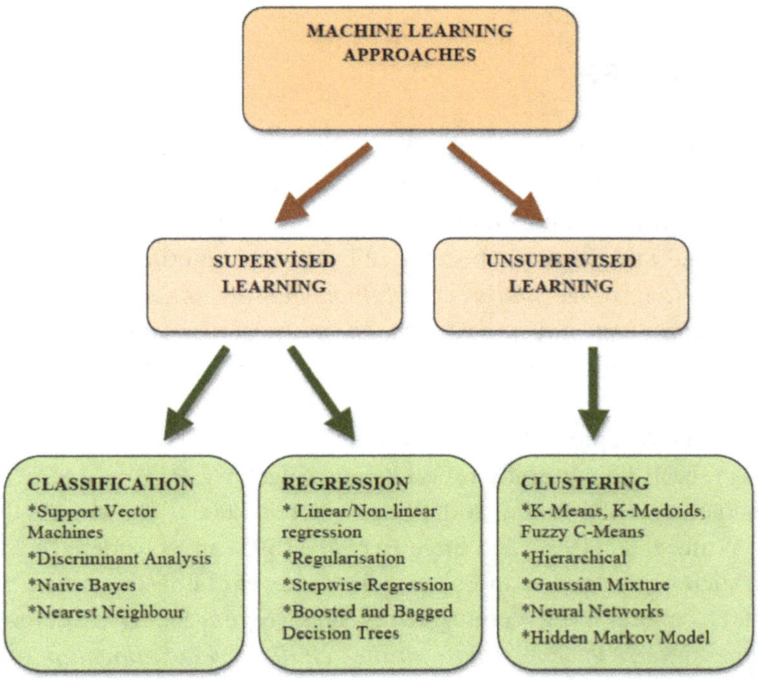

Verburg et al. (2004) argue that there is a weak connection between spatially explicit land-use models and social sciences; and therefore, human behavior is neglected in many spatially explicit land-use models.

In order to address some of these issues, hybrid models of land-use/cover change that are based on an estimation model followed by an addition of a simulation model have been introduced. Markov models are such type of models that estimate land-use changes as a function of a variety of explanatory variables and based on these estimations, they predict changes in spatial patterns of landscapes (Irwin and Geoghegan, 2001; Wu et al., 2006; Valdez et al., 2016). From the review of Irwin and Geoghegan (2001), examples of hybrid models include the LUCAS model (Berry et al., 1996; Hazen and Berry, 1997) and the CLUE model (Veldkamp and Fresco 1996a; 1997; Verburg and Overmars, 2007; Han et al., 2015). Both models estimate the impact of various explanatory variables (e.g. slope, soil, aspect, elevation, location) on different types of land-use/cover change and using the estimated models, they simulate the impact of different policy scenarios on land-use/cover change. A more recent study by Puertas et al. (2014) developed an integrated model where land-use change is modelled by applying logistic regression model, Markov chain and cellular automata. A similar research by Samardzic-Petrovic et al. (2017) utilized three different machine learning approaches including decision trees, neural networks and support vector machines for short-term land-use change modeling. Other examples of hybrid models are: land use scanner (LUS) (Hilferink and Rietveld, 1999), ProLand and UPAL (Sheridan et al., 2007), and the SLUDGE model (Verburg et al., 2006b). Gibon et al. (2010) stated that socio-ecological processes need to be considered in the modeling approach and that the scenarios need to be elaborated using hybrid or integrated approach for the examination of alternative futures of land-use change.

CELLULAR APPROACHES

Cellular approaches simulate changes over time by combining probability maps with spatial interaction effects. Regarding spatial interaction effects, cellular automata (CA) approach focuses on the interaction between land-use in a specific location, conditions at that location and land-use types in the surrounding locations. These spatial interactions between a cell and its neighbors are defining characteristics of cellular automata. *...The long-range spatial correlations between the system's states in different time periods are shown to generate structure and pattern. These correlations can be quantified and compared to a random configuration and possible states to yield a measure of relative order in a system...* (Irwin and Geoghegan, 2001, p. 10).

The transition rules in CA represent the land-use dynamics which are explained by urban theories, and mainly based on the well-known study by White and Engelen (1993) (see also White and Engelen, 1997; 2000; Straatman et.al. 2004; White, 2005; Engelen et.al. 2007). The models developed by White and Engelen (2000), White et al. (1997) and Engelen et al. (1995) are considered as constrained cellular automata models. In such a constrained approach, a higher level of constraint is utilized to regulate the quantity of change at the cellular level (Verburg et al., 2004) (Figure 5). *...The socio-economic demands computed at the macro level are the constraints for the cellular level of allocation of land uses with cellular automata, which in turn determines the actual land-use patterns. A feedback to the regional level is achieved through the influence of land-use densities and suitabilities at the cellular level on the regional demand in the next iteration...* (Verburg et al., 2004). The factors determining the spatial dynamics in CA modeling are presented in Figure 6.

Figure 5. The top-down allocation approach
Source: Verburg et al. (2004)

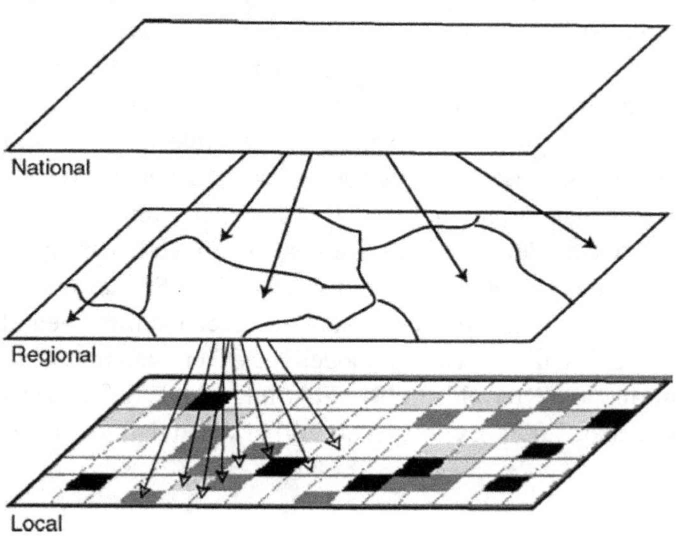

Figure 6. An example of the factors determining the spatial dynamics in cellular modeling
Source: Convery and Shahumyan (2009)

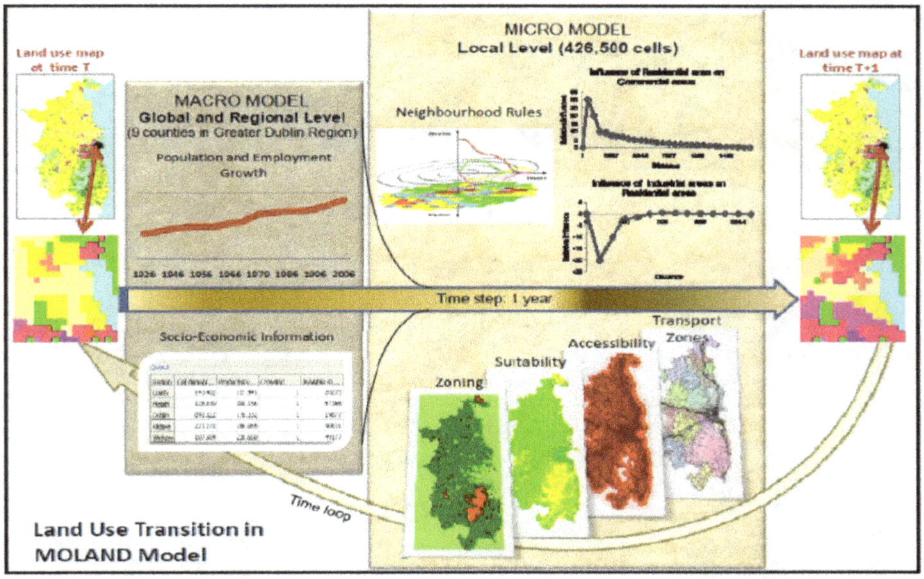

The CA modeling of land-use change is under a developing process since new developments take place in different directions. Following White and Engelen (1993), some other examples of CA models have emerged: A spatial dynamic system is developed integrating socio-economic and environmental factors at variety of scales in Engelen et al. (1995). Likewise, Batty (1998) utilized an extended cellular automation -in which demand for location drives the system responding through the supply of actual development- to model spontaneous development in an urban area e.g. regeneration of inner cities and location of edge cities. Barredo et al. (2003) applied dynamic approach of the CA for modeling future urban land-use scenarios while Vliet et al. (2009) focused on a variable grid CA to allow incorporation of more spatial information by aggregating land-uses at greater distances. Furthermore, there are joint CA-Markov models combining the Markov technique with the CA modeling as in Balzter et al. (1998); Almeida et al. (2003); Caruso et al. (2005) and Yang et al. (2016). An example of a joint CA-Markov modeling framework is provided in Figure 7 from Yang et al. (2016). In this example, the CA is combined with a Markov model to control the transition quantity; and this is linked to GIS where the local rules were used to simulate the future land-use change (Figure 7). Other examples of cellular automata models that are used to explain urban growth patterns are Wu and Webster (1998), Li and Yeh (2000), Silva and Clarke (2005), Petrov et al. (2009), Alexander et al. (2016) and Gaunaridis et al. (2018). There are also hybrid models that involve combinations of several techniques such as statistical techniques with the cellular models and system models (Aspinall, 2004; Rutherford et al., 2008) or cellular models integrated into the systems dynamics model (Southworth et al. 1991; Parker et al. 2003).

Spatial heterogeneity is related to a variety of features across landscape, which will drive actual changes in land-use pattern. To test the hypothesized growth rules in the cellular automata approach, an empirical model is required pointing to the identification problem that arises in the identification of the interaction effects from other elements of landscape heterogeneity (e.g. zoning, environmental features).

Figure 7. An example of an integrated modeling of Markov and CA
Source: Yang et al. (2016)

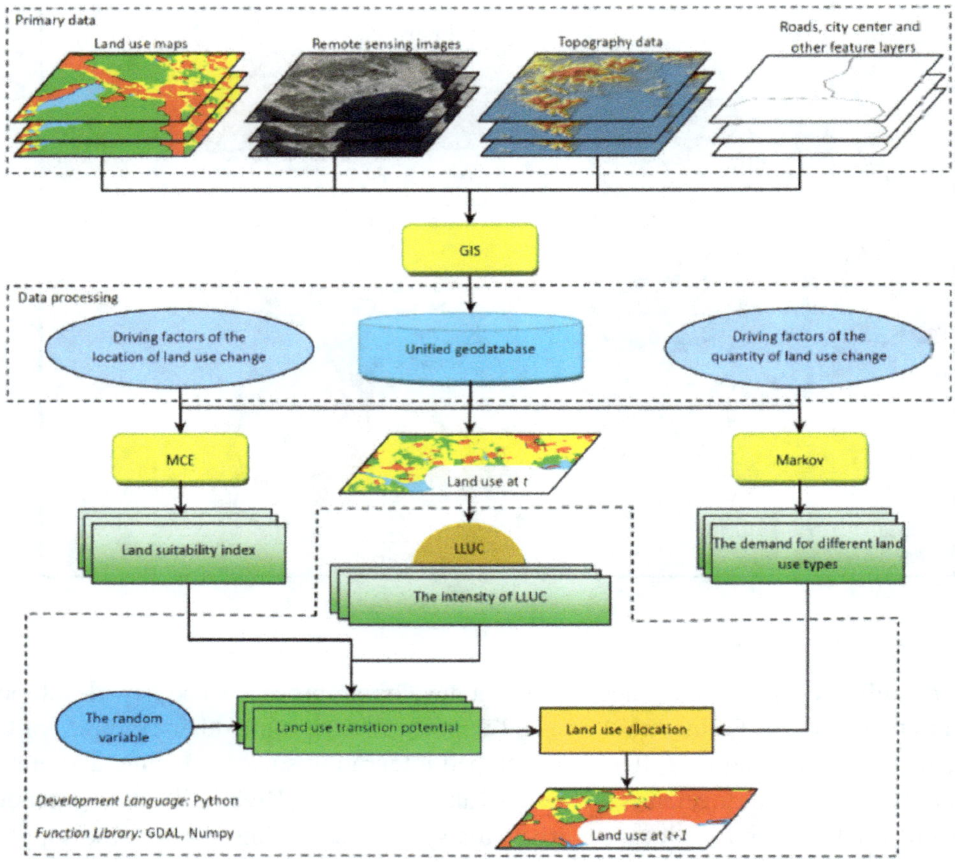

However, CA models are based on a simple spatial landscape and were not developed to *differentiate the interaction effects from a variety of exogenous variables that may also generate same patterns of development* (Irwin and Geoghegan, 2001, p.10). The simple decision rules embedded by the CA models make it difficult to examine the impacts of policy interventions on the individual agents that make land-use change decisions (Brown et al., 2004). A final shortcoming of the CA models is that these models are lacking an economic rationale. Examination of land-use change patterns under alternative policy scenarios should be based on *an understanding of how individual landowners will react under different policy scenarios* (Irwin and Geoghegan, 2001). This requires an integration of the economic theory to the spatial process of land-use change.

There are recent improvements to the CA models regarding the explained shortcomings. To incorporate the impact of multiple agents on land-use change, there are studies focusing on integrated multi-agent and cellular automata modeling. Some examples are Moreno et al. (2007), Zhao and Peng (2012) and Tan et al. (2015). Further to this, there have been increasing attempts to integrate economic models with the CA land-use change models. Some examples of such integrated models are provided in Van Delden and McDonald (2010) and Harvey et al. (2019).

SECTOR-BASED ECONOMIC MODELS

These models are often based on macroeconomic theory and structural modeling approaches. The models mostly focus on inputs and outputs, and trade among regions and sectors, and describe the equilibrium flows of population, employment or other economic parameters to identify demand for different land-uses. As an example, Jun (2005) proposed a model where land-use demand is determined by inter-sectorial and interspatial relations of production, income generation and consumption through application of input-output multipliers (Figure 8). In the model, residential land-use demand projection is based on projected population and the household size (which is correlated to projected employment determined by generated output from the input-output model) and the space required by each household. Non-residential land-use demand is derived from employment in different sectors and the land requirement per employee working in a particular sector. The model is eventually simulated based on different scenarios to examine the impact of government policy on regional and metropolitan economic activity (see also Jin and Wilson, 1993; Jun, 2004).

Input-output approach is also utilized in general equilibrium models which provides integrated assessment of spatial and inter-temporal interactions among different socio-economic and physical forces that drive land-use/land cover change. As an example, readers can refer to LUC model of IIASA that was developed for China (see Fischer and Sun, 2001). LUC model is aimed at assessing spatial and temporal interactions among different socio-economic and biophysical factors that drive land-use/cover change. The model is based on a general equilibrium modeling, which uses input-output tables and applies a dynamic welfare optimization model. In the optimization model, the optimum levels of resources and transformations are provided as a function of the initial state of the economy and resources, of consumer preferences and production relations, and of exogenously specified dynamics and constraints such as population and climate change (Verburg et al., 2004). The model has coarse spatial resolution and is highly data demanding due to the large number of sectors of the economy included in the model. A more recent

Figure 8. An example of an input-output model
Source: Jun (2005)

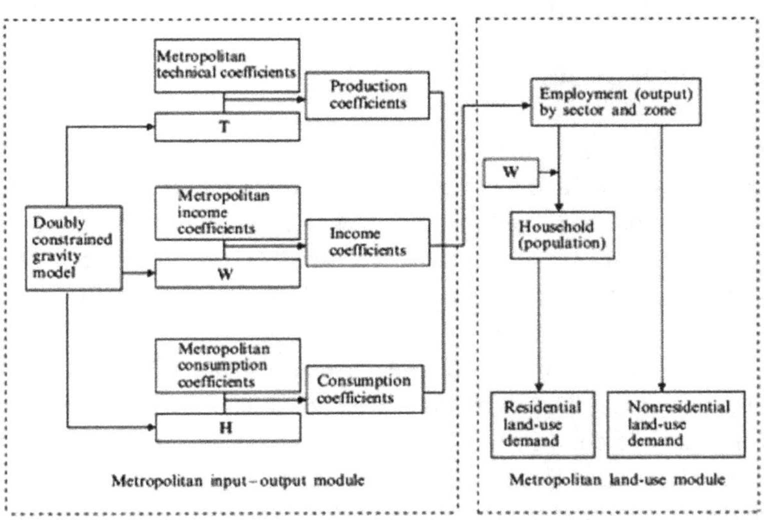

example is the newly developed dynamic spatial general equilibrium model of European Commission i.e. RHOMOLO. RHOMOLO simulates policy scenarios related to the EU cohesion policy and for the analysis of the regional impacts or impacts on EU member states. The Model takes into consideration the location choice of different agents and captures the connection between agglomeration and dispersion forces in determining the spatial equilibrium (Brandsma et al., 2015). The Model is dynamic considering that it utilizes time-series information on factors of production, human capital and technology.

The sector-based models are criticized due to the difficulties that arise from scaling these models (Verburg et al., 2004). The problems of scale are sound when this type of models is used at coarse scales or higher aggregation levels (Jansen and Stoorvogel, 1998). Another shortcoming of macroeconomic models is that these models do not consider land as spatially explicit; rather land is generally represented as a factor of production (Alcamo et al., 2006). Such modeling examples are GTAP, GEM-E3 and the NEMESIS (Brecard et al., 2006; Capros et al., 2013; Corong et al., 2017). The first two are general equilibrium models, and the latter is a macro sectoral econometric model. These models deal with land-use change by focusing on the entire economy and the interactions between economic sectors across regions. For a detailed review of the macroeconomic models, refer to Bockstael and Irwin (2000), and Wiedmann et al. (2007).

SPATIALLY DISAGGREGAED ECONOMIC MODELS

These models are spatially explicit models that are mainly based on the microeconomic theory to examine the changes in land-use and cover and the equilibrium land-use patterns as an outcome of individual decisions. Economic theory on urban location suggests that there is competition for scarce space among different activities resulting in exploitation of best sites by the highest bidders. This process of competition for space is represented by the bid-rent theory which is a central concept in urban location theories. The non-spatially explicit models are mainly based on the bid-rent theory that is linked with a limited spatial extent. Spatially explicit models, on the other hand, are based on integration of the economic theory to the spatial modeling of land-use/cover change. This modeling work is significant considering that the modeling framework has a theoretical background where the underlying theory is linked to the empirical model. In the following sections, we review some examples of the models under this category.

Non-Spatially Explicit Models

The models are based on individuals who make land-use decisions with the aim of maximizing expected returns or utility derived from the land and use microeconomic theory for the model development including the choice of functional form and the variables (Ruben et al., 1998; Verburg et al., 2004). In explaining the spatial structure of land-use patterns in an urban-regional setting, a theoretical perspective had been developed by various key researchers including Von Thunen (1826); Marshall (1920); Losch (1954); and Alonso (1964). At the basis of the theoretical framework of the microeconomic models is the bid-rent model that was initially developed by Alonso (1964) and others (Mills, 1967; Muth, 1969). The bid-rent model presumes that there is a trade-off between land rents and transportation costs in an urban area that is composed of concentric rings of residential development around the urban center, where residential density decreases as distance from the urban center increases (Figure 9). Therefore, households optimize their location decisions by trading off land rents and the accessibility to the urban

Figure 9. Illustration of Alonso's (1964) bid-rent theory

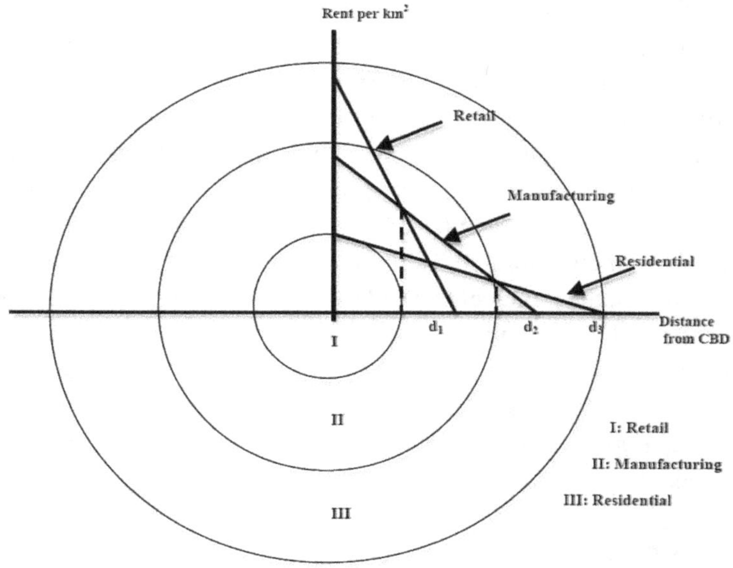

center and bid up higher land rents for locations close to the city center (Figure 9). There are studies in the literature that criticized the bid-rent model by claiming that the model's predictions are limited (compared to actual land-use patterns) to explain the complexity of spatio-temporal patterns of urban development (Anas et al., 1998; Irwin and Geoghegan, 2001).

A common criticism of bid-rent models such as Alonso (1964) is based on three central points: The first is the focus on transportation and location within a perfect equilibrium market and the strong assumption of complete availability of information (Anas et al. 1998; Arthur, 2005). The second is related to the exclusion of heterogeneous agents i.e. households with varying incomes or tastes, and the agglomeration of activities in a single central business district-CBD (Wheaton, 2004). In fact, the models assuming heterogeneity of households in terms of income and preferences have been developed (Sasaki, 1987). Lastly, it can be suggested that the principle urban economic theories lack the input of historical influences and dynamics in explaining the spatial preference rigidities. For instance, most European cities had already set their patterns based on this criterion. In addition, the impact of individuals' decisions and social background on the movement and selection of locations is underrepresented (Anas et al. 1998; Parker et al. 2003).

Ricardian land rent theory provides an improvement to the bid-rent model through explaining differentiation in land rents due to the differences in land quality that stems from a heterogeneous landscape. However, the Ricardian theory fails to consider relative locations of land-uses that form the spatial structure. The literature has extended the bid-rent model through empirical studies that focus on evolution of employment sub-centers (Anderson and Bogart, 2001; McMillen and Smith, 2003), applications based on multi-centric city models (McMillen, 2004; Lee, 2007; Behrens et al. 2009); and developments of more sophisticated models based on a general equilibrium setting (Fujita et al., 1997; Berliant and Wang, 2008; Burnett, 2012). More recent literature has focused on developing general equilibrium models of urban land-use which do not have a priori assumption of location of employment or households. In other words, the model will determine the spatial organization of economic or all other activities endogenously, which were previously given as exogenous in the bid-rent models.

Regarding the general equilibrium models, we can give examples from the research performed by Fujita et al. (1997), Lucas and Rossi-Hansberg (2002), Cavailhes et al. (2007), and Berliant and Wang (2008). The common point of these studies is that they are based on non-monocentric general equilibrium models where the location of households and firms are endogenously determined without specifying the location as a priori. The interdependence among individual households and/or firms is related to a variety of factors *e.g. demand and supply linkages between households and firms, knowledge spillovers among firms, or congestion effects among residential land-uses* (Irwin and Geoghan, 2001). Depending on the parameters of the model, the monocentric and multicentric patterns can also be determined within the model. Because of the endogenous process introduced by these models, they are more robust than the traditional bid-rent models. However, as explained by Irwin and Geoghegan (2001), the equilibrium solution of these models cannot fully capture much of the actual heterogeneity of the landscape. Therefore, these models provide *a fairly abstract description of the landscape pattern based on equilibrium conditions* (Irwin and Geoghegan, 2001, p. 13).

Spatially Explicit Models

These models are based on integration of economic theory to the spatial modeling of land-use/cover change. Bockstael (1996) argued that many economic phenomena are related to fundamental spatial processes. For instance, the economic process of commodity and services exchange has a spatial component considering that much of the exchange takes place in the urban center and/or close to employment centers (Bockstael, 1996). Other examples of spatial processes are the diffusion of information and adoption of technology, and the effects of externalities. In Bockstael's (1996) explanation, the former process is dynamic and is often associated with spatial processes considering that diffusion of knowledge and technology takes place either geographically or over more abstract surfaces. In a similar vein, externalities are based on spatial processes as these are linked to *the motion of spatially dynamic domino effect* (Bockstael, 1996). There is vast literature on the integration of economic theory to the land-use modeling framework in the landscape ecology literature. For instance, Chomitz and Gray (1996) developed a model of deforestation where the optimal use of landscape is determined by the highest rents, which is the outcome of landowners' profit maximization. This model uses economic theory to model landscape change by using remotely sensed data and spatial GIS data. The likelihood of conversion from forest use to agriculture was modeled as a function of soil quality, and input and output prices in a given location. Similar to the bid-rent model, the spatial variation of prices depends on the location of land-uses and transportation costs to market centers. Other examples of economic and spatial modeling of deforestation can be found in Müller and Zelner (2002), Echeverria et al. (2008), Rosa et al. (2013), and Dezecache et al. (2017).

There are also studies in the urban land modeling literature aiming to modeling development and redevelopment in an urban area through application of logistic regression functions that describes either the probability of particular land-use type occurring or of the location transitioning from one land-use type to another. This approach has been used for modeling urban development (Landis, 1995; Landis and Zhang, 1998a; 1998b) and along urban-rural gradients (Wear and Bolstad, 1998). Regarding urban development, the logic behind the choice of explanatory variables is based on economic theory so that the impact of different policies that affects returns from different land-uses can be predicted from the model. Examples of explanatory variables include initial land-use, various drivers of demand, distance and accessibility measures, costs of development, and returns to alternative uses. As argued by Irwin and Geoghegan (2001), these models fail to link the unit of observation and the decision maker, and to capture the impacts of variables on land-use change over a longer time period. Therefore, these models can be improved through development of a *dynamic model of land-use change in which the model integrates the economic theory, and the spatially distributed land-use decisions can be aggregated to describe the predicted changes in regional pattern* (Irwin and Geoghegan, 2001, p. 14). The dynamic economic land-use models that incorporate spatial heterogeneity are under development (see Epple et al., 2010).

AGENT-BASED MODELS

In Brown et al.'s (2013, p. 453) explanation, agent-based models are structural considering that these models are developed by the modeler to represent the processes and interactions that are linked to the observed land-use changes (see also Parker et al., 2003; Matthews et al., 2007). Agent-based models focus on the agents, the complexity emerging as a result of the interaction among agents, and their decision-making in relation to the urban environment (Evans and Manson, 2007) (Figure 10). Agents are accepted to be autonomous and rational decision-making entities implying that they have control over their actions, they are goal oriented and they are capable of interacting with each other for achieving set of goals in perfect rationality framework (Parker et al., 2003). Agent-based models are built on economic theory and these *models show flexibility to incorporate any theory decision-making and land-use change* (Brown et al., 2013, p. 453). Given the focus on the decision-making process, agent-based models can be considered as a process-based bottom-up approach that is based on actual land-use decisions of the agents at the local level (Robinson et al., 2007).

Figure 10. Components in an agent-based land-use/cover change model.
Source: Huigen (2003)

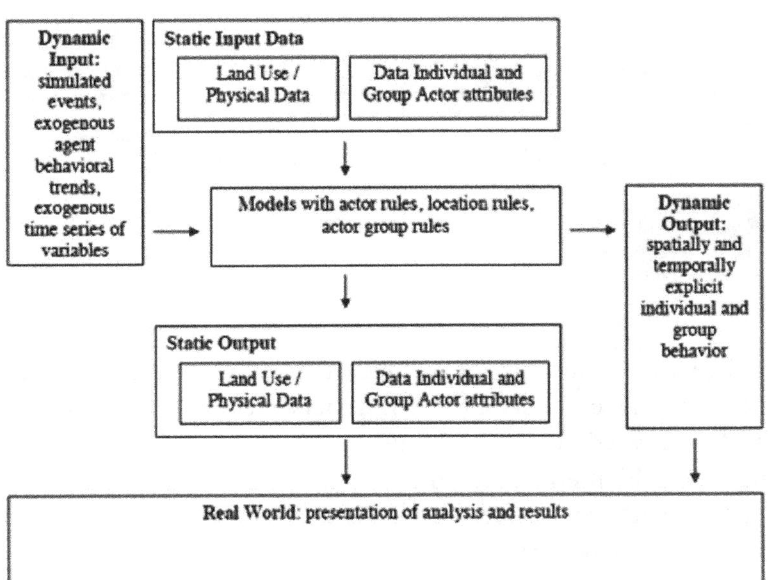

Figure 11. An example of the agricultural land-use changes through Agent-based Modeling
Source: Beckers et al. (2018)

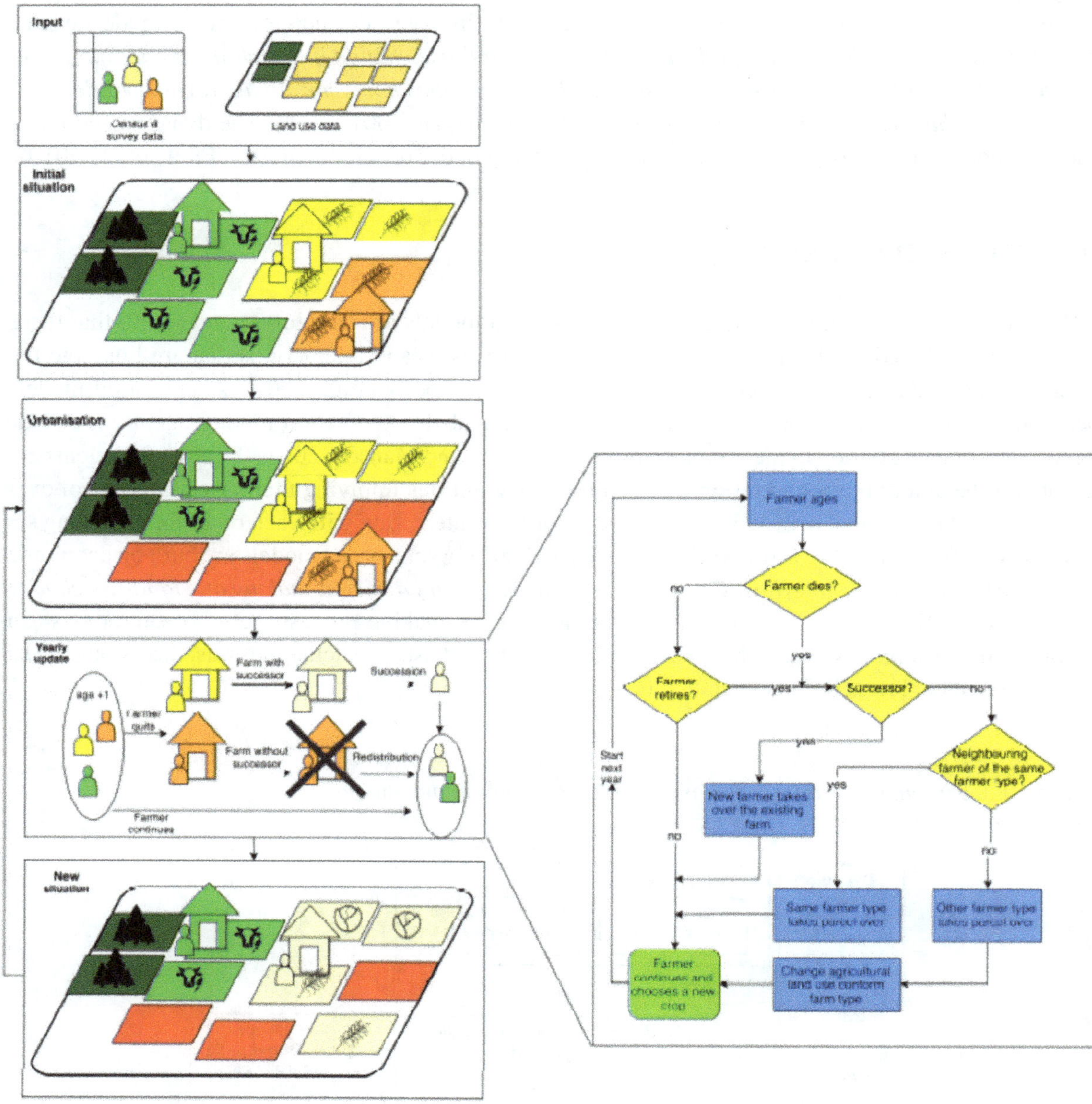

Some examples of agent-based models are Kollman et al. (1997); Batty (2001); Ronald et al. (2007); Murray-Rust et al. (2014); Grinberger et al. (2015), Andersen et al. (2017); and Beckers et al. (2018). The illustration of the Agent-based Model (ABM) used for modeling the agricultural land-use dynamics is from Beckers et al. (2018) (Figure 11). A further example of an ABM for the simulation of urban dynamics after a hypothetical earthquake is provided in Figure 12 (Grinberger et al. 2015). In these models, the outcome of an action to every perfectly rational agent depends on both his actions and those of other agents, ending up a recursive programming problem, prioritizing the use of game theory to explain the

Figure 12. An example of an agent-based simulation model of urban dynamics
Source: Grinberger et al. (2015)

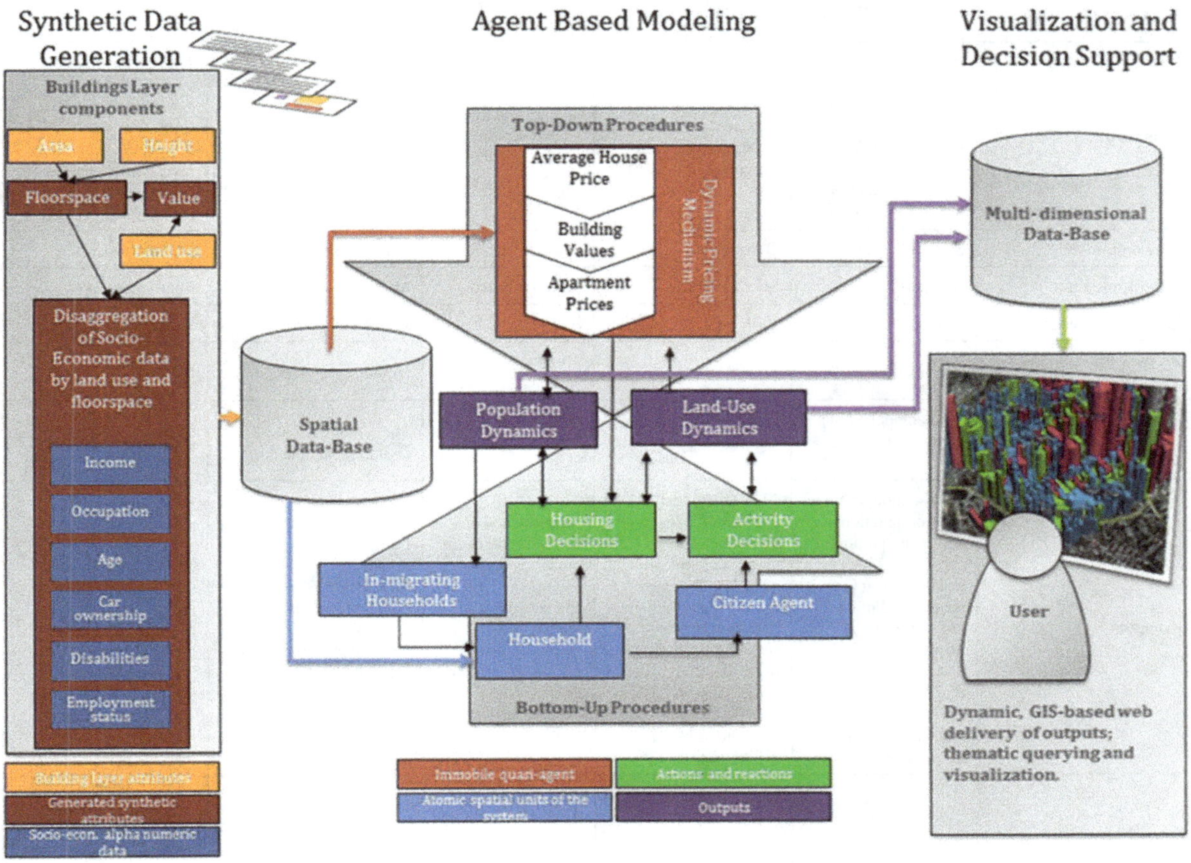

decision-making mechanisms in agent-based models. Applications of game theory in agent-based models can be found in Tesfatsion (2001), Parsons et al. (2002) and Pyka and Fagiolo (2007). A detailed review of multi-agent models can be seen in Parker et al. (2003), which focused on agent-based models as tools for representing human decision-making and simulating the aggregate outcomes that result from decisions made by many individuals. According to Verburg et al. (2004), the simulated individual behavior at aggregate levels could lead to the development of new theories linking individual behavior to collective behavior. …*Such studies indicate how individual people interact to form groups and organize collective action, and how such collective decisions vary with group size, collective social capital etc.…* (Verburg et al., 2004, p. 310). An alternative view suggests that combining agent-based models, to represent individual decision-making, with cellular models, to represent biophysical landscape change, offers a promising approach for future model development in land-use change literature (Parker et al., 2003).

As highlighted by Verburg et al. (2004), multi-agent models are highly data demanding regarding socio-economic data obtained from extensive field surveys and questionnaires, and it can be difficult to obtain sufficient data at the individual/household level to develop a well-structured decision-making model. Except few studies that utilized agent-based approach for modeling large regions (Fontaine and Rounsevell, 2009; Valbuena et al., 2010), many of the currently available multi-agent models can simu-

Table 2. Main characteristics of different land-use change modeling frameworks

Characteristics	Machine Learning Approaches	Cellular Models	Sector-based Economic Models	Spatially Disaggregated Economic Models	Agent-based Models
Formalization	Focus on projection of patterns based on theory and methods including artificial neural networks, classification and regression trees and logistic regression	These models represent processes of change based on locational information	These models focus on inputs and outputs and trade among regions and sectors to compute demand of land for different uses	These models are based on micro-economic theory and focus on changes in land use as an outcome of individual decisions	These models focus on transitional dynamics that arise from individual agents' decisions and interactions and conditions that evolve the system over time
Time dynamics	Static/Dynamic	Dynamic	Static/Dynamic	Static/Dynamic	Static/Dynamic
Geographic scale	All scales	Micro to meso	Macro to meso	Micro to meso	Micro to meso
Spatial complexity	Medium-high	High	Low	High-medium	High
Decision making complexity	Low-medium	Low-medium	Low	High	High
Temporal complexity	Static models: low; Dynamic models: high	Medium-high	Static models: low; Dynamic models: high	Static models: low; Dynamic models: high	High
Strengths	*These models have fast computational speed and predictive capacity. *They can operate at high spatial resolutions over large study areas. *They are easy to transfer to different urban environments and require limited expert knowledge	*The CA models use simple rules to model behavior of greatest complexity *The CA approach involves integration of economic and environmental modeling *The functioning of CA modeling can be easily understood by non-expert users	*General and partial equilibrium models incorporate endogenous prices and substitution effects *Input-output models track intre-industry linkages and they are easy to implement	*The theoretical framework of these models is well established *They include the individual-level of factors in developing regional land-use patterns *They have the ability to incorporate multiple sources of spatial heterogeneity	*They represent the interactions between the agents and the spatial outcome. *They are built on economic theory and have the flexibility to integrate any theory of decision making and land-use
Weaknesses	*These models mainly focus on correlations between spatial indicators and land-use classes at one point in time and do not consider spatiotemporal dynamics of land-use change *A further limitation is related to their restricted application to change detection/prediction of categorical land cover information	*The linkages to collaboration and competition among human agents are not included *There are no theoretical basis in developing the neighborhood functions *Function settings in the CA approaches are based on developers' knowledge and hardly documented *Irregularity of land-uses are not fully covered in many of the CA approaches	*In most cases, these models are not spatially explicit and individual behavior is not related to price mechanism *Partial equilibrium models lack feedbacks with other sectors *The mathematical functions used in the modeling process are difficult to calibrate with real data *In input-output models, prices are fixed and no substitution effects. For general equilibrium models, there is high data requirement	*Unlikely to cover all the structural parameters of the behavioral process *The partial equilibrium models may miss important feedbacks and cannot fully capture spatial dynamics; they are restricted to short-term	*They have extensive data requirements *Lack of standardization *Large number of parameters *Model complexity

Source: the authors

late simplified hypothetical landscapes considering that the number of interacting agents and the factors influencing land-use change are still large to develop comprehensive models (Kanaroglou and Scott, 2001; Verburg et al., 2004). SIMPOP model is such an example of an agent-based model that simulated hypothetical urban development and transition (Sanders et al., 1997). A further problem is the scale and aggregation issues for the use of the model at large regional scales that resulted in the use of agent-based approaches for the modeling of local area processes. The potential for analyzing policy impacts is a strength of the agent-based models; however, the models are applied at spatial scales that may not be relevant for the scales of analysis where policy and land-use plans are developed (Rounsevell et al., 2012). Verburg et al. (2004, p. 312) claim that further efforts are undertaken to develop operational multi-agent models that can simulate realistic land-use changes (Bousquet et al., 1998; Berger, 2001; see Verburg et al., 2004 for further references). Such an approach requires extensive data at the individual/household level to build a structured and validated decision-making model and observed spatial data of land-use/ land cover change is not sufficient to validate such a model (Verburg et al., 2004, p. 312). Model validation is based on the comparison of model outcomes for a historical period with the observed changes of the land-use. The methods for validation for the land-use change models are provided in Pontius (2000).

A final example is the combination of cellular models with the agent-based models i.e. the multi-agent systems for land-use/cover modeling. In these models, *cellular modeling represents physical and ecological aspects of a system while an agent-based system represents human decision-making* (Parker et al. 2003, p. 318). An example is Ligtenberg et al. (2001) which developed an integrated multi-agent and CA simulation approach by utilizing a bottom-up modeling concept to represent individual actor behavior. The main goal of this approach is to aggregate local scale conditions and actors from the bottom-up in such a way that actors interact with each other locally-having their own preferences and budgets- and follow the social decision rules of the related neighborhoods (see also Holland and Miller, 1991; Batten, 2001; Elsner, 2010). Another approach considers path dependence in integrated multi-agent and CA models and explains existing urban patterns as a result of the positive or negative feedback coming from the historical events. Some examples are Batty (2005); Brown et al. (2005); and Sohl et al. (2007). A comprehensive review of agent-based models is provided in Groeneveld et al. (2017). Following the review of five key types of land-use change modeling approaches, we provide main characteristics of these different approaches in Table 2. And, the overall land-use change modeling process is presented in Figure 13.

FUTURE RESEARCH DIRECTIONS: MODEL INTEGRATION

As the review of the literature suggests, there are a wide variety of approaches that have been built on different theories of decision-making, economy and natural sciences, and providing assessment frameworks for different concepts in relation to policy support, planning and decision-making processes (Table 2). In this view, the tools and approaches should be selected and designed to complement each other in order to provide coherent outputs. Such tools and approaches could be effectively used considering the complexity and uncertainties that increase over time. Land-use change dynamics is related to cross-scale processes, such as exchanges of energy, matter and information, and spatio-temporal factors within socio-ecological systems (Brown et al., 2013). This implies that there are interdependent parts within land-use systems, which are characterized by strong *interactions between the parts, complex feedback loops that make it difficult to distinguish cause from effect, and significant time and space lags discontinuities, thresholds, and limits* (Verburg et al, p. 318 referred to: Costanza and Wainger, 1993).

This complexity makes the integration of different feedback mechanisms in the land-use system an important issue in the land-use change literature. The integration of different approaches enables modelers to represent processes from different disciplines and provide information for planners and decision makers that has additional value compared to information provided from individual modeling approaches (Rounsevell et al., 2012). A major challenge for developing integrated modeling frameworks is providing linkages between different modeling systems. Verburg et al. (2004) asserted that for the models where individual systems are not fully integrated and feedbacks are not developed, the modeling framework can be complicated and may not adequately represent the observed complexity of the land-use system. Therefore, land-use models are required to integrate different drivers and components of land-use change into the modeling framework to provide significant improvements to land-use modeling in the future. In fact, complicated nature of integrated modeling approaches makes it difficult to deal with feedbacks and quantify how individual model components interact (Sohl and Claggett, 2013). There is a body of literature suggesting that integration of different model components may not result in improved model performance or be not more useful for policy-makers (Pontius and Spencer, 2005; Van Delden et al., 2011). This literature argues that basic software coupling of models is more practical and easier to understand compared to the conceptual framework developed through the linking of independent models (Van Delden et al., 2011).

Verburg et al. (2004) described two approaches for the integration of different sub-systems that vary in the degree of integration (Figure 14). The first approach analyzes and models the different sub-systems separately. In this approach, *interactions and feedbacks between the sub-systems are either assumed to be negligible or must be clearly defined* (Verburg et al., 2004, p. 318). The second approach, on the other hand, gives more importance to the interactions between the sub-systems. The approach depends on the endogeneity of variables, which are functions of the interactions between the system components (Verburg et al., 2004). An example of this approach is the top-down hierarchy built in a global model where constrained land areas are allocated to more detailed scales through the cellular automata (Verburg et al., 2008). Most integrated modeling systems are related to climate change and its policy dimensions (Verburg et al., 2004). Some examples of integrated models are provided in Verburg et al. (2004) including: IMAGE2 (Alcamo et al., 1998); IIASA-LUC (Fischer and Sun, 2001); and Patuxent Landscape model (Geoghegan et al., 1997). The literature on integrated land-use models has been growing recently. These include CLUE-CR (Veldkamp and Fresco, 1996b), UrbanSim (Waddell, 2002), land use modeling system (LUMOS) (Beurden et al., 2007), MAS model (Loibl et al., 2007), and the LANDSHIFT modeling framework (Schaldach and Koch, 2009) (references are from: Azadi et al., 2017). From this literature, it is highlighted that to assess and integrate different feedbacks in the land-use system through developments in theory, data and integrated modeling applications is an important issue for the future land-use modeling applications.

Figure 14. Interactions between the land-change model and its sub-systems
Source: Adapted from: Pozoukidou et al. (2017); Verburg et al. (2004)

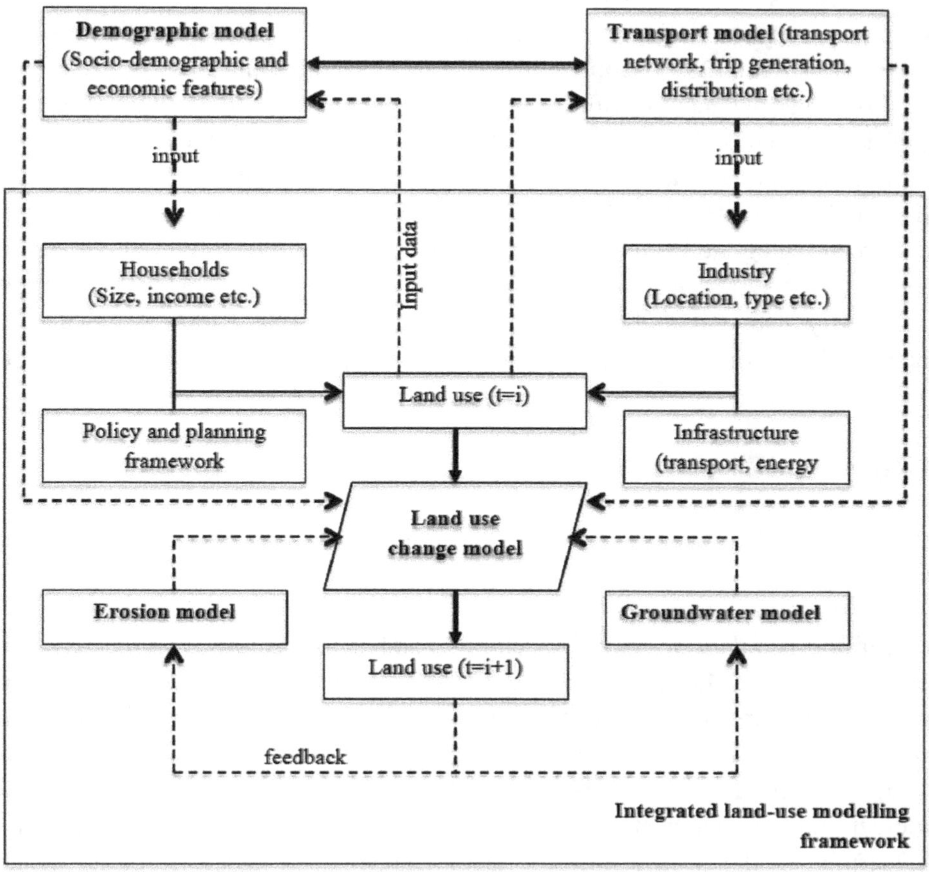

CONCLUSION

General Conclusions

A variety of approaches has been developed to model land-use change ranging from pattern-based approaches to approaches focusing on environmental structure and human decision processes. Following Brown et al. (2013), we have reviewed five key types of modeling approaches that are widely used and applicable at regional to global scales and are useful in policy evaluation and impact assessment of land-use change. However, there are a number of issues that should be addressed in the land-use modeling framework. For instance, many of the land-use models are limited in addressing the feedbacks between individual interconnections and the environment. The understanding of impacts of actors and driving factors on land-use change has come to the forefront. The aspects of land-use management are also important, but these are not fully covered in many of the land-use change models. The link between land cover and the function provided by the land is not easy to uncover as it depends on local and contextual factors (Van Soesbergen, 2016). There are land-use models that are capable of incorporating some of these processes, but the development of such models is still in progress. As stated by Ramankutty et al. (2008), many of the issues that need to be covered in land-use modeling framework are related to limitations of data; particularly the limitations of the consistent land cover data for different time periods. According to Van Soesbergen (2016), applications of new algorithms and classifications to the existing data for different time periods may provide an improvement to address the data limitations.

Though significant progress has been achieved in land-use modeling theory and applications, more aspects are required to be incorporated for future development of the land-use models. This is important for the land-use modeling research in that models can represent better the multi-scale characteristics of land-use system and integrate better the different components of the system. Integration of different feedbacks of the land-use system is a multi-disciplinary concept that is related to the integration of different modeling theory, data and applications. Such a multi-disciplinary framework can help to better understand the complexity of land-use systems and assist in policy decisions and evaluations on the impacts of land-use change.

Specific Issues

We have listed a number of concluding remarks regarding the specified modeling approaches we have covered in this review.

- Machine learning is a data-driven approach that has been successfully implemented in different geo-science disciplines. Machine learning approaches comprise computational techniques utilized from identifying patterns of data. In such an approach, machine learning algorithms statistically learn patterns based on present correlations defined by a training set of data and develops learned mapping between predictor variables and a target variable (Bishop, 2006). These models have been widely used to predict changes in land cover for a given case study area or region. Statistical machine learning methods have provided advancement over physically based models through their fast-computational speed and better predictive capacity (Rogan et al., 2008). Therefore, they can operate at high spatial resolutions and over large study areas. A commonly accepted limitation of machine learning approaches is their restricted application to change detection/prediction of categorical land cover information (Patil et al., 2017). Many of the land-use change models that are based on machine learning approaches focus on the correlations between spatial indicators and land-use classes at the same time point but they do not estimate land-use change to predict near future. Machine learning approaches that are based on static land-use modeling are easy to transfer to different urban environments and would require limited involvement of expert knowledge. The informative attributes' (predictors) selection process may provide benefits including improvement of model accuracy, reducing noisy inputs and model complexity and reducing the time required for training (Samardzic-Petrovic et al., 2015).
- Cellular approaches integrate economic and environmental processes and thus provide an improvement over purely geographical and economic land-use models. In most cases, integrated models are based on a combination of separate economic and environmental processes capable of spatially explicit modeling of different land-uses. Hasbani (2008) listed several reasons that make CA models suitable for land-use modeling: 1) Information from other models can be integrated to CA models to spatially allocate land-use changes; 2) The cell state is typically a land-use but also

can be specified differently (such as population density etc.) (White and Engelen, 2000); 3) The models can be constrained in different ways to reflect the territorial and local tendencies (Jenerette and Wu, 2001); 4) There is the possibility to specify the quantity of land that is changed one time step to another; 5) The CA models are simple considering that they incorporate a limited set of equations and transition rules but may also include complexity (White and Engelen, 2000); 6) The model inputs include mostly the raster images that can be easily developed by using the GIS software. We note that the CA models are useful for capturing patterns of land-use changes rather than precise location of land-uses at the local/regional scales (Jenerette and Wu, 2001). There are also limitations of CA approaches as they do not include conflict resolving rules (Junfeng and Boerboom, 2006). This is related to the collaboration and competition between human agents to change the urban area for their own purposes. Verburg et al. (2004) pointed to the function settings in the CA approach as they are hardly documented and mostly based on developers' knowledge and some calibration. Further to this, the authors stated that the theoretical basis of the quantification of neighborhood functions is poor as such quantification is based on expert knowledge (Verburg et al., 2004). Therefore, it is recommended to develop more sophisticated approaches to define the neighborhood effects and to improve the CA modeling approach further. Another shortcoming of the CA models is the use of regular raster grids. In this, it is assumed that there is homogeneous land-use, but it may contain heterogeneity in reality. Therefore, a regular grid cell cannot represent irregular land-uses in urban environments. In most CA models, when a cell begins to develop, it is completely developed in the next time step. However, in reality not all the different land-uses develop at the same speed and at the same time. This variability in spatio-temporal dynamics needs to be involved in the modeling process (Liu and Anderson, 2004). Though the need for using irregular structures is introduced in the literature long time ago (Couclelis, 1985), there have been few implementations in reality. According to Stevens and Dragicevic (2007, p. 709), the reasons are three folds. First, *computationally intensive modeling is required to search irregular neighborhoods*. Second, *there is the overhead associated with the vector data structure to represent irregular shapes*. A further reason is the *complexity involved in defining some of the basic components such as neighbor and irregular grid*. Due to the complexities in implementing such irregularities in traditional CA, most CA modelers use the custom-built software (White et al., 1997). This has created challenges related to *high degree of specialization and time required to develop such CA modeling tools, problems of data compatibility and lack of GIS functions in the custom applications* (Stevens and Dragicevic, 2007, p. 710).

- Sector-based economic models are mostly equilibrium models that allocate land demand through demand and supply structures of land intensive economic sectors where the model equates demand and supply under exogenous constraints (Heistermann et al., 2006). These models are generally suitable for macro to meso-scale analysis as they consistently address demand, supply and trade through price mechanisms for the sectors specified in the analysis. Data availability appears to be an issue to develop modeling frameworks at the regional or more local scales. Beaussier et al. (2019, p. 418) pointed to multi-disciplinary work and technical skills required to build model interfaces and a further risk of *model coupling that increases complexity leading to a black box and loss of replicability*. They are generally dynamic models that take into account the timing of the factors considered in the analysis, which makes them more computationally intensive, compared with static models. They are limited though for two reasons: First, individual behavior is not related to price mechanisms and the model's incorporation of impact of demand on actual land-use change process is limited. Second, many of these models are not developed as land-use models as they only consider the market process and include land as a factor of production rather than allocate land spatially. De Rosa et al. (2016) mentioned that sector-based economic models address land competition and transformations through mathematical functions that are difficult to calibrate with the real data. In their explanation, the assumptions determining those functions are difficult to detect and involves the risk that they may remain unknown to both policy-makers and land-use modeling practitioners. Further to this, the effect of uncertainties caused by multiple parameters in the modeling process may be difficult to reveal (see De Rosa et al., 2016).

- Spatially disaggregated economic models of land-use change derive from economic models of individual land-use decisions. The theoretical framework of these models is well established in urban economics science. For the improvement of these models, *improved datasets, improved methods and improved theory are required* (Irwin and Geoghegan, 2001). The increasing availability of spatial social data has contributed to improved applications of spatially explicit modeling. In Irwin and Geoghegan's (2001) words, obtaining better data to be applied in the spatially explicit models in association with more sophisticated spatial-economic theories requires improved empirical methods such as those used in the field of spatial econometrics. Despite there are advances achieved in the field of spatial econometrics, this field requires *further improvement to catch up with the advances achieved in its dynamic counterpart i.e. time series econometrics* (Irwin and Geoghegan, 2001). In spatially explicit economic modeling framework, the spatial dimension and geographical interaction of economic agents are key drivers of economic processes and the resultant spatial outcome. Therefore, these interactions between the agents require incorporating such interconnections in the modeling process. The improvements in the theoretical models of interacting agents and the interconnection between them have started in the recent years to shift the emphasis from the individual behavior of agents to interconnections among them (Ertur and Koch, 2007). This has provided new insights to examine neighborhood effects, spatial spillovers and network effects (Manski, 2000). These models can be applied at every scale ranging from applications in local areas to more macro levels of applications as in the case of regional economic models. A major challenge is the aggregation of individual choices to regional outcomes. Bockstael and Irwin (2000) stated that aggregating individual land-use choices to regional market outcomes is more complicated considering the existing spatial heterogeneity. The development of the theoretical literature pointing to these issues has been in progress, and as highlighted by Irwin and Geoghegan (2001), new theoretical and empirical improvements about the evolution of urban land-use patterns are needed.
- Agent-based modeling provides an improvement in the land system science since the models couple social and environmental models by focusing on agents' decision making in relation to the urban environment and examine the collective responses of the agents in environmental management policies. These models can incorporate different theories and different scientific disciplines. An important question, however, is when should agent-based approaches be used in a research or policy context? Agent-based land-use/cover change models can represent the physical system well considering that physical environment has been researched for a relatively long period of time; and therefore, relevant datasets and the methods for data development are well established (Huigen, 2003). As highlighted by Huigen (2003), the main difficulties occur in the specification of the agent decision rules, the quantification of decision rules and qualitative variables, and specification of the model's boundary. The decision to include the factors in the model either

exogenously or endogenously and the specification of the feedbacks that should be integrated to the model are controversial (Huigen, 2003). The literature points to the heterogeneity of land-use systems as they indicate a high degree of heterogeneity particularly at the individual farm level. Matthews et al. (2007) noted that this heterogeneity is the main characteristics of its own right that influences the dynamics of such systems. Therefore, using agent-based models is beneficiary in modeling systems where the heterogeneity between the agents and the interactions between them influence land-use system. The literature has shown that agent-based approaches are also useful to evaluate social norms or social structure where the modeling approach incorporates social processes and non-monetary influences on decision making (Kohler and Gumerman, 2000; Matthews et al., 2007).

Concluding Remarks

Land-use modeling is complex considering the data requirements, methodological developments and application process. The land-use modeling requires specific expertise and capacity to specify drivers, obtain spatial data, and to perform and evaluate analysis and statistical relationships. The choice of land-use model to be applied in a case study depends on scale and extent of analysis, purpose and scope, and the datasets that are available to be used in the analysis. As recommended by Van Soesbergen (2016), for larger projects, it may be useful to participate land-use developers considering that many of these models have not been provided for public use. However, some of the freely available or low-cost models may be useful for small-scale projects. Finally, Van Soesbergen (2016) pointed to existing datasets of modeled land-use under different future scenarios that can be incorporated and used in future projects.

We have highlighted the importance of integration of different disciplines in the modeling framework. The techniques and methods that are developed in different disciplines may assist in developing better simulation algorithms. The methodological integration necessitates the researchers to move forward their disciplinary traditions (Verburg et al., 2004). The literature has shown that the integrated modeling approaches result in more improved results compared to traditional modeling approaches (Walsh et al., 1999; Walsh and Crews-Meyer, 2002; the references are from: Verburg et al., 2004). Therefore, the inter-disciplinary approaches highlight the strengths of multi-disciplinary approaches in land-use modeling and assist in developing a better understanding of complex land-use systems and providing an improved communication between the different agents that are influential in land-use change.

ACKNOWLEDGMENT

The study presented here was funded by The Scientific and Technological Research Council of Turkey (TUBİTAK) linked to BIDEB-2232 Programme, Project No. 118C002. We thank to three anonymous reviewers for their helpful comments and suggestions on the earlier versions of this chapter.

REFERENCES

Acheampong, A. R., & Silva, E. A. (2015). Land use-transport interaction modeling: A review of the literature and future research directions. *Journal of Transport and Land Use*, *8*(3), 11–13.

Adams, D. M., Alig, R. J., Callaway, J. M., McCarl, B. A., & Winnett, S. M. (1996). *The Forest and Agricultural Sector Optimization Model (FASOM): Model structure and policy applications*. Research Paper PNW-RP-495. Portland, OR: U.S. Department of Agriculture, Pacific Northwest Research Station.

Adams, R. (1995). Assessing the performance of the NELUP hydrological models for river basin planning. *Journal of Environmental Planning and Management*, *38*(1), 53–76. doi:10.1080/09640569513110

Agarwal, C., Green, G. L., Grove, M., Evans, T., & Schweik, C. (2000). *A review and assessment of land-use change models: Dynamics of space, time, and human choice*. Bloomington, IN: Centre for the Study of Institutions, Population and Environmental Change, Indiana University.

Alcamo, J., Kok, K., & Busch, G. (2006). Searching for the future of land: scenarios from the local to global scale. In E. F. Lambin & H. J. Geist (Eds.), *Land use and land-cover change: local processes and global impacts* (pp. 137–156). New York: Springer. doi:10.1007/3-540-32202-7_6

Alcamo, J., Leemans, R., & Kreileman, E. (1998). *Global change scenarios of the 21st Century. Results from the IMAGE 2.1 Model*. London: Elsevier.

Alexander, P. J., Fealy, R., & Mills, G. M. (2016). Simulating the impact of urban development pathways on the local climate: A scenario-based analysis in the greater Dublin region, Ireland. *Landscape and Urban Planning*, *152*, 72–89. doi:10.1016/j.landurbplan.2016.02.006

Almeida, M. C., Batty, M., Monteiro, A. M. V., Câmara, G., & … . (2003). Stochastic cellular automata modeling of urban land use dynamics: Empirical development and estimation. *Computers, Environment and Urban Systems*, *27*(5), 481–509. doi:10.1016/S0198-9715(02)00042-X

Alonso, W. (1964). *Location and land use*. Cambridge, MA: Harvard University Press. doi:10.4159/harvard.9780674730854

Alsharif, A. A. A., & Pradhan, B. (2014). Urban sprawl analysis of Tripoli Metropolitan City (Libya) using remote sensing data and multivariate logistic regression model. *Photonirvachak (Dehra Dun)*, *42*(1), 149–163. doi:10.100712524-013-0299-7

An, L., & Brown, D. G. (2008). Survival analysis in land change science: Integrating with GIScience to address temporal complexities. *Annals of the Association of American Geographers*, *98*(2), 323–344. doi:10.1080/00045600701879045

Anas, A., Arnott, R., & Small, K. A. (1998). Urban spatial structure. *Journal of Economic Literature*, *36*, 1426–1464.

Andersen, L. E., Groom, B., Killick, E., Ledezma, J. C., Palmer, C., & Weinhold, D. (2017). Modeling land use, deforestation, and policy: A hybrid optimisation-heterogeneous agent model with application to the Bolivian Amazon. *Ecological Economics*, *135*, 76–90. doi:10.1016/j.ecolecon.2016.12.033

Anderson, N. B., & Bogart, W. T. (2001). The structure of sprawl: Identifying and characterizing employment centres in polycentric metropolitan areas. *American Journal of Economics and Sociology*, *60*(1), 147–169. doi:10.1111/1536-7150.00058

Arthur, W. B. (2005). Out-of-equilibrium economics and agent-based modeling. In K. Judd, & L. Tesfatsion (Eds.), Handbook of Computational Economics, Vol. 2: Agent-Based Computational Economics. North-Holland: Elsevier.

Aspinall, R. (2004). Modeling land use change with generalized linear models-A multi-model analysis of change between 1860 and 2000 in Gallatin Valley, Montana. *Journal of Environmental Management*, *72*(1-2), 91–103. doi:10.1016/j.jenvman.2004.02.009 PMID:15246576

Azadi, H., Barati, A. A., Rafiaani, P., Taheri, F., Gebrehiwot, K., Witlox, F., & Lebailly, P. (2017). Evolution of land-use change modeling: Routes of different schools of knowledge. *Landscape and Ecological Engineering*, *13*(2), 319–332. doi:10.100711355-016-0311-9

Baker, W. L. (1989). A review of models of landscape change. *Landscape Ecology*, *2*(2), 111–133. doi:10.1007/BF00137155

Balzter, H., Braun, P. W., & Kohler, W. (1998). Cellular automata models for vegetation dynamics. *Ecological Modelling*, *107*(2/3), 113–125. doi:10.1016/S0304-3800(97)00202-0

Barredo, J. I., Kasanko, M., McCormick, N., & Lavalle, C. (2003). Modeling dynamic spatial processes: Simulation of urban future scenarios through cellular automata. *Landscape and Urban Planning*, *64*(3), 145–160. doi:10.1016/S0169-2046(02)00218-9

Barreteau, O., & Bousquet, F. (2000). SHADOC: A multi-agent model to tackle viability of irrigated systems. *Annals of Operations Research*, *94*(1/4), 139–162. doi:10.1023/A:1018908931155

Basse, M. R., Omrani, H., Charif, O., Gerber, P., & Bodis, K. (2014). Land use changes modeling using advanced methods: Cellular automata and artificial neural networks. The spatial and explicit representation of land cover dynamics at the cross-border region scale. *Applied Geography (Sevenoaks, England)*, *53*, 160–171. doi:10.1016/j.apgeog.2014.06.016

Batten, D. (2001). Complex landscapes of spatial interaction. *The Annals of Regional Science*, *35*(1), 81–111. doi:10.1007001680000032

Batty, M. (1998). Urban evolution on the desktop: Simulation with the use of extended cellular automata. *Environment & Planning A*, *30*(11), 1943–1967. doi:10.1068/a301943

Batty, M. (2001). Agent-based pedestrian modeling. *Environment & Planning B*, *28*(3), 321–326. doi:10.1068/b2803ed

Batty, M. (2005). *Cities and complexity: Understanding cities with cellular automata, agent-based models and fractals*. Cambridge, MA: The MIT Press.

Beaussier, T., Caurla, S., Bellon-Maurel, V., & Loiseau, E. (2019). Coupling economic models and environmental assessment methods to support regional policies: A critical review. *Journal of Cleaner Production*, *216*, 408–421. doi:10.1016/j.jclepro.2019.01.020

Behrens, K., Gaigne, C., & Thisse, J. F. (2009). Industry location and welfare when transport costs are endogenous. *Journal of Urban Economics*, *65*(2), 195–208. doi:10.1016/j.jue.2008.11.003

Berger, T. (2001). Agent-based spatial models applied to agriculture: A simulation tool for technology diffusion, resource use changes and policy analysis. *Agricultural Economics*, *25*(2-3), 245–260. doi:10.1111/j.1574-0862.2001.tb00205.x

Berliant, M., & Wang, P. (2008). Urban growth and subcenter formation: A trolley ride from the Staples Centre to Disneyland and the Rose Bowl. *Journal of Urban Economics*, *63*(2), 679–693. doi:10.1016/j.jue.2007.04.009

Berry, M., Hazen, B., MacIntyre, R. L., & Flamm, R. (1996). LUCAS: A system for modeling land-use change. *IEEE Computational Science & Engineering*, *3*(1), 24–35. doi:10.1109/99.486758

Beurden, J. B., Bakema, A., & Tijbosch, H. (2007). A land use modeling system for environmental impact assessment: recent applications of the LUMOS toolbox. In E. Koomen, J. Stillwell, A. Bakema, & H. J. Scholten (Eds.), *Modeling land use change progress and applications* (pp. 281–296). Dordrecht: Springer. doi:10.1007/1-4020-5648-6_16

Bilsborrow, R. E., & Okoth Ogondo, H. W. O. (1992). Population-driven changes in land use in developing countries. *Ambio*, *21*, 37–45.

Bishop, C. M. (2006). *Pattern recognition and machine learning*. New York: Springer.

Blumler, M., MacDonald, G., & Millington, A. (2011). Biocomplexity. In M. Blumler, G. MacDonald, A. Millington, & U. Schickhoff (Eds.), *Handbook of Biogeography* (pp. 469–487). London: Sage Publications.

Bockstael, N. E. (1996). Modeling economics and ecology: The importance of a spatial perspective. *American Journal of Agricultural Economics*, *78*(5), 1168–1180. doi:10.2307/1243487

Bockstael, N. E., & Irwin, E. G. (2000). Economics and the land use-environment link. In H. Folmer & T. Tietenberg (Eds.), *The international yearbook of environmental and resource economics 1999/2000* (pp. 1–54). Northampton, MA: Edward Elgar Publishing.

Bommel, P., Becu, N., Le Page, C., & Bousquet, F. (2016). Cormas: An Agent-Based simulation platform for coupling human decisions with computerized dynamics. In T. Kaneda, H. Kanegae, Y. Toyoda, & P. Rizzi (Eds.), *Simulation and gaming in the network society. Translational Systems Sciences* (Vol. 9). Singapore: Springer. doi:10.1007/978-981-10-0575-6_27

Bousquet, F., Bakam, I., Proton, H., & Le Page, C. (1998). Cormas: Common-pool resources and multiagent systems. *Lecture Notes in Artificial Intelligence*, *1416*, 826–837.

Brandsma, A., Kancs, A., Monfort, P., & Rillaers, A. (2015). RHOMOLO: A dynamic spatial general equilibrium model for assessing the impact of cohesion policy. *Papers in Regional Science*, *94*(S1), S197–S221. doi:10.1111/pirs.12162

Brecard, D., Fougeyrollas, A., Le Mouel, P., Lemiale, L., & Zagamé, P. (2006). Macroeconomic consequences of European research policy: Prospects of the NEMESIS model in the year 2030. *Research Policy*, *35*(7), 910–924. doi:10.1016/j.respol.2006.03.001

Briassoulis, H. (2000). Analysis of land use change: theoretical and modeling approaches. In S. Loveridge (Ed.), *The web book of regional science*. Morgantown, WV: West Virginia University.

Britz, W. (Ed.). (2004). *CAPRI modeling system documentation, final report of the FP5 shared cost project CAP-STRAT: Common Agricultural Policy strategy for regions, agriculture and trade, QLTR-2000-00394*. Bonn University.

Brown, D. G., Page, S. E., Riolo, R. L., Zellner, M., & Rand, W. (2005). Path dependence and the validation of agent-based spatial models of land use. *International Journal of Geographical Information Systems*, *19*(2), 153–174. doi:10.1080/13658810410001713399

Brown, D. G., Verburg, P. H., Pontius, R. G. Jr, & Lange, M. D. (2013). Opportunities to improve impact, integration, and evaluation of land change models. *Current Opinion in Environmental Sustainability*, *5*(5), 452–457. doi:10.1016/j.cosust.2013.07.012

Brown, D. G., Walker, R., Manson, S., & Seto, K. (2004). Modeling land use and land cover change. In G. Gutman, A. C. Janetos, C. O. Justice, & ... (Eds.), *Land Change Science* (pp. 395–409). Kluwer Academic Publishers.

Bryant, D., Nielsen, D., & Tangley, L. (1997). *The last frontier forests: Ecosystems and economies on the edge*. Washington, DC: World Resources Institute.

Burnett, P. (2012). Urban industrial composition and the spatial expansion of cities. *Land Economics*, *88*(4), 764–781. doi:10.3368/le.88.4.764

Capros, P., Van Regemorter, D., Paroussos, L., & Karkatsoulis, P. (2013). *GEM-E3 Model documentation. No. EUR26034 EN*. European Commission Joint Research Centre. Retrieved from ftp://139.191.159.82/pub/EURdoc/JRC83177.pdf

Caruso, G., Rounsevell, M., & Cojocaru, G. (2005). Exploring a spatio-dynamic neighbourhood-based model of residential behavior in the Brussels peri urban area. *International Journal of Geographical Information Science*, *19*(2), 103–123. doi:10.1080/13658810410001713371

Cavailhes, J., Gaigne, C., Tabuchi, T., & Thisse, J. F. (2007). Trade and the structure of cities. *Journal of Urban Economics*, *62*(3), 383–404. doi:10.1016/j.jue.2006.12.002

Ceuclelis, H. (1985). Cellular worlds: A framework for modeling micro-macro dynamics. *Environment & Planning A*, *17*(5), 585–596. doi:10.1068/a170585

Cheong, S., Brown, D. G., Lopez-Carr, D., & Kok, K. (2012). Mixed methods in land-change research: Towards integration. *Transactions of the Institute of British Geographers*, *37*(1), 8–12. doi:10.1111/j.1475-5661.2011.00482.x

Chomitz, K., & Gray, D. (1996). Roads, land use and deforestation: A spatial model applied to Belize. *The World Bank Economic Review*, *10*(3), 487–512. doi:10.1093/wber/10.3.487

Clarke, K. C., & Gaydos, L. J. (1998). Loose-coupling a cellular automaton model and GIS: Long-term urban growth prediction for San Fransisco and Washington/Baltimore. *International Journal of Geographical Information Science*, *12*(7), 699–714. doi:10.1080/136588198241617 PMID:12294536

Clifton, K., Ewing, R., Knaap, G.-J., & Song, Y. (2008). Quantitative analysis of urban form: A multidisciplinary review. *Journal of Urbanism*, *1*(1), 17–45. doi:10.1080/17549170801903496

Convery, S., & Shahumyan, H. (2009). *Applications of GIS and Urban Modeling for Environmental Policy*. Presentation in UEP Events. Retrieved from: www.uep.ie/downloads/

Corong, E., Hertel, T., McDougall, R., Tsigas, M., & van der Mensbrugghe, D. (2017). The Standard GTAP Model, Version 7. *Journal of Global Economic Analysis*, *2*(1), 1–119.

Costanza, R., Voinov, A., Boumans, R., Maxwell, T., Villa, F., Wainger, L., & Voinov, H. (2002). Integrated ecological economic modeling of the Patuxent River watershed, Maryland. *Ecological Monographs*, *72*(2), 203–231. doi:10.1890/0012-9615(2002)072[0203:IEEMOT]2.0.CO;2

Costanza, R., Wainger, L. A., & Folke, C. (1993). Modeling complex ecological economic systems. *Bioscience*, *43*(8), 545–556. doi:10.2307/1311949

Davidson, M. D. (2017). Equaty and the conservation of the global ecosystem services. *Sustainability*, *9*(3), 339. doi:10.3390u9030339

De Groot, R. S., Alkemade, R., Braat, L., Hein, L., & Willemen, L. (2010). Challenges in integrating the concept of ecosystem services and values in landscape planning, management and decision making. *Ecological Complexity*, *7*(3), 260–272. doi:10.1016/j.ecocom.2009.10.006

De Groot, R. S., Wilson, M. A., & Boumans, R. M. J. (2002). A typology for the classification, description and valuation of ecosystem functions, goods and services. *Ecological Economics*, *41*(3), 393–408. doi:10.1016/S0921-8009(02)00089-7

De Nijs, T. C. M., de Niet, R., & Crommentuijin, L. (2004). Constructing land use maps of the Netherlands in 2030. *Journal of Environmental Management*, *72*(1-2), 35–42. doi:10.1016/j.jenvman.2004.03.015 PMID:15246572

De Rosa, M., Trydeman Knudsen, M., & Hermansen, J. E. (2016). A comparison of land use change models: Challenges and future developments. *Journal of Cleaner Production*, *113*, 183–193. doi:10.1016/j.jclepro.2015.11.097

Dendoncker, N., Rounsevell, M., & Bogaert, P. (2007). Spatial analysis and modeling land use distributions in Belgium. *Computers, Environment and Urban Systems*, *31*(2), 188–205. doi:10.1016/j.compenvurbsys.2006.06.004

Dezecache, C., Salles, J. M., Vieilledent, G., & Hérault, B. (2017). Moving forward socio-economically focused models of deforestation. *Global Change Biology*, *23*(9), 3484–3500. doi:10.1111/gcb.13611 PMID:28055134

Echeverria, C., Coomes, D. A., Hall, M., & Newton, A. C. (2008). Spatially explicit models to analyze forest loss and fragmentation between 1976 and 2020 in southern Chile. *Ecological Modelling*, *212*(3-4), 439–449. doi:10.1016/j.ecolmodel.2007.10.045

EEA. (2008). *Modeling environmental change in Europe: Towards a model inventory* (SEIS/Forward). EEA Technical Report No. 11/2008. Copenhagen: European Environmental Agency.

Elsner, W. (2010). The process and a simple logic of 'Meso'. Emergence and the co-evolution of institutions and group size. *Journal of Evolutionary Economics*, *20*(3), 445–477. doi:10.100700191-009-0158-4

Engelen, G., Lavalle, C., Barredo, J. I., Meulen van der, M., & White, R. (2007). The Moland modeling framework for urban and regional land-use dynamics. In E. Koomen, J. Stillwell, A. Bakema, & H. J. Scholten (Eds.), Modeling Land-Use Change. Springer.

Engelen, G., White, R., Uljee, I., & Drazan, P. (1995). Using cellular automata for integrated modeling of socio-environmental system. *Environmental Monitoring and Assessment*, *34*(2), 203–214. doi:10.1007/BF00546036 PMID:24202047

Epple, D., Gordon, B., & Sieg, H. (2010). 'Drs. Muth and Mills meet Dr. Tiebout: Integrating location specific amenities into multi-community equilibrium models. *Journal of Regional Science*, *50*(1), 381–400. doi:10.1111/j.1467-9787.2009.00649.x

Ertur, C., & Koch, W. (2007). Growth, technological interdependence and spatial externalities: Theory and evidence. *Journal of Applied Econometrics*, *22*(6), 1033–1062. doi:10.1002/jae.963

Etter, A., McAlpine, C., Wilson, K., Phinn, S., & Possingham, H. (2006). Regional patterns of agricultural land use and deforestation in Colombia. *Agriculture, Ecosystems & Environment*, *114*(2-4), 369–386. doi:10.1016/j.agee.2005.11.013

Evans, T. P., & Manson, S. (2007). Guest editorial: Space, complexity, and agent-based modeling. *Environment & Planning B*, *34*(2), 196–199. doi:10.1068/b3402ed

FAO. (2016). *State of the World's forests 2016. Forests and agriculture: Land-use challenges and opportunities*. Online Report. Rome: FAO. Retrieved from http://www.fao.org/3/a-i5588e.pdf

Fischer, G., & Sun, L. X. (2001). Model based analysis of future land-use development in China. *Agriculture, Ecosystems & Environment*, *85*(1-3), 163–176. doi:10.1016/S0167-8809(01)00182-7

Fitz, H. C., De Bellevue, E. B., Costanza, R., Boumans, R., Maxwell, T., Wainger, L., & Sklar, F. H. (1996). Development of a general ecosystem model for a range of scales and ecosystems. *Ecological Modelling*, *88*(1-3), 263–295. doi:10.1016/0304-3800(95)00112-3

Fontaine, C. M., & Rounsevell, M. D. A. (2009). An agent-based approach to model future residential pressure on a regional landscape. *Landscape Ecology*, *24*(9), 1237–1254. doi:10.100710980-009-9378-0

Fujita, M., Thisse, J. F., & Zenou, Y. (1997). On the endogenous formation of secondary employment centres in a city. *Journal of Urban Economics*, *41*(3), 337–357. doi:10.1006/juec.1996.2002

Gaunaridis, D., Chorianopoulos, I., & Koukoulas, S. (2018). Exploring prospective urban growth trends under different economic outlooks and land-use planning scenarios: The case of Athens. *Applied Geography (Sevenoaks, England)*, *90*, 134–144. doi:10.1016/j.apgeog.2017.12.001

Geist, H. J., & Lambin, E. F. (2002). Proximate causes and underlying driving forces of tropical deforestation. *Bioscience*, *52*(2), 143–150. doi:10.1641/0006-3568(2002)052[0143:PCAUDF]2.0.CO;2

Geoghegan, J., Wainger, L. A., & Bockstael, N. E. (1997). Spatial landscape indices in a hedonic framework: An ecological economics analysis using GIS. *Ecological Economics*, *23*(3), 251–264. doi:10.1016/S0921-8009(97)00583-1

Gertsev, V., & Gertseva, V. (2004). Classification of mathematical models in ecology. *Ecological Modelling*, *178*(3-4), 329–334. doi:10.1016/j.ecolmodel.2004.03.009

Gibon, A., Sheeren, D., Monteil, C., Ladet, S., & Balent, G. (2010). Modeling and simulating change in reforesting mountain landscapes using a social-ecological framework. *Landscape Ecology*, *25*(2), 267–285. doi:10.100710980-009-9438-5

Grinberger, A. Y., Lichter, M., & Felsenstein, D. (2015). Simulating urban resilience: Disasters, dynamics and (synthetic) data. In S. Geertman, J. Ferreira Jr, R. Goodspeed, & J. Stillwell (Eds.), *Planning Support Systems and Smart Cities. Lecture Notes in Geoinformation and Cartography*. Cham: Springer. doi:10.1007/978-3-319-18368-8_6

Groeneveld, J., Müller, B., Buchmann, C. M., Dressler, G., Guo, C., Hase, N., ... Schwarz, N. (2017). Theoretical foundations of human decision-making in agent-based land use models-A review. *Environmental Modelling & Software*, *87*, 39–48. doi:10.1016/j.envsoft.2016.10.008

Han, H., Yang, C., & Song, J. (2015). Scenario simulation and the prediction of land use and land cover change in Beijing, China. *Sustainability*, *7*(4), 4260–4279. doi:10.3390u7044260

Harvey, E. P., Cardwell, R. C., McDonald, G. W., Van Delden, H., Vanhout, R., Smith, N. J., ... van den Belt, M. (2019). Developing integrated models by coupling together existing models; land use, economics, demographics and transport in Wellington, New Zealand. *Computers, Environment and Urban Systems*, *74*, 100–113. doi:10.1016/j.compenvurbsys.2018.07.004

Hasbani, J.-G. (2008). *Semi-automated calibration of a cellular automata model to simulate land-use changes in the Calgary region* (Ms thesis). University of Calgary, Department of Geomatics Engineering, Canada.

Hazen, B. C., & Berry, M. W. (1997). The simulation of land-cover change using a distributed computing environment. *Simulation Practice and Theory*, *5*(6), 489–514. doi:10.1016/S0928-4869(96)00026-2

Heistermann, M., Muller, C., & Ronneberger, K. (2006). Land in sight? Achievements, deficits and potentials of continental to global scale land use modeling. *Agriculture, Ecosystems & Environment*, *114*(2-4), 141–158. doi:10.1016/j.agee.2005.11.015

Herold, M., Couclelis, H., & Clarke, K. C. (2005). The role of spatial metrics in the analysis and modeling of urban land use change. *Computers, Environment and Urban Systems*, *29*(4), 369–399. doi:10.1016/j.compenvurbsys.2003.12.001

Hersperger, A. M., Gennaio, M. P., Verburg, P. H., & Burgi, M. (2011). Feedback loops in conceptual models of land change: Lost in complexity? *Ecology and Society*, *16*(2), r1. doi:10.5751/ES-04204-1602r01

Hilferink, M., & Rietveld, P. (1999). Land use scanner: An integrated GIS based model for long term projections of land use in urban and rural areas. *Journal of Geographical Systems*, *1*(2), 155–177. doi:10.1007101090050010

Holland, J., & Miller, J. H. (1991). Artificial adaptive agents in economic theory. *The American Economic Review*, *81*(2), 65–71.

Huigen, M. G. A. (2003). *Agent-based modeling in land use and land cover change studies*. Interim Report IR-03-044, International Institute for Applied Systems Analysis-IIASA, Austria.

Irwin, G. E., & Geoghegan, J. (2001). Theory, data and methods: Developing spatially explicit economic models of land use change. *Agriculture, Ecosystems & Environment*, *85*(1-3), 7–23. doi:10.1016/S0167-8809(01)00200-6

Jankowski, P., & Richard, L. (1994). Integration of GIS-based suitability analysis and multicriteria evaluation in a spatial decision support system for route selection. *Environment & Planning B*, *21*(3), 323–340. doi:10.1068/b210323

Jansen, H., & Stoorvogel, J. J. (1998). Quantification of aggregation bias in regional agricultural land use models: Application to Giocimo County, Costa Rica. *Agricultural Systems*, *58*(3), 417–437. doi:10.1016/S0308-521X(98)00039-0

Jenerette, D. G., & Wu, J. (2001). Analysis and simulation of land-use change in the central Arizona-Phoenix region, USA. *Landscape Ecology*, *16*(7), 611–626. doi:10.1023/A:1013170528551

Jin, Y., & Wilson, A. (1993). Generation of integrated multispatial input-output models of cities (GIMI-MoC): Initial stage. *Papers in Regional Science*, *72*(4), 351–367. doi:10.1007/BF01576848

Jorgensen, S. (2008). Overview of the model types available for development of ecological models. *Ecological Modelling*, *215*(1-3), 3–9. doi:10.1016/j.ecolmodel.2008.02.041

Jun, M.-J. (2004). A metropolitan input-output model: Multisectoral and multispatial relations of production, income formation, and consumption. *The Annals of Regional Science*, *38*(1), 131–147. doi:10.100700168-003-0138-9

Jun, M.-J. (2005). Forecasting urban land-use demand using a metropolitan input-output model. *Environment & Planning A*, *37*(7), 1311–1328. doi:10.1068/a3723

Junfeng, J., & Boerboom, L. (2006). Transition rule elicitation methods for urban cellular automata models. In J. van Leeuwen & H. Timmermans (Eds.), *Innovation in design & decision support systems in architecture and urban planning*. Springer.

Kanaroglou, P., & Scott, D. (2001). Integrated urban transportation and land-use models for policy analysis. In M. Dijst, W. Schenkel, & I. Thomas (Eds.), *Governing cities on the move*. Aldershot, UK: Avebury.

Kohler, T. A., & Gumerman, G. J. (2000). *Dynamics in human and primate societies: Agent based modeling of social and spatial processes. Sante Fe Institute Studies in the Sciences of Complexity*. Oxford, UK: Oxford University Press.

Kollman, K., Miller, J. H., & Page, S. E. (1997). Political institutions and sorting in a Tiebout Model. *The American Economic Review*, *87*, 977–922.

Koomen, E., Stillwell, J., Bakema, A., & Scholten, H. J. (Eds.). (2007). *Modeling land use change: Progress and applications*. Springer.

Lambin, E. F. (1997). Modeling and monitoring land-cover change processes in tropical regions. *Progress in Physical Geography*, *21*(3), 375–393. doi:10.1177/030913339702100303

Lambin, E. F., Rounsevell, M. D. A., & Geist, H. J. (2000). Are agricultural land-use models able to predict changes in land-use intensity? *Agriculture, Ecosystems & Environment*, *82*(1-3), 321–331. doi:10.1016/S0167-8809(00)00235-8

Landis, J. (1995). Imagining land use futures: Applying the California urban futures model. *Journal of the American Planning Association*, *61*(4), 438–457. doi:10.1080/01944369508975656

Landis, J., & Zhang, M. (1998a). The second generation of the California urban futures model. Part 1: Model logic and theory. *Environment & Planning A*, *25*(5), 657–666. doi:10.1068/b250657

Landis, J., & Zhang, M. (1998b). The second generation of the California urban futures model. Part 2: Specification and calibration results of the land-use change submodel. *Environment & Planning A*, *25*(6), 795–824. doi:10.1068/b250795

Landis, J. D., Monzon, J. P., Reilly, M., & Cogan, C. (1998). *Development and pilot application of the California Urban and Biodiversity Analysis (CURBA) Model*. Paper presented at the 1998 ESRI International User Conference.

Lavalle, C., Baranzelli, C., & Batista e Silva, F. (2011). A high resolution land use/cover modeling framework for Europe: Introducing the EU-ClueScanner100 Model. In B. Murgante et al. (Eds.), ICCSA-Part I (pp. 60-65). Springer.

Lee, B. (2007). Edge or edgeless cities? Urban spatial structure in US metropolitan areas, 1980 to 2000. *Journal of Regional Science*, *47*(3), 479–515. doi:10.1111/j.1467-9787.2007.00517.x

Lee, D. B., Jr. (1973, May). Requiem for large-scale models. *AIP Journal*, 163–177.

Leh, M., Bajwa, S., & Chaubey, I. (2013). Impact of land use change on erosion risk: An integrated remote sensing, geographic information system and modeling methodology. *Land Degradation & Development*, *24*, 409–421.

Li, S., Verburg, P. H., Shihai, L., Wu, J., & Xiaobing, L. (2012). Spatial analysis of the driving factors of grassland degredation under conditions of climate change and intensive use in Inner-Mongolia, China. *Regional Environmental Change*, *12*(3), 461–474. doi:10.100710113-011-0264-3

Li, X., & Yeh, A. G. (2000). Modeling sustainable urban development by the integration of constrained cellular automata and GIS. *International Journal of Geographical Information Science*, *14*(2), 131–152. doi:10.1080/136588100240886

Ligtenberg, A., Bregt, A. K., & van Lammeren, R. (2001). Multi-actor-based land use modeling: Spatial planning using agents. *Landscape and Urban Planning*, *56*(1-2), 21–33. doi:10.1016/S0169-2046(01)00162-1

Liu, X., & Anderson, C. (2004). Assessing the impact of temporal dynamics on land-use change modeling. *Computers, Environment and Urban Systems*, *28*(1-2), 107–124. doi:10.1016/S0198-9715(02)00045-5

Loibl, W., Tötzer, T., Köstl, M., & Steinnocher, K. (2007). Simulation of polycentric urban growth dynamics through agents: model concept, application, results and validation. In E. Koomen, J. Stillwell, A. Bakema, & H. J. Scholten (Eds.), *Modeling land use change progress and applications* (pp. 219–235). Dordrecht: Springer.

Losch, A. (1954). *The economics of location* (W. H. Woglom & W. F. Stolper, Trans.). New Haven, CT: Yale University Press.

Lucas, E. R., & Rossi-Hansberg, E. (2002). On the internal structure of cities. *Econometrica*, *70*(4), 1445–1476. doi:10.1111/1468-0262.00338

Manski, C. F. (2000). Economic analysis of social interactions. *The Journal of Economic Perspectives*, *14*(3), 115–136. doi:10.1257/jep.14.3.115

Marshall, A. (1920). *Principles of economics*. London: McMillan.

Mas, J.-F., Kolb, M., Paegelow, M., Camacho Olmedo, M. T., & Houet, T. (2014). Inductive pattern-based land use/cover change models: A comparison of four software packages. *Environmental Modelling & Software*, *51*, 94–111. doi:10.1016/j.envsoft.2013.09.010

Matthews, R., Gilbert, N., Roach, A., Polhill, J., & Gotts, N. (2007). Agent-based land-use models: A review of applications. *Landscape Ecology*, *22*(10), 1447–1459. doi:10.100710980-007-9135-1

McMillen, D. P. (2004). Employment densities, spatial autocorrelation, and subcenters in large metropolitan areas. *Journal of Regional Science*, *44*(2), 225–243. doi:10.1111/j.0022-4146.2004.00335.x

McMillen, D. P., & Smith, S. C. (2003). The number of subcenters in large urban areas. *Journal of Urban Economics*, *53*(3), 321–338. doi:10.1016/S0094-1190(03)00026-3

Mertens, B., & Lambin, E. (1997). Spatial modeling of deforestation in southern Cameroon. *Applied Geography (Sevenoaks, England)*, *17*(2), 143–162. doi:10.1016/S0143-6228(97)00032-5

Meyfroidt, P., Lambin, E. F., Erb, K.-H., & Hertel, T. W. (2013). Globalization of land use: Distant drivers of land change and geographic displacement of land use. *Current Opinion in Environmental Sustainability*, *5*(5), 438–444. doi:10.1016/j.cosust.2013.04.003

Michener, W. K., Baerward, T. J., & Firth, P. (2001). Defining and unraveling biocomplexity. *Bioscience*, *51*(12), 1018–1024. doi:10.1641/0006-3568(2001)051[1018:DAUB]2.0.CO;2

Mills, E. S. (1967). An aggregative model of resource allocation in a metropolitan area. *The American Economic Review*, *57*, 197–210.

Moreno, N., Quintero, R., Ablan, M., Barros, R., Dávila, J., Ramírez, H., ... Acevedo, M. F. (2007). Biocomplexity of deforestation in the Caparo tropical forest reserve in Venezuela. An integrated multi-agent and cellular automata model. *Environmental Modelling & Software*, *22*(5), 664–673. doi:10.1016/j.envsoft.2005.12.022

Müller, D., & Zeller, M. (2002). Land use dynamics in the central highlands of Vietnam: A spatial model combining village survey data with satellite imagery interpretation. *Agricultural Economics*, *27*(3), 333–354. doi:10.1111/j.1574-0862.2002.tb00124.x

Murray, A. T. (2010). Advances in location modeling: GIS linkages and contributions. *Journal of Geographical Systems*, *12*(3), 335–354. doi:10.100710109-009-0105-9

Murray-Rust, D., Brown, C., van Vliet, J., Alam, S. J., Robinson, D. T., Verburg, P. H., & Rounsevell, M. (2014). Combining agent functional types, capitals and services to model land use dynamics. *Environmental Modelling & Software*, *59*, 187–201. doi:10.1016/j.envsoft.2014.05.019

Muth, R. (1969). *Cities and Housing*. Chicago: University of Chicago Press.

O'Callaghan, J. R. (1995). NELUP: An Introduction. *Journal of Environmental Planning and Management*, *38*(1), 5–20. doi:10.1080/09640569513084

Overmars, K. P., Verburg, P. H., & Veldkamp, T. (2007). Comparison of a deductive and an inductive approach to specify land suitability in a spatially explicit land use model. *Land Use Policy*, *24*(3), 584–599. doi:10.1016/j.landusepol.2005.09.008

Parker, D. C., Manson, S. M., Janssen, M. A., Hoffmann, M. J., & Deadman, P. (2003). Multi-agent systems for the simulation of land-use and land-cover change: A review. *Annals of the Association of American Geographers*, *93*(2), 314–337. doi:10.1111/1467-8306.9302004

Parsons, S., Gmytrasiewicz, P., & Wooldridge, M. (2002). *Game theory and decision theory in agent-based systems*. Kluwer Academic Publishers. doi:10.1007/978-1-4615-1107-6

Patil, S. D., Gu, Y., Dias, F. S. A., Stieglitz, M., & Turk, G. (2017). Predicting the spectral information of future land cover using machine learning. *International Journal of Remote Sensing*, *38*(20), 5592–5607. doi:10.1080/01431161.2017.1343512

Pazur, R., & Bolliger, J. (2017). Land changes in Slovakia: Past processes and future directions. *Applied Geography (Sevenoaks, England)*, *85*, 163–175. doi:10.1016/j.apgeog.2017.05.009

Petrov, L., Lavalle, C., & Kasankob, M. (2009). Urban land use scenarios for a tourist region in Europe: Applying the MOLAND model to Algarve, Portugal. *Landscape and Urban Planning*, *92*(1), 10–23. doi:10.1016/j.landurbplan.2009.01.011

Pijanowski, P. C., Brown, D. G., Shellito, B. A., & Manik, G. A. (2002). Using neural nets and GIS to forecast land use changes: A land transformation model. *Computers, Environment and Urban Systems*, *26*(6), 553–575. doi:10.1016/S0198-9715(01)00015-1

Pontius, R. G. (2000). Quantification error versus location error in comparison of categorical maps. *Photogrammetric Engineering and Remote Sensing*, *66*, 1011–1016.

Pontius, R. G. Jr, Cornell, J. D., & Hall, C. A. S. (2001). Modeling the spatial pattern of land-use change with GEOMOD2: Application and validation for Costa Rica. *Agriculture, Ecosystems & Environment*, *85*(1-3), 191–203. doi:10.1016/S0167-8809(01)00183-9

Pontius, R. G. Jr, & Spencer, J. (2005). Uncertainty in extrapolations of predictive and-change models. *Environment & Planning B*, *32*(2), 211–230. doi:10.1068/b31152

Pozoukidou, G., Gavanas, N., & Verani, E. (2017). Land use models and sustainable urban mobility plans: An integrative approach for strategic planning. *International Journal of Sustainable Development and Planning*, *12*(5), 867–882. doi:10.2495/SDP-V12-N5-867-882

Puertas, O. L., Henriquez, C., & Meza, F. J. (2014). Assessing spatial dynamics of urban growth using an integrated land use model: Application in Santiago Metropolitan Area, 2010-2045. *Land Use Policy*, *38*, 415–425. doi:10.1016/j.landusepol.2013.11.024

Pyka, A., & Fagiolo, G. (2007). Agent-based modeling: A methodology for neo-schumpeterian economics. In H. Hanusch & A. Pyka (Eds.), *The Elgar companion to neo-schumpeterian economics*. Cheltenham, UK: Edward Elgar. doi:10.4337/9781847207012.00037

Ramankutty, N., & Foley, J. A. (1999). Estimating historical changes in global land cover: Croplands from 1700 to 1992. *Global Biogeochemical Cycles*, *13*(4), 997–1027. doi:10.1029/1999GB900046

Randolph, J. (2004). *Environmental land use planning and management*. Washington, DC: Island Press.

Rindfuss, R. R., Entwisle, B., Walsh, S. J., An, L., Badenoch, N., Brown, D. G., ... Verburg, P. H. (2008). Land use change: Complexity and comparisons. *Journal of Land Use Science*, *3*(1), 1–10. doi:10.1080/17474230802047955 PMID:20354590

Robinson, D. T., Brown, D. G., Parker, D. C., Schreinemachers, P., Janssen, M. A., Huigen, M., ... Barnaud, C. (2007). Comparison of empirical methods for building agent-based models in land-use science. *Journal of Land Use Science*, *2*(1), 31–55. doi:10.1080/17474230701201349

Rocha, J., Ferreira, J. C., Simoes, J., & Tenedorio, J. A. (2007). Modeling coastal and land use evolution patterns through neural network and cellular automata integration. *Journal of Coastal Research*, *50*, 827–831.

Rogan, J., Franklin, J., Stow, D., Miller, J., Woodcock, C., & Roberts, D. (2008). Mapping land-cover modifications over large areas: A comparison of Machine Learning Algorithms. *Remote Sensing of Environment*, *112*(5), 2272–2283. doi:10.1016/j.rse.2007.10.004

Ronald, N., Sterling, L., & Kirley, M. (2007). An agent-based approach to modeling pedestrian behavior. *International Journal of Simulation*, *8*(1), 25–38.

Rosa, I. M. D., Purves, D., Souza, C. Jr, & Ewers, R. M. (2013). Predictive modeling of contagious deforestation in the Brazilian Amazon. *PLoS One*, *8*(10), e77231. doi:10.1371/journal.pone.0077231 PMID:24204776

Rouchier, J., Bousquet, F., Requier-Desjardins, M., & Antona, M. (2001). A multi-agent model for describing transhumance in North Cameroon: Comparison of different rationality to develop a routine. *Journal of Economic Dynamics & Control*, *25*(3-4), 527–559. doi:10.1016/S0165-1889(00)00035-X

Rounsevell, M. D. A., Pedroli, B., Erb, K. H., Gramberger, M., Busck, A. G., Haberl, H., ... Wolfslehner, B. (2012). Challenges of land system science. *Land Use Policy*, *29*(4), 899–910. doi:10.1016/j.landusepol.2012.01.007

Ruben, R., Moll, H., & Kuyvenhoven, A. (1998). Integrating agricultural research and policy analysis: Analytical framework and policy applications for bio-economic modeling. *Agricultural Systems*, *58*(3), 331–349. doi:10.1016/S0308-521X(98)00034-1

Rutherford, G. N., Bebi, P., Edwards, P. J., & Zimmermann, E. N. (2008). Assessing land-use statistics to model land cover change in a mountainous landscape in the European Alps. *Ecological Modelling*, *212*(3-4), 460–471. doi:10.1016/j.ecolmodel.2007.10.050

Samardzic-Petrovic, M., Dragicevic, S., Bajat, B., & Kovacevic, M. (2015). Exploring the decision tree method for modeling urban land use change. *Geomatica*, *69*(3), 313–325. doi:10.5623/cig2015-305

Samardzic-Petrovic, M., Kovacevic, M., Bajat, B., & Dragicevic, S. (2017). Machine learning techniques for modeling short term land-use change. *International Journal of Geo-Information*, *6*(12), 387. doi:10.3390/ijgi6120387

Sanders, L., Pumain, D., Mathian, H., Guerin-Pace, F., & Bura, S. (1997). SIMPOP: A multiagent system for the study of urbanism. *Environment & Planning B*, *24*(2), 287–305. doi:10.1068/b240287

Sasaki, K. (1987). A comparative static analysis of urban structure in the setting of endogenous income. *Journal of Urban Economics*, *22*(1), 53–72. doi:10.1016/0094-1190(87)90049-0

Schaldach, R., & Koch, J. (2009). Conceptual design and implementation of a model for the integrated simulation of large-scale land-use systems. In Information technologies in environmental engineering. Environmental science and engineering, Part 2 (pp. 425-438). Academic Press. doi:10.1007/978-3-540-88351-7_32

Seto, K. C., Güneralp, B., & Hutyra, L. R. (2012). Global forecasts of urban expansion to 2030 and direct impacts on biodiversity and carbon pools. *Proceedings of the National Academy of Sciences of the United States of America*, *109*(40), 16083–16088. doi:10.1073/pnas.1211658109 PMID:22988086

Sheridan, P., Schroers, J. O., & Rommelfanger, E. (2007). GIS-based modeling of land use systems. In E. Koomen, J. Stillwell, A. Bakema, & H. J. Scholten (Eds.), *Modeling land use change progress and applications* (pp. 375–389). Dordrecht: Springer.

Shooshtari, S. J., & Gholamalifard, M. (2015). Scenario-based land cover change modeling and its implications for landscape pattern analysis in the Neka Watershed, Iran. *Remote Sensing Applications: Society and Environment*, *1*, 1–19. doi:10.1016/j.rsase.2015.05.001

Silva, E. A., & Clarke, K. C. (2005). Complexity, emergence and cellular urban models: Lessons learned from applying Sleuth to two Portuguese Metropolitan Areas. *European Planning Studies*, *13*(1), 93–116. doi:10.1080/0965431042000312424

Sohl, T. L., & Claggett, P. R. (2013). Clarity versus complexity: Land use modeling as a practical tool for decision-makers. *Journal of Environmental Management*, *129*, 235–243. doi:10.1016/j.jenvman.2013.07.027 PMID:23954777

Sohl, T. L., Sayler, K. L., Drummond, M. A., & Loveland, T. R. (2007). The FORE-SCE Model: A practical approach for projecting land cover change using scenario-based modeling. *Journal of Land Use Science*, *2*(2), 103–126. doi:10.1080/17474230701218202

Southworth, F., Dale, V. H., & O'Neill, R. V. (1991). Contrasting patterns of land use in Rondonia, Brazil: Simulating the effects on carbon release. *International Social Science Journal*, *130*, 681–798.

Stevens, D., & Dragicevic, S. (2007). A GIS-based irregular cellular automata model of land-use change. *Environment & Planning B*, *34*(4), 708–724. doi:10.1068/b32098

Straatman, B., White, R., & Engelen, G. (2004). Towards an automatic calibration procedure for constrained cellular automata. *Computers, Environment and Urban Systems*, *28*(1-2), 149–170. doi:10.1016/S0198-9715(02)00068-6

Sui, D. Z. (1998). GIS-based urban modeling: Practices, problems and prospects. *International Journal of Geographical Information Science*, *12*(7), 651–671. doi:10.1080/136588198241581 PMID:12294534

Sun, B., & Robinson, D. T. (2018). Comparisons of statistical approaches for modeling land-use change. *Land (Basel)*, *7*(4), 144. doi:10.3390/land7040144

Tan, R., Liu, Y., Zhou, K., Jiao, L., & Tang, W. (2015). A game-theory based agent-cellular model for use in urban growth simulation: A case study of the rapidly urbanizing Wuhan area of Central China. *Computers, Environment and Urban Systems*, *49*, 15–29. doi:10.1016/j.compenvurbsys.2014.09.001

Tesfatsion, L. (2001). Introduction to the special issue on agent-based computational economics. *Journal of Economic Dynamics & Control*, *25*(3/4), 281–293. doi:10.1016/S0165-1889(00)00027-0

Turner, B. L., Lambin, E. F., & Reenberg, A. (2007). The emergence of land change science for global environmental change and sustainability. *Proceedings of the National Academy of Sciences of the United States of America*, *104*(52), 20666–20671. doi:10.1073/pnas.0704119104 PMID:18093934

Turner, M. G., Wear, D. N., & Flamm, R. O. (1996). Land ownership and land-cover change in the Southern Appalachian Highlands and the Olympic Peninsula. *Ecological Applications*, *6*(4), 1150–1172. doi:10.2307/2269599

Valbuena, D., Verburg, P. H., Bergt, A. K., & Ligtenberg, A. (2010). An agent-based approach to model land-use change at a regional scale. *Landscape Ecology*, *25*(2), 185–199. doi:10.100710980-009-9380-6

Valdez, V. C., Ruiz-Luna, A., & Berlanga-Robles, C. A. (2016). Effects of land use changes on ecosystem services value provided by coastal wetlands: Recent and future landscape scenarios. *Journal of Coastal Zone Management*, *19*(1), 1000418. doi:10.4172/jczm

Van Delden, H., & McDonald, G. (2010). *Towards the integration of economic and land use change models*. 5th International Congress on Environmental Modeling and Software, Ottawa, Ontoria, Canada.

Van Delden, H., van Vliet, J., Rutledge, D. T., & Kirkby, M. J. (2011). Comparison of scale and scaling issues in integrated land-use models for policy support. *Agriculture, Ecosystems & Environment*, *142*(1-2), 18–28. doi:10.1016/j.agee.2011.03.005

Van Keulen, H., Van Ittersum, M. K., & de Ridder, N. (2000). New approaches to land use planning. In R. P. Roetter, H. van Keulen, A. G. Laborte, C. T. Hoanh, & H. H. van Laar (Eds.), *Systems research for optimising future land use in South and Southeast Asia*. IRRI.

Van Soesbergen, A. (2016). *A review of land use change models*. UNEP and John D. and Catherine T. Mac Arthur Foundation.

Varga, M., Balogh, S., & Csukas, B. (2016). GIS based generation of dynamic hydrological and land patch simulation models for rural watershed areas. *Information Processing in Agriculture*, *3*(1), 1–16. doi:10.1016/j.inpa.2015.11.001

Veldkamp, A., & Fresco, L. (1997). Exploring land use scenarios: An alternative approach based on actual land use. *Agricultural Systems*, *55*(1), 1–17. doi:10.1016/S0308-521X(95)00079-K

Veldkamp, A., & Fresco, L. O. (1996a). CLUE: A conceptual model to study the conversion of land use and its effects. *Ecological Modelling*, *85*(2-3), 253–270. doi:10.1016/0304-3800(94)00151-0

Veldkamp, A., & Fresco, L. O. (1996b). CLUE-CR: An integrated multiscale model to simulate land use change scenarios in Costa Rica. *Ecological Modelling*, *91*(1–3), 231–248. doi:10.1016/0304-3800(95)00158-1

Veldkamp, A., & Lambin, E. F. (2001). Predicting land use change. *Agriculture, Ecosystems & Environment*, *85*(1-3), 1–6. doi:10.1016/S0167-8809(01)00199-2

Verburg, P., Soepboer, W., Veldkamp, A., Limpiada, R., Espaldon, V., & Mastura, S. S. A. (2002). Modeling the Spatial Dynamics of Regional Land Use: The CLUE-S Model. *Environmental Management*, *30*(3), 391–405. doi:10.100700267-002-2630-x PMID:12148073

Verburg, P. H., Eickhout, B., & van Meijl, H. (2008). A multi-scale, multi-model approach for analysing the future dynamics of European land use. *The Annals of Regional Science*, *42*(1), 57–77. doi:10.100700168-007-0136-4

Verburg, P. H., & Overmars, K. (2007). Dynamic simulation of land-use changes trajectories with the Clue-S Model. In E. Koomen, J. Stillwell, A. Bakema, & H. J. Scholten (Eds.), *Modeling land-use change. The GeoJournal Library* (Vol. 90). Dordrecht: Springer.

Verburg, P. H., & Overmars, K. P. (2009). Combining top-down and bottom-up dynamics in land-use modeling: Exploring the future of abandoned farmlands in Europe with the Dyna-CLUE model. *Landscape Ecology*, *24*(9), 1167–1181. doi:10.100710980-009-9355-7

Verburg, P. H., Rounsevell, M. D. A., & Veldkamp, A. (2006b). Scenario-based studies of future land use in Europe. *Agriculture, Ecosystems & Environment*, *114*(1), 1–6. doi:10.1016/j.agee.2005.11.023

Verburg, P. H., Schot, P. P., Dijst, M. J., & Veldkamp, A. (2004). Land use change modeling: Current practice and research priorities. *GeoJournal*, *61*(4), 309–324. doi:10.100710708-004-4946-y

Verburg, P. H., Schulp, C. J. E., Witte, N., & Veldkamp, A. (2006a). A downscaling of land use change scenarios to assess the dynamics of European landscapes. *Agriculture, Ecosystems & Environment*, *114*(1), 39–56. doi:10.1016/j.agee.2005.11.024

Verburg, P. H., Veldkamp, A., de Koning, G. H. J., Kok, K., & Bouma, J. (1999). A spatial explicit allocation procedure for modeling the pattern of land use change based upon actual land use. *Ecological Modelling*, *116*(1), 45–61. doi:10.1016/S0304-3800(98)00156-2

Vliet, J., White, R., & Dragicevic, S. (2009). Modeling urban growth using a variable grid cellular automaton. *Computers, Environment and Urban Systems*, *33*(1), 35–43. doi:10.1016/j.compenvurbsys.2008.06.006

Von Thunen, H. (1826). Der isolierte staat. In *Von Thunen's isolated state, 1968* (C. M. Warterberg, Trans.). Pergamon.

Waddell, P. (2002). UrbanSim; modeling urban development for land use, transportation and environmental planning. *Journal of the American Planning Association*, *68*(3), 297–314. doi:10.1080/01944360208976274

Wagner, D. F. (1997). Cellular automata and geographic information systems. *Environment & Planning B*, *24*(2), 219–234. doi:10.1068/b240219

Walsh, S. J., & Crews-Meyer, K. A. (Eds.). (2002). *Linking people, place, and policy. A GIScience approach*. Boston: Kluwer Academic Publishers. doi:10.1007/978-1-4615-0985-1

Walsh, S. J., Evans, T. P., Welsh, W. F., Entwisle, B., & Rindfuss, R. R. (1999). Scale dependent relationships between population and environment in Northeastern Thailand. *Photogrammetric Engineering and Remote Sensing*, *65*, 97–105.

Wear, D. N., & Bolstad, P. (1998). Land-use changes in Southern Appalachian landscapes: Spatial analysis and forecast evaluation. *Ecosystems (New York, N.Y.)*, *1*(6), 575–594. doi:10.1007100219900052

Webster, C. J. (1993). GIS and the scientific inputs to planning. Part 1: Description. *Environment & Planning B*, *20*(6), 615–748. doi:10.1068/b200709

Webster, C. J. (1994). GIS and the scientific inputs to planning. Part 2: Prediction and prescription. *Environment & Planning B*, *21*(2), 145–157. doi:10.1068/b210145

Wegener, M. (2004). Overview of land-use transport models. In D. A. Hensher & K. Button (Eds.), *Transport geography and spatial systems* (pp. 127–146). Kidlington: Pergamon/Elsevier. doi:10.1108/9781615832538-009

Wheaton, W. (2004). Commuting, congestion, and employment dispersal in cities with mixed land use. *Journal of Urban Economics*, *55*(3), 417–438. doi:10.1016/j.jue.2003.12.004

White, R. (2005). Modeling multi-scale processes in a cellular automata framework. In J. Portugali (Ed.), *Complex artificial environments. Verlag*. Springer.

White, R., & Engelen, G. (1993). Cellular automata and fractal urban form: A cellular modeling approach to the evolution of urban land use patterns. *Environment & Planning A*, *25*(8), 1175–1199. doi:10.1068/a251175

White, R., & Engelen, G. (1997). Cellular automata as the basis of integrated dynamic regional modeling. *Environment & Planning B*, *24*(2), 235–246. doi:10.1068/b240235

White, R., & Engelen, G. (2000). High-resolution integrated modeling of the spatial dynamics of urban and regional systems. *Computers, Environment and Urban Systems*, *24*(5), 383–400. doi:10.1016/S0198-9715(00)00012-0

White, R., Engelen, G., & Uijee, I. (1997). The use of constrained cellular automata for high-resolution modeling of urban land-use dynamics. *Environment & Planning B*, *24*(3), 323–343. doi:10.1068/b240323

Wicke, B., Verweij, P., Meijl, H. V., Vuuren, D. P. V., & Faaij, A. P. C. (2012). Indirect land use change: Review of existing models and strategies for mitigation. *Biofuels*, *3*(1), 87–100. doi:10.4155/bfs.11.154

Wiedmann, T., Lenzen, M., Turner, K., & Barrett, J. (2007). Examining the global environmental impact of regional consumption activities-Part 2: Review of input-output models for the assessment of environmental impacts embodied in trade. *Ecological Economics*, *61*(1), 15–26. doi:10.1016/j.ecolecon.2006.12.003

Wilbanks, T. J., & Kates, R. W. (1999). Global changes in local places: How scale matters. *Climatic Change*, *43*(3), 601–628. doi:10.1023/A:1005418924748

Witzke, H. P., & Zintl, A. (2003). *CAPSIM-Complete documentation*. Bonn: Final Report to Eurostat.

Wu, F., & Webster, C. J. (1998). Simulation of land development through the integration of cellular automata and multicriteria evaluation. *Environment & Planning B*, *25*(1), 103–126. doi:10.1068/b250103

Wu, Q., Li, H.-Q., Wang, R.-S., Paulussen, J., He, Y., Wang, M., ... Wang, Z. (2006). Monitoring and predicting land use change Beijing using remote sensing and GIS. *Landscape and Urban Planning*, *78*(4), 322–333. doi:10.1016/j.landurbplan.2005.10.002

Yang, J., Su, J., Chen, F., Xie, P., & Ge, Q. (2016). A local land use competition cellular automata model and its applications. *International Journal of Geo-Information*, *5*(7), 106. doi:10.3390/ijgi5070106

Zhao, L., & Peng, Z.-R. (2012). LandSys: An agent-based Cellular Automata model of land use change developed for transportation analysis. *Journal of Transport Geography*, *25*, 35–49. doi:10.1016/j.jtrangeo.2012.07.006

Zhu, J., Zheng, Y., Carroll, A. L., & Aukema, B. H. (2008). Autologistic regression analysis of spatial-temporal binary data via Monte Carlo maximum likelihood. *Journal of Agricultural Biological & Environmental Statistics*, *13*(1), 84–98. doi:10.1198/108571108X273566

Zurek, M. B., & Henrichs, T. (2007). Linking scenarios across geographical scales in international environmental assessments. *Technological Forecasting and Social Change*, *74*(8), 1282–1295. doi:10.1016/j.techfore.2006.11.005

Chapter 10
Space Syntax Theory and Its Contribution to Urban Design

Ahmed M. Refaat
Assiut University, Egypt & Effat University, Saudi Arabia

ABSTRACT

This chapter presents the theory of space syntax and its contributions to urban design. The theory, introduced in 1984 by Hillier and Hanson, refers to a set of principles and techniques to investigate and reveal the characteristics of urban settlements via concrete measures. In this regard, the current chapter is divided into several sections. The first section demonstrates the space syntax theory at both conceptual and methodological levels. In the second section, the measures of connectivity, integration, choice, and intelligibility are discussed, and linked to the traditional urban values such as accessibility and legibility the traditional urban values. In the third section, types of space syntax analysis are discussed, in addition to their importance and their application potentials. This chapter demonstrates the contributions of space syntax theory to urban design and architecture. This finding provides a potential mechanism for explaining how the theory succeeded in presenting numerical and concrete measures to reveal what have been intangible values of urban settlements such as legibility and accessibility.

INTRODUCTION

Space syntax represents one of the seminal theories in both urban design and architectural realms. The theory focuses on analyzing the spatial structure in all its diverse forms, cities, settlements and buildings, to investigate the relationship between human societies and space organization. Despite that some of the theory foundational concepts were envisaged in the mid-1970s by Hillier and Leaman, the theory was formally introduced in 1984 by Hillier and Hanson in their book "The Social Logic of Space". The authors were motivated by the desire to find a concrete methodology to describe, represent and quantify the spatial structure of settlements or buildings that map the relation between society and its urban and architectural forms. In their book, Hillier and Hanson demonstrated the main concepts, paradigms and measures of the theory, which were developed later by Hillier (1996a) in his seminal book "Space is the Machine". In this sense, the theory can be defined as a set of techniques to represent, quantify and

DOI: 10.4018/978-1-5225-9238-9.ch010

interpret the spatial structure - termed as a spatial configuration in the theory literature – of both urban and architectural forms, so their underlying social logic can be enunciated (Bafna, 2003).

According to the theory, by analyzing the spatial structure, which in itself carries social information and content for a given settlement, the morphological features of its urban form should be disclosed, and the social relations can be mapped (Hillier and Hanson, 1984; Hillier et al., 1987). Also, the configuration of the spatial structure of a given settlement might influence the way its inhabitants and visitors use and perceive it (Penn, 2003; Hillier, 1996a). In this regard, the theory has the competence of mapping the impact that spatial structure, or spatial configuration, has on social life, encountering and movements in urban and architectural spaces (Netto, 2016).

SPACE SYNTAX THEORY AT THE CONCEPTUAL LEVEL

The departure point of space syntax theory is to build up a conceptual model that investigates the society-space relationship on the basis of the social content of spatial patterning and the spatial content of social patterning (Hillier and Hanson, 1984). This base or paradigm denies the simple notion of space-as-form and society-as-content. However, this base stresses on the social structure as inherently spatial, and the spatial structure has a fundamental social logic (Bafna, 2003). Therefore, the space syntax paradigm bridges the static narrative of society and space as discrete entities to a more dynamic discourse: each mutates and reforms the other. To tackle this dynamic discourse, two main challenges had to be addressed: identifying the nature of the spatial structure at the conceptual level, and how to describe, quantify and interpret its latent socio-spatial qualities at the methodological and operational levels. Here, the author discusses how the theory addressed the first challenge while the second challenge is discussed in the next section.

To identify the nature of spatial structure at the conceptual level, the theory mapped out two important premises. First, the spatial structure in all its forms, cities, settlements or buildings, is not a by-product of some extraneous determinative factors such as topography or climate, or by society kinship systems or mythologies (as reported by structural anthropologists) or by the biological desire of individual to identify a clear "territory" (as the Territoriality theory states) (Hillier and Hanson, 1984). Rather, the theory conceptualizes the spatial structure as a discrete entity that holds within itself both the construction rules (the genotype of the pattern) and final form (the phenotype of the structure). In this regard, the theory denies the previous approaches such as territoriality theory for Newman and cognitive studies of Lynch; arguing that these studies failed to offer an operational description and focused on the behavioral reaction of individuals towards their built environment rather than on the space as an autonomous entity (Hillier and Hanson, 1984).

Moreover, the theory argues that it is only the urban grid configuration which can influence the patterns of movement and the distribution of important buildings as reported in the natural movement theory coined by Hillier et al. (1993). As shown in Figure 1, the configuration of the urban grid (C) influences both movement patterns (M) and attractors (A). Alternatively, movement patterns and attractors have no effect on the grid configuration since the grid configuration existence precedes theirs. Thus, when agreement occurs among configuration, movement and attraction, the configuration is assigned to be the primary cause for such agreement. As a result, 'natural movement' was defined as the proportion of pedestrian movement of each street or axial line resulting from the urban grid configuration itself rather

Figure 1. The effect of urban grid configuration on movement and buildings distribution.
Source: (Hillier et al., 1993)

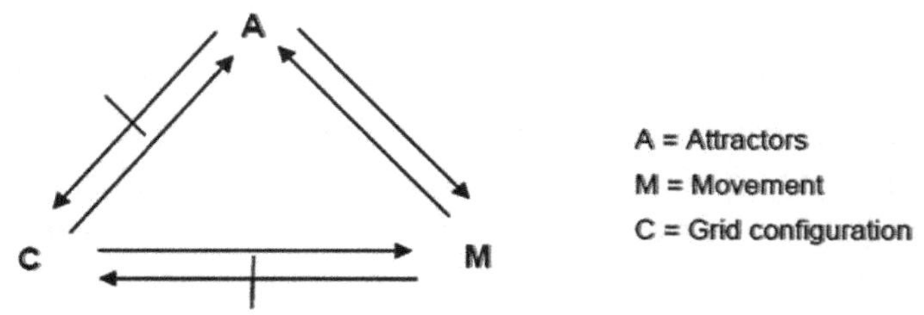

than from the presence of any attractors (Hillier, 1996b; Hillier et al., 1993). Attractors, however, can work as a multiplier for the movement and increase the effects of the urban grid configuration.

To verify this theory, several studies, (Hillier and Iida, 2005; Penn et al., 1998; Hillier, 1996b; Croxford et al., 1996; Penn and Dalton, 1994; Hillier et al., 1993), compared the distribution of movement patterns in urban areas with their integration patterns and confirmed the dominant influence of a spatial configuration on movement patterns. Furthermore, the influence of the grid configuration extends to other aspects of urban form (Hillier, 1999a; Hillier, 1996a; Hillier, 1996b). As a result of the grid configuration, mixed land-uses and high building densities tend to locate themselves on lines which have a high proportion of 'through' movement. These lines are those which have the highest accessible (choice measure) values and consequently represent the lines habitually used by people when travelling between different origins and destinations in the system. In this sense, the movement through these lines was considered as a by-product movement of different trips in the system (Hillier, 1996b). The distribution of this by-product movement is termed 'movement economy' since this distribution determines the patterns of land-uses and stimulates urban life (Hillier, 1996b). This distribution of by-product movement, which is based mainly on the grid configuration itself, is responsible for providing cities with their livability and functionality (Mohamed, 2009). On the other hand, the relative importance of different areas in a given city depends on their spatial structure features and their ability to anchor themselves in the whole structure of the city (Hillier, 1996a). If an area has a cluster of lines in the system with the highest integration, it is attracted more by-product movement than other areas. As a result, as shown above, these areas are attracted higher densities of development and uses to benefit from this by-product movement. Thus, those areas are expected to be the centers of urban functionality and livability within the system (Kubat et al., 2001).

The second is the spatial structure that should be recognized as a system. Influenced by structuralism epistemology, the theory conceptualizes the spatial structure as a global form (as a spatial configuration) produced via local rules which govern the growth of its local cells (individual spaces). In this sense, the local rules represent the genotype of the structural pattern while the global form represents the phenotype of the structure. This last premise allows the theory to have the capabilities to interpret different morphological patterns. In other words, by identifying and quantifying these rules (the genotype), it is possible to interpret different urban and architectural morphological forms (the phenotype). In this regard, the theory presents its seminal contribution: the ability to address different morphological forms. Hence, the second main challenge emerges: how to describe, quantify and interpret the spatial structure at the methodological and operational level to reveal its latent rules or genotype pattern?

SPACE SYNTAX THEORY AT THE METHODOLOGICAL AND OPERATIONAL LEVEL

The above challenge represents the departure point of the theory at the methodological and operational level. This challenge includes three concerns should be addressed: the description of spatial structure in a precise and concrete method that can reflect the unique signature of different patterns, the technique that can transform this description to measurable quantitative values, and, finally, the approach that can bridge the phenotypes qualities of the spatial structure to reveal its genotypes rules. To respond to these three concerns, respectively, the theory introduced the axial and convex map representations to describe different settlements, adopted graph theory techniques to express the axial and convex representations in a quantitative form, and depended on the topological approach to reveal the genotypes rules.

Regarding the description of the spatial structure and motivated by the desire to describe the spatial structure by its own term, the theory introduced two different, but complementary, methods to represent open spaces: axiality and convexity (Hillier and Hanson, 1984). In the axial map representation, the urban and architectural layouts are covered by the fewest and longest set of accessible and sight lines that cover the whole layout Figure 2. In the convex map representation, the urban and architectural layouts are divided into a set of the fewest and fattest convex spaces Figure 2. In this sense, axiality refers to spaces in terms of their one-dimensional organization. In other words, it associates with one-dimensional or linearly extended spaces. Alternatively, convexity refers to the spaces in terms of their two-dimensional organization or the "fattest" spaces (Hillier and Hanson, 1984; Brown, 1990). Accordingly, axiality is seen as "the maximum global extension of the system of spaces unified linearly", while convexity is seen as "the maximum local extension of the system of spaces unified two-dimensionally" (Hillier and Hanson, 1984). In social terms, axiality is related to the movement of individuals through the system and their awareness of their locations in this system. Conversely, convexity is more related to the presence of individuals in space and their awareness of the features and events that occur in this space. Thus, axiality is more related to global events in the system, while convexity is more related to the local events and informal interpersonal encounters in the system (Seamon, 2007; Seamon, 1993).

The advantage of axial and convex representations is their ability to be represented graphically and, consequently, to be measured quantitatively which address the second concern (Hillier and Hanson,

Figure 2. The axial and convex maps of Gassin, France
Source: (Hillier and Hanson, 1984)

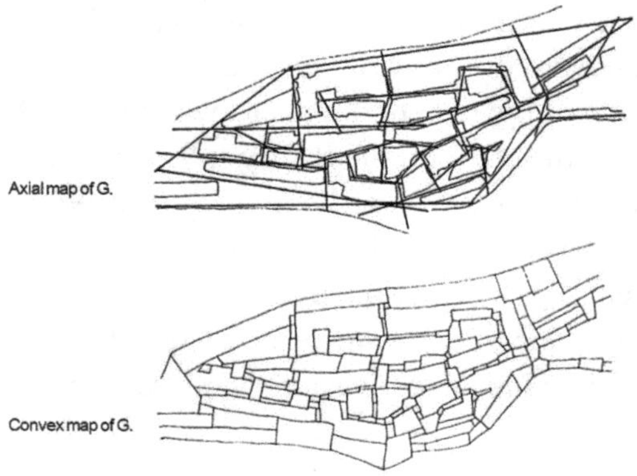

1984). Accordingly, the theory adopted graph theory techniques. At the operational level, this can be done through drawing axial and convex maps of the layouts under investigation. Then, axial and convex maps are connected into graphs. For axial maps, axial lines are represented as nodes or vertices in the morphological graph while each line in the graph represents an intersection between two axial lines (two nodes) in the spatial network (AL-Sayed et al., 2014). Similarly, in convex maps, convex spaces are represented as nodes or vertices in the graph while accessible connections between spaces are seen as lines or links between these nodes as shown in Figure 3 (Turner et al., 2005; Bafna, 2003).

To address another concern, identifying the proper approach that can bridge the phenotypes qualities of the spatial structure to reveal its genotypes rules, the analysis of the resultant graph representations adopted the topological approach. In this approach, the graph representation is analyzed according to the adjacency and conjunction relations of its nodes (whether these nodes represent axial lines or convex spaces) rather than physical distance or metric measures. Consequently, the topological analysis is a free metric analysis (Hillier and Hanson, 1984). The proposition underlying the use of topological abstraction is that topological descriptions can reveal the sociological aspects of the spatial configuration of both urban and architectural forms (Bafna, 2003). This proposition has a widespread acceptance within space syntax literature based on common-sense observation rather than theoretical arguments. However, at the methodological level, topological descriptions are more robust and reliable in analyz-

Figure 3. Urban and architectural layouts (a), axial and convex maps representation (b), axial and convex representation by a graph (c), the graph connectivity is highlighted in (d &e)
Source: (Al_Sayed et al., 2014)

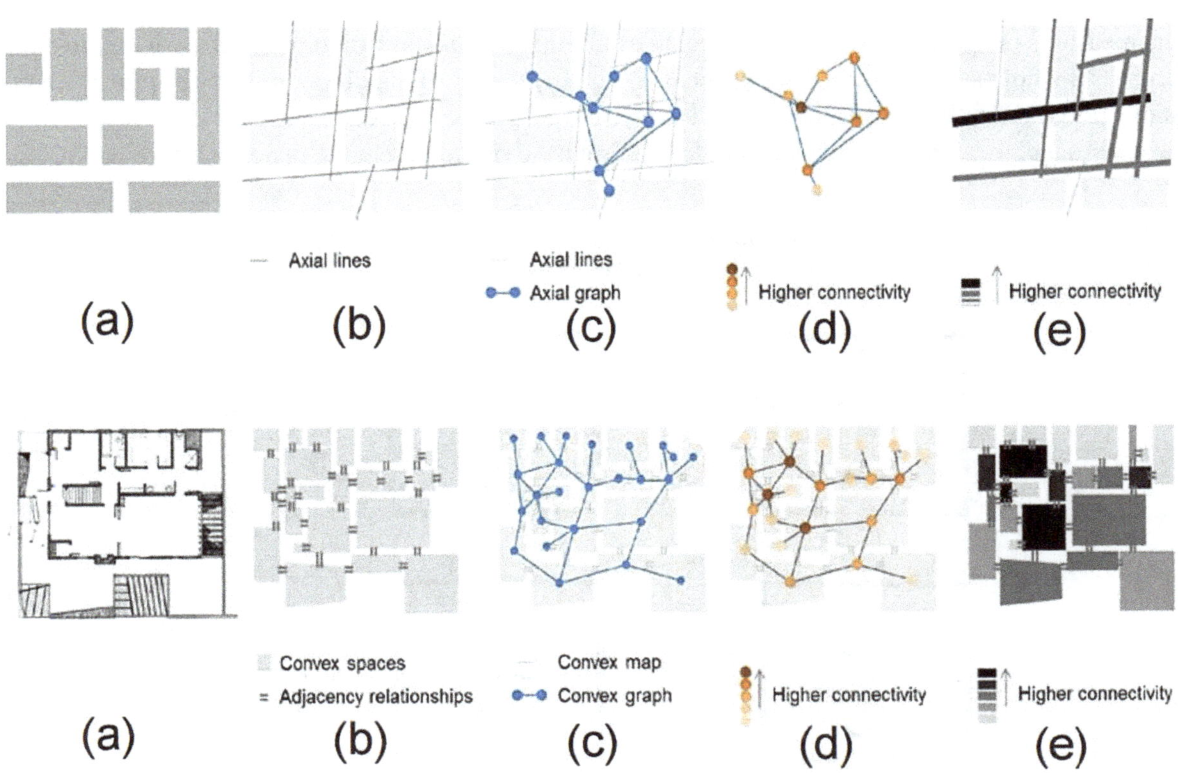

ing spatial structures. Topological descriptions can evidently be expressed in quantitative tools from applied discrete mathematics and can offer interpretations that go in accordance with the structuralism orientation of space syntax theory.

Nevertheless, the significant aspect of topological descriptions is that they: "allow researchers a systematic way of disregarding small, circumstantial, and generally sociologically irrelevant geometrical differences between configured spaces, thus enabling several different spaces to be classified together under a broader topological category" (Bafna, 2003). For instance, despite the similarity in shape and number of spaces in the four layouts in Figure (4), the topological graph analysis offers significant differences in terms of space hierarchy, controllability and segregation. In this sense, the topological approach contributes to the ultimate objective of space syntax theory: identifying the abstract rules (genotypes) underlying spatial forms (phenotypes) rather than spatial forms themselves (Hillier and Hanson, 1984).

However, recent studies (Tavroulaki et al., 2017; Turner, 2007; Hillier, et al., 2007; Hillier and Iida, 2005) have reported that using geometric and metric approaches in addition to the topological one can capture more precisely the structure-function relations in urban context such as natural movement patterns and land-use distribution. In this regard, Hillier et al. (2007) stated that measures which express the geometric and topological properties – termed as topo-geometric - of the urban context are more precisely to map out the structure-function relations at the global scale (termed as the foreground networked). Meanwhile, the measurements express metric properties (physical distance) of the urban context are more accurate in capturing structure-function relations at the local scale (termed as the background networked).

MAIN MEASURES OF SPACE SYNTAX THEORY AND THEIR MEANINGS

The measures of space syntax theory focus on examining the spatial structure of urban or architectural forms as a whole system; examining its local and global features in addition to their reciprocal effects. At the global scale, the measures of global integration and global choice represent the core measures. On the other hand, local integration, local choice and connectivity represent the major measures at the local

Figure 4. The role of the topological graph representation in revealing the inherited socio-spatial features (http://homes.chass.utoronto.ca/~banning/ANT%20417/417review.htm)

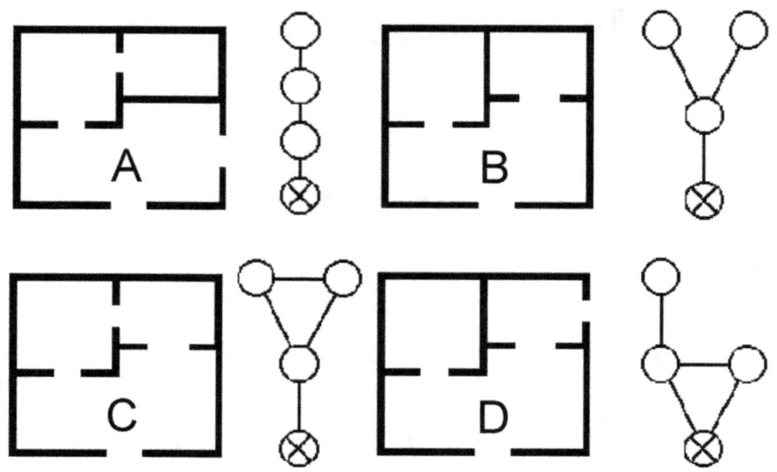

scale. Alternatively, the intelligibility measure, which evaluates the degree of legibility of the system under investigation, represents the main measure which examines the reciprocal relationship between local and global features of urban and architectural forms. In the following section, the notion of depth is introduced, because of its importance, in addition to a brief demonstration of four important measures: connectivity, integration, choice and intelligibility.

Depth

The depth of one space from another is defined by the number of intervening spaces between these two spaces (Bafna, 2003). The depth measure achieves its importance when comparing two or more spaces in relation to a third space (termed the root space) usually represent the entrance in architectural buildings Figure 4. This importance increases when this comparison is conducted between the whole spatial structure of architectural and urban forms. For example, as shown above in Figure (4), there are two main types of spatial configuration of architectural and urban forms: deep and shallow structures. Deep structures, such as A in Figure 4, are distinguished by passing through a sequential chain of spaces (commonly form a linear structure) (Dovey, 1996). On the other hand, shallow structures, such as D in Figure 4, offer less control on movements and social encounter in addition to showing less hierarchy. Alternatively, at the individual scale, spaces B and C, as shown in Figure 7, have different depth values in relation to C. in this regard, space B, which has a high depth value, shows a high degree of privacy and segregation.

Connectivity

Connectivity is a local measure for each node (axial line or convex space) in the graph representation. Connectivity value for a specific node represents the number of nodes that are directly connected to this node (Bafna, 2003). Connectivity measure, in addition to local integration (see below), is commonly used to reveal the syntactic features of axial lines or convex spaces in their local context or their adjacent surrounding. Highly connected spaces are expected to catch the majority of movement volume in their local context, while the lowest ones are expected to have low volume of movement and a high degree of privacy. Also, the measure of connectivity is used, in addition to global integration, to identify the degree of legibility of architectural and urban forms (see intelligibility measure below).

Integration

The measure of integration or relative asymmetry (RA) means the average depth of a node, which can be a convex space or an axial line, to all other nodes in the system. Integration, in this sense, represents a global measure which reflects the relationship between a given space and all other spaces in the system (Klarqvist, 1993). RA can be calculated from the following equation:

$RA = 2\left(MD - 1\right) / \left(K - 2\right)$ (Hillier and Hanson, 1984)

Where MD is the mean depth of space from other spaces and K is the number of spaces in the system. The resulting value is between 0 and 1. Spaces which have low integration values are likely to be integrated with other spaces in the system, while spaces with high values are likely to be segregated from other spaces in the system[1] (Hillier and Hanson, 1984). The integrated spaces are those spaces which have less depth from other spaces in the system, while segregated spaces are those which have more depth from other spaces in the system (Hillier, 1996a). When the integration value of a given space is calculated in relation to all spaces in the system, it is termed 'global integration' (R=n). The term 'local integration' applies when the integration value of this space is calculated in relation to all spaces located three steps away from it (R=3) (Hillier, 1996a). The distribution of the most integrated spaces in the system is termed the 'integration core'. This distribution can be mapped as a specific proportion, such as the top 5% or top 10%, of the most integrated spaces in the system (Hillier and Hanson, 1984). Therefore, the integration core identifies the most accessible spaces in the system and affects the distribution of movement, land-uses and the encounters between inhabitants and strangers (Hillier, 1989). The shape of the integration core in a given settlement represents a significant feature of its spatial structure (Klarqvist, 1993; Hillier, 1989; Hillier and Hanson, 1984).

Regarding types of movement, the integration measure is linked to the "to-movement" type rather than "through-movement" one. Like this, integrated spaces represent the milieus of important destinations with the spatial structure of urban layouts. Also, integration measures are the most appropriate predictor for the patterns of both pedestrian and vehicular movements. Highly integrated lines are those which have high rates of movement and vice versa (Hillier, 1996b). Furthermore, local integration (R=3) is found to be more associated with pedestrian movement, while global integration (R=n) is more associated with vehicular movement (Hillier, 1996a; Hillier, 1996b). However, in several studies, vehicular rather than pedestrian movement showed a higher and more constant correlation with integration measures (Dawson, 2003; Penn, 2003). Alternatively, the net road width was reported as a secondary factor affecting the vehicular movement on the scale of the main road network (Krishnamurthy & Arasan, 2012; Tang et al., 2012; Penn et al., 1998; Hillier, 1996b).

Choice

The 'choice' is a dynamic measure that describes the movement flow through urban space or a group of them. Figure 5 shows that a space has a high choice value when a lot of shortest paths, connecting all spaces to all spaces of a system, pass through this space (Klarqvist, 1993). The choice is linked to the "through-movement" type and, consequently, it is commonly used to measure the degree of accessibility. The measure of choice experienced several developments to address not only topological properties of the urban network but also both geometric and metric ones (Hillier, et al., 2007; Hillier and Iida, 2005). Furthermore, it was reported that the measure of choice that geometrically (least angle change) or topologically (fewest turns) weighted is a good predictor of movement (Hillier, et al., 2007).

Intelligibility

Another important measure is intelligibility. Intelligibility is a characteristic of the whole structure of a settlement. It refers to the ability to predict the global spatial structure through the local properties of its individual spaces or "[t]he degree to which what can be seen and experienced locally in the system allows the large-scale system to be learnt without conscious effort" (Hillier, 1996a). In this sense, in an intelligible system, well-connected spaces are well-integrated spaces, while in an unintelligible system, well-connected spaces are not well-integrated (Hillier, 1996a). Intelligibility can be measured by examining the correlation between connectivity and global integration. The resulting correlation value is between 0 and 1. High correlation values indicate that the system is likely to be intelligible and vice versa.

ANALYTICAL TYPES OF SPACE SYNTAX THEORY

Based on the paradigm of convexity and axiality, topological and graph theory approached in addition to metric and angular ones, and the core measures of the theory (integration, choice and connectivity), four types of analysis have been produced to address the spatial configuration and visual qualities of both urban and architectural forms. These types included convex analysis, isovist analysis and visibility analysis (VGA), axial analysis, and segment analysis. These analyses vary in their importance of application in relation to the architectural and urban disciplines as shown in Figure (6) whereas convex analysis has more applications in the architecture realm, while segment analysis is commonly applied in urban contexts. The following section discusses the aim and importance of each type.

Convex Analysis

Convex analysis is mainly conducted in the case of investigating the arrangement of buildings or urban layout and their generative types or rules (Bafna, 2003). The analysis depends on the convex map representation in which the spatial complexity of architectural and urban layouts is reduced to the fewest and fattest convex spaces. Then, as mentioned earlier, each convex space is represented by a node in graph representation and each permeable connection between two spaces is represented by an edge (Figure 7). This generic graph representation reveals the spatial relationships between the convex spaces of the whole layout in quantitative measures such as connectivity, integration and choice to identify the most connected and accessible spaces within the layout. However, a specific graph representation, termed as justified graph or j-graph, has been developed to reveal more features of the architectural and urban spatial structures. In this, j-graph, convex spaces of the spatial structure are ordered from a specific space, commonly the entrance, termed the root space. All spaces that are one step away from the root space are put in the first level, while all spaces that are two steps away from the root space are put in the

Figure 5. Choice measure illustration
Source: http://www.ia.arch.ethz.ch/wp-content/uploads/2016/10/slides.pdf

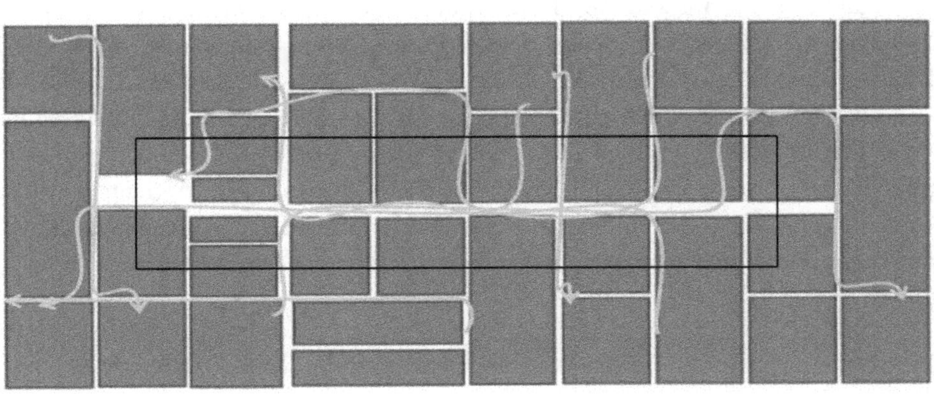

Figure 6. The relation between space syntax analyses and Architecture and urban disciplines.
Source: the author

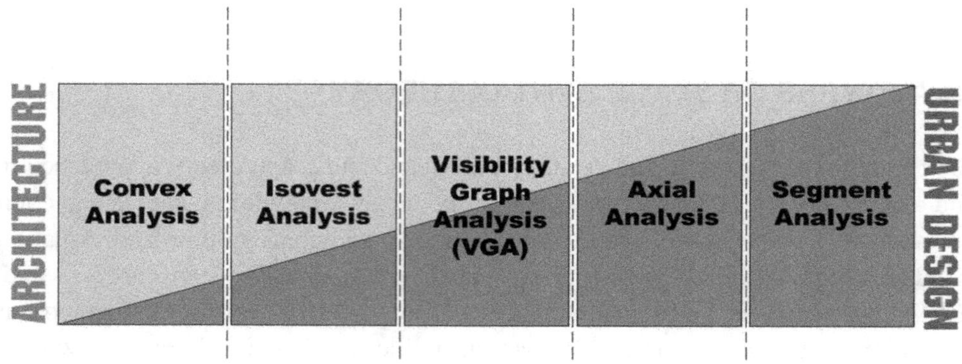

Figure 7. Convex spaces and their associated graphs.
Source: (Dawes and Ostwald, 2013)

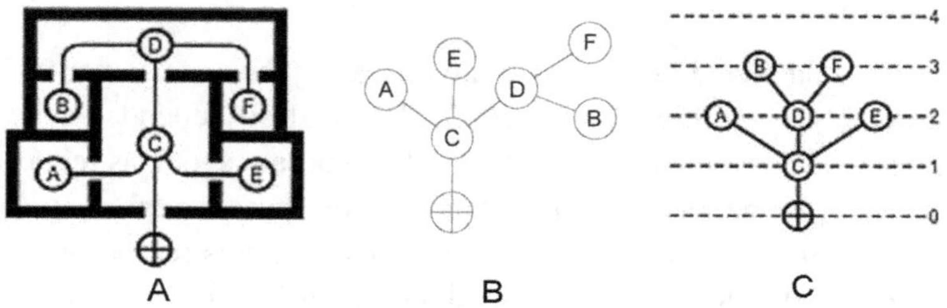

second level and so on. J-graph importance lies in addressing the notion of depth as mentioned earlier. j-graph is commonly used to identify the deep vs. shallow structures on one hand, and to reveal the social hierarchy within architectural and urban layouts on the other hand.

In this sense, convex analysis, thru using both generic graph representation and j-graph, aims to reveal the socio-spatial structure of architectural and urban layouts: the genotype of architectural and urban forms. This genotype represents the essential rules and inherited properties that govern the resultant architectural and urban forms and their future transformations. In this regard, Dovey describes that genotype as, "Genotypes are not formally types as often debated in architecture, but institutionally and epistemologically embedded clusters of spatial segments together with syntactic rules of sequence and adjacency. In a given cultural context, buildings types such as houses, schools, offices, and so on are generated from a limited number of spatial genotypes" (Dovey, 1996).

At the application and research level, convex analysis has been used to identify the socio-spatial genotypes of residential and public buildings in several cultural contexts (Bai & Nasu, 2018; Mohamed, 2014), and to evaluate the functional and circulation efficiency of several buildings such as hospitals (Cocina, 2017).

Isovist Analysis and Visibility Graph Analysis (VGA)

The Isovist term refers to the area of the spatial environment directly visible from a specific point within the space as shown in Figure 8. The term was first introduced by Benedikt in 1979. Isovist analysis aims to examine the visual field properties, in terms of area and perimeter, which can be produced from a specific point in the architectural and urban layouts. Thus, there is a possibility to quantify the space and identify what our perception of space might be and the potential of its use (Turner et al., 2001). Isovist can also be constructed from an area (such as convex space) to show the visible field from this area. Moreover, a set of isovists can be constructed from a path at regular intervals to display how the user who walk through this path experience space from an origin point, such as entrance, to a particular destination.

At the application level, isovist analysis can be used to investigate the visual experience the users of architectural and urban form should have, to compare the visual fields properties among main squares in cities or main spaces within the buildings, and to evaluate, visually, the location of main service or public transportation hubs at urban scale, and the location of main functions and circulation cores at the architectural scale.

To examine the interrelation between isovist fields constructed from all points within a specific spatial layout, space syntax research developed the isovist analysis to a more advanced analysis known as visibility graph analysis or VGA (Turner et al., 2001). In this analysis, each point within the layout, which has its own isovist value according to its visibility field, is represented as a node in the graph representation. Hence, common quantitative measures such as connectivity and integration can be calculated and visual relationships between different points (nodes) in the layout can be mapped Figure (9). The significance of visibility graph analysis lies in its relation to the manifestation of spatial perception such as way-finding, movement, and space utilization (AL-Sayed, 2014).

Figure 8. An Isovist field, incorporating the visible area from a generating location.
Source: (Turner et al., 2001)

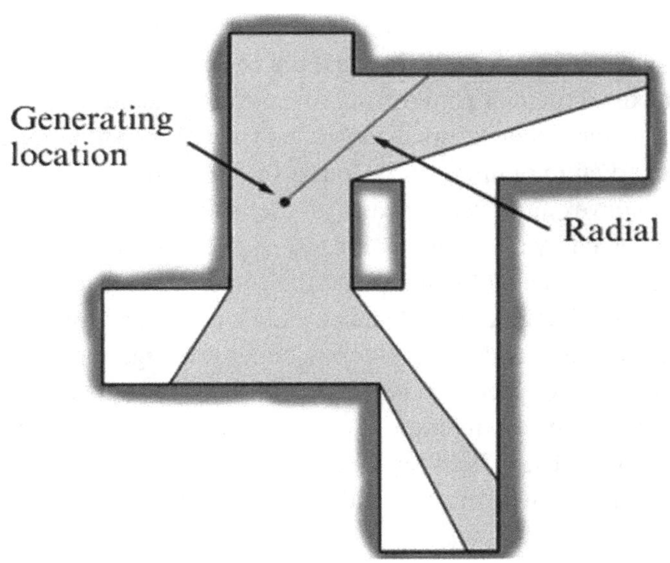

Figure 9. The visibility graph analysis of Effat University Library (visual connectivity measure).
Source: the author

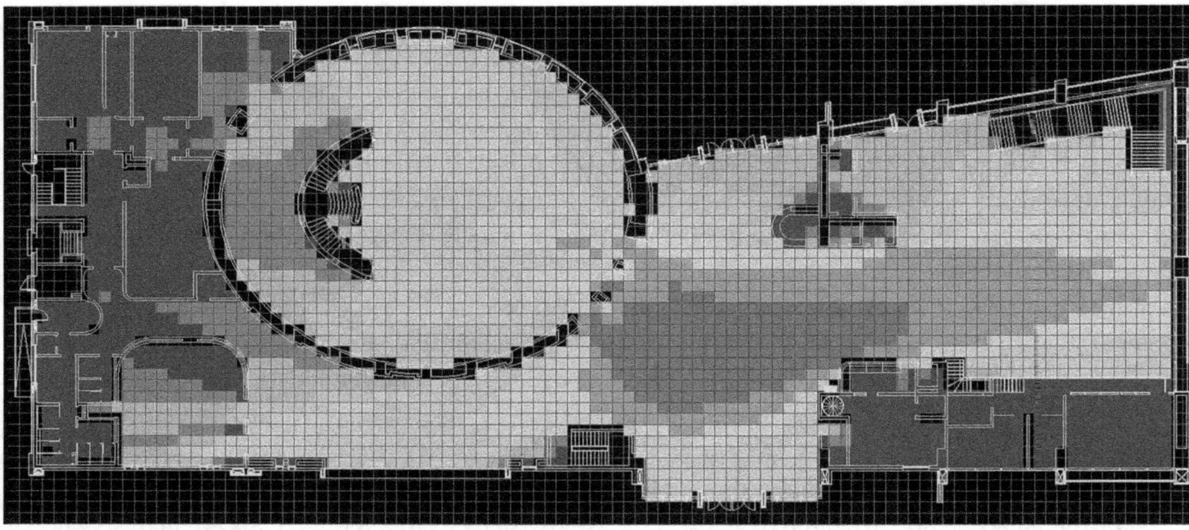

In application and academic studies, visibility graph analysis (VGA) has been used to examine its validity to predict pedestrian movement flow at the urban scale. For example, two important studies reported that, through the comparison between axial map analysis and visibility graph analysis, VGA shows a significantly higher correlation with the pedestrian movement in urban contexts than axial analysis measures (Desyllas & Duxbury, 2001; Turner and Penn, 1999). At the architectural scale, several empirical studies reported that visibility analyses are helpful to forecast spatial behavior and affective qualities of indoor spaces (Franz, von der Heyde & Bülthoff, 2005; Wiener & Franz, 2005).

Axial Analysis

The axial analysis represents the most important and widespread analysis in space syntax literature and studies. The analysis is commonly used to understand the behavioral characteristics and encounter patterns within spatial contexts, to predict patterns of movements, to identify the strategic destinations or spaces in urban and architectural context (most integrated spaces), and to highlight the most accessible spaces within the spatial structure of urban and buildings layouts (spaces that have high choice values) (Bafna, 2003). Similar to convex analysis, axial analysis depends on axial maps representation in which the spatial layout is represented by the minimal, longest and accessible set of axial lines. In graph representation, axial lines are represented by nodes and their intersections represented by links or edges. The measures of connectivity, integration and choice can be calculated for axial analysis to reveal the inherited features of the urban and architectural forms under investigation. In addition to the common syntactic measures introduced by space syntax theory, the axial analysis offers an important measure called synergy. Synergy examines the relation of different parts of a city to its whole structure via a scatter plot examining the correlation between local integration (R=3) and global integration (R=n). Through this analysis, the most sensitive areas within the city structure can be identified and, accordingly, future development to enhance the spatial structure performance can be achieved with minimum interventions, in addition to the possibilities to evaluate different future development plans (Mohamed and Brown, 2009).

Segment Analysis

Segment analysis represents an important development in space syntax realm at both the conceptual and the methodological levels (Tavroulaki et al., 2017). At the conceptual level, segment analysis bridges the topological approach to include geometric configuration of urban and architectural space (metric and angular approaches). In this sense, the theory responds to the concerns of axial map representation (Hillier and Iida, 2005; Hillier, 1999a; Hillier, 1999b; Turner, 2001; Steadman, 2004; Dalton, 2001) and the adoption of topological approach only to reveal the intrinsic attribute of architectural and urban forms, overlooking the geometrical features of urban and architectural configuration in terms of metric and angular aspects. At the methodological level, segment analysis offered the use of Road-center-line maps, largely available in GIS-based segment maps, in space syntax analysis allowing the integration between space syntax analysis and GIS software (Turner, 2007; Dalton et al., 2003). This opens new possibilities to compare between syntactic measures, such as connectivity, integration, and choice, and new variables such as density, buildings height, land use types, and street width.

At the operational level, segment analysis is based on segment map representation in which an axial map is broken into segments. These segments represent the inter-junction lines between points of intersection in axial map. Thus, the segment map increases the representation resolution of street networks that, recently, bring the segment analysis to the fore in the space syntax studies. The analysis of these segments considers not only the topological connections but also consider their metric and angular connections. In all types, each segment is represented by a node in graph representation and its connections by edges or links. Subsequently, syntactic measures such as connectivity, integration and choice can be calculated. However, these measures vary in segments analysis based on the three aspects of analysis: topological, metric and angular. The integration measure has three forms: topological integration, metric integration and angular integration.

Moreover, the analysis can produce mixed measures such as angular integration with metric radius and angular depth with topological radius. In this regard, and according to several academic studies, the most critical measures of segment analysis are integration and choice measures that adopt angular depth with metric radius. These two measures can reveal the degree of accessibility of urban layouts and identify "to-movement" and "through-movement" potentials. Furthermore, segment analysis has been considered a powerful predict tool for social and economic activity patterns in an urban context (Al_Sayed et al., 2014). At the application and research level, segment analysis has been used to capture the vehicular and pedestrian movement patterns in urban layouts (Turner, 2007; Paul, 2012).

IMPACTS OF SPACE SYNTAX THEORY ON URBAN DESIGN REALM

This section discusses and outlines the impacts of space syntax theory and its contribution to the urban design realm. The first contribution was restoring the vital role of spatial structure, terms as spatial configuration, within urban contexts. This was achieved through the theory techniques (convexity and axiality) which allowed for describing open spaces via their own terms to reveal their local and global features. In this sense, the urban designers and researchers are capable to examine the spatial structure qualities as an isolated variable and trace the effect of these qualities on other aspects of urban context (social, functional, visual and perceptual). The second contribution of the theory is its capability to measure intangible qualities presented in the urban design literature such as legibility and accessibility quantitatively. Furthermore, it allows tackling the reciprocal effect between local and global spatial qualities of the urban context and tracing the effect of this mutual impact on the social performance and functionality of the urban settlement as a whole (via intelligibility and synergy measures and integration core shape). The third contribution is the ability to examine the reciprocal effect between spatial structure features and other dimensions of urban design. It allows examining, in a concrete and measurable methodology, the reciprocal effect between spatial configuration (spatial dimension) and other dimensions of urban design: functional, social, visual, perceptual and morphological. The following section discusses the key finding regarding these reciprocal effects between spatial configurations and perceptual and social dimensions of urban design, as the effect of spatial configuration on functional dimension was addressed in section one where the movement economy theory and land use distribution were discussed

Regarding perceptual dimensions, several studies have focused on the relation between the urban grid configuration and the cognition of the built environment (Hölscher et al., 2007; Long et al., 2007; Yun & Kim, 2007; Kim & Penn, 2004; Penn, 2003; Zimring & Dalton, 2003; Kim, 2001). Some studies compare the integration values of Hampstead Garden Suburb with the integration values produced from the digitized sketch maps of the same area that had been drawn by its residents (Kim, 2001). Kim (2001) has proved that there is a significant correlation between the integration values in reality and those of the sketch maps. The comparison of local integration (R=3) of the built environment and the sketch maps, however, shows a higher correlation than the comparison of their global integration. This shows that local integration is "a good predictor of the cognitive representations of spatial configuration". Also, the results showed that the local integration is more associated than the global measures with the frequency of elements and streets appearing in the sketch maps (Kim, 2001). In other words, the frequency of elements drawn in sketch maps is expected to show a higher correlation with the local integration rather than the global one. Also, several studies reported that the top 10% of the most integrated axial lines (the integration core) of a given area matches the key paths appearing in the sketch maps of this area (Montello, 2007; Dalton & Bafna, 2003).

Kim (2001) reported that the intelligibility of the built environment has a decisive role in its cognition. The more intelligibility the built environment, the more accurate and complex its cognition representations are and vice versa. This role of intelligibility was confirmed later by Penn (2003). However, one of the key findings of Penn is that the cognition of the built environment depends on topological relations rather than metric ones (2003). This concurs with another study which reported that it is the geometrical and topological properties which affect the cognition of the built environment (Hillier & Iida, 2005).

Regarding the social dimension and based on the original concept of space syntax theory, a number of studies focused on the relation between the urban grid configuration and its corresponding social patterns. Hillier argues that it is the grid configuration which generates the opportunities for social interactions, encounters between individuals, and community life or what he termed as 'the virtual community' (1996a). Also, Hillier reports that the comparison between syntactic measures and the movement patterns of a given area reflects its virtual community (1989).

On the other hand, several studies focused on the relation between the integration and movement patterns and the crime patterns (Lopez & Nes, 2007; Farooq, 2007; Şahbaz & Hillier, 2007; Hillier, 1999). These studies reported that well integrated and busy areas and spaces are safer than segregated

& empty ones. In this sense, space syntax studies support Jane Jacobs (1961) and Zelinka and Brennan (2001) who argued for connected and mixed-use streets to achieve safe environments (Hillier and Şahbaz, 2008). Several studies reported that attaining safety in urban areas is an inter-dependency issue, which involves other aspects besides spatial factors, such as density, dwelling type and socio-economic variables (Hillier & Şahbaz, 2008; Farooq, 2007; Hillier, 2004). In addition, Hillier and Şahbaz reported that, contrary to the claims of traditional studies, not all crimes are associated with segregated streets. In this, street robbery is found to be linked with the most integrated streets (2008).

CONCLUSION

This paper concludes by arguing that space syntax is one of the theories that has been so controversial and has evoked different responses from scholars and researchers ranging from full acceptance to prior rejection. Actually, although the theory cannot cover all aspects and relationship in terms of space-society relations, space syntax has addressed unexplored areas in terms of socio-spatial relations and opened new prospects to analyze, investigate and evaluate a wide range of morphological forms at both urban and architectural scale. Space syntax theory has produced concrete and operational methodology to represent, quantify, analyze and interpret the built environment based on its spatial configuration. It offers a wide range of analytical types to addresses several objectives such as identifying the socio-spatial genotypes of urban and architectural form (convex analysis), recognizing users' perception of the built environment and their visual experience (isovist and visibility graph analysis), capturing the patterns of movement and encounter in spatial settings, evaluating the degree of accessibility and intelligibility of architectural and urban layouts, and identifying the most vibrant areas or spaces within cities and buildings (axial and segment analysis). Also, it facilitates the connections between the analysis of the urban grid and other aspects of urban form (functional, perceptual and social) accounting on the social content of space and the spatial content of society via a concrete and objective methodology.

The concluded remarks are in line with Penn (2003) through describing space syntax as a social theory that tends to look at how an individual is constituted by all others in a group or society. Space syntax as an analytic methodology looks at how space is constituted by all other spaces in a configuration. From both theoretical and methodological perspectives, space syntax takes an objective rather than subjective epistemological stance. This provides a good starting point for discussion and further research.

To sum up, Space Syntax substantially supports the multiple urban design stages. For instance, in the pre-design stage, the space syntax offers concrete measures, such as integration, choice, and intelligibility, to evaluate the existing urban context and to select the appropriate locations for different land uses. In the proposal development stage, space syntax offers different techniques, such as axial, convex and visibility analyses, all used to evaluate design alternatives and to conduct a rational alternatives comparison. In the post-design stage, the theory offers different measures and techniques to evaluate the final design proposal, in a digital setting, to test its efficiency and effectiveness prior to the implementation stage.

REFERENCES

Al Sayed, K., Turner, A., Hillier, B., Iida, S., & Penn, A. (2014). *Space Syntax Methodology* (4th ed.). London: Bartlett School of Architecture, UCL.

Bafna, S. (2003). Space syntax: A brief introduction to its logic and analytical techniques. *Environment and Behavior*, *35*(1), 17–29. doi:10.1177/0013916502238863

Bai, L., & Nasu, S. (2018). Common Space Spatial Layout Transition in Japanese Nursing Home - By Space Syntax point of view. *Asian Journal of Environment-Behaviour Studies*, *3*(7), 19–30. doi:10.21834/aje-bs.v3i7.256

Benedikt, M. L. (1979). To take hold of space: Isovists and isovist fields. *Environment & Planning B*, *6*(1), 47–65. doi:10.1068/b060047

Brown, F. E. (1990). Comment on Chapman: Some Cautionary Notes on the Application of Spatial Measures to Prehistoric Settlements. In The Social Archaeology of Houses. Edinburgh, UK: Edinburgh University Press.

Cocina, G. Giulia (2017). Responding to User Cultural Needs in Hospitals with the Support of SPACE Syntax Analysis. In *Proceedings of the 11th Space Syntax Symposium* (pp. 64.1-64.10). Lisbon, Portugal: Instituto Superior Técnico, Departamento de Engenharia Civil, Arquitetura e Georrecursos.

Croxford, B., Penn, A., & Hillier, B. (1996). Spatial Distribution of Urban Pollution: Civilizing Urban Traffic. *The Science of the Total Environment*, *189-190*, 3–9. doi:10.1016/0048-9697(96)05184-4

Dalton, N. (2001). Fractional configurational analysis and a solution to the Manhattan problem. *Proceedings 3rd International Space Syntax Symposium.*

Dalton, N., Peponis, J., & Dalton Conroy, R. (2003). To tame a TIGER, one has to know its nature 1: Extending weighted angular integration analysis to the description of GIS road-centerline data for large scale urban analysis. In *Proceedings of the 4th International Space Syntax Symposium.* London: University College London.

Dalton, R. C., & Bafna, S. (2003). The Syntactical Image of the City: A Reciprocal Definition of Spatial Elements and Spatial Syntaxes. *Proceedings of the 4th International Space Syntax Symposium.*

Dawes, M., & Ostwald, M. J. (2013). Precise Locations in Space: An Alternative Approach to Space Syntax Analysis using Intersection Points. *Architecture Research*, *3*(1), 1–11.

Dawson, P. C. (2003). Analysing the Effects of Spatial Configuration on Human Movement and Social Interaction in Canadian Arctic Communities. *Proceedings of the 4th international space syntax symposium.*

Desyllas, J., & Duxbury, E. (2001). Axial maps and visibility graph analysis. In *Proceedings of the 3rd International Symposium on Space Syntax* (pp. 27.1–27.13). Atlanta, GA: Georgia Institute of Technology.

Dovey, K. (1996). Social Theory and Space Syntax: Exploring Power and Spatial Programming. In *Evolving Environmental Ideals - Changing Way of Life, Values and Design Practices: IAPS 14 Conference Proceedings* (pp. 149-158). Stockholm, Sweden: Royal Institute of Technology.

Farooq, A. (2007). Social Malice by Housing Type. In *Proceedings of the 6th International Space Syntax Symposium*. Istanbul: ITU Faculty of Architecture.

Franz, G., Von der Heyde, M., & Bülthoff, H. H. (2005). Predicting experiential qualities of architecture by its spatial properties. In B. Martens & A. G. Keul (Eds.), *Designing Social Innovation: Planning, Building, Evaluating* (pp. 157–166). Cambridge, MA: Hogrefe and Huber.

Franz, G., & Wiener, J. M. (2005). Exploring isovist-based correlates of spatial behavior and experience. In A. van Nes (Ed.), *Proceedings of the 5th International Space Syntax Symposium*. Delft, The Netherlands: TU Delft Press.

Hillier, B. (1989). The Architecture of the Urban Object. *Ekistics*, *56*, 5–21.

Hillier, B. (1996a). *Space Is the Machine: A Configurational Theory of Architecture*. Cambridge, UK: Cambridge University Press.

Hillier, B. (1996b). Cities as Movement Economy. *URBAN DESIGN International*, *1*(1), 41–60. doi:10.1057/udi.1996.5

Hillier, B. (1999). The Common Language of Space: A Way of Looking at the Social, Economic and Environmental Functioning of Cities on a Common Basis. *Journal of Environmental Sciences (China)*, *11*, 344–349.

Hillier, B. (1999b). Centrality as a Process: Accounting for Attraction Inequalities in Deformed Grid. *URBAN DESIGN International*, *4*(3-4), 107–127. doi:10.1057/udi.1999.19

Hillier, B., & Hanson, J. (1984). *The Social Logic of Space*. Cambridge, UK: Cambridge University Press. doi:10.1017/CBO9780511597237

Hillier, B., Hanson, J., & Graham, H. (1987). Ideas Are in Things: An Application of Space Syntax Method to Discovering House Genotypes. *Environment and Planning. B, Planning & Design*, *14*(4), 363–385. doi:10.1068/b140363

Hillier, B., & Iida, S. (2005). Network and Psychological Effects in Urban Movement. In A. G. Cohn & D. M. Mark (Eds.), Lecture Notes in Computer Science: Vol. 3693. *Spatial Information Theory. COSIT 2005*. Berlin: Springer. doi:10.1007/11556114_30

Hillier, B., Penn, A., Hanson, J., Grajewski, T., & Xu, J. (1993). Natural Movement: Or, Configuration and Attraction in Urban Pedestrian Movement. *Environment and Planning. B, Planning & Design*, *20*(1), 29–66. doi:10.1068/b200029

Hillier, B., & Sahbaz, O. (2008). *An Evidence-Based Approach to Crime and Urban Design or, Can We Have Vitality, Sustainability and Security All at Once?* Retrieved from https://spacesyntax.com/wp-content/uploads/2011/11/Hillier-Sahbaz_An-evidence-based-approach_010408.pdf

Hillier, B., Turner, A., Yang, T., & Park, H. T. (2007). Metric and Topo-Geometric Properties of Urban Street Network: Some Convergences, Divergences and New Results. In *Proceedings of the 6th International Space Syntax Symposium*. Istanbul: ITU Faculty of Architecture.

Hölscher, C., Dalton, R. C., & Turner, A. (2007). *Proceedings of the Workshop on Spatial Cognition: Space Syntax and Spatial Cognition*. Bremen: Universität Bremen.

Jacobs, J. (1961). *The Death and Life of Great American Cities*. New York: Random House.

Kim, Y. O. (2001). The Role of Spatial Configuration in Spatial Cognition. *Proceedings of the 3rd International Space Syntax Symposium*.

Kim, Y. O., & Penn, A. (2004). Linking the Spatial Syntax of Cognitive Maps to the Spatial Syntax of the Environment. *Environment and Behavior*, *36*(4), 483–504. doi:10.1177/0013916503261384

Klarqvist, B. (1993). Space Syntax Glossary. *Nordisk Arkitekturforskning*, *2*, 11–12.

Krishnamurthy, K., & Arasan, V. T. (2012). Effect of road width and traffic volume on vehicular interactions in heterogeneous traffic. *Journal of Advanced Transportation*. doi:10.1002/atr.196

Kubat, A. S., Istek, C., & Asami, Y. (2001). Characterization of the Street Networks in the Traditional Turkish Urban Form. *Environment and Planning. B, Planning & Design*, *28*(5), 777–795. doi:10.1068/b2718

Long, Y., Baran, P. K., & Moore, R. (2007). The Role of Space Syntax in Spatial Cognition: Evidence from Urban China. In *Proceedings of the 6th International Space Syntax Symposium*. Istanbul: ITU Faculty of Architecture.

Lopez, M. J. J., & Nes, A. V. (2007). Space and Crime in Dutch Built Environments: Macro and Micro Scale Spatial Conditions for Residential Burglaries and Thefts from Cars. In *Proceedings of the 6th International Space Syntax Symposium*. Istanbul: ITU Faculty of Architecture.

Mohamed, A. M. R. M. (2014). *The Spatial Configuration of Formal Housing Types in New Assiut City, Egypt. In Environment and Design*. Istanbul, Turkey: Re-Thinking Residential Environments.

Mohamed, A. M. R. M., & Brown, F. (2009). The Urban Grid and its Consequences Assiut: An Egyptian Case-Study. In *Proceedings of the 7th International Space Syntax Symposium*. Stockholm: KTH.

Montello, D. R. (2007). The Contribution of Space Syntax to a Comprehensive Theory of Environmental Psychology. In *Proceedings of the 6th International Space Syntax Symposium*. Istanbul: ITU Faculty of Architecture.

Netto, V. (2016). 'What is space syntax not?' Reflections on space syntax as sociospatial theory. *Urban Des Int.*, *21*(1), 25–40. doi:10.1057/udi.2015.21

Paul, A. (2012). Unit-segment Analysis: A Space Syntax Approach to Capturing Vehicular Travel Behavior Emulating Configurational Properties of Roadway Structures. *Journal of Transport & Infrastructure Research*, *12*(3), 275–290.

Penn, A. (2003). Space Syntax and Spatial Cognition or, Why the Axial Line? *Environment and Behavior*, *35*(1), 30–65. doi:10.1177/0013916502238864

Penn, A., & Dalton, N. (1994). The Architecture of Society: Stochastic Simulation of Urban Movement. In Simulating Societies: The Computer Simulation of Social Phenomena. London: UCL Press.

Penn, A., Hillier, B., Banister, D., & Xu, J. (1998). Configurational Modelling of Urban Movement Networks. *Environment and Planning. B, Planning & Design*, *25*(1), 59–84. doi:10.1068/b250059

Şahbaz, Ö., & Hillier, B. (2007). The Story of the Crime: Functional, Temporal and Spatial Tendencies in Street Robbery. In *Proceedings of the 6th International Space Syntax Symposium*. Istanbul: ITU Faculty of Architecture.

Seamon, D. (1993). The Life of the Place: A Phenomenological Commentary on Bill Hillier's Theory of Space Syntax. *Environmental & Architectural Phenomenology Newsletter*, *4*, 10–19.

Seamon, D. (2007). A Lived Hermetic of People and Place: Phenomenology and Space Syntax. In *Proceedings of the 6th International Space Syntax Symposium*. Istanbul: ITU Faculty of Architecture

Steadman, P. (2004). Developments in space syntax. *Environment and Planning. B, Planning & Design*, *31*(4), 483–486. doi:10.1068/b3104ed

Tang, T. Q., Yang, X. B., Wu, Y. H., Caccetta, L., & Huang, H. J. (2012). A multilane traffic flow model with lane width and the number of lanes. *Chinese Physics Letters*, *29*(9), 098903-1, 098903–098904. doi:10.1088/0256-307X/29/9/098903

Tavroulaki, I., Marcus, L., & Berghauser Pont, M. (2017). Representations of street networks in space syntax towards flexible maps and multiple graphs. *Proceeding of the 11th International Space Syntax Symposium*.

Turner, A. (2001). Angular analysis. *Proceedings of the 3rd international symposium on space syntax*, 30-1.

Turner, A. (2007). From axial to road-centre lines: A new representation for space syntax and a new model of route choice for transport network analysis. *Environment and Planning. B, Planning & Design*, *34*(3), 539–555. doi:10.1068/b32067

Turner, A., Doxa, M., O'Sullivan, D., & Penn, A. (2001). From isovists to visibility graphs: A methodology for the analysis of architectural space. *Environment and Planning. B, Planning & Design*, *28*(1), 103–121. doi:10.1068/b2684

Turner, A., & Penn, A. (1999). Making Isovists Syntactic: Isovist Integration Analysis. *Proceeding of the second space syntax symposium*.

Turner, A., Penn, A., & Hillier, B. (2005). An Algorithmic Definition of the Axial Map. *Environment and Planning. B, Planning & Design*, *32*(3), 425–444. doi:10.1068/b31097

Yun, Y. W., & Kim, Y. O. (2007). The Effect of Depth and Distance in Spatial Cognition. In *Proceedings of the 6th International Space Syntax Symposium*. Istanbul: ITU Faculty of Architecture.

Zelinka, A., & Brennan, D. (2001). *Safescape: Creating Safer, More Livable Communities through Planning and Design*. Chicago: Planners Press.

Zimring, C., & Dalton, R. C. (2003). Linking Objective Measures of Space to Cognition and Action. *Environment and Behavior*, *35*(1), 3–16. doi:10.1177/0013916502238862

ADDITIONAL READING

Griffiths, S. (2017). To Go with the Flows or to Flow with the Nodes? An Exploration of 'post-disciplinary' Theories of Movement in Space Syntax and Mobilities Research. In: Heitor, T and Serra, M and Silva, JP and Bacharel, M and Da Silva, LC, (eds.) *Proceedings of the 11th Space Syntax Symposium.* (pp. 64.1-64.10). Portugal: Lisbon. Instituto Superior Técnico, Departamento de Engenharia Civil, Arquitetura e Georrecursos.

Hillier, B. (2002). A Theory of the City as Object: Or, How Spatial Laws Mediate the Social Construction of Urban Space. *URBAN DESIGN International*, *7*(3-4), 153–179. doi:10.1057/palgrave.udi.9000082

Hillier, B., & Netto, V. (2001). Society Seen through the Prism of Space: Outline of a Theory of Society and Space. IN Hanson, J. (Ed.) *Proceedings of the 3rd International Space Syntax Symposium.* Atlanta GA.

Hillier, B., & Penn, A. (2004). Rejoinder to Carlo Ratti. *Environment and Planning. B, Planning & Design*, *31*(4), 501–511. doi:10.1068/b3019a

Hillier, B., Turner, A., Yang, T., & Park, H. T. (2007). Metric and Topo-Geometric Properties of Urban Street Network: Some Convergences, Divergences and New Results. IN Kubat, A. S., Ö. Ertekin, Y. I. Güney, & E. Eyüboğlu (Eds.) *Proceedings of the 6th International Space Syntax Symposium.* Istanbul: ITU Faculty of Architecture.

Marshall, S. (2005). *Streets and Patterns.* London: Spon Press.

Peponis, J., Allen, D., Haynie, D., Scoppa, M., & Zhang, Z. (2007). Measuring the Configuration of Street Networks: The Spatial Profiles of 118 Urban Areas in the 12 Most Populated Metropolitan Regions in the Us. IN Kubat, A. S., Ö. Ertekin, Y. I. Güney, & E. Eyüboğlu (Eds.) *Proceedings of the 6th International Space Syntax Symposium.* Istanbul: ITU Faculty of Architecture.

Ratti, C. (2004). Space Syntax: Some Inconsistencies. *Environment and Planning. B, Planning & Design*, *31*(4), 487–499. doi:10.1068/b3019

Ratti, C. (2004). Rejoinder to Hillier and Penn. *Environment and Planning. B, Planning & Design*, *31*(4), 513–516. doi:10.1068/b3019b

ENDNOTE

1 In Depthmap and Mindwalk software, the integration value is calculated as equal to (1/RA) so the high values refer to the highly integrated spaces and vice versa.

Chapter 11
Space Syntax Approach for Articulating Space and Social Life

Abdelbaseer A. Mohamed
Al-Azhar University, Egypt

ABSTRACT

This chapter sets out to provide a detailed description of the relationship between space and society. It begins by discussing how people co-live in spaces and how such spaces co-live as communities. Understanding the relationship between space and society requires shedding light on how (1) communities emerge and work and (2) people build their social network. The chapter's main premise is that spatial configuration is the container of activities and the way we construct our cities influences our social life. Therefore, the urban environment should be analyzed mathematically using urban models in order to evaluate and predict future urban policies. The chapter reviews a space-people paradigm, Space Syntax. It defines, elaborates, and interprets its main concepts and tools, showing how urban space is modelled and described in terms of various spatial measures including connectivity, integration, depth, choice, and isovist properties.

INTRODUCTION

The connection between the physical urban environment and society is obvious and whilst sometimes neglected; it is something that we have to consider when planning for people. The urban system allows societal processes to develop in a conditional sense. Meanwhile, the urban system is mostly a product of socic-cultural, economic, and administrative processes (Klaasen, 2003). Put differently, society shapes buildings and the way we organize these buildings influences our social life situations. However, the literature on social theory shows that historical changes have received more attention over space and geography. In fact, space was virtually absent in social affairs (Netto, 2007). Differentiation between traditional society and modern society of the nineteenth century was the core on which many paradigmatic approaches to social aspects were built. Near the end of the twentieth century, the role of space

DOI: 10.4018/978-1-5225-9238-9.ch011

as a key element in (re)producing a society with its various socioeconomic conditions has been asserted by many experts in sociology, geography, anthropology, architecture and urban planning.

Among the question underlying discussions about space and society, three are perhaps key: (1) How does the type of society influence one's behavior? (2) How do certain social choices relate to architecture and design? Or how can we understand our societies as spatial systems? (3) How can we measure the local and global properties of urban space? What follows is a selective discussion of theories on space and society, starting with theories from sociologists like Karl Marx and Max Weber, and ending with those from urban morphology pioneers like Bill Hillier and Julienne Hanson.

The Impact of Society on Space

This section reviews paradigmatic approaches that discuss the issue of social cohesion. Reviewing approaches from classic to more recent works show that scholars understand society-space interrelations differently.

Solidarity, Conflict, and Space

Abd El-Rahman Ibn Khaldun (1332-1406), a Muslim scholar, contrived a theory of social conflict based on his studies of nomadic and sedentary societies. At the basic of this theory is the concept of 'group feeling' or solidarity (asabiyyah) (Giddens, 2009, p. 22). According to Ibn Khaldun, groups with strong social bonds, Nomadic Bedouin tribes, dominate and rule those with weaker internal ties, sedentary town-dwellers (Elshater, 2015, p. 26). Accordingly, the Bedouin's lifestyles turn more urbanized with weaker solidarity leaving them vulnerable to attack from external enemies. This circular system of rising and decline explains states-formation (Giddens, 2009). Another earlier initiative for explaining how societies work was introduced by Auguste Comte (1798- 1875) who was concerned with social changes after the French revolution. Comte tried to reconstruct the French society arguing that sociology like physics and chemistry can use scientific techniques to explain the social world. He postulated that inequalities and social fragmentation caused by industrialization could be overcome by producing a new moral consensus that would enhance bonding society (Giddens, 2009).

Progressively, Spencer (1872) conceptualized the theory of organic analogy which states that society is analogous to an organism. In this model, the various organs in the body integrate together to keep the whole system functioning and regulated. Likewise, "the various parts of society (the economy, the polity, health care, education, etc.) work together to keep the entire society functioning and regulated" (Stolley, 2005: 23).

Emile Durkheim (1858- 1917) was influenced by Spencer's work on social evolution and the organic analogy, and by Comte's ideas. Nevertheless, he described Comte's work as vague and speculative, arguing that Comte failed to establish sociology scientifically (Giddens, 2009). In the book, *The Division of Labor in Society*, Durkheim (1893) differentiates between two types of solidarity, mechanical and organic solidarities, according to societal changes through the growing division of labor (Durkheim, 1893; Vaughan, 1999; Magda, 2003; Stolley, 2005; Netto, 2007). This approach is known as functionalism. In it, Durkheim suggested that the type of solidarity in pre-industrial society, traditional society, was mechanical, where societies share the same identity, values, and beliefs. In this model, strong bonds emerge vividly within clearly recognizable boundaries where society is small and organized around kinship affiliations. Here, solidarity is derived from the collective conscience. The other model of social bonds, based on specialization, is organic solidarity, shifted from mechanical solidarity, which arises from the process of trade demands (Vaughan, 1999). This is a natural result of the transition from traditional society, rural, to modern society, cities, where the social network is transpatial (independent of the space). Here, the society gets complex, and social bonds based on conscience collective is no longer possible because people are more individuated, Durkheim postulates that cities are collections of non-indigenous groups —individual immigrants (Vaughan, 1999). Craib provides a good description of this solidarity: *In contemporary society I am dependent on other people- many of whom live on the other side of the world- to grow my food, make my clothes, supply me with warmth and light, print the books I read, service the word processor I write on, and so on. I could have or do none of these things without society.* (Craib,1997, p. 67).

In this sense, it is common to see different occupations interdependent within the societal system. Nevertheless, this organic solidarity can result in **anomie,** a period of chaos, which implies a lack of moral guidance due to social changes caused by individuals' specialized roles (Stolley, 2005). In other words, anomie is an intervening period between an old society, where values breakdown, and a new society, where new values need to emerge. So, anomie implies individualism that is destructive. This individualism should be evaded through reassuring moral order within social entities such as the education system (Dingley, 2008). Another problem which could prevent organic solidarity is forced division of labor where jobs are not distributed based on occupants' skills but their social class. This inequality should be removed to reintegrate individuals into society.

American sociologist Talcott Parsons, (1902-1972) was influenced by Durkheim's functionalist approach. Parson reshaped Durkheim's scheme of social solidarity. According to him, the fundamental function of solidarity is to integrate 'norms and values in the social system' (Palacios, 2007). While mechanical, traditional, solidarity derives from the common values of the society, organic, modern, solidarity is expressed through norms of economy –property, market relations, etc. In other words, both mechanical and organic solidarity draw upon common values.

Furthermore, mechanical solidarity, normative integration, seems to precede organic solidarity, functional integration. However, organic solidarity is not subordinated to mechanical solidarity, as they are both at the same level of institutionalization. This differentiation of Parson's understanding of solidarity led him to identify another type of solidarity, 'diffuse solidarity,' which mediates mechanical and chemical solidarity. Diffuse solidarity refers to "the common matrix out of which *both* the others have emerged by a process of differentiation" (Parsons,1985, p. 209; Palacios, 2007, p. 42). It is "based on common belongingness" (Cohen and Arato, 1994, p. 131).

Another contemporary of Durkheim who gave attention to social ties is the German Sociologist Ferdinand Toennies (1855-1936). Toennies differentiated between two types of communities: *Gemeinschaft*, community, and *Gesellschaft,* association. The former refers to traditional society, rural community, which reflects the importance of kinship and religion, while the latter refers to urban life where individuality and impersonality are at the core. Alongside Toennies is Georg Simmel (1858-1918) who was interested in studying the influence of the size of groups. According to him, people tend to create certain responses in larger cities to accommodate the various demands of urban life. This makes individuals cope with situations intellectually. For example, people could guide a wayfarer but ignore a beggar (Stolley, 2005).

Unlike Durkheim who was concerned with social harmony, solidarity, Karl Marx (1818-1883) and Max Weber (1864-1920) were interested in social conflicts. According to Marx, social conflict is a result of growing capitalism, which reflects inequality in society. Marx observed the escalating tension between the factory owners and workers where differences in social classes are highlighted. Marx defines a social class as "a group of people who stand in a common relationship to the means of production –the means by which they gain a livelihood." (Giddens, 2009, p. 439). Followers of Marxism shifted the focus from class and economic relations into wider areas such as inequality in gender. This was the foundation for feminist theory (Stolley, 2005). As for Max Weber, the focus of his studies was the correlation between individuals and society. In this sense, his concern was again totally different from Durkheim who dealt with society as a whole. According to Weber, social action is a result of the rationality of human behavior. This trend of *rationalization of society* stems from the rise of capitalism and modern bureaucracy. In other words, Weber argued that the distinctiveness of modern society was a result of transformations in the way of thinking (Netto, 2007; Giddens, 2009). He, also, asserted that we should understand the meanings behind the actions.

While Durkheim and Marx see that social structure exists out of individuals, Weber argues that it is social action which counts. He states that cultural ideas and values alongside economic factors have a direct impact on social changes. Accordingly, Weber proposes that stratification is shaped by three aspects: class, i.e. differences in economic conditions; status, that is, differences in 'the social honor or prestige'; and party which refers to a defined group of people who work together due to the common features they have (Giddens, 2009). In this sense, he is again different from Marx, who confines the causes of stratification on economic factors. So, according to Weber, people from aristocratic families continue to enjoy the prestige and honor even when their wealth has been lost (ibid.). Eventually, Weber's theory, in spite of its significance in interpreting human action, failed to grasp the levels of both the social life and of society (Magda, 2003, p. 9).

The Impact of Space on Human Behavior: A Brief Review

The literature on social theory shows that historical changes have received more attention over space and geography; space was virtually absent in social affairs (Netto, 2007). Approaching the end of the twentieth century, the role of space as a key element in (re)producing a society with its various socioeconomic conditions has been asserted by many experts in sociology, geography, anthropology, architecture, urban planning and so on.

Theorists differ in their views about the notion of the implications of space on society. Erving Goffman, for example, recognizes the space as the background of activities with no relation to society. Nevertheless, the urban space, according to him, plays an important role in understanding face-to-face situations (Magda, 2003). Likewise, Levitas (1978) states that social structures cannot be explained by urban form since many streets are very similar looking but the activities occurring on such streets are very different. In contrast with Goffman and Levitas, some theorists believe that space influences society. Whyte (1943), for example, stated that the street space enhances social interaction between residents. He found that certain spatial and societal patterns emanated spontaneously in a certain urban order (Whyte, 1980).

Whyte's view can also be seen in the review of *The Death and Life of Great American Cities* by Jane Jacobs (1961), who objected to the planning ideals of modernism. She pointed out the importance of sidewalks on human social behavior. According to Jacobs, secure and safe sidewalks can be created by separating public and private spaces from each other and by creating a constant flow of people on the streets (Jacobs 1961; Pongsmas, 2004). In other words, she encourages open and accessible public spaces, in which strangers and inhabitants are a source of safety (Hillier, 2004). If a city fails to create safe sidewalks, problems like social segregation will emerge because people will avoid walking which in turn will impede positive interaction. Jacob demonstrated that *primary mixed uses, small blocks,* and *aged blocks* are all generators of *diversity* that is the key element of a successful city (Jacob, 1961; Pongsmas, 2004). However, the physical properties that meet Jacobs' principles are ill-defined (Peponis, 1989).

In *the Image of the City*, Kevin Lynch (1960) reveals the influence of urban environment on people's orientation. According to him, building the image of an environment is a two-way process, and is the consequence of an interface between the viewer and the settings in which he or she lives. The procedure is not just how users realise objects and others, but also how others comprehend them as a part of the environment. Consequently, such process reflects the influence of the observer and environment on each other. Lynch says, "nothing is experienced by itself, but in relation to its surroundings, the sequences of events leading up to it" (Lynch, 1960, p. 1).

In *A City Is Not A Tree*, Christopher Alexander (1966) postulates that a city is a complex object, an abstract structure which is not a tree but rather a semi-lattice. A semi-lattice structure of a city overlaps with a semi-lattice social structure. In other words, social relations in modern society are transpatial in order to be congruent with the spatial arrangement of the city. Alexander also differentiates between natural cities, or cities that arise spontaneously, natural cities, and artificial ones, or cities that are created by architects and planners (Alexander, 1966). He rejected modern town planning principles, particularly single-use zoning, for producing non-overlapping hierarchies. However, this rejection led him to devote more attention to the spatial distribution of activity patterns than to the morphology of space (Peponis, 1989).

Another proponent of space implication on society is Michelson (1970), who suggests that social solidarity is promoted by the pattern of streets. Similar to Michelson, the social anthropologist Pierre Bourdieu suggests that social solidarity is strengthened by close proximity which is likely to create encounters (Michelson, 1970). In the book *'Life Between Buildings'* Gehl (1971) studied the relationship between physical environment and social activities. He categorized outdoor activities into three types: (1) necessary (e.g. going to school or work), (2) optional (i.e. activities reckoning on outdoor physical conditions), (3) social or resultant (i.e. communal activities such as conversations, greetings, and barbeque). For instance, activities like sitting around and enjoying sunbathing only take place when conditions of place and weather are optimal. Gehl (1971) found that poor physical environments impede the occurrence of optional and social activities, hence the diminishing of opportunities for social interaction (Gehl, 1987). Nevertheless, Gehl's findings need to be verified mathematically using an objective tool.

Architect and city planner Oscar Newman (1972) claims that there is a direct relationship between crime and space. In his book *Defensible Space* in 1972, he states that defensible space is a socio-physical phenomenon where space and society are both parts of a successful space. He asserts the role of physical features as follows: *Regardless of the social characteristics of inhabitants, the physical form of housing was shown to play an important role in reducing crime and in assisting residents in controlling behavior in their housing environments (Newman, 1996, p. 25).*

According to the theory of defensible space, safer places will be achieved if the residents of an area feel belonging to and responsible for their territory. The number of people living in a territory is influenced by the physical form of housing since crime rates will be lower in buildings with few families having their own individual entry than buildings with more families sharing a common entry. This is based on the principle of natural surveillance where the residents' ability to monitor what is happening is closely linked to the physical characteristics of an area (Newman, 1996). Newman encourages closed and inaccessible spaces such as cul-de-sac for a safer environment, in which residents are the primary source of natural surveillance while outsiders are the source of the threat (Hillier, 2004). This seems contradictory to Jacobs's views. Nevertheless, the differences between the two schools of thoughts should be acknowledged, while the presence of a stranger in a public street is ordinary and vital for natural policy, it would be unexpected and hence the danger in semi-private spaces (Hillier, 2004).

Bill Hillier and Simon Shu (1999) questioned the theory of defensible space and gave evidence for that from a study of crime vulnerability in several housing types, in order to 'cover a wide range of social classes', in England: *The findings from this research provide empirical evidence for scepticism about the idea of "territoriality" and "defensible space" put forward by Oscar Newman (Newman, 1972), and suggest that, other things being equal, property crimes tend to cluster in those globally or locally segregated areas, particularly in cul-de-sac footpaths and rear dead-end alleys, but also in those segregated short cul-de-sac carriageways which Newman considered to be the key to increase local surveillance and hence to exclude casual intrusion by non-residents. Positive features which make spaces safer are integrated through roads with front entrances on both sides, exactly those anonymous spaces Newman considers more prone to crime* (Shu, 2000).

Besides, Hillier and Hanson (1984) contradict Newman's thoughts by suggesting that crime is prevented through the co-presence of people on the streets: *Unlike Oscar Newman's 'defensible space' theory, which emphasises inhabitants policing space and excluding strangers, our research has led us to conclude that strangers police space and inhabitants police strangers, thus generating 'automatic' control in the area... and so reducing certain street crimes. The interaction and accessibility for strangers and inhabitants are profoundly influenced by the convex and axial organization of an urban area and its interface with the buildings (Hillier et al, 1983, p. 52).*

In this sense, Hillier and Hanson deal with the urban space not as an object with a passive role, but rather as (re)producer of society. Both Hillier and Hanson founded the theory of *space syntax* where the relationship between the morphological construction of man-made surroundings and social constructions or events is investigated. Hillier and Hanson (1984) propose the concept of *intelligibility* as an indicator of 'sense of place'. The intelligibility is well-defined as the grade to what can be realised and experienced locally in the scheme allows the large–scale arrangement to be learned without conscious efforts" (Hillier, 1996b, p. 171). In this sense, the degree of intelligibility is defined by the degree of accessibility in a street system rather than by a sense of belonging to a territory. This means that the possible social networks are shaped through the morphology of urban space that is the 'virtual community' of the potential co-presence of people. Moreover, an intelligible system is one in which well-connected spaces also tend to be well-integrated ones and vice versa. Put differently, in an intelligible domain the association between local and global possessions of space is perfect, so the whole can be interpreting from the part.

In another pioneering study, Arnis Siksna (1997) compared centers of twelve North American and Australian cities in terms of block size and form. He found that particular block forms perform better than others for making city center more livable because of the effect they have on movement rates, frontages potentiality, and the degree of continuity of urban fabrics (Siksna, 1997). On the other hand, Hernandez et al. (2007) argue that mobility is one of the key indicators of place attachment —people usually conform to the constraints of spatial settings.

While traffic can dilute and challenge everyday social life, pedestrians and so well-functioning streets are typically seen as crucial for good city life by the virtue of fostering natural surveillance and therefore feeling of safety. Published in 1981, Donald Appleyard's *Livable Streets* defends the street as a place with key social functions. Appleyard studied the possible influence of different traffic volumes on the amount of social relationships that develop on the streets. He concluded that streets with light or no traffic are a fundamental aspect of livable environments (Appleyard, 1981). In the 21st century, the concept of livability received much more attention and had become more prominent.

While the word livability is broadly used in the USA, other terms such as sustainability, placemaking and life quality are more common elsewhere. Livability stresses soft values in policies and plans at cities and national level, whereas sustainability broadens the focus from social aspects to include economic and environmental domains. Differently, New Urbanism, found in 1993, concentrates primarily on design (Gehl &Svarre, 2013).

In *Great Streets*, architect and city planner Allan Jacobs (1993) studied the connection between design qualities and urban life. Jacobs believed that systematic observation is especially important for learning human behavior. Based on field observations of a wide array of street types all over the world, he emphasized the importance of physical factors in creating great streets and so livable communities (Jacobs, 1993).

In about 2000, urban life studies became mainstream. By using dialogue-based manual methods (e.g. interviews and workshops) as well as new technologies (e.g. cameras and GPS senders), countless works on the connections between public social life and form have been conducted (Gehl, 2010; Tonkiss, 2013). Later in the early 2010s, the Arab world has witnessed a revolutionary wave, The Arab Spring, of peaceful demonstrations and revolts to end long-term regimes. Public spaces such as Cairo's Tahrir Square in Egypt and Istanbul's Taksim Square in Turkey were the magnet for these revolts. This attests that while social media and communication technologies can initiate civil uprising, the public realm is still substantial as the place where people can meet up to protest. It is also the stage for public speeches and real-life interactions. Bein in public enhances social interactive skills that cannot be taken only from one's family. If so, design decisions should not be overlooked through dealing solely with static drawings or maps.

From the above, it is well-defined that there are various theories and approaches to the notion of space and society. For reasons of scope, the following sections are devoted to reviewing, with more detail, the space syntax approach. As will be shown, space syntax is a scientific and human-focused theory and method that treats spatial layout a social, economic and political problem. The principles upon which the method is built are derived from real-life observations of human behavior. This is originally conceived to help architects and city planners improve the design of built environments and guide land use planning.

SPACE SYNTAX THEORY AND METHOD

Various early initiatives attempted to analyze the spatial dimension in order to interpret the social pattern. One of these initiatives is Lionel Martin's study (1975) in which he claims that street network is the container of activities and that the way we construct our cities influences our social life. In his view, the urban environment should be analyzed mathematically. In line with this, Bullock and Dickens (1975) suggest using urban models in order to evaluate and predict future urban policies. Further, Krüger (1979) suggests splitting the urban layout into a pattern of linkages in a way similar to planets and stars (Vaughan, 1999).

Space syntax is originated and developed in the 1970s at the Bartlett Unit for Architectural Studies at University College London (Hillier et al., 1983; Hillier & Hanson1984; Space Syntax Ltd., 2004), aims at analyzing space configuration in order to reveal social patterns. The purpose is to make the configuration seen by people as a non-discursive thing used intuitively in everyday life, a discursive thing

(Westin, 2011: 229). Space syntax attempts to define society, different social solidarities identified by Durkheim, through space. In other words, space syntax is a graph-based theory which defines people's relations through relations between street segments in an urban grid. The virtual community is produced based on spatial configuration, the relation between two spaces taking into account at least third space and at most all other spaces in the system (Hillier & Hanson, 1984).

Space syntax research found consistency between the pattern of social relations and the pattern of relations created by the spatial formations (Netto, 2007). In their book *the social logic of space*, Hillier and Hanson (1984) argue: *why should these patterns [of space] be different in different societies? Here we found the general sociology of Durkheim... profoundly suggestive. Durkheim had distinguished between two fundamentally different principles of social solidarity or cohesion: 'organic' solidarity based on interdependence through differences, such as those resulting from the division of labour; and a 'mechanical' solidarity based on integration through similarities of belief and group structure. This theory was profoundly spatial: organic solidarity required an integrated and dense space, whereas mechanical solidarity preferred a segregated and dispersed space... In the work of Durkheim, we found the missing component of a theory of space (Hillier and Hanson 1984, p.18).*

As can be seen from above, space syntax expands beyond the Durkeimian approach, since the idea is that social information is embedded in the configuration. Importantly, the arrangement of social networks in space can take three potential patterns according to the degree of socio-spatial correspondence. When aspects of the social structure are projected into space, then the produced society is called 'correspondence society', where local identities tend to be strengthened. On the other hand, when the 'social fabric' crosses space, then we get what is known as 'non-correspondence society' where global solidarity tends to construct local transpatial identities. Another kind of society, which is not just created according to spatial or transpatial forms, is that model of 'differential solidarities' where the social fabric is projected in different configurations based on class or gender differences (Hillier and Hanson, 1984; Hanson and Hillier, 1987; Hillier, 2009c; Al Ghatam, 2012). It is worth mentioning that these three kinds of society are developed from Durkheim's ideas about societal solidarities.

From above, it is clear that Hillier and Hanson are concerned with the implications of the physical dimension of space. They criticized the naïve descriptions of the city depending on simplified linguistic concepts such as hierarchies and regularized geometries (Hillier, 2009a). These naïve descriptions may have resulted from the lack of the means to describe the complexities of the urban environments. For example, while the two plans shown in Figure 1 are similar in adjacency and the number of units, they

Figure 1. Two architectural plans with different spatial configuration
Source: the author

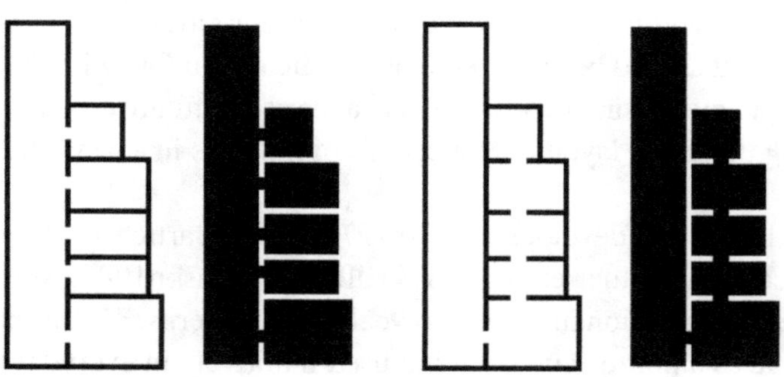

are completely unlike in terms of permeability. The position of doors in each plan influences the use of spaces according to how we relate such spaces to each other. Then, the question is how to describe the arrangement of spaces, spatial configuration, in each plan objectively?

To answer the question above, Hillier and Hanson, in their book *The Social Logic of Space*, (1984) developed three iconic ideas to represent space: Axiality (longest axial sight), when people move in it, convexity (in which all points are visible from each other), when they interact within it; and isovist which refers to the conventional settings of all points in the field of eye-vision that are visible through a given vantage point in the urban space and with regard to an setting (Benedikt, 1979, p. 47) (Figure 2).

According to Hillier and Hanson (1984) movement is generated naturally by the spatial configuration of the urban environment. In other words, 'natural movement' is generated automatically depending on the underlying degree of accessibility (Hillier et al 1993).

Natural movement –and so the spatial configuration itself– influences land use pattern since spaces with high movement rates will attract commercial uses, while non-movement seeking activities (e.g. residence and the like) will migrate to locations with low co-presence (Hillier, 1996a; Greene, 2002; Topçu et al, 2007). The attracted uses produce 'multiplier effect' on movement since they increase the importance of the locations themselves and, in turn, encourage further uses. In other words, the spatial configuration may influence both movement and attractors, but the latter two cannot change the configu-

Figure 2. Axiality, convexity, and isovist
source: the author

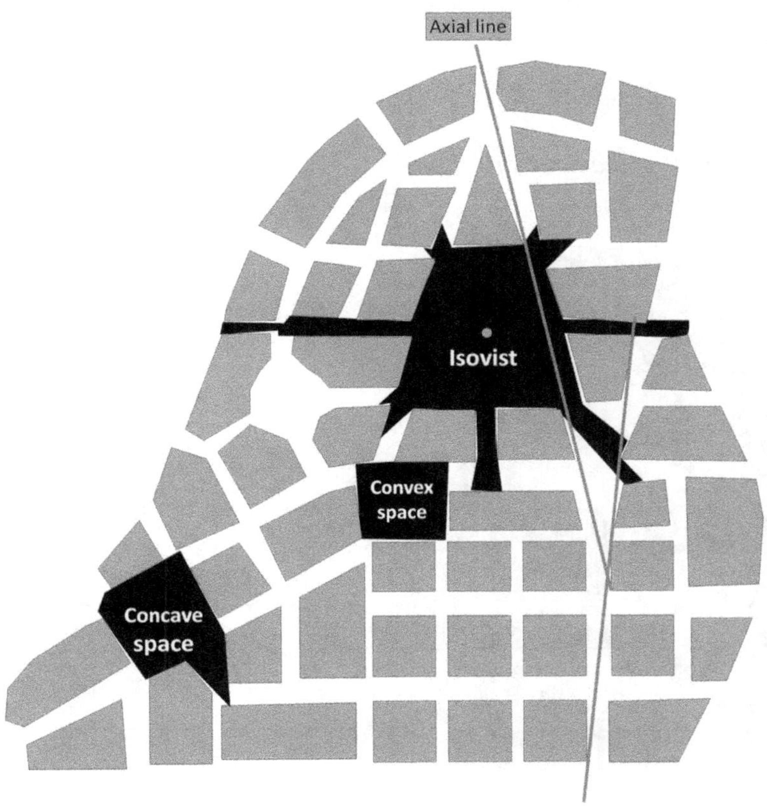

ration. Rather, movement and attractors can influence each other. This dynamic process of configuration, movement, and attraction is what Hillier has called 'movement economy' (Hillier et al, 1993).

REPRESENTING URBAN SPACE

Urban space can be represented either as a set of convex space or as a map of axial lines. A convex map consists of the largest and fattest convex spaces that cover the area (Hillier et al., 1983). Likewise, if the street network of an urban environment is covered with the fewest and longest axial lines, then we get what Hiller and Hanson called 'axial map' (figure 3).

The axial map can be generated automatically using UCL Depthmap. The software creates all potential axial lines. Then, all line map is reduced to the fewest line map. Remarkably, the manually drawn axial map in figure 3 is relatively different from that produced automatically in Figure 4.

A closer look to the urban grid of any city will show that it is made up of a dual network –the foreground network, made up of the few numbers of longer lines representing centers and sub-centers, and the background network, made up of larger number of shorter lines representing residential space (Hillier, 2009a; 2009b). Here the space acts either generatively, in the foreground network, in order to maximize co-presence creating new relationships and giving a city a universal spatial structure or conservatively, in the background network, so as to reproduce cultural patterns through restraining and preserving existing movement pattern, which gives the city its spatial identity (Hillier 1999; Wang, 2012; Psarra, 2009; 2010; Sailer et al, 2012). For example, Figure 5 shows the city of Cairo has a universal spatial form of a 'deformed wheel' in the foreground network and a larger number of shorter lines (or segments) in the background network.

Figure 3. Convex map (left) and axial map (right)
Source: the author

Figure 4. All line map (left) and fewest line map (right)
Source: The author

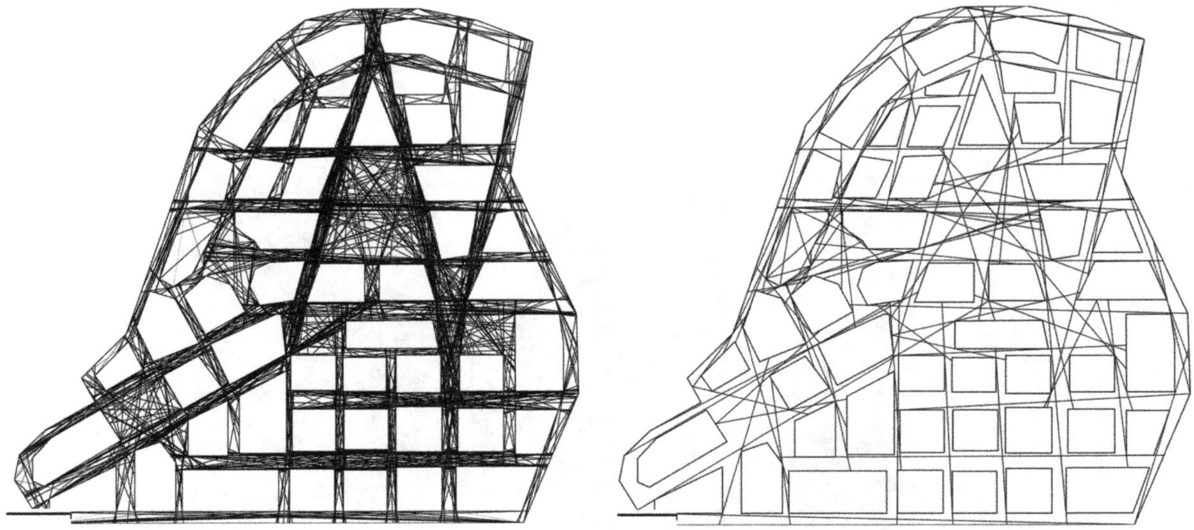

As a matter of fact, axial lines reflect the degree of accessibility from each line to all other lines in the system. The axial map was progressively developed into a segment map where the street segment between junctions is the spatial element (Hillier and Stonor, 2010; van Nes, 2011). The structure of the urban grid, then, shows the potential movement of people since spaces with high syntactic values will generate co-presence and interaction higher than spaces which are syntactically segregated.

CONFIGURATIONAL ATTRIBUTES

Integration (Closeness)

The built environment is the container of various activities such as eating, gathering, interacting, etc. Undoubtedly, there is an expected relation between space and its use inherent in the relation between the configuration of people and configuration of space. Public life studies show that most people prefer simplest, minimum directional change paths over complex, many directional changes paths -even if the complex path is the shorter (Figure 6).

If the longest and fewest axial lines of an urban system are drawn, then the product is the axial map of that system. Once obtained, the relationships between the axial lines comprising the axial map can be examined and interpreted. In terms of topological distance (fewest turns), if the interest is to see the number of turns that must be crossed to get from an origin to a potential destination within the system, then the calculated value is called depth. It shows how a line relates to all other lines in the system. That is the degree of spatial accessibility or closeness where maximum depth means a minimum number of turns, hence maximum integration, also called *to-movement* (Figure 7). Put another way, depth and integration are in contrast with each other.

Figure 5. Cairo has a foreground network of a 'deformed wheel' pattern while having a background structure of shorter street segments reflecting the spatial dimension of cultural context
Source: The author

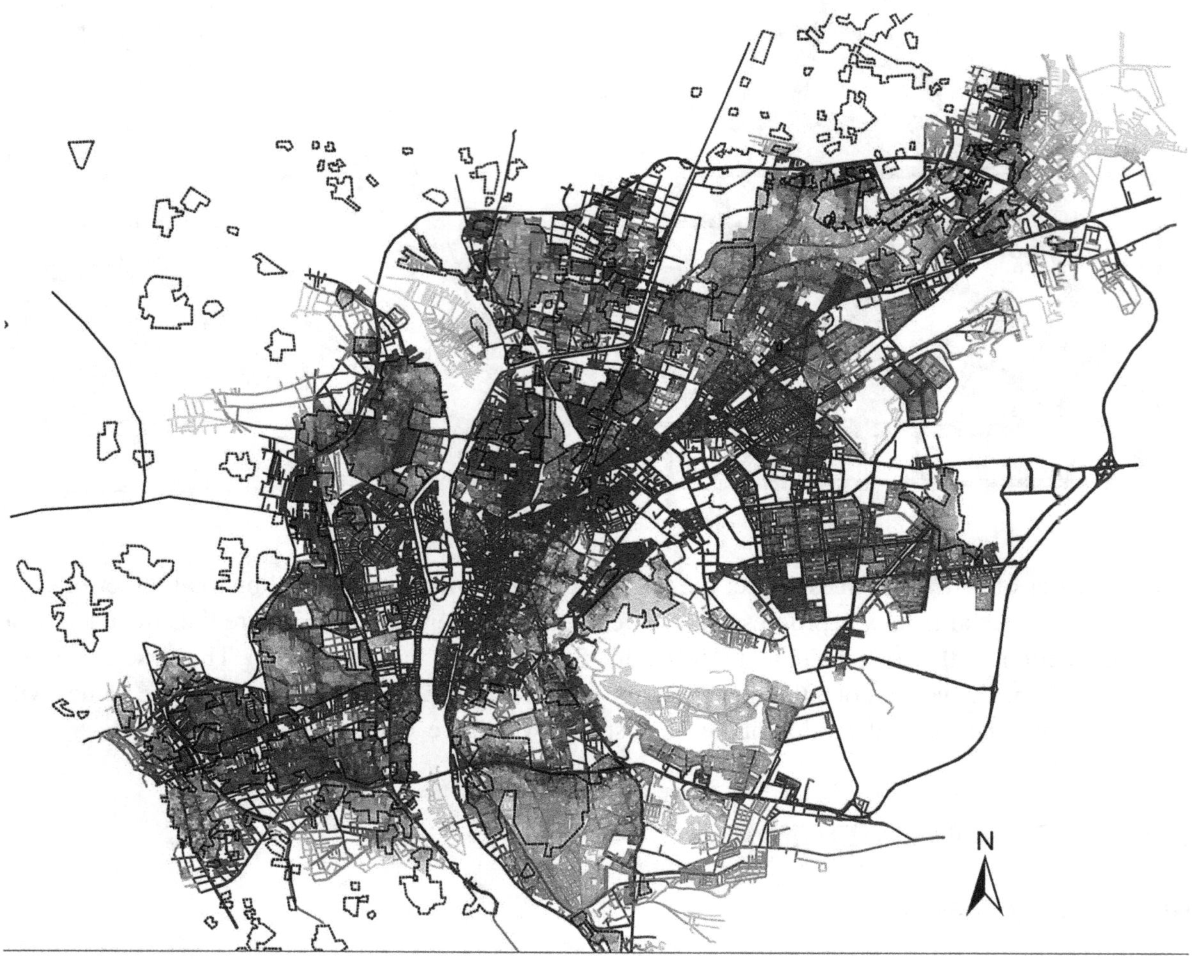

To illustrate, assume a very simple urban system of only 7 spaces or streets. If the axial map of that system is drawn then it is possible to select any axial line and set all lines that are directly connected with it on a separate level until we pass through all spaces in the system, then the number of levels, also called steps or point depth (now step depth), required to reach all lines in the system can be specified and values of integration, and depth can be measured. Step depth measures how far a line or a segment is from the selected root line or segment in terms of topological, metric, or geometric distance (Turner, 2004, pp. 25, 28). As shown in Figure 8, by starting at line 7, we find that it is connected to line 1 and 6. Then line 1 is linked to line 3, 4 and 5, while line 6 joins line 5 and 4; further, line 3 is connected to line 2. By setting each direct connection on a separate level, we notice that for line 7 we have to join up four levels or steps to cover the whole system. The produced graph is called justified graph and can take various patterns (e.g. diamond shape, tree) according to the branching of the considered space and according to the syntactic size of the system itself. As to line 1, we can see that it's directly linked to lines

Figure 6. Most people prefer simple paths
Source: The author

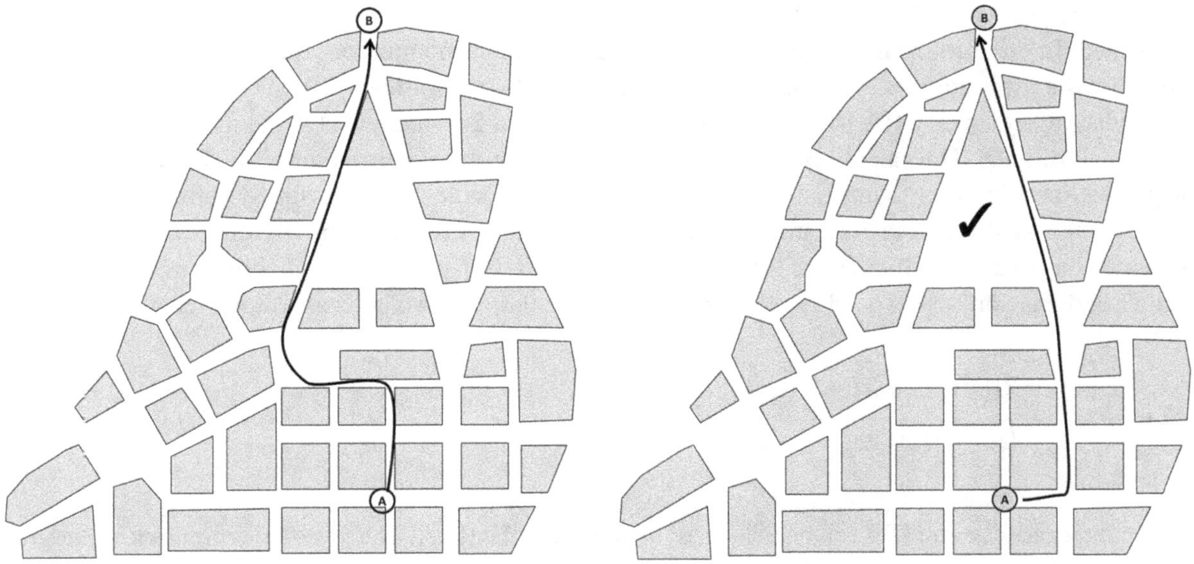

Figure 7. Integration/ to-movement reflects the degree of spatial accessibility from an origin to all potential destinations.
Source: the author

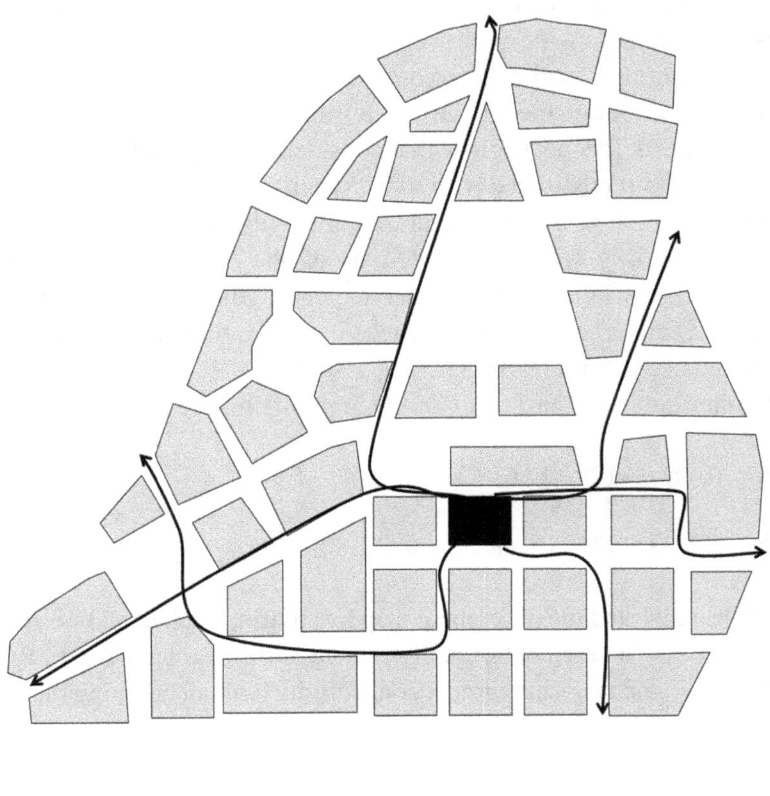

3, 4, 5 and 7. Then line 5 is connected to line 2 and 6. So we find that we need 3 steps to pass through the whole system. Apparently, line 7 is more segregated than line 1 because line 7 needs a higher number of levels (3). As a rule of thumb, lines with many levels in relation to all others seem to be spatially segregated. In order to measure the mean depth of a particular line multiply the number of the lines on a level by the level number, sum across levels and divide by the number of lines (Hillier et al., 1983). Then relative asymmetry RA (relative depth) = 2(MD -1)/(n-2), where MD is the mean depth and n is the number of lines. The calculated value ranges between 0 and 1, where a high value indicates the least integrated lines. In order to normalize the measure and be capable of comparing systems with different sizes, we use Real Relative Asymmetry (RRA), which can be calculated by dividing RA by D-value, diamond shape value (Hillier et al., 1983). The D-value is sensitive to the number of lines in the system (Hillier and Hanson, 1984, p. 112). The formula for calculating D-value is as follows (Shafiei, 2013):

$$Dn = \frac{2\left\{n\left[\log 2\left(\left(n+2\right)/3\right)-1\right]+1\right\}}{\left(n-1\right)\left(n-2\right)}$$

The integration value ITG can be measured by inverting the RRA. The low value implies low integration and vice versa. The following simple equation is used in calculating integration:

ITG = 1/RRA

It is worth mentioning that the integration value of a line varies according to the number of considered steps; if we consider the total number of steps or direction changes in a system, then the obtained measure is at a global scale, whereas considering only a particular number of levels in calculation gives more local measure. For instance, counting how deep or shallow each street axis is from all lines is called global integration Rn, whereas limiting calculations to only the first three levels gives radius-3 integration or local integration R3; so, all lines outside the third step from the line for which the integration R3 is being calculated are not included. It is noteworthy that considering only the first level gives 'connectivity' (the number of lines that are directly joined with it), so specifying the type of integration depends up on radius–n integration (Hillier, 1997). In our example, connectivity of line 1 equals 4 because it is directly connected to four lines (3, 4, 5, 7), whereas connectivity of line 7 is 2. The relationship between global integration Rn and connectivity indicates the degree of 'intelligibility'. The intelligibility is defined as "the degree to which what can be seen and experienced locally in the system allows the large–scale system to be learned without conscious efforts" (Hillier, 1996b, p. 171). The value of intelligibility ranges between 0 (for the unintelligible system) to 1 (for the high intelligible urban system).

Space 1 mean depth = ((4*1) + (2*2)) / 6 = 1.3

Space 7 mean depth = ((2*1) + (3*2) + (1*3)) / 6 = 1.83

The axial map can be converted into a segment map by splitting each axial line at the locations where it intersects with other lines. The element of analysis here is the segment between intersections rather than the whole axial line. So, topological segment connectivity is about counting the number of segments

Figure 8a. A hypothetic model of seven axial lines and their topological relationship: The simple model of axial lines
Source: the author

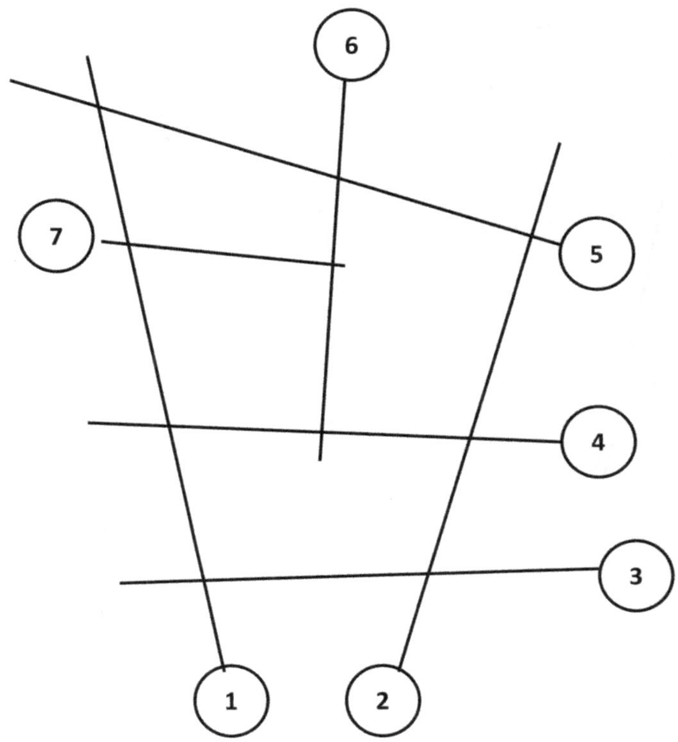

Figure 8b. A hypothetic model of seven axial lines and their topological relationship: The simple justified graphs
Source: the author

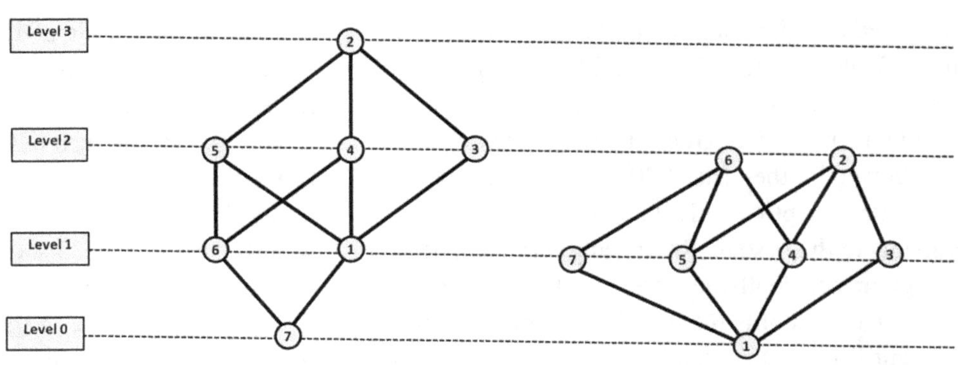

that are directly connected to a certain segment (Figure 9). In other words, it shows how a segment is connected to other segments in its direct vicinity (van Nes, 2011).

In terms of geometric distance (i.e. least angle change), the total angular depth value is calculated by summing all angular turns from a segment to another via the lowest angular deviation cost (Turner,

Figure 9. Calculation of topological segment connectivity
Source: the author

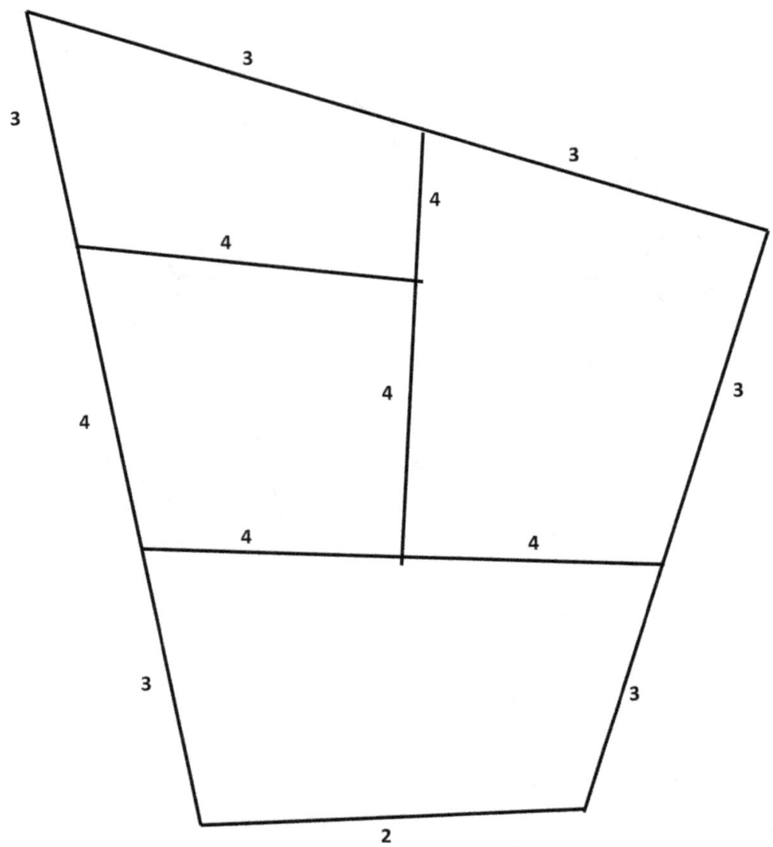

2005). Weights are assigned according to the degree of directional change. For instance, an angle of 45 degrees is weighted as 0.5, an angle of 90 degrees is assigned a value of 1, and an angle of 180 degrees takes a value of 2 and so forth (Ibid). To illustrate that imagine a simple model comprised of four segments connected to one another at different angles (figure 10). The depth from segment A to B is 0.5, since they form a turn of 45 degrees. The depth from segment A to C is 0.833, the turn of 45 degrees from A to B followed by the turn of 30 degrees to C (Turner, 2005, p. 148).

The angular mean depth is calculated by dividing the total depth by the node count, the number of nodes or segments in the system. So, the angular mean depth of A = [(B) 0.5+ (C) 0.833 + (D) 0.833] / (3) = 0.722 Angular step depth measures the shallowest angular path from the root segment (or segments) within the system and records the path length on the segment. So, as mentioned before in the previous example, segment B has a step depth of 0.5 from A, while the segment C has a step depth of 0.833.

It is worth stating that there are three weights of distance: the metric distance which defines a street network by the shortest passes, the topological distance which calculates a street network in terms of fewest turns, and finally the geometrical distance which measures a street network regarding least angle change. The radius of analysis can be topological in terms of the number of steps or metric where segments located within a particular metric distance are only considered. For example, radius n considers

Figure 10. A hypothetic model of four street segments and their angular relationship
Source: (Turner, 2005, p. 148)

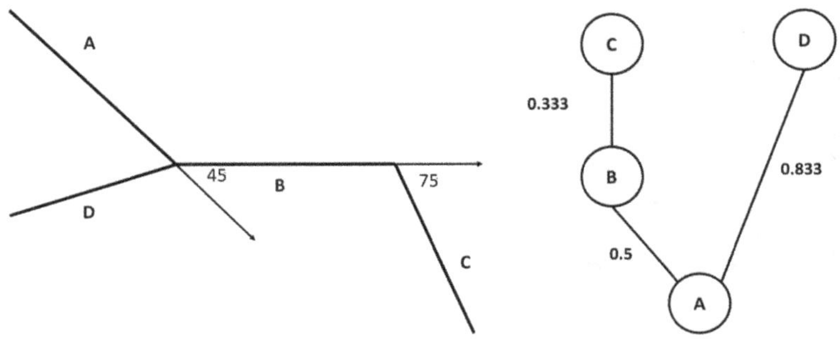

all segments within a certain system, while radius 2000 limit the syntactic calculations to only segments that are located at a distance up to 2000 meters from a particular segment, so all segments outside the distance of 2000 meters are excluded from the analysis. The higher the radius is, the more global the measure is. Radii of 400 (5-minute walk), 800 (10-minute walk) and 1200 meters (15-minute walk) are proper for analyzing pedestrian movement rate at the local scale, while radius 2000 meters and up are suitable at a more global scale. As to vehicular movement, a 5-minute car-driving can be represented by 1200m radius analysis and so on.

In a nutshell, the syntactic measures of each segment can be applied in relation to all other segments in the system, on a global scale or specified to local distance. In other words, the type of syntactic attributes, whether global or local, will vary according to various radii, where a large radius indicates more of a global extent of the measures. In order to understand the relationship between syntactic measures at different radii, the correlation between the local angular integration (connectivity) / to-movement, and the angular global integration is known as segment-intelligibility, whereas the correlation between the local angular choice/ through-movement, and the global choice is expressed as Space-Crossing measure (Al-Ghatam, 2012). Both segment-intelligibility and space-crossing measures are considered as an indicator of the quality of an environment. Put another way, if the correlation between local and global properties of space is perfect, then the whole can be read from the part and vice versa.

Normalizing the measures enables comparing urban systems with different sizes with each other. Consequently, a neighborhood could be compared with the whole city and with smaller urban systems as well. Actually, it is more important to normalize on the local scale than it is on the global scale (radius n), that is because radius n is a constant. Normalization can be problematic in areas where there are huge variations between sparse and dense grid. Normalized angular integration (NAIN) is measured as:

```
NAIN = NC^1.2/TD
In UCL Depthmap, NAIN can be calculated as follows:
value ("T1024 Node Count RXXX metric") ^ 1.2 /
(value ("T1024 Total Depth RXXX metric") +2) (Al_Sayed et al, 2014)
```

Figure 11. Global topological (left) and angular (right) integration Rn with darker segment lines for higher values.
Source: the author

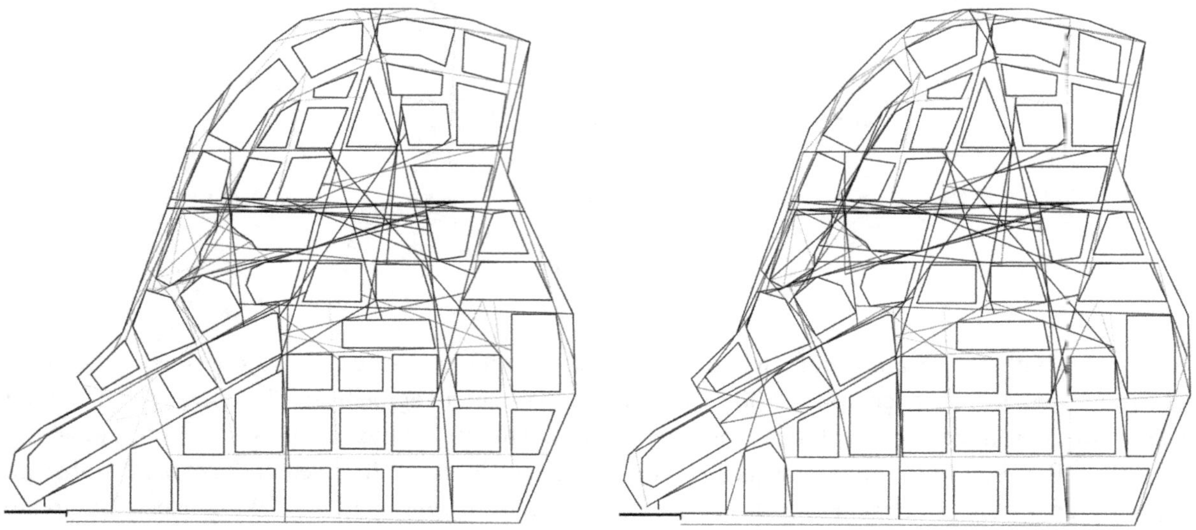

Integration Core

The integration core is the band of the most integrating spaces (top 25% integrating lines is recommended for small spatial systems of less than 100 lines while the top 10% is suggested for the integration core of larger ones) that tend to cluster in a particular area within the system (Hillier et al., 1983; Space Syntax Ltd., 2004). The integration core usually overlaps with the planned centers (Space Syntax Ltd., 2004) and constitutes the main structure of the cognitive map (Mohamed, 2010). In well-used urban areas, the integration core penetrates into the whole system reducing journey lengths between the live center and all other parts (Hillier et al., 1983). Nevertheless, the shape of the core differs from a city to another. It can take the form of the regular grid, a 'spiky potato', or a 'deformed wheel'. In addition, it can have a clustered or a fragmented structure (Klarqvist, 1993). Figure 12 shows the integration core of Cairo metropolitan. The city has the pattern of a 'deformed wheel', darker integrated lines comprising a hub in the center linked to the edge (the rim) by axial integrators representing spokes.

Choice (Betweenness)

Choice or *through-movement* shows how likely space is crossed with respect to all other pairs of segments (Hillier et al., 1983; Klarqvist, 1993; Turner, 2005, 2007) (Figure 13). Thus, if the shortest angle paths between each potential origin and destination pairs of segments in the system are constructed, then the total number of times each node or segment will be passed through as a part of journeys' routes is called choice (Turner, 2005; 2007). In other words, choice "calculates the number of shortest paths overlap between all nodes in the graph" (Varoudis et. al, 2013, pp. 57-61). If the number of overlaps of a segment or space is high, then it will get a strong choice value and vice versa. Cul-de-sacs and dead-end streets are assigned a choice value of 0 because they are not crossed by the shortest passes joining

Figure 12. Cairo has the pattern of a 'deformed wheel'
Source: The author

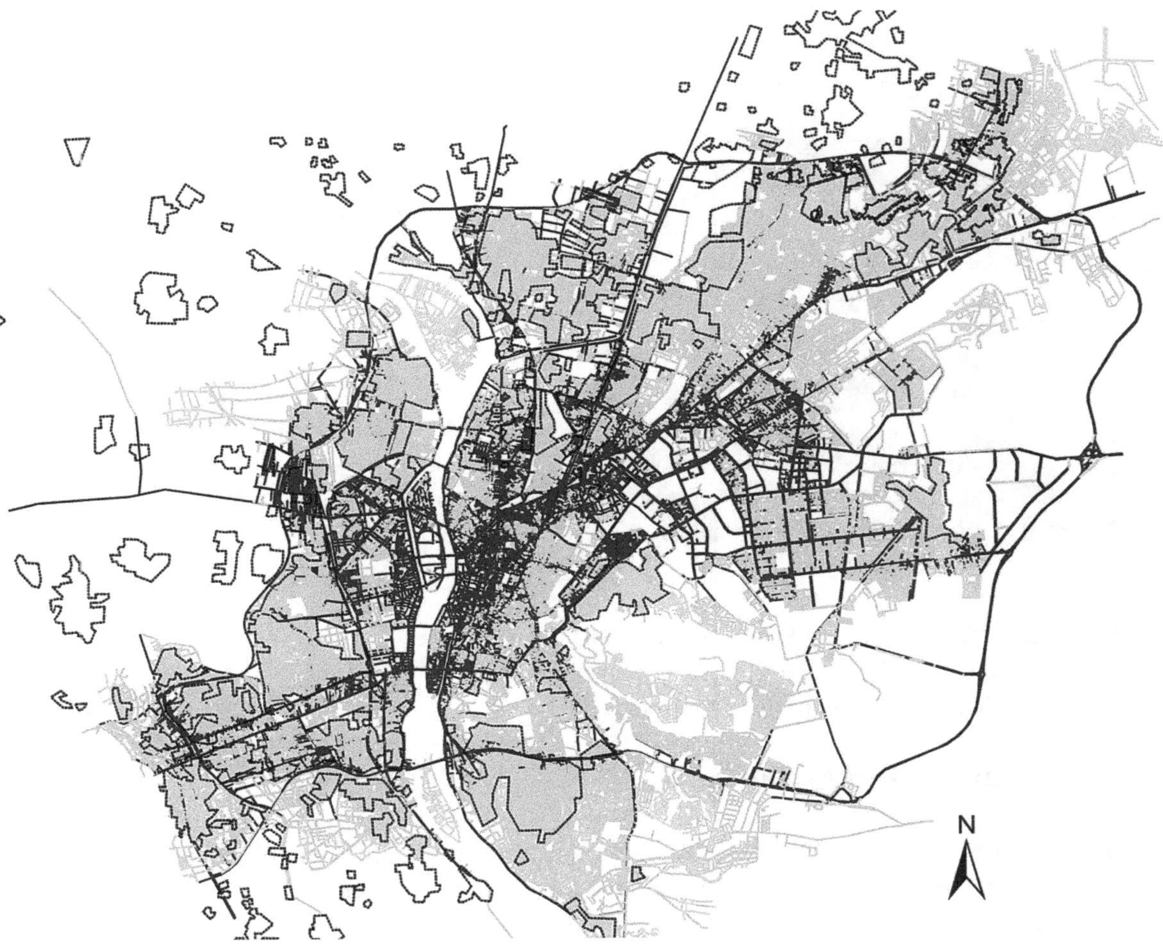

all segments to all segments of a system. In other words, the flow through cul-de-sacs and the like is 0. Finally, it is worth mentioning that global choice corresponds to centers, whilst local choice corresponds to local sub-centers.

1. Overlapping higher choice segments as a part of the shortest path between potential origin and destination pairs of buildings
2. Calculated Angular Choice Rn with darker lines for higher values

The normalizing angular choice was introduced by Tao Yang in Hillier et al. (2012). NACH has proven to be independent of the size of urban systems. The formula for normalizing angular choice (NACH) is as follows:

- NACH = logCH+1/logTD+3
- NACH is calculated in UCL Depthmap software as follows:

Figure 13. The concept of choice/ through-movement
Source: The author

- log (value ("T1024 Choice RXXX metric") +1)/
- log (value ("T1024 Total Depth RXXX metric") +3)
- Where XXX is the metric radius of interest (Al_Sayed et al, 2014)

VGA and Agent-Based Analysis

Visibility Graph Analysis (VGA) is based on the concept of isovist, "the field of view within which the entire space boundaries can be defined through the transportation of observer's eye without geometric obstacle" (Mohamed, 2010, p. 89). To make a visibility graph of a particular space we have to set a grid of equal squared cells. Then, the analysis will show how a cell in the grid is related to all other cells in term of mutual visibility. It is noteworthy that not all co-visible spaces are also co-accessible (e.g. a glass wall is visible but not accessible) (Turner, 2003; Can, 2012). To say it differently, the pattern of VGA depends upon the degree of permeability between spaces (Figure 14). VGA analysis captures local properties such as the perimeter of the view field, occlusivityand compactness and global properties such as visual integration and visual entropy. Compactness reflects the degree to which the isovist takes the shape of an ideal circle, whereas occlusivity gives an idea of the length of the "open" edges.

Agent-based modelling is a 'pedestrian simulation' based on the spatial settings of the urban environment (Figure 15). Understanding the behaviour of pedestrian movement is needed for identifying locations of congestion. Micro-simulation model of human movement can be done using UCL Depthmap software (Turner, 2004).

Figure 14. Visual integration for two different plans. Note: lighter areas have a higher degree of visual integration
Source: the author

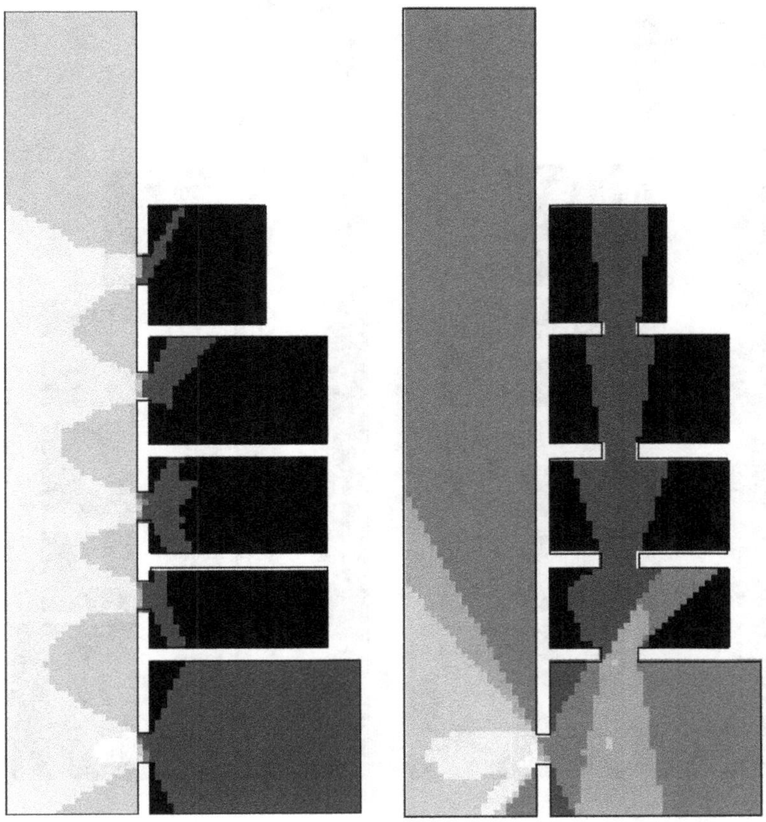

CONCLUSION

This chapter has reviewed theories on issues relevant to the interrelations of space and society with a special focus on space syntax methodology. Vast literature emphasized how industrialization was associated with social changes. Social bonds are stronger in pre-industrial or traditional society than in modern society. While traditional society is organized around kinship and religion affiliations, modern society is based on individuality and specialization. Boundaries of space and community in the former overlap with each other, whereas the relations among communities in the later are transpatial, independent of space boundaries; thus, social groups have multiple memberships. The third type of community could have both spatial and transpatial links as a certain social group can maintain its solidarity and at the same time can integrate into employment and education with the host society; in this sense, the concentration may not equal segregation. In fact, the notion of solidarity was the main paradigm on which many theorists built their thoughts about society constitution.

On the other hand, architects and city planners like Oscar Newman and journalists like Jane Jacobs have grasped the notion of solidarity and social responsibility in different views: while Newman suggests the notion of local territories through encouraging enclosed spaces with limited access, Jacobs

Figure 15. Agent-based analysis for two different plans. Note: lighter parts have a higher movement rate
Source: the author

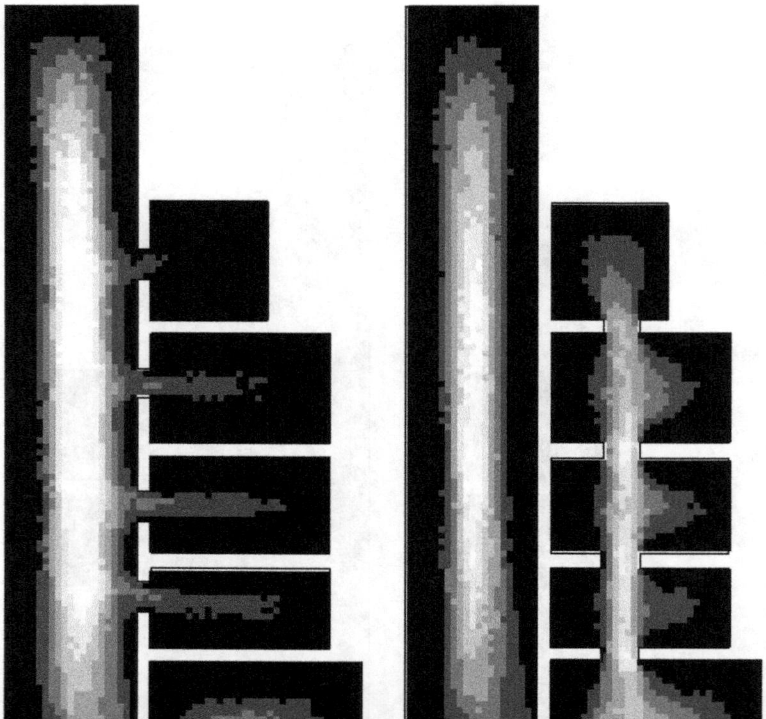

advocates large scale integration of urban spaces. Both schools of thoughts stress the role of the human dimension in increasing security in cities. Nevertheless, Oscar Newman's concept of urban territoriality was rejected by philosophers like Bill Hillier who described Newman's theory of defensible space as normative, narrow and static.

Whilst some scholars like Erving Goffman depict the space as a background of activities, others — like Hillier and Hanson— believe that social information is implanted in the spatial configuration of the urban layout. Hillier argues that space is a physical entity. According to him, while space and society are mutually related, they must be separated from each other and then bridged through studying how they affect each other. Hillier tried to explain the cause' and 'effect' of space in an objective way using a human-focused approach.

Space syntax is a set of robust theories and tools for mapping the likely social effects of spatial arrangements. The methodology can mathematically measure local and broader global physical properties of urban environments such as connectedness and intelligibility. Space syntax describes space in terms of abstract properties of relational patterns. Unlike traditional public life studies, space syntax technique observes life indirectly through predicting movements. With the help of computer programes scripted with selected principles of human behaviour, space syntax depicts where and how people will probably walk around in urban spaces. Processed mathematical models show how accessible a space is from all other spaces of the system and how many times the space is used as a part of the shortest paths between other spaces. The more accessible the space, the higher the potential for movement in the area.

On the basis of movement rate, movement-seeking activities (e.g. commercial uses) will be located along accessible routes, whilst less-movement activities (e.g. residential uses) will migrate to the rest parts of the urban system. Well-used street environments are expected to be richly accessible and informally supervised by people, while lifeless suburbs will have lower degrees of spatial accessibility. In recent years, the space syntax methodology has shifted from exclusive concern with movement and functions to other implications of a spatial configuration such as criminal activity and building density.

Although widely used and published since the 1970s in a number of countries, space syntax has various shortcomings. First, processed maps are difficult for non-specialists to read without understanding the theory and the calculations in the method itself. Second, there is a great deal of simplification in space syntax studies when looking only at the two-dimensional spatial patterns. Furthermore, spaces syntax models focus exclusively on movement and social interfaces; however, social practice and human interaction are more complex and not easily predictable. In other words, space syntax as a socio-spatial theory is described by some sociologists and public life scholars as environmental or spatial deterministic. If so, experiencing the city is mandatory for constructing more accurate models and to get a grip on the interaction between public life and form. This does not mean that space syntax technology and other new methods will not contribute significantly to future social life studies. Ideally, the methods and tools should not be taken for granted; rather, they should often be deeply informed by observations as well as a prerequisite knowledge about buildings and city districts.

REFERENCES

Al-Ghatam, W. (2012). Cultural movement patterns and social implications in space of villages absorbed by cities in Bahrain. In Proceedings of the 8th International Space Syntax Symposium (8210:01-31). Santiago: Pontificia Universidad Católica.

Al Sayed, K., Turner, A., Hillier, B., & Iida, S. (2014). *Space Syntax Methodology*. London: University College London Press.

Alexander, C. (1966). A city is not a tree. *Design Magazine*, *206*, 46–55.

Appleyard, D. (1981). *Livable Streets*. Berkeley, CA: University of California Press.

Benedikt, M. L. (1979). To take hold of space: Isovists and isovist fields. *Environment & Planning B*, *6*(1), 47–65. doi:10.1068/b060047

Bullock, N., & Dickens, P. (1975). Use of models in planning. In L. Martine & L. March (Eds.), *Urban space and structures* (pp. 97–112). Cambridge, UK: Cambridge University Press.

Can, I. (2012). *In-between space and social interaction: A case study of three neighborhoods in Izmir* (Doctoral thesis). University of Nottingham.

Cohen, J., & Arato, A. (1994). *Civil society and political theory*. Cambridge, MA: The MIT Press.

Craib, I. (1997). *Classical Social Theory: An Introduction to the Thought of Marx, Weber, Durkheim, and Simmel*. Oxford, UK: Oxford University Press.

Dingley, J. (2008). *Nationalism, Social Theory and Durkheim*. Hampshire, UK: Palgrave. doi:10.1057/9780230593107

Durkheim, E. (1893). The Division of Labor in Society. New York: Macmillan Publishing Company.

Elshater, A. (2015). Urban Design Redux: Redefining a professional practice of specialization. *Ain Shms Engineering Journal*, *6*(1), 25–39. doi:10.1016/j.asej.2014.08.004

Gehl, J. (1987). Life between buildings: Using public space (J. Koch, Trans.). New York: Van Nostrand Reinhold.

Gehl, J. (2010). *Cities for People*. Washington, DC: Island Press.

Gehl, J., & Svarre, B. (2013). *How to study public life*. Washington, DC: Island Press. doi:10.5822/978-1-61091-525-0

Gehl Architects. (2007). *Public Spaces, Public Life*. Sydney: City of Sydney.

Giddens, A. (2009). *Sociology*. Oxford, UK: Wiley.

Greene, M. (2002). *From Informal Settlements to Sustainable Neighbourhoods: The role of spatial configuration in the urban system* (Unpublished doctoral dissertation). University of London.

Hanson, J. & Hillier, B. (1987). The architecture of community: some new proposals on the social consequences of architectural and planning decisions. *Architecture et Comportment/Architecture and Behaviour, 3*(3), 251 - 273.

Hernandez, B., Hidalgo, M. C., Salazar-Laplace, M. E., & Hess, S. (2007). Place Attachment and Place Identity in Natives and Non-natives. *Journal of Environmental Psychology*, *27*(4), 310–319. doi:10.1016/j.jenvp.2007.06.003

Hillier, B. (1996a). Cities as movement economies. *URBAN DESIGN International*, *1*(1), 41–60. doi:10.1057/udi.1996.5

Hillier, B. (1996b). *Space is the Machine*. Cambridge, UK: Cambridge University Press.

Hillier, B. (1997). Cities as movement economies. In P. Droege (Ed.), *Intelligent Environments: Spatial Aspects of the information revolution* (pp. 295–344). Elsevier B.V. doi:10.1016/B978-044482332-8/50020-2

Hillier, B. (1999). Centrality as a process: accounting for attraction inequalities in deformed grids. In *Proceedings of the 2nd International Space Syntax Symposium* (vol. 1, pp. 06.1: 06.20). Brasilia.

Hillier, B. (2004). Can streets be made safe? *URBAN DESIGN International*, *9*(1), 31–45. doi:10.1057/palgrave.udi.9000079

Hillier, B. (2009a). Spatial sustainability in cities: organic patterns and sustainable forms. In *Proceedings of the 7th International Space Syntax Symposium* (pp. k01.1-k01.20). Stockholm: KTH.

Hillier, B. (2009b) The genetic code for cities – is it simpler than we thought? In *Proceedings of Complexity in Cities conference*. Delft: TU Delft Faculty of Architecture.

Hillier, B. (2009c) What do we need to add to a social network to get a society? Answer: something like what we have to add to a spatial network to get a city. In: D. Koch, L. Marcus, & J. Steen (Eds.) *Proceedings of the 7th International Space Syntax Symposium* (pp. p. 42). Stockholm: Royal Institute of Technology (KTH).

Hillier, B., & Hanson, J. (1984). *The Social Logic of Space*. Cambridge, UK: Cambridge University Press. doi:10.1017/CBO9780511597237

Hillier, B., Hanson, J., Peponis, J., Hudson, J., & Burdett, R. (1983). Space Syntax: A different urban perspective. *The Architects Journal*, *178*, 47–63.

Hillier, B., Penn, A., Hanson, J., Grajewski, T., & Xu, J. (1993). Natural Movement: Or, configuration and attraction in urban pedestrian movement. *Environment & Planning B*, *20*(1), 29–66. doi:10.1068/b200029

Hillier, B., & Shu, S. (1999). Design for secure space. *Planning in London*, *29*, 36–38.

Hillier, B. & Stonor, T. (2010) Space Syntax- Strategic Urban Design: City Planning Review. *The City Planning Institute of Japan*, *59*(3), 7-11.

Hillier, B., Yang, T., & Turner, A. (2012). Normalising least angle choice in Depthmap and how it opens up new perspectives on the global and local analysis of city space. *Journal of Space Syntax*, *3*(2), 155–193.

Jacobs, A. (1993). *Great Streets*. Cambridge, MA: MIT Press.

Jacobs, J. (1961). *The Death and Life of Great American Cities*. Singapore: Penguin Books.

Klaasen, I. T. (2003). *Knowledge-based design: developing urban & regional design into a science*. Delft: Delft University Press/Techne Press.

Klarqvist, B. (1993). A space syntax glossary. *Nordic Journal of Architectural Research*, *2*, 11–12.

Krüger, M. J. T. (1979). An approach to build-form connectivity. *Environment and Planning. B, Planning & Design*, *6*(1), 67–88.

Levitas, S. (1978). Anthropology and the sociology of the streets. In S. Anderson (Ed.), *On streets* (pp. 225–240). Cambridge, MA: The MIT Press.

Lynch, K. (1960). *The Image of the City*. Cambridge, MA: The MIT Press.

Magda, M. (2003). *An investigation of the relation of space to society: a discussion on Giddens, H. Lefebvre and space syntax* (M.Sc. thesis). University College London.

Martin, L. (1975). The grid as generator in urban space and structures. In M. Carmona & S. Tiesdell (Eds.), *Urban design reader* (pp. 70–82). Architectural Press.

Michelson, W. (1970). *Man, and his urban environment: A sociological approach.* Reading, MA: Addison-Wesley.

Mohamed, A. (2010). *Understanding urban navigation: way-finding within the city* (Unpublished M.Sc. thesis). Ain Shams University.

Netto, V. M. (2007). *Practice, communication and space: A reflection on the materiality of social structures* (Unpublished doctoral dissertation). University of London.

Newman, O. (1972). *Defensible space: crime prevention through urban design*. New York: Macmillan.

Newman, O. (1996). *Creating defensible space*. Washington, DC: U.S. Dept. of Housing and Urban Development, Office of Policy Development and Research.

Palacios, J. (2007). *The Catholic Social Imagination: Activism and the Just Society in Mexico and the United States*. Chicago: The University of Chicago Press. doi:10.7208/chicago/9780226645025.001.0001

Parsons, T. (1985). *Talcott Parsons on Institutions and Social Evolution: Selected Writings (Heritage of Sociology Series)*. Chicago: The University of Chicago Press.

Peponis, J. (1989). Space, culture and urban design in late modernism and after. Ekistics, 56(334/335), 93-108.

Pongsmas, N. (2004). *Configuration of Urban Space and Social Sustainability of Urban Neighborhood: A Case Study on the City of San Diego at the Dawn of the Twentieth-First Century* (Unpublished doctoral thesis). Texas Tech University.

Project for Public Spaces, Inc. (2000). *How to Turn a Place Around: A Handbook for Creating Successful Public Spaces*. New York: Project for Public Spaces, Inc.

Psarra, S. (2009). The ghost of conceived space – what kind of work does or should space syntax perform for architecture. In *Proceedings of the 7th international space syntax symposium* (089:01-089:10). Stockholm: KTH.

Psarra, S. (2010). The Ghost of Conceived Space – What Kind of Work Does or Should Space Syntax Perform for Architecture. *The Journal of Space Syntax*, *1*(1), 17–29.

Sailer, K., Pomeroy, R., Raheem, A., Budgen, A., & Lonsdale, N. (2012). The generative office building. In *Proceedings of the 8th International Space Syntax Symposium* (pp. 8010:01-8010:25). Santiago: Pontificia Universidad Católica.

Shafiei, K. (2013). *The Fragmented Morphology of Spontaneous Settlements: The case study cities of Jeddah (Saudi Arabia) and Zahedan (Iran)* (Unpublished doctoral thesis). University of London.

Shu, C. F., (2000). Housing layout and crime vulnerability. *Urban Design International*, *5*(3-4), 177-188.

Siksna, A. (1997). The effects of block size and form in North American and Australian city centres. *Urban Morphology*, *1*, 19–33.

Space Syntax Limited. (2004). *Agora Cities for People 2004, D3.1 Technology Platform Report*. London: Space Syntax Limited.

Spencer, H. (1872). *Principles of psychology* (Vol. 2). London: Williams and Norgate.

Stolley, K. (2005). *The basics of sociology*. Westport, CT: Greenwood Press.

Tonkiss, F. (2013). *Cities by Design, The Social Life of Urban Form*. Cambridge, UK: Polity Press.

Topcu, M., Topcu, K., & Kabut, A. (2007), Movement economy dependent on urban design. In *Proceeding of the 6th International Space Syntax Symposium* (pp. 117-01: 117-06). Istanbul: Istanbul Teknik Üniversitesi.

Turner, A. (2003). Analysing the visual dynamics of spatial morphology. *Environment and Planning. B, Planning & Design*, 30657–30676.

Turner, A. (2004). *Depthmap 4: A Researcher's Handbook*. London: Bartlett School of Graduate Studies, UCL.

Turner, A. (2005). Could a road-centre line be an axial line in disguise? In *Proceeding of 5th International Space Syntax Symposium* (pp. 145-159). Delft: TU Delft, Tech. Press.

van Nes, A. (2011). The One- and Two-Dimentional Isovists Analysis in Space Syntax. In S. Nijhuis, R. Lammeren, & F. Hoeven (Eds.), *Exploring the Visual Landscape: Advances in Physiognomic Landscape Research in The Netherlands* (Vol. 2, pp. 163–183). Research in Urbanism Series.

Vaughan, L. (1999). *Clustering, Segregation and the Ghetto: The Spatialisation of Jewish Settlement in Manchester and Leeds in the 19th Century* (Unpublished doctoral dissertation). University College London.

Wang, M. (2012). *Extending Geographic Information systems to urban morphological analysis with a space syntax approach* (Unpublished M.Sc. thesis). University of Gävle (HIG).

Westin, S. (2011). The life and form of the city: An interview with Bill Hillier. *Space and Culture*, *14*(2), 227–237. doi:10.1177/1206331211404391

Whyte, W. (1980). *The Social Life of Small Urban Spaces*. New York: Ingram.

Chapter 12

Space Syntax Beyond Cairo Street Markets:
An Application of a Theory on Socio-Spatial Contemporary Practices

Amany Ramadan Arisha
https://orcid.org/0000-0002-9903-7350
Arab Academy for Science, Technology, and Maritime Transport, Egypt

Nancy Mostafa Abd El-Moneim
Arab Academy for Science, Technology, and Maritime Transport, Egypt

ABSTRACT

Street vending is a growing controversial phenomenon in urban environments. It is a survival strategy and an economic opportunity for countless numbers of marginalized vendors. However, the temporal presence of vendors is portrayed as the source of substantial urban issues, which detract from the quality of the urban public space and the public life of individuals. This chapter aims to propose a practical approach to understand the impact of vendors' temporal presence on the quality of urban space and social life. By space syntax theory, this study utilizes pragmatic methods in the fields of social and human sciences; to analyze the socio-spatial and temporal attributes of the vending phenomenon in relation to urban users' movement in a case study street market at Cairo. The findings introduce a syntactic methodology that highlights the profound relationship between users and informal urban markets to be applied in diverse contexts.

INTRODUCTION

Cairo Street Markets are a day-to-day reality in all countries especially the developing countries (Brown, 2006). It is a common occupation of selling or retailing merchandise and services in urban spaces including streets, alleyways, avenues, boulevards, parking lots, street intersections, sidewalks, and public parks (Bromley, 2000; Yatmo, 2008).

DOI: 10.4018/978-1-5225-9238-9.ch012

The public trading phenomenon is a representative of an activity mostly initiated by unemployment, rapid urbanization, disproportionate resource allocation, and lack of sustainable development (Kafafy, 2016; Li et al., 2018; Sarpong & Nabubie, 2015). It is the most effective opportunity for income generation in cities for countless numbers of marginalized vendors, who encounter limited chances for getting a formal work (Bhowmik, 2005; Donovan, 2008; Forkuor et al., 2017; Roever & Skinner, 2016). This income generation has positioned trading in urban environments as a survival strategy and a mean for a fair livelihood against the gross inequality and exploitation of the socioeconomic system in cities of hardship (Abusaada & Elshater, 2019).

Moreover, street vendors are the bedrock of the urban economy where each successive layer above utilizes vendors' presence to their advantage (Brown, 2006; Rajagopal, 2010a). In the sense that, vendors adapt their trading activities to meet the urban users' socio-economic needs (Bhowmik, 2005; Rajagopal, 2010b). Vending operations varies depending on the events in the space, the weather, and time of the day (Bromley, 2000), as well as potential customers' flow (Reid et al., 2010; Wongtada, 2014). Thus, vending activity is a practical approach to cater for seasonal, sporadic, and unusual demands of customers (Sarpong & Nabubie, 2015).

The street vending practice constitutes a crucial component of the informal economy (Roever & Skinner, 2016). This economic informality implies that even though vendors are associated with the offer of legitimate or socially acceptable merchandise and services, they lack the presence of any formal registration, taxation or licensing (Bouhali, 2018; Kafafy, 2017). In this specific situation, the term 'informal' seems, by all accounts, to be depicted as 'illegal' (Cross, 2000; Graaff & Ha, 2015). Missing the fact that the vendors' utilization of urban environments for their own livelihoods is a grassroots initiative that represents a remarkable case of self-improvement for those confronting hardship to comply with the unreasonable bureaucracy through their meager personal assets (De Soto, 1989).

Furthermore, vending operations take place on public or private property that is not legitimately assigned for trading activities, and at least in part outside business regulation, planning codes, or other regulatory requirements (Brown et al., 2010). Yet, not only the constraints of having no formal access to a trading space that constitutes an urban issue, but also the concentration of vendors in a particular location creating a market-place that is called 'urban market' is the source of substantial urban problems (Bromley, 2000; Hays-Mitchell, 1994; Recio & Gomez, 2013). For instance, city image degradation, traffic congestion, pedestrian movement blockage, urban space deterioration, disorder, low levels of cleanness, vandalism, and so on (Anjaria, 2006; Bromley & Mackie, 2009; Crossa, 2009; Donovan, 2008), this has positioned street vending as an urban threat to the quality of the urban public space and the social life of individuals.

A street vendor is conceptualized as an individual who offers merchandise accessible to be purchased by the public people without having a permanent built-up structure or entity (Bhowmik, 2005). Therefore, street trading operations could be stationary or may be mobile (Bhowmik, 2005; Bromley, 2000).

In this manner, vendors are a temporal urban feature that could appear and vanish at any time and location beyond the initially intended planning, so detracting from the city form and its image (Yatmo, 2008). This situation is caused by the vendor physical and visual appearance in the urban environment (Yatmo, 2009). This temporal presence may turn into a violation of the functional zoning code and public order, creating conflicts between the temporary and the fixed activities as the temporal dimension of the urban experience is usually slighted in urban planning and design. Consequently, public trading is revealed as a contested activity between various social groups over the utilization of urban environments.

The above overview illustrates the contradiction and dualism in views towards this contemporary urban practice; which has negatively overshadowed the potentials of the vending phenomenon. Public trading is not necessarily a problem, but rather a functional and resilient component of cities. In the sense that, vendors adapt the spatial urban environment to meet the users' socio-economic needs; which increase the vendor income level in return. Although the capacity of adaptation of this activity induces a process of transformations in urban environments. Yet, the relation between the sociological and locational potential of the vending activity is slightly neglected in the existing literature. The problem presented in this research is that the transformation of the urban space into a temporal informal trading organization changes the socio-spatial relations in the area; which impacts the vitality of urban and social contexts.

Rather than merely contributing to the overwhelming debate on the vendors' role in urban spaces. The following research seeks to acknowledge street vending not as an economic practice or a mechanism to cope with the socio-economic or political injustices, but as a socio-spatial activity to produce a vital urban space. The authors of the current work argue that applying the principles of contemporary digital technology to the field of street vending could bring a fresh perspective to understand the complex attributes of the informal urban marketplace.

The objective of this chapter is to propose a practical approach to understand the impact of vendors' temporal presence on the quality of urban public space and social life. To achieve these, this study analysis an urban public space that is utilized as a temporal market by 'space syntax theory' in a case study at Cairo, where vendors have become an essential aspect of the Egyptian urban scenario.

Space syntax—as a theory, approach, and method—is concerned with associating the physical aspects of urban environments with its functional, social, and behavioral attributes (Hillier, 2008; Koohsari et al., 2014; Penn, 2008). Over the past two decades, this discipline has been adequately contributing to various issues in cities (Lerman & Lebendiger, 2017; Marcus et al., 2010); such as vehicular movement planning and public transportation (De Koning et al., 2017; Lerman et al., 2014; Penn et al., 1998), urban sustainability and resilience for hazards (Maureira & Karimi, 2017), environmental and spatial cognition (Conroy-Dalton, 2003; Marcus et al., 2016; Osman & Suliman, 1994), land uses and urban economic development (Bolton et al., 2017; Penn & Turner, 2004), sprawl and slum upgrading (Karimi & Parham, 2012), in addition to crime and social segregation (Hillier, 2002; Vaughan, 2007).

Besides, space syntax software 'Depthmap' as an open source makes it easy to provide computational support for the development of urban morphological studies (Turner, 2011; Varoudis, 2012). Accordingly, 'space syntax' is ready to hit the mainstream as one of the significant fields that can enhance the dialogue between urban design theory, method, and empirical research (Griffiths, 2014; Karimi, 2018), even in the face of growing challenges for cities (e.g. Chen, 2017; Karimi, 2016; Serra & Pinho, 2013).

The uniqueness of this research is that it breaks new grounds by analyzing the socio-spatial and temporal attributes of street vending activity, utilizing a syntactical approach. This contribution could be a step in linking this practical application with informal street markets field in Egypt; which in return will enhance public space utilization, users' social life, and vendors' economic performance.

The next section discusses the role of public trading as a socio-spatial activity and the vitality of marketplaces as temporal public spaces. Followed by an interpretation of 'space syntax' concepts and measures, and the potentialities of this analytical approach from previous studies and real-life projects. The section afterward proposes a syntactic research methodology that highlights the profound relationship between users and informal urban markets to be applied in future urban design projects and regulation policies. The final section discusses the vending phenomenon in the Egyptian context, presenting a selected street market as an analytical case study.

URBAN MARKETS: A SOCIO-SPATIAL CONTEMPORARY PRACTICE

The marketplace is considered an essential part of urban traditions and cultural life throughout history (Tangires, 2003). Historical scholars have described marketplaces as the fundamental feature and the central constituency of the city (e.g., Al-Maqrizi, 1908; Braudel, 1979; Ibn-Khaldun, 1967; Mumford, 1961; Raymond, 2007; Wiet, 1964). Marketplace trading is the earliest and most dominant form of public activity in cities (Unlu-Yucesoy, 2013), as the first free enterprise zone (McMillian, 2002).

Ostensibly, the marketplace or street vending area represents spaces where consumers and vendors meet for commercial purposes; either at daily or periodic markets (Morales, 2011). It has a variety of names, such as public markets, traditional markets, farmers markets, craft markets, urban markets, and square or street markets (Bromley & Mackie, 2009; Unlu-Yucesoy, 2013). Marketplaces are owned and operated by various types of official or unofficial organizations, not just city governments. Yet, they are distinguished by the kind of merchandise and services offered - whether general or specialist, regardless of where and when transactions take place and whether the market's operations are formal or informal (Bromley, 1998).

Predominately, the presence of the urban marketplace relies upon its function within the city trading hierarchy, as shown in Figure 1. Cities and urban markets are inherently associated and reliant on one another (Schappo & Van Melik, 2017). Cities provide social demand, and location for urban markets, while urban markets create sustenance, profit, and culture to urban communities (Hiebert et al., 2015; PPS, 2003). Like this, urban markets are multifaceted places serving a variety of purposes; which are appropriate to their specific context and participants (Morales, 2009), and depend on the spatial, demographic, social, economic, and political structure of the city.

Regardless of the different names, functions, governance, and settings for marketplaces; dealing with the market as an economic event, and the marketplace as a mere container of commercial transactions

Figure 1. The main types of urban markets or street vending areas
Source: The authors

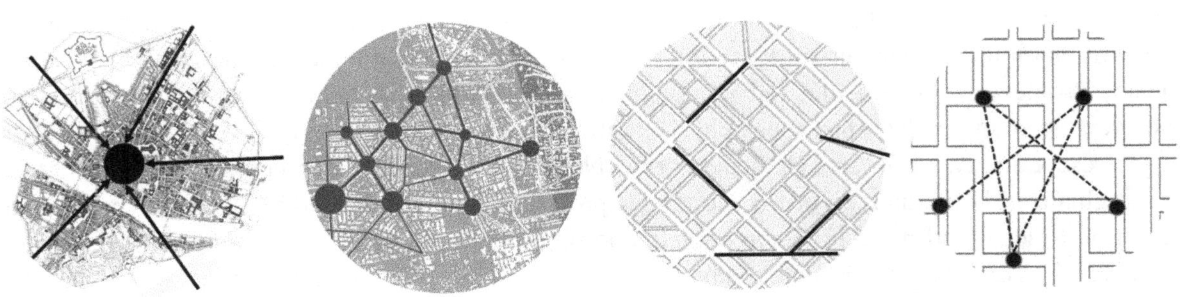

City Market: usually at the center of the city, constituting the main trading zone which sells general and specialist merchandise.	**Neighborhood Market:** usually serving each neighborhood in the district separately within the convenient walking distance of consumers, selling day-to-day goods.	**Street Market:** serving the passing consumers, its operations could be daily or temporally on specific times of day/week/year, type of merchandise sold varies according to times of operation.	**Hub Markets:** serving busy pedestrian locations such as: a public transportation node, an administrative building, a touristic spot, a public leisure garden, or at street intersection.

underestimates the role of marketplaces as a place-producing, recreational facility, and community development tool in regional and urban scope (Aliyah et al., 2017; Al-Maimani et al., 2014).

Marketplaces have a long history in integrating communities: economically, socially, culturally, and politically (Hiebert et al., 2015), as well as enhancing the health of marginalized populaces (Morales, 2011). At the macro level in the urban scope, urban trading activities are a link in the international/ local merchandise supply chain; through providing a cheap sales outlet for local and regional products (Rajagopal, 2010a). Urban markets are long-term investments as an essential part of food distribution infrastructure; which stabilizes the regional food systems and achieves local sustainability (Hedden, 1929). At the same time, the existence of any urban market impacts other regional land use planning, buildings utilization, street pattern, movement pattern, and will in the long run influence the urban development (Aliyah et al., 2017).

On the other hand, at the micro level, marketplaces are destinations of informal associations in which they unite disparate groups, reconstituting public spaces by producing fluid places that promote interaction, upward social mobility, encourage experimentation, and renew community spirit (PPS, 2003; Watson, 2009). More precisely, marketplaces are a conviviality addition to urban life as bright and inclusive public spaces creating a unique character to the city (Rajagopal, 2010a). Urban markets are adaptable spatial and temporal organizations that could be key drivers of progress as an avenue for economic improvement, regeneration, and sustainable living (Schappo & Van Melik, 2017). Given this context, marketplaces can be tools for diversifying business activity and creating multi-use and multi-scaled socio-economic environments which integrate activities, individuals, and spaces (Morales, 2011).

Contested Contemporary Practice

In the contemporary scene, with the enormous expansion of the urban region, the relationship between the city and the urban trading territory has altered (Unlu-Yucesoy, 2013). The rapid demographic growth of city populaces, rapid urbanization and migration, shrinkage of formal appropriate employment opportunities, and political restructuring, over and above that the adoption of economic liberal, and globalized policies since the 1980s and 1990s have forced increasing numbers of individuals into the informal sector, especially urban trading (Bhowmik, 2003; Sarpong & Nabubie, 2015). Indeed, The situation increase the competition for utilizing the urban space for livelihoods of countless numbers of marginalized vendors, causing problems with vehicular traffic, complicating the public life of urban residents, and placing considerable strain on the urban space physical infrastructure (Crossa, 2009).

Correspondingly, a principal challenge for the New Urban Agenda is to recognize the contributions of marginalized individuals to urban economies prosperity by promoting their access or progression transition to legal and more productive livelihood opportunities, side by side to spatial development strategies. These strategies promote the inclusivity, accessibility, and quality of urban public spaces as pillars on which sustainable urbanization could occur (UN General Assembly, 2016). Given the expanding number of individuals entering this occupation, few cities adequately balance the need to support livelihoods with the need to manage urban environments (Brown et al., 2010; Lyons & Snoxell, 2005). Consequently, the lack of economic and regulatory formal recognition anticipated a negative image towards vending operations as a non-urban and deviant pattern of behavior and attitude.

The majority of arguments that are marshaled against the urban vending activity are founded on vendors' utilization of urban public spaces, which are initially unintended for trading purposes (Onyango et al., 2012; Roever & Skinner, 2016). This issue has positioned their presence as a violation of functional zones that have been determined by urban planners and regulators. More precisely, vendors are a potential threat to existing socio-spatial order that is desired by hegemonic social groups.

Although urban public space is widely recognized as a common property resource to which all individuals have equal and free rights to access and use (Abd El-Moneim & Galal, 2017; Harvey, 2008, 2012; Lefebvre, 2003; Lynch, 1960). With regards to public trading, this is not the situation, as vendors' utilization of urban space is invariably contested including many competing and mutually contradicted interests (Boonjubun, 2017; Graaff & Ha, 2015; Maneepong & Walsh, 2013).

In the new millennium, urban public space is a commodity in which cities use to compete and promote themselves and their power (Graaff & Ha, 2015; Low & Smith, 2005). More precisely, cities' image is defined by the spatial attributes of the public space and by the social activities it accommodates, reflecting many aspects of the urbanizing world. Subsequently, the existence of traditional unregulated livelihoods practices in urban environments is a symbol of disorder, backwardness, chaotic, sinful (Reyes, 2013); which is contrasted against the modern urban agenda, whose main aspects are aesthetics and public order (Setsabi & Leduka, 2008). Vendors are considered, in this manner, a hindrance to drawing in both domestic and international investors and tourists (Crossa, 2009; Oz & Eder, 2012).

This chapter departs from the mainstream negative assumptions underlying past research on public trading activity. Provisionally speaking, vendors occurrences and the urban issues or conflicts that may pursue can be tended by revisiting the vending phenomenon from a fresh perspective that appreciates their role in the vibrancy of urban spaces. This highlights the need to explore informal urban marketplaces through the manifestation of three main issues: public spaces, public life, and vitality of urban or social contexts.

Street Markets as Public Spaces

Urban public space is portrayed as the physical space and spatial configurations which could impact individuals' movement, experience, and activities, moreover, the social relations that determine the creation, utilization, and control of that space within the city public realm (Brown, 2006; Carmona et al., 2003). It includes all the spaces that have accepted communal access or use rights, which could be formal parks, squares, streets, and vacant lands, also the areas between buildings and on pavements edge (Drummond, 2000; Gehl, 2011). Jacobs (1961) mentioned that *streets and their sidewalks are the principal public places of cities, as the most essential organs. Think about a city and what comes to mind? It is the street. If the city's streets look fascinating, the city looks interesting; if streets look dull, the city looks dull.*

Unlike any other urban space, streets are the foundation of cities which determine its spatial and visual characteristics with two primary functions: mobility and place (Tandon & Sehgal, 2017; Von Schonfeld & Bertolini, 2017), as shown in Figure 2. Regarding mobility function, streets host an essential role of transportation in terms of movement, flow, and accessibility of individuals and vehicular. While, regarding place function, these urban avenues retain a remarkable identity by creating a sense of place and offer social or recreational activities, that are functional and psychological importance to individual's lives (Gehl, 2010; Jacobs, 1993). More precisely, streets are quintessential social public spaces of cities (Mehta, 2013). Also, streets can be vibrant with a variety of economic attractions, therefore, keep the city alive through supporting the urban economy, as well as urban sociability (Garvin, 2016).

On the set of public trading, street markets epitomize the inconveniences and opportunities that accompany the utilization of urban environments for informal vending. Focusing on the positive potentials

Figure 2. The primary functions of urban streets
Source: The authors

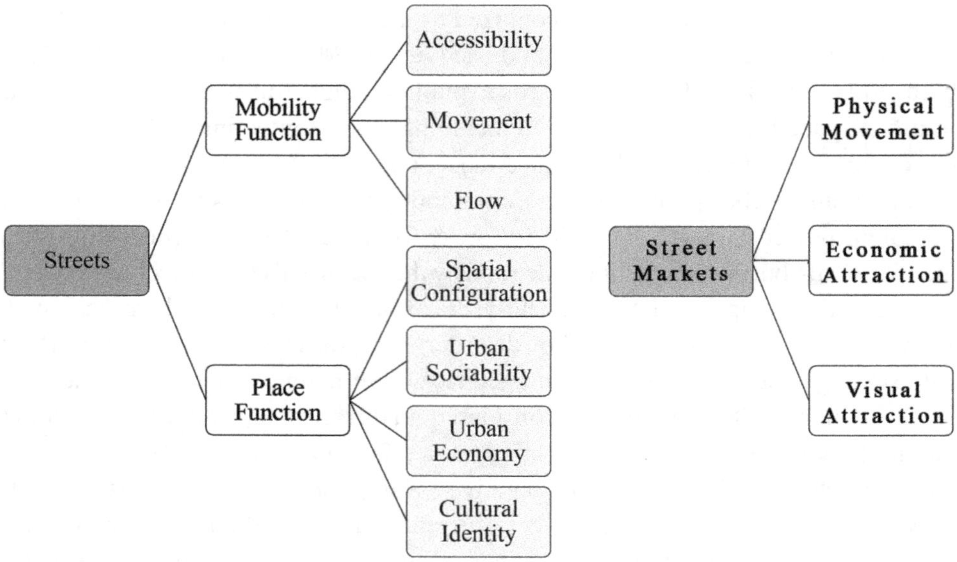

of this phenomenon, the presence and the configuration of street vendors in urban streets contribute to the quality of life and enhance the public safety as 'eyes on the street' (Abd El-Moneim & Galal, 2017; Carmona et al., 2003; Duneier, 1999; Jacobs, 1961; Watson, 2009). Bromley (1998) and Whyte (1980) describe the presence of vending activities in urban streets as a vibrant third place, which can revitalize the life of dull streets and flourish a lively social life. In the meantime, regulated and ordered vendors upgrade the visual image of streets with their colorful stalls, costumes, and merchandise, and serve as living signs to indicate where economic and social activity occurs (Yatmo, 2009).

In the street market context, there is a functional relationship between vending pattern, social pattern, and urban economy (Aliyah et al., 2017). Referring to urban space definition, informal markets can be vital public spaces when it integrates both the spatial dimension of the urban area, the temporal dimension of the urban experience, the social dimension of vendors and urban users, and the vending operations pattern (Schappo & Van Melik, 2017; Unlu-Yucesoy, 2013; Watson, 2009), as shown in Figure 3.

Socio-Spatial Attributes of Street Vending Activity

Leading to the next aspect of the chapter; urban markets represent a lively everyday arena in the city, constituting a crucial part of public social life (Efroymson, 2015; Rajagopal, 2010b). This life is consistent with the argument of departure in this section, that street vending is a resilient occupation, where a social system of norms governs urban spatial transformation process. Consequently, users' affinity or animosity relationship with informal street markets depends on an intricate socio-spatial and socio-economic attributes. Yet, there is a significant variation in the attributes that could influence the vendors' adaptation; which varies considerably according to the economic, social, regulatory, political structures, or even the degree of the official control within a specific spatial context (Forkuor et al., 2017).

Figure 3. The four dimensions of informal street markets as vital urban public spaces
Source: The authors

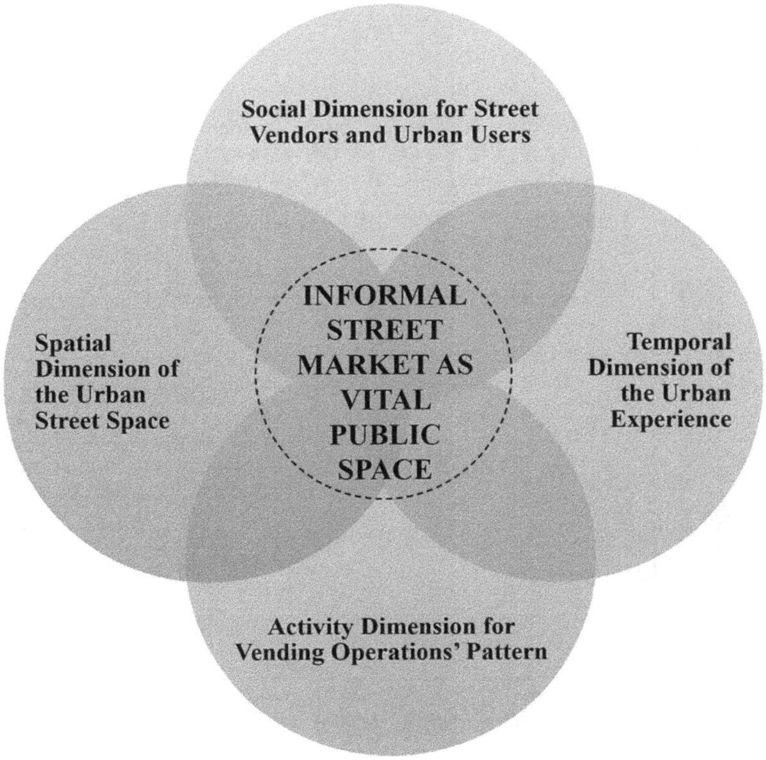

On the spatial level, the viability of informal street markets is featuring two main spatial levels, as shown in Figure 4. First, the street market as an 'urban entity'; which is connected to more extensive geographical areas that influence the vehicular traffic, pedestrian movement, buildings utilization, and regional land use planning, as well as business development (Aliyah et al., 2017; Schappo & Van Melik, 2017). In this context, the street market location and its proximity to other territories impact the urban development and enhances the locational utility for users. The convenience of the market location to urban consumers is a crucial determinant to predict its accessibility (Wongtada, 2014). For instance, the existence of street markets in a contested place within the proper walking distance that is compatible with the shopping process; can be tolerated as it reduces the miles traveled by inhabitants to fulfill their necessities (Bromley, 2000). Moreover, the proximity to surrounding regional and urban spaces which draw in or create a flow of pedestrian movement, e.g., residential regions, institutional zones, commercial centers, public transportation services. These issues are a vital determinant factor for vendors.

Second, the street market as a 'place-setting,' in terms of users' movement within the urban space and their accessibility to the building entrances. In this context, movement factors are related to the street physical features in terms of street width and type (e.g. shared street, pedestrian-only, and vehicular-lane), street network hierarchy (e.i. primary, secondary, and service), and streetscape elements (e.g. potential seating locations, shaded or not, and nearby activities). Moreover, the spatial configuration and distribution pattern, are essential determinants of the vending activity that either complicate or upgrade

Figure 4. The spatial attributes of the vending activity
Source: The authors

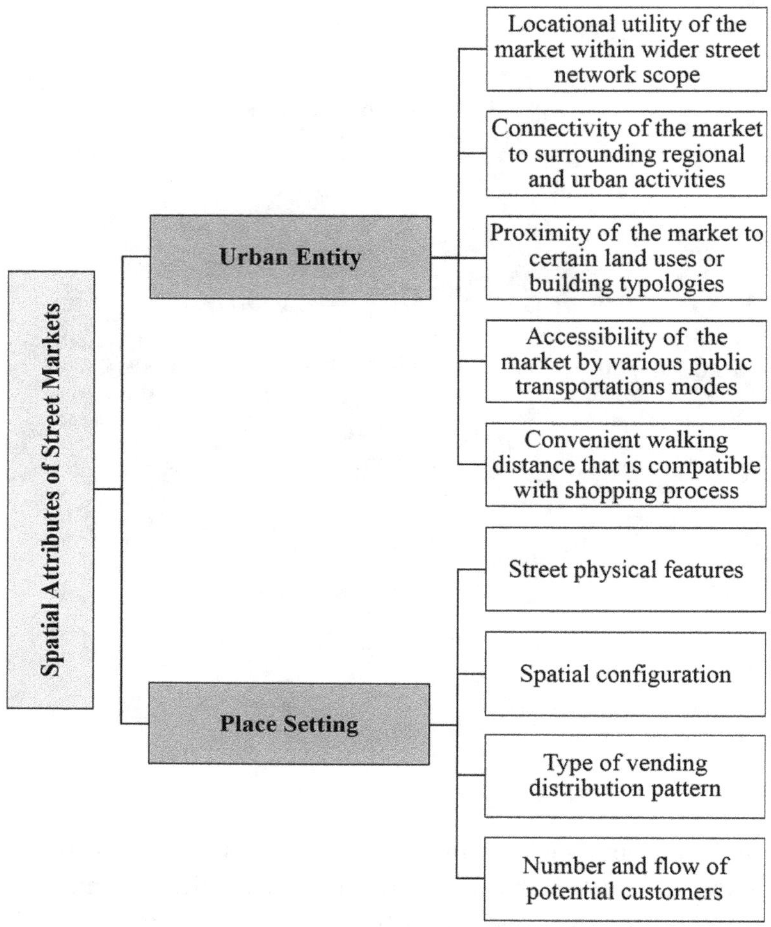

the vitality of public life for urban users (Hays-Mitchell, 1994; Rajagopal, 2010b). Both are considered a product of a series of social agreements and negotiations between street vendors and urban residents living or working in urban space (Efroymson, 2015). In this situation, contemporary vending activities are stimulated through interactions with prospecting customers to achieve the economic advantage of the vendor (Reid et al., 2010; Wongtada, 2014). So, the number and flow of potential urban consumers are other crucial factors that determine the locational preference of vendors (Hays-Mitchell, 1994).

On the social level, street vendors adapt their public trading activities regarding the scale of operation, location, time, number of vendors, terms of workforce, gender percentage, prices, and types of merchandise that sometimes are not available in off-street commercial stores; according to the needs of urban communities (Bhowmik, 2005; Bromley, 2000; Hays-Mitchell, 1994; Rajagopal, 2010b). This adaptation is a versatility that contributes to vending activity survival and prosperity. Moreover, the interaction between the various urban users (e.g. local residents, visitors, and passerby) and vendors, as well as the communication between vendors themselves; are essential factors to determine the user's reactions to the marketplace (e.g. feeling of safety, feeling of comfort, rate of visiting, land level of interest).

Regarding the vending activity pattern, vendors adjust the market processes and non-trading operations according to the socio-spatial attributes of urban users and urban space, in terms of on-site storage facilities, garbage collection, solid waste disposal, vendors' public toilets and sanitation services, as well as electricity, gas, and water supply networks, and so on. Vendors manage the informal marketplace and non-trading operations by their own, creating conflicts or coexistences between the various social groups over the utilization of the urban space. Similar to the pattern of customers flow, the market process pattern changes the rhythm of the existing socio-spatial relations in the street market.

Temporality Versus Vitality

The temporal dimension of the urban experience is an essential attribute; which could contribute to the livability and vitality of urban spaces. Vitality is one of the performance measurements of urban design; which deals with the socio-spatial success of urban environments (Lynch, 1960). It is a result of both the visual and physical quality of the urban space and the diversity of activities (Jalaladdini & Oktay, 2012), as well as the extent to which a place adapts to the temporal attributes.

In the context of informal street markets, one of the most crucial determinants of vitality is the temporal patterns of the vending activity (Hays-Mitchell, 1994; Li et al., 2018; Yatmo, 2009). This issue refers to timing and duration of vending operations, some vendors, the density of distribution, and the flow of pedestrians or vehicles across the day-night and during peak hours, as well as the presence of various activities. In this level, authors of the present work suggest four axial criteria interrelating socio-economic and tempo-spatial attributes of this phenomenon, to investigate the impact of informal vendors presence on the vitality of urban space and public life, as illustrated in Figure 5.

Space Syntax Theory

In the early seventies, a team led by Professor Bill Hiller developed 'Space Syntax' as an overarching paradigm that investigates society-space relation in the field of urban planning and architecture design (Hillier & Hanson, 1984), through a set of theoretical approaches that examine movement, visual perception, and human occupation. For instance, the theory of order and structure (Hanson, 1989), natural movement theory (Hillier et al., 1993), and movement economy theory (Hillier & Penn, 1996), as well as the theory of centrality as a process (Hillier, 2001). Furthermore, a set of analytical methods to investigate the spatial configuration influence on urban communities and vice versa (Hillier et al., 1983; Hillier & Hanson, 1998), such as axial analysis, visual graph analysis, and segment-angular analysis (Hillier & Hanson, 1984; Turner, 2003).

Space syntax is an arrangement of techniques for the representation, quantification, and analyzing any spatial configuration in buildings or urban settlements. Configuration is characterized as, at least, the relation between two spaces considering a third, and, at most, as the relations among spaces in a complex considering every single space in the complex. (Hillier et al., 1987)

The critical point for 'space syntax' development is that any social structure is inherently spatial, and contrarily that the configuration of any occupied space has on the basic level a social logic (Bafna, 2003; Hillier et al., 1993; Hillier, 2008). This 'configuration-function' or 'space-society' paradigm qualifies

Figure 5. The proposed four axial criteria to understand and develop a solution for the informal vending activity
Source: The authors

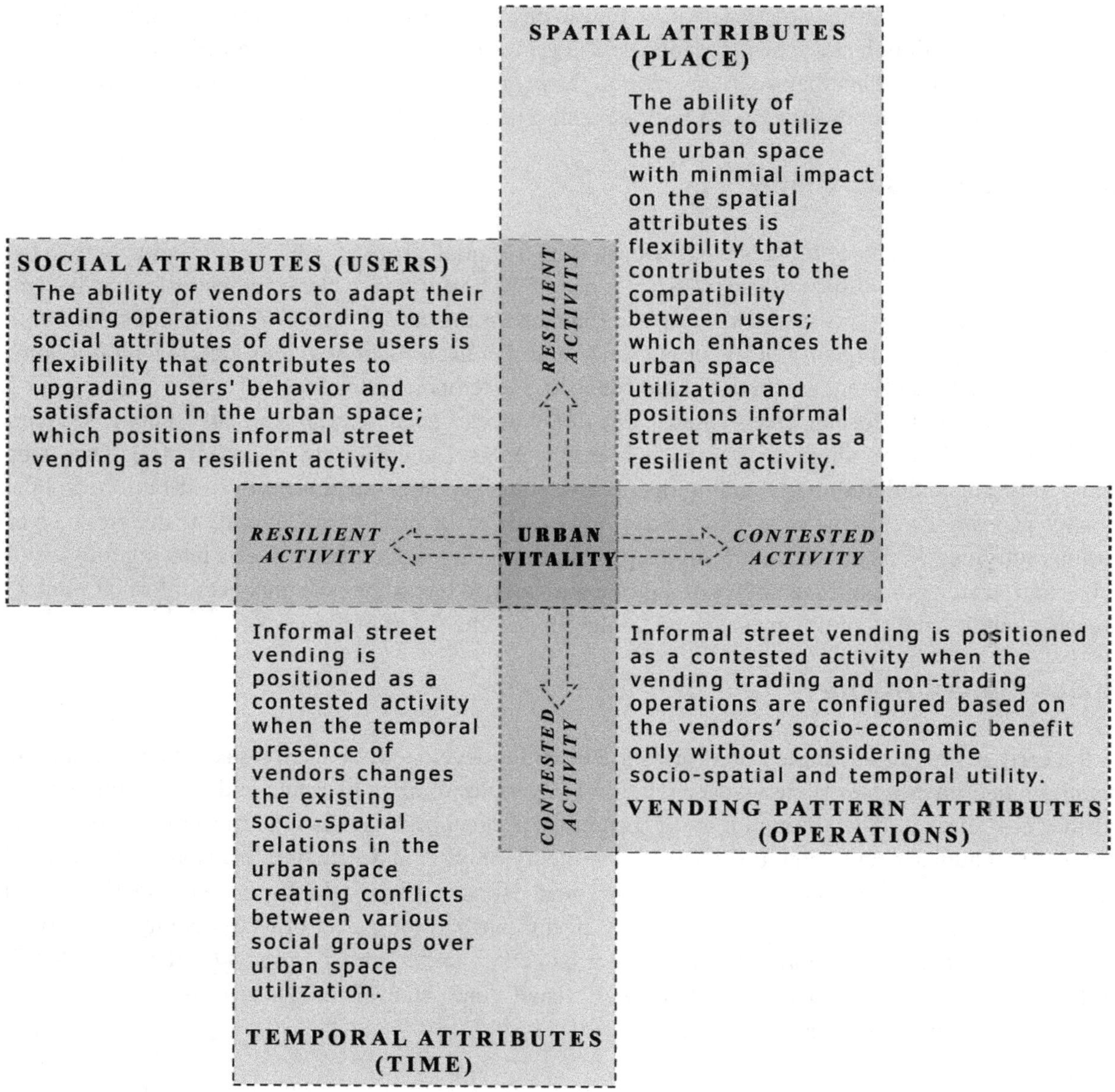

any physical structure utilizing a set of spatial descriptors, such as 'axial maps,' or 'convex maps,' or 'isovist maps,' as shown in Figure 6.

Syntactic maps are used to create a pattern of spatial relationships in a graph representation utilizing 'AGRAPH' software application, to be quantified mathematically by an analytical computer software 'Depthmap' (Turner, 2011; Varoudis, 2012), based on some syntactic measures, as summarized in Table 1.

Figure 6. Types of syntactic maps
Source: The authors

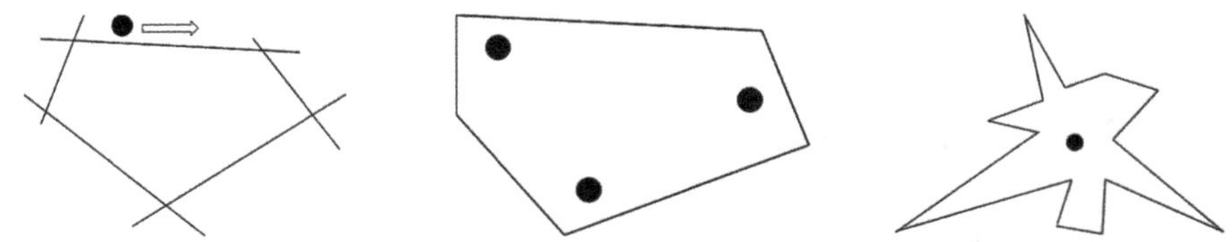

Axial Map:
Depicts the longest and the least number of 'axial lines' that connect all the spaces in a network layout with each other by links. It represents the way people move, along a line.

Convex Map:
Depicts the least number of convex spaces that fully cover a layout and the connections between them. It represents where people see each other and interact.

Isovist Map:
Depicts the visual and physical permeability of any convex space or axial line to all others. It represents the visibility field that has various shapes depending on the observer location.

Table 1. Syntactic measures

Syntactic Measures	
Graph	A figure that represents the relationship between all the linear 'axial lines' or nodal 'convex spaces' in the spatial layout.
Syntactic Step	The direct connection between one step and its immediate neighborhoods, or the change of direction from one axial line to another.
Justified Graph	A restructured graph with the goal that particular space is set at the base called 'the root space,' at that point all spaces that are one syntactical step far from it are placed on the first level above, spaces that are two syntactic steps far on the second level, and so on.
Degree of Depth	The essential measure extracted from a justified graph, which speaks to the relationship of a node to the graph that contains it, through estimating the sum of the connections (steps) that it could be passed by to move from space (node) to another on the spatial layout.
Connectivity	This variable estimates the number of immediate spaces that are one step away from each convex space or axial line. It is a local measure to count the number of connections that each space has to its neighborhood spaces.
Control Value	A dynamic local value to express the extent to which a specific space controls access to its immediate neighborhoods, considering alternative connections that each space has.
Local & Global Integration	A static local or global value that estimates the average depth of a particular space to all other spaces in the layout. Spaces can be positioned from the most integrated to the most isolated.

Source: The authors

Syntactic measures can be identified with social attributes, to test socio-spatial hypotheses or comprehend the social effect of spatial layouts or configurations (Al-Sayed, 2012; Hillier et al., 1993; Penn, 2008).

The Axial Analysis

The most fundamental syntactic approach for urban studies is 'axial maps'; which is an uncomplicated model of the spatial network. Axial maps depict the longest and the least number of 'axial lines' that connect all the spaces in a network layout with each other by links (Hillier & Hanson, 1984). Axial lines are the visibility lines (perception) or the movement lines (navigation) of the individuals and vehicles. Axial lines are 'scale-free' objects (Griffiths, 2014), which could be treated as continuous entities or segments to calculate metric distances (how far to travel), topological distances (how many changes in direction), and angular distances (what degree of angular shift). In the case of axial analysis, the spatial network is translated into a topological graph, where straight lines are represented as nodes and their intersections as edges.

Potentialities of Space Syntax About Vending Activity

Referring to the vending dilemma mentioned earlier, authorities and urban regulators guide their undertakings to eliminate or relocate the vending activities in urban spaces, which is usually portrayed as a phase towards enhancing the quality and visual image of the urban space. Taking into consideration that the location is a crucial determinant for vendors' income, eviction or relocation campaigns deprive vendors of their main livelihood (Roever & Skinner, 2016; Setsabi & Leduka, 2008). Subsequently, relocated vendors abandon the formalized markets and return to the same or alternative unplanned public trading locations (Boonjubun, 2017; Bromley & Mackie, 2009; Crossa, 2009; Donovan, 2008; Taylor & Song, 2016).

However, the existing literature on public trading has utilized diverse research methods (e.g., on-site observations, interviews, and questionnaires). Unfortunately, these conventional methods have been inconvenient to reveal that these elimination campaigns comprise an exclusionary socio-economic process, yet, improve vendors working conditions. Except for John et al. (2016) who utilized space syntax methods, revealing that vendors relocation from public spaces in downtown in Porto Alegre, Brazil to a formalized market building does not compile with the vendors' trading performance or customers' movement. 'Space syntax' can propose a practical approach to understand the impact of informal street markets location and spatial configuration on potential customers; which could enhance the activity pattern, individuals' social life, and vendors economic performance, with minimum negative impacts on the urban spatial experience, and surrounding street network.

A SYNTACTIC METHODOLOGY

A syntactic methodology is conducted to understand the impact of the transformation of urban streets into informal markets, as illustrated in Figure 7. In so doing, the study analyzes a street that is utilized as a temporal market; comparing pre, and during the vending operations. This chapter deals with the first spatial level of urban markets, as an 'urban entity.' Taking into consideration that 'space syntax' is a method to explore cities, not to assign universal meanings to particular syntactic measures (Griffiths, 2014). In this chapter, by the criteria mentioned earlier, the syntactic measures interrelate the socio-spatial and temporal attributes; to investigate the impact of the market presence on a wider network of urban streets and the users' movement in the urban environment.

STREET VENDING PHENOMENON IN THE EGYPTIAN CONTEXT

Cairo city the capital of Egypt, the largest city in Africa and the Middle East, is considered a standout political and cultural focal point in the region (Sims, 2012). This cosmopolitan city has experienced various intervals that rolled out radical changes in its social and economic life, which is reflected in its urban fabric (Abd El-Moneim & Galal, 2017; Nagati & Stryker, 2013). It is a mega-city, inhabited with

Figure 7. A proposed syntactic methodology and steps of the study
Source: The authors

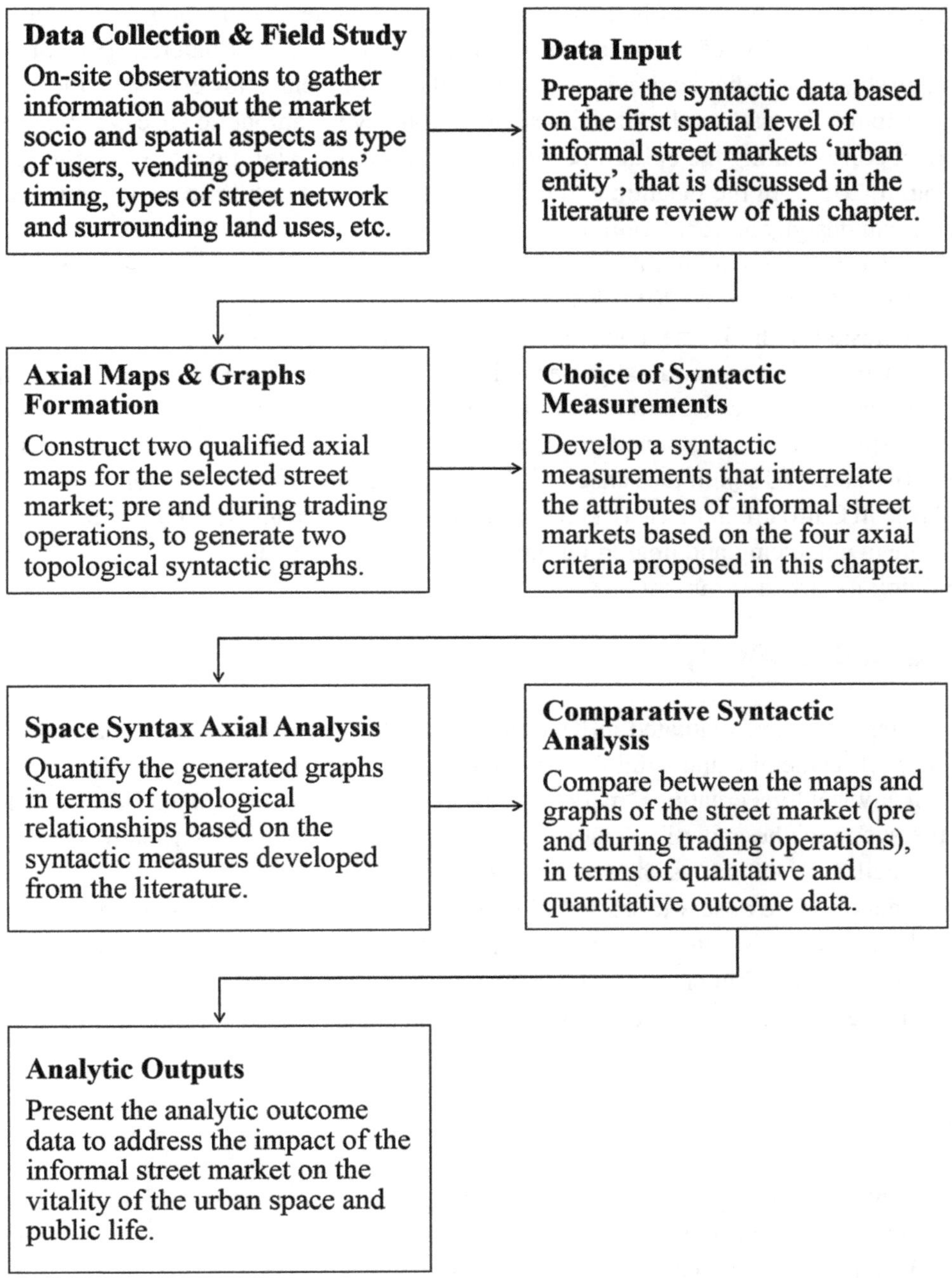

millions of individuals, with an increase in the demographic growth and urban agglomeration rate day by day (Abd Elrahman, 2016). With a high unemployment rate that is estimated to be 12.5% by the end of 2016 (CAPMAS, 2018), Cairo urban spaces are swallowed by the growing number of marginalized vendors as a necessary survival strategy (Kafafy, 2016).

With regards to that, Cairo is a city that holds an extensive variety of informality in housing and working urban environments that may influence the livability for its individuals (Stryker & Nagati, 2016). This image has situated as a 'city of hardship' with an abnormal level of difficulty that causes inhabitants to suffer in their daily lives from the absence of order (Abusaada & Elshater, 2019). Although Cairo meets the requirements of urbanization in its integrated environments, however, lack the knowledge to manage the attributes of vital public life. Consequently, many urban public spaces are not doing their essential role in consolidating the cultural, economic, and political significance of public spaces of the city (El Fayoumi, 2017). In the meantime, the social production of public space is regularly neglected by Egyptian urban design and regulation policies.

After the January 25 revolution, informal vending activities has known a dramatic spatial expansion in Cairo city centers (Brown et al., 2017; Kafafy, 2017). Correspondingly, urban environments for Cairo city become contested with diverse, conflicting activities, leading to a prevalent decline in the use of urban public spaces. Although vendors' presence is portrayed as an urban disaster in the Egyptian context, it represents an economical solution for several unemployed citizens (Bouhali, 2018; Kafafy, 2016). However, these spontaneous, unplanned practices participate in making sense of place in the Egyptian urban context (Elshater, 2014). In the sense that the misallocations of urban uses and activities generate the vendors' role in Cairo urban fabric (El Fayoumi, 2017). In this manner, vendors are actors who have an essential role in enhancing and upgrading the city functioning (Elshater, 2014), creating a system of sustainable living for urban communities.

The Analytical Case Study

Ismail Abaza Street Market is situated in Khairat, El-Sayeda Zainab District in Cairo, within an overcrowded district of residential and administration buildings. This strategic location offers an advantage that draws in many vendors, creating a marketplace called 'Souq El-Mowazafen' or 'Souq El-Wezarat.' Therefore, the scope of the analytical study is within walking distance from main streets and public transportation facilities (300 m), as shown in Figure 8.

The preliminary on-site observations reveal that Ismail Abaza is a morning market only, as shown in Figure 9, where vendors transform the public street into a pedestrian market as a response to the employees' demand for affordable products at a convenient distance from their workplace as well as their work rhythm. Therefore, the vending activity pattern is adaptable to one social group of urban users.

FINDINGS

Based on the syntactic investigation, the chapter aims to explore the impact of the transformation of Ismail Abaza Street to an informal morning market. This section presents comparatively the results of the socio-spatial and temporal analysis; pre and during the vending operations.

Axial Representation of a Spatial Urban Entity

The initial step is to construct two axial maps for the selected market; pre and during the vending operations. Formally speaking, this spatial representation reveals that Ismail Abaza St. as an urban marketplace segregates the urban fabric and the surrounding street network in the selected layout.

Figure 8. The selected street market and surrounding street network
Source: The authors

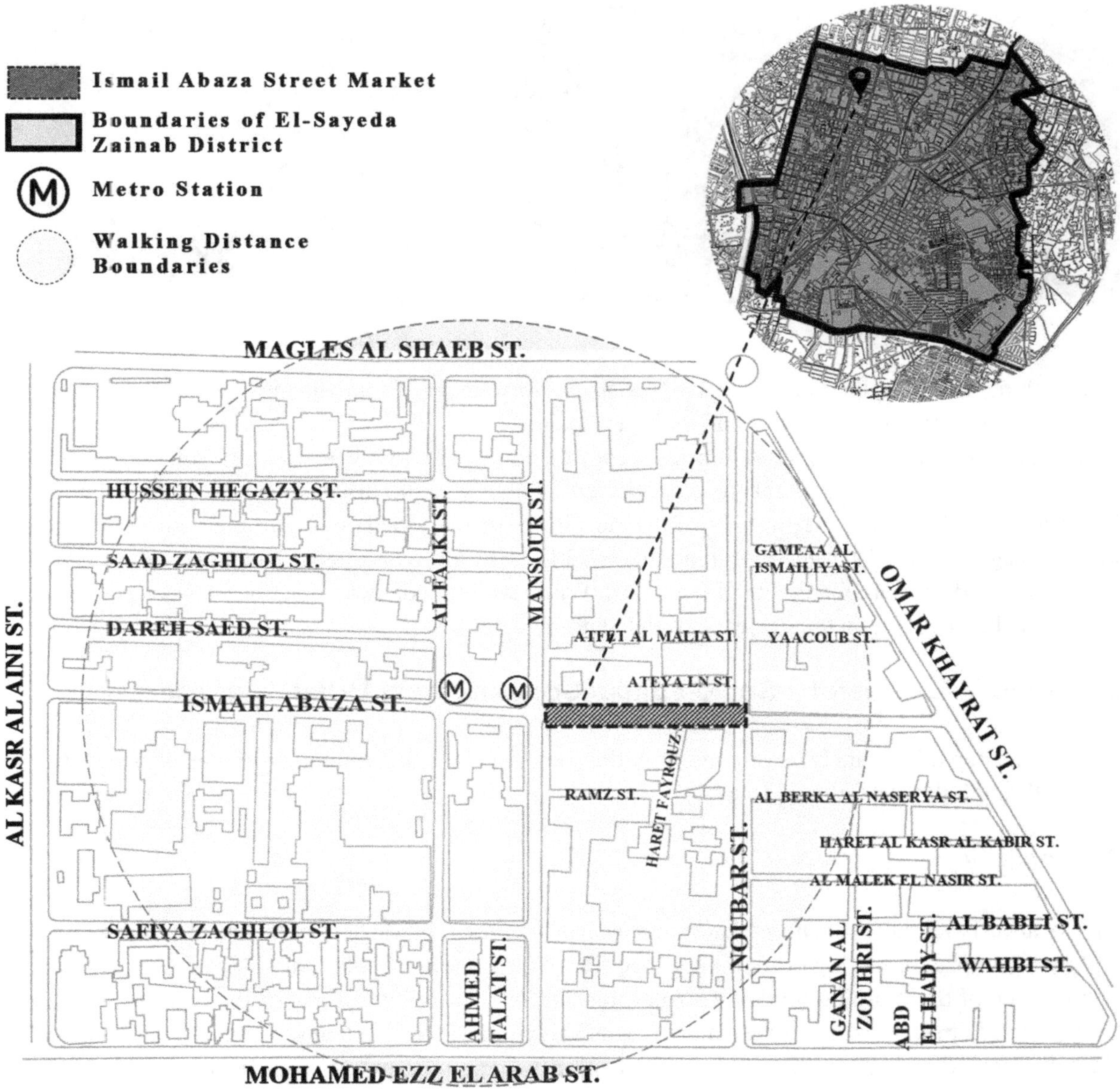

Ismail Abaza St. (segment 6), Magles Al Shaeb St. (segment 2), and Mohamed Ezz El Arab St. (segment 8); are three direct connection axes between Al Kasr Al Aini Street and Omar Khayrat Street. However, the location of segment (2) and (8) as the border lines for the selected urban area, has situated segment (6) as the only continuous movement path cutting through the whole spatial layout. Hence, the presence of the morning urban market at Ismail Abaza St. has not only sectioned the street network into two discrete urban environments but also increased the number of street segments (represent the change

Figure 9. The transformation of Ismail Abaza Street to an informal morning market
Source: The authors

of direction), as highlighted in Figure 10. Increasing the number of segments influences the immediate connections between urban spaces and their permeability.

The connections between axial lines are represented through the formation of syntactic graphs. In Figure 11, (a) shows that segment (6) is connected with another six urban streets (1, 9, 11, 13, 14, and 28). While, (b) shows that the presence of the morning market divided this continuous segment into two sporadic segments (6, 19), where segment (6) is still connected with three urban streets (1, 9, and 11), and segment (19) is combined with (14, and 29). While the immediate connection between segment (6) and (13) is blocked due to the presence of the vending activities.

Syntactic Measures of a Socio-spatial Contemporary Urban Practice

The second step of this analysis is to quantify the socio-spatial and temporal attributes of Ismail Abaza Market through the syntactic measures.

Connectivity

This local variable shows the immediate connections between urban spaces or the change of direction, to explore the impact of linking or unlinking specific streets from the whole spatial layout. Predominately, the presence of the morning market does not impact the connectivity of any of the other streets in the spatial layout directly, except for segment (6), as shown in Figure 12. Consequently, Ismail Abaza St. Market is considered an additional syntactic step, influencing the users' movement in and around the urban space. In the sense that the blockage of this part of the street shifts the traffic flow towards other streets in the spatial layout, and increases the distance traveled by the inhabitants or employees to reach their destinations, as well as decreases the permeability of surrounding buildings and urban spaces.

The Degree of Depth

Depth is a primary measure to investigate movement and accessibility within an urban space that is extracted from justified graphs, to represent the degree to which individuals must traverse to go from the root space to all the other spaces in the spatial layout. Accordingly, depth is useful to investigate the

Figure 10. Illustration of the market and surrounding street network as axial maps
Source: The authors

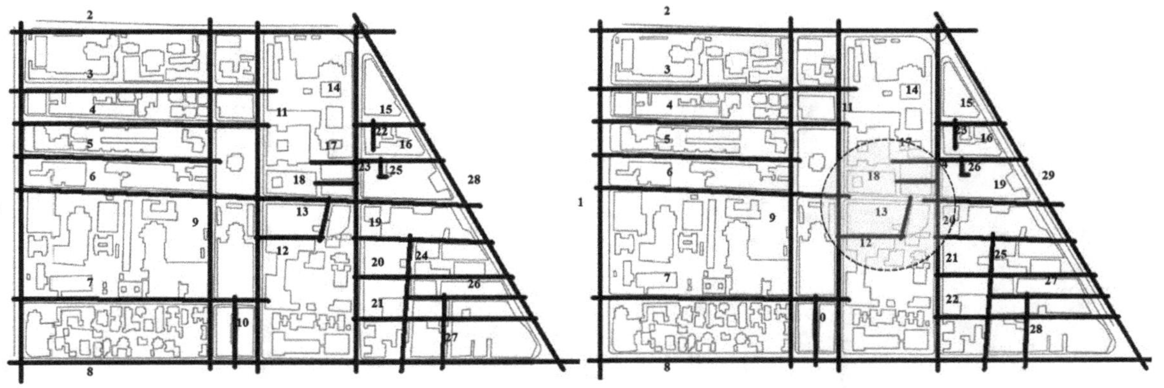

(a) The axial map of Ismail Abaza urban context, at night before the vending operations where the number of segments=28.

(b) The axial map of Ismail Abaza urban context, during the morning street market where the number of segments=29.

Figure 11. Syntactic graphs for the selected urban context
Source: The authors

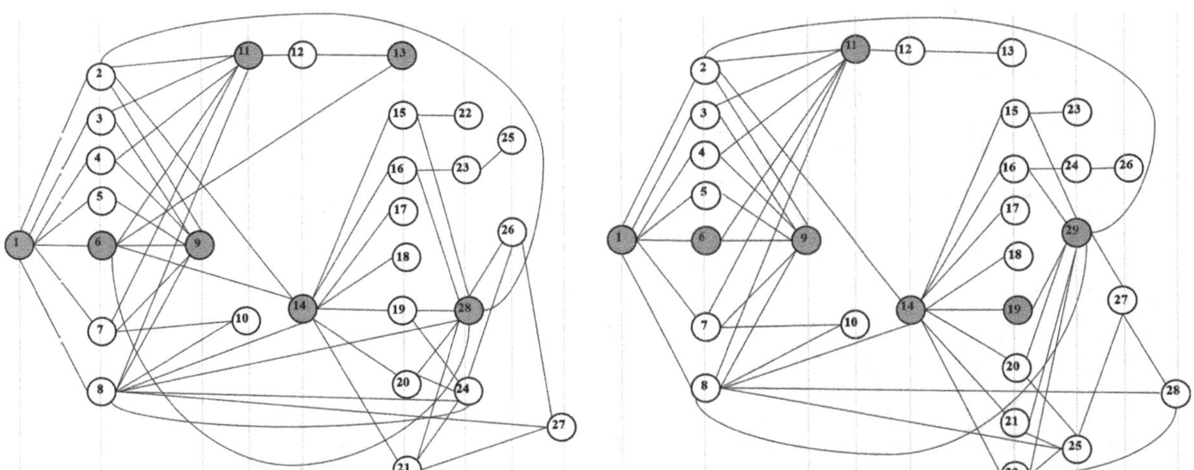

(a) The connectivity graph representing the syntactic relationships of the urban context segments, at night before the vending operations.

(b) The connectivity graph representing the syntactic relationships of the urban context segments, during the morning street market.

locational utility of the market for various users. Considering the morning market as the destination, it impeded the urban inhabitants' direct-access to their buildings as segment (6) is the only direct pathway to the residential area. While, for temporal employees and market visitors, the market is situated two steps away from the administration area and one step away from the metro station, as shown in Figure 13.

Figure 12. The number of immediate connections for each street in the layout
Source: The authors

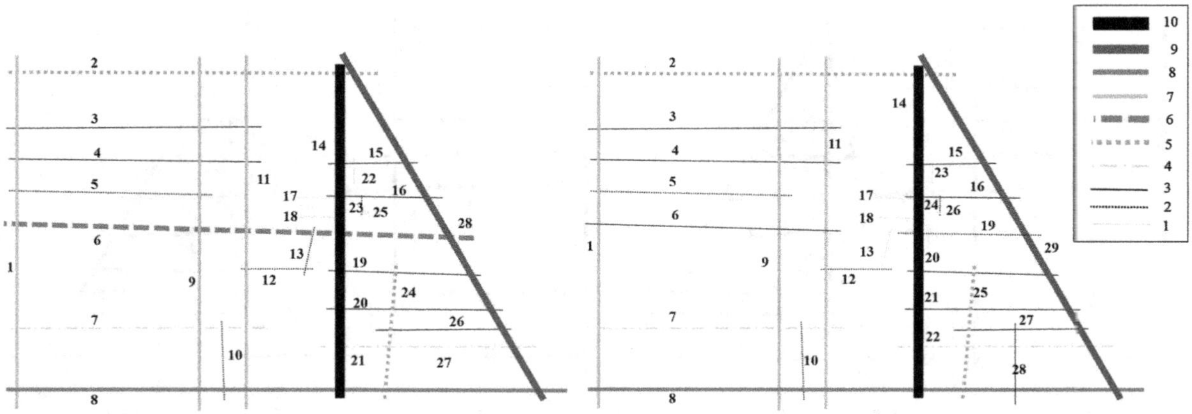

(a) The number of immediate connections that are one step away from each other; for each urban street in the layout, at night before the vending operations.

(b) The number of immediate connections that are one step away from each other; for each urban street in the layout, during the morning street market.

Figure 13. The degree of depth from segment (1)
Source: The authors

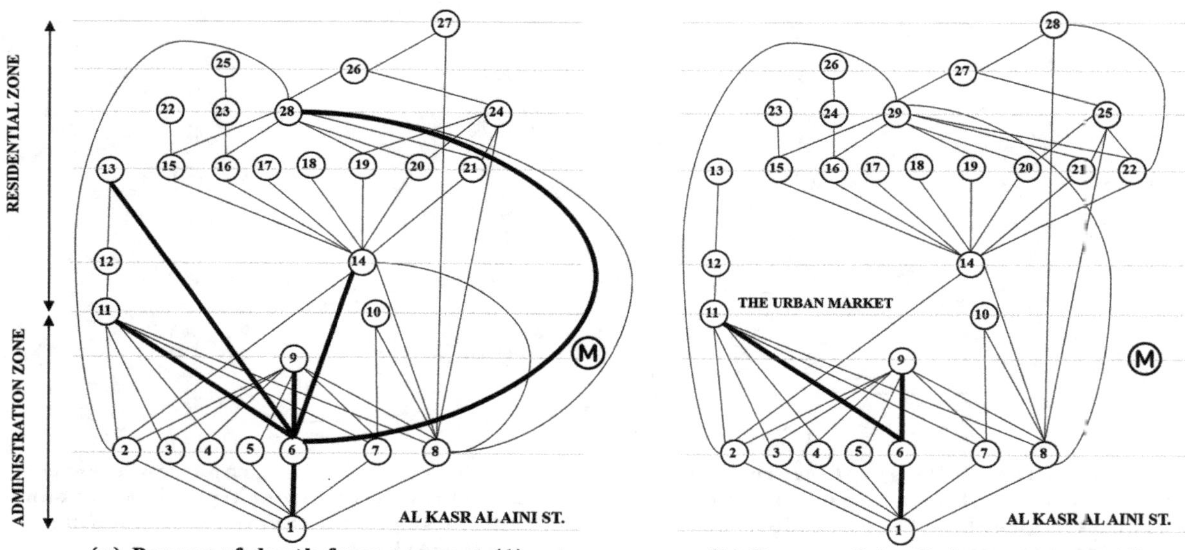

(a) Degree of depth from segment (1), at night before the vending operations.

(b) Degree of depth from segment (1), during the morning market.

On the contrary to segment (1), the justified graph from segment (2) as the root space shows a more integrated and accessible spatial relationship towards all the other urban spaces in the layout, as shown in Figure 14. Even when the morning market is operating, urban inhabitants, employees, and market visitors can reach their destinations efficiently.

Figure 14. The degree of depth from segment (2)
Source: The authors

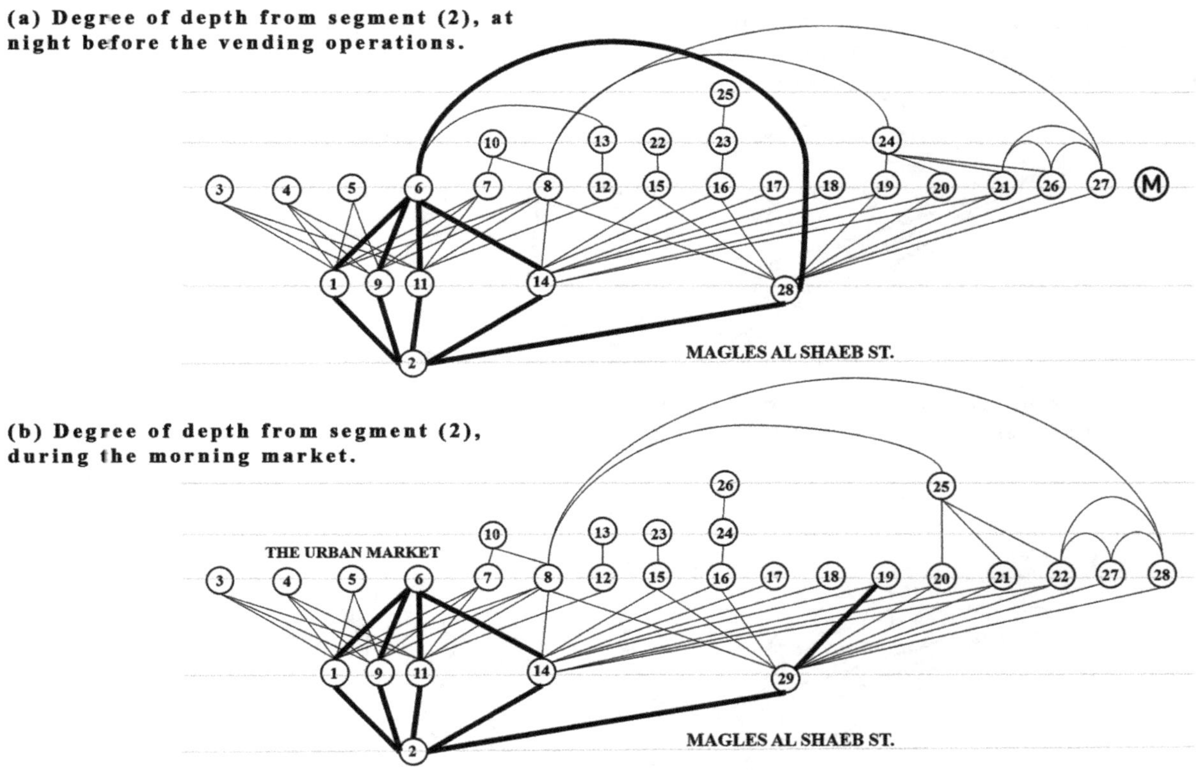

Similarly, the justified graph from segment (8) shows a high level of integration and accessibility for the various users, as shown in Figure 15. From segment (8) urban inhabitants can reach their residential buildings without passing by the urban market from segment (14), and urban customers can enter the street market from segment (11).

Predominately, the presence of the morning market increases the degree of depth of the spatial layout for the permanent urban inhabitants, while at the same time decreases the degree of depth for the temporal employees and potential customers, as shown in Figure 16. In this manner, the temporal presence of Ismail Abaza vendors is considered a response to the social demand of particular type of users at a convenient topological distance and specific times.

Control Value

The control value is the sum of the inverse values of the connectivity parameter of all alternative connections. Table 2 reveals that the market decreases the extent to which the surrounding streets control direct access to the residential area. Subsequently, inhabitants are experiencing a kind of hardship that is different from those visiting this urban space for work or services.

Figure 15. The degree of depth from segment (8)
Source: The authors

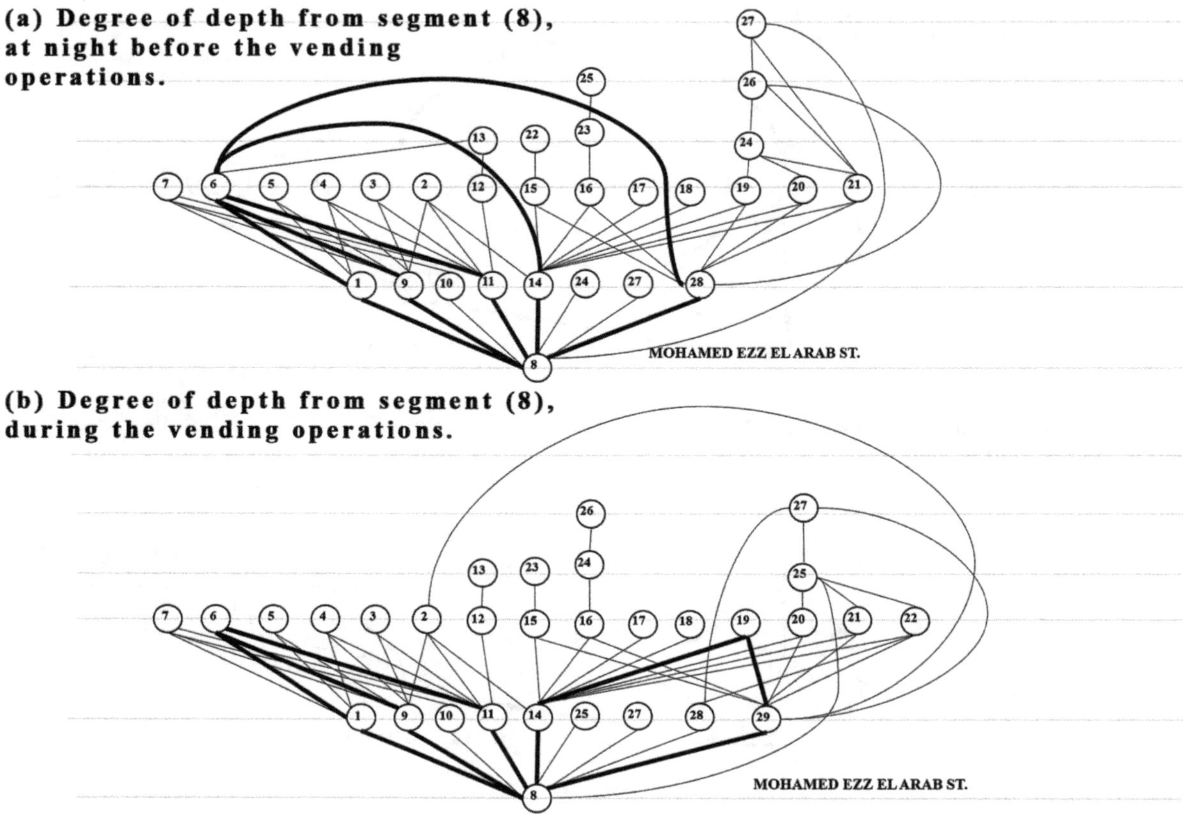

Integration

Similarly, the integration value is the inverse of the mean depth of each street in the layout. Hence, streets with less depth are more integrated into the whole spatial layout, while streets with more depth are the most segregated, as shown in Table 3.

DISCUSSION

After analyzing and comparing the axial maps and syntactic measures of Ismail Abaza St. Market; pre and during the vending operations, it is clear that the social demand of customers and the economic advantage of vendors play a vital role in the proximity of the market location to the administration zone and the metro station.

On the one hand, the social attributes determine the vending activity spatial properties. On the other hand, even when the urban layout lies within the same metric distance, the accessibility of the market and the residential area is influenced by different topological distances. This reveals that the spatial properties of the urban space determine how individuals utilize it in return.

Figure 16. The degree of depth from a segment (28, 29)
Source: The authors

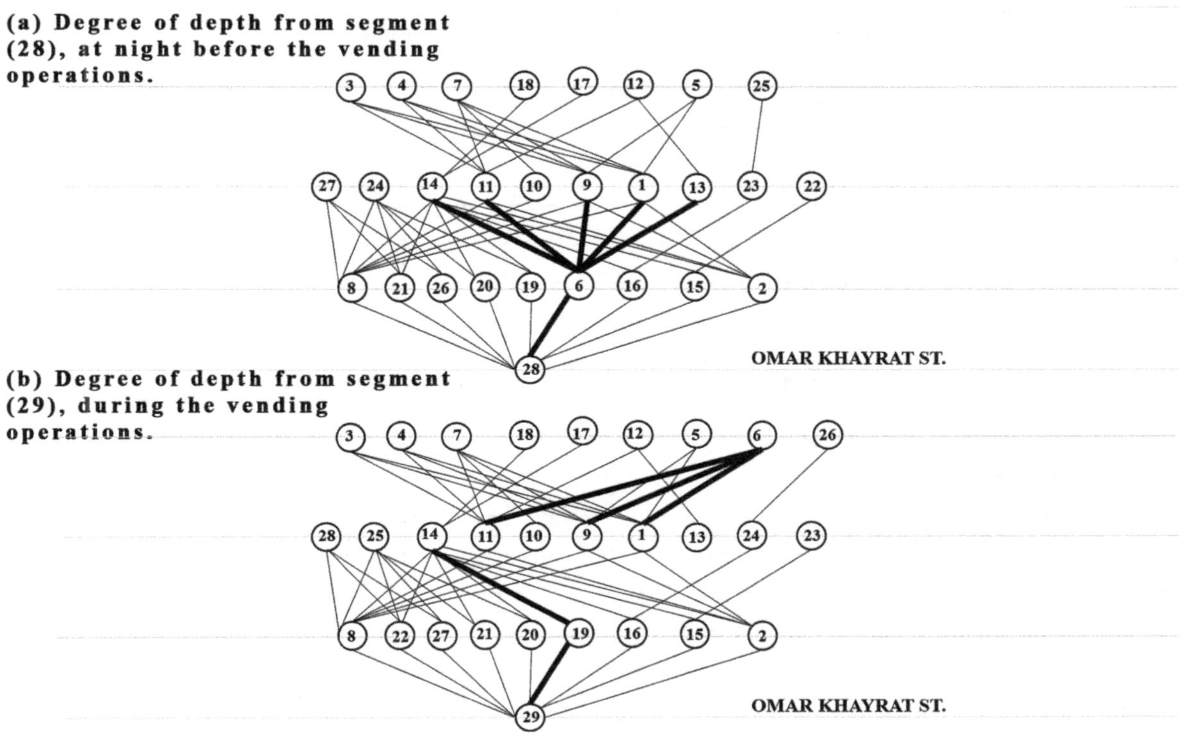

Referring to the main crux of urban planning to create vital urban space, the presence of the market as a response to the temporal and social dimension of employees needs regarding scheduling and location, reveals that Ismail Abaza St. Market is a vital urban public space for temporal users, while creates a kind of hardship for permanent inhabitants. However, this chapter agrees with Bromley (2000) and Gehl (2010) that there is difficulty in closing off public streets and utilizing it as marketplaces, which needs to be managed and appropriately regulated on various spatial and social levels.

Taking into consideration the diversity of users, where each social group is experiencing a different urban experience in terms of physical movement or social activities. However, digital research methods could not successfully address the social aspect of the urban experience. In this manner, a complementary traditional research tool of observing public life is needed in mapping the behavior and needs of various individuals within and around the informal street market. The key, therefore, is to interrelate socio-economic and tempo-spatial attributes in the area of informal street markets in Egypt, through utilizing traditional and digital research methods.

FUTURE RESEARCH DIRECTIONS

Space syntax can offer several significant advantages to understand the complexity of public trading, through investigating the market as an 'urban-entity' that is connected to a wider urban spatial network,

Table 2. Control values

Axial Line	Connectivity	Control Value
Surrounding Streets and the Market Pre Vending Operations		
#1	7	1.715
#2	5	1.309
#6	6	1.386
#8	8	1.488
#9	7	1.715
#11	7	1.595
#14	10	1.186
#28	9	1.110
Surrounding Streets and the Market During Vending Operations		
#1	7	1.715
#2	5	1.308
#6	3	1.202
#8	8	1.484
#9	7	1.715
#11	7	1.595
#14	10	1.065
#19	2	0.106
#29	9	0.987

Source: The authors

or as a 'place-setting' that is related with configuration and social behavior. Utilizing this analytic method in various social and urban contexts (historical areas, new urban settlements, and rural areas) could be an opportunity to tackle the negative implications of these contemporary practices on the urban experience. Furthermore, the authors of the current work are conducting follow-up research using traditional methods to reveal the informal street markets vitality and show how traditional and digital techniques can be complementary.

CONCLUSION

This study acknowledged the street market as an urban setting, which could be a vital public space when fulfilling the social needs of permanent and temporal users. This chapter extracted criteria that interrelate socio-economic and tempo-spatial attributes of vending activity as a step towards developing a practical approach for managing public trading in public spaces and upgrading the vitality of public life. Besides, the syntactic methodology conducted in this investigation can be valuable in the field of informal street markets as a practical research method; which clarifies the impact of the spatial properties of a particular space in determining how inhabitants utilize it and vice-versa, although it needs to be used side-by-side with traditional research tools for a more effective outcome.

Table 3. Integration value

Axial Line	Local Depth (steps=3)	Global Depth	Mean Depth	Integration
Surrounding Streets and the Market Pre Vending Operations				
#1	15	109	4.037	0.248
#2	52	56	2.074	0.482
#6	51	55	2.037	0.491
#8	48	61	2.259	0.443
#9	50	63	2.333	0.429
#11	50	63	2.333	0.429
#14	51	51	1.889	0.529
#28	53	53	1.963	0.509
Surrounding Streets and the Market During Vending Operations				
#1	15	114	4.071	0.246
#2	51	59	2.107	0.475
#6	35	87	3.107	0.322
#8	51	59	2.107	0.475
#9	51	68	2.429	0.412
#11	53	66	2.357	0.424
#14	52	56	2	0.5
#19	46	79	2.821	0.354
#29	56	56	2	0.5

Source: The authors

REFERENCES

Abd El-Moneim, N. M., & Galal, O. M. (2017). Cairene Public Realm: An Analysis of the City Public Spaces and its Quality. *1st International Conference on Towards a Better Quality of Life*. 10.2139srn.3163037

Abd Elrahman, A. S. (2016). Tactical Urbanism: A Pop-Up Local Change for Cairo's Built Environment. *Prodecia-Social and Behavioral Science*, *216*, 224–235. doi:10.1016/j.sbspro.2015.12.032

Abusaada, H., & Elshater, A. (2019). From Chaos to Order: Articulating the Urban Policies for Cities of Hardship. In U. G. Benna (Ed.), *Industrial and Urban Growth Policies at the Sub-National, National, and Global Levels* (pp. 41–64). IGI Global. doi:10.4018/978-1-5225-7625-9.ch003

Al-Maimani, A., Salama, A. M., & Fadli, F. (2014). Exploring Socio-Spatial Aspects of Traditional Souqs: The Case of Souq Mutrah, Oman. *Archnet-IJAR: International Journal of Architectural Research*, *8*(1), 50–65. doi:10.26687/archnet-ijar.v8i1.356

Al-Maqrizi, T.A. (1908). *Histoire d'Egypte de Makrizi* (E. Blochet, Trans.). Paris: E. Leroux.

Al-Sayed, K. (2012). A Systematic Approach Towards Creative Urban Design. *Design Computing, and Cognition*, *12*, 1–20.

Aliyah, I., Setioko, B., & Pradoto, W. (2017). Spatial Flexibility in Cultural Mapping of Traditional Market Area in Surakarta: A Case Study of Pasar Gede in Surakarta. *City, Cultura e Scuola, 10*, 41–51.

Anjaria, J. S. (2006). Street Hawkers and Public Space in Mumbai. *Economic and Political Weekly, 41*(21), 2140–2146.

Bafna, S. (2003). Space Syntax: A Brief Introduction to its Logic and Analytical Techniques. *Environment and Behavior, 35*(1), 17–29. doi:10.1177/0013916502238863

Bhowmik, S. (2003). National Policy for Street Vendors. *Economic and Political Weekly, 38*(16), 1543–1546.

Bhowmik, S. (2005). Street Vendors in Asia: A Review. *Economic and Political Weekly, 40*(22/23), 2256–2264.

Bolton, T., Francis, N., & Froy, F. (2018). The Impact of Space Syntax on Urban Policy Making: Linking Research into UK Policy. *11th International Space Syntax Symposium.*

Boonjubun, C. (2017). Conflicts over streets: The eviction of Bangkok street vendors. *Cities (London, England), 70*, 22–31. doi:10.1016/j.cities.2017.06.007

Bouhali, A. (2018). Negotiating Streets and Space in Transnational Trade Marketplaces in Oran (Algeria) and Cairo (Egypt): "Place Struggle" in the Commercial City. *Articulo-Journal of Urban Research,* (17/18), 1-21.

Braudel, F. (1979). *The Wheels of Commerce: Civilization and Capitalism, 15th-18th Century* (Vol. 2). New York, NY: Harper & Row.

Bromley, R. (2000). Street Vending and Public Policy: A Global Review. *The International Journal of Sociology and Social Policy, 20*(1/2), 1–29. doi:10.1108/01443330010789052

Bromley, R. D. F. (1998). Market-place Trading and the Transformation of Retail Space in the Expanding Latin American City. *Urban Studies (Edinburgh, Scotland), 35*(8), 1311–1333. doi:10.1080/0042098984376

Bromley, R. D. F., & Mackie, P. K. (2009). Displacement and the New Spaces for Informal Trade in the Latin American City Centre. *Urban Studies (Edinburgh, Scotland), 46*(7), 1485–1506. doi:10.1177/0042098009104577

Brown, A. (Ed.). (2006). *Contested Space: Street Trading, Public Space, and Livelihoods in Developing Countries*. Rugby, UK: ITDG Publications. doi:10.3362/9781780444703

Brown, A., Kafafy, N., & Hayder, A. (2017). Street Trading in the Shadows of the Arab Spring. *Environment and Urbanization, 29*(1), 283–298. doi:10.1177/0956247816673559

Brown, A., Lyons, M., & Dankoco, I. (2010). Street Traders and the Emerging Spaces for Urban Voice and Citizenship in African Cities. *Urban Studies (Edinburgh, Scotland), 47*(3), 666–683. doi:10.1177/0042098009351187

CAPMAS. (2018). *Egypt in Figures: 2018 (Statistical Yearbook, Ref. No: 71-01112-2018)*. Arab Republic of Egypt.

Carmona, M., Heath, T., Oc, T., & Tiesdell, S. (2003). *Public Places- Urban Spaces: The Dimensions of Urban Design*. New York, NY: Architectural Press.

Chen, X. (2017). Research on the Application of Space Syntax in Urban Planning, *World. Journal of Engineering Technology*, *5*, 29–35.

Conroy-Ddalton, R. (2003). The Secret is to follow your nose: Route path selection and angularity. *Environment and Behavior*, *35*(1), 107–131. doi:10.1177/0013916502238867

Cross, J. (2000). Street Vendors, Modernity and Postmodernity: Conflict and Compromise in the Global Economy. *The International Journal of Sociology and Social Policy*, *20*(1/2), 29–51. doi:10.1108/01443330010789061

Crossa, V. (2009). Resisting the Entrepreneurial City: Street Vendors' Struggle in Mexico City's Historic Center. *International Journal of Urban and Regional Research*, *33*(1), 43–63. doi:10.1111/j.1468-2427.2008.00823.x

De Koning, R. E., Van Nes, A., Ye, Y., & Roald, H. (2017). Strategies for Integrated Densification with Urban Qualities: Combining Space Syntax with Building Density, Land Usage, Public Transport and Property Rights in Bergen City. *11th International Space Syntax Symposium*.

De Soto, H. (1989). *The Other Path: The Invisible Revolution in the Third World*. New York, NY: Harper and Row.

Donovan, M. G. (2008). Informal Cities and the Contestation of Public Space: The Case of Bogota's Street Vendors, 1988-2003. *Urban Studies (Edinburgh, Scotland)*, *45*(1), 29–51. doi:10.1177/0042098007085100

Drummond, L. B. W. (2000). Street Scenes: Practices of Public and Private Space in Urban Vietnam. *Urban Studies (Edinburgh, Scotland)*, *37*(12), 2377–2391. doi:10.1080/00420980020002850

Duneier, M. (1999). *Sidewalk*. New York, NY: Farrar, Status, and Giroux.

Efrcymson, D. (2015). *Urban Menace or Valuable Asset? The Social and Economic Role of Street Vendors in Cities*. Health Bridge.

El Fayoumi, M. A. (2017). Street Vendors' Role in Main Squares Utilization as a Type of Tactical Urbanism Application in G.C.R.: A Criticism Review on Ramses Square. *Procedia Environmental Sciences*, *37*, 153–163. doi:10.1016/j.proenv.2017.03.031

Elshater, A. (2014). Tactical Urbanism: A Method of Community Empowerment in Cairo Neighborhoods. World Barcelona SB14.

Forkuor, J. B., Akuoko, K. O., & Yeboah, E. H. (2017). Negotiation and Management Strategies of Street Vendors in Developing Countries: A Narrative Review. *SAGE Open*, *7*(1), 1–13. doi:10.1177/2158244017691563

Garvin, A. (2016). *What Makes a Great City?* Washington, DC: Island Press. doi:10.5822/978-1-61091-759-9

Gehl, J. (2010). *Cities for People*. Washington, DC: Island Press.

Gehl, J. (2011). *Life Between Buildings: Using Public Spaces*. Washington, DC: Island Press.

Graaff, K., & Ha, N. (Eds.). (2015). *Street Vending in the Neoliberal City: A Global Perspective on the Practices and Policies of a Marginalized Economy*. New York, NY: Berghahn Books.

Griffiths, S. (2014). Space Syntax as Interdisciplinary Urban Design Pedagogy. In M. Carmona (Ed.), *Explorations in Urban Design: An Urban Design Research Primer* (pp. 157–167). London, UK: Ashgate Publishing.

Hanson, J. (1989). *Order and Structure in Urban Space: A Morphological History of the City of London*. London, UK: UCL.

Harvey, D. (2008). The Right to the City. *New Left Review*, *53*, 23–40.

Harvey, D. (2012). *Rebel Cities: From the Right to the City to the Urban Revolution*. London, UK: Verso.

Hays-Mitchell, M. (1994). Street Vending in Peruvian Cities: The Spatio-Temporal Behavior of Ambulantes. *The Professional Geographer*, *46*(4), 425–438. doi:10.1111/j.0033-0124.1994.00425.x

Hedden, W. P. (1929). *How great cities are fed*. New York, NY: Heath and Company.

Hiebert, D., Rath, J., & Vertovec, S. (2015). Urban Markets and Diversity: Towards a Research Agenda. *Ethnic and Racial Studies*, *38*(1), 5–21. doi:10.1080/01419870.2014.953969

Hillier, B. (2001). Centrality as a Process: Accounting for Attraction Inequalities in Deformed Grids. *URBAN DESIGN International*, *4*(3), 107–127.

Hillier, B. (2002). *Can Streets Be Made Safe?* Oxford, UK: University of Oxford.

Hillier, B. (2008). Space and Spatiality: What the Built Environment Needs from Social Theory. *Building Research and Information*, *36*(3), 216–230. doi:10.1080/09613210801928073

Hillier, B., & Hanson, J. (1984). *The Social Logic of Space*. Cambridge, UK: Cambridge University Press. doi:10.1017/CBO9780511597237

Hillier, B., & Hanson, J. (1998). Space Syntax as a Research Programme. *Urban Morphology*, *2*(2), 108–111.

Hillier, B., Hanson, J., & Graham, H. (1987). Ideas are Things: An Application of the Space Syntax Method to Discovering House Genotypes. *Environment and Planning. B, Planning & Design*, *14*(4), 363–385. doi:10.1068/b140363

Hillier, B., Hanson, J., Peponis, J., Hudson, J., & Burdett, R. (1983). Space Syntax: A Different Urban Perspective. *Architectural Journal*, *30*, 47–63.

Hillier, B., & Penn, A. (1996). Cities as Movement Economies. *URBAN DESIGN International*, *1*(1), 49–60. doi:10.1057/udi.1996.5

Hillier, B., Penn, A., Hanson, J., Ggrajewski, T., & Xu, J. (1993). Natural Movement or Configuration and Attraction in Urban Pedestrian Movement. *Environment and Planning. B, Planning & Design*, *20*(1), 29–66. doi:10.1068/b200029

Ibn-Khaldun, A. (1967). The Muqaddimah: An Introduction to History (Rosenthal, F. Trans., and Dawood, N.J. Ed.). London, UK: Routledge & Kegan Paul Publisher.

Jacobs, A. (1993). *Great Streets*. Cambridge, MA: MIT Press.

Jacobs, J. (1961). *The Death and Life of Great American Cities*. London, UK: Peregrine Books.

Jalaladdini, S., & Oktay, D. (2012). Urban Public Spaces and Vitality: A Socio-Spatial Analysis in the Streets of Cypriot Towns. *Procedia: Social and Behavioral Sciences, 35*, 664–674. doi:10.1016/j.sbspro.2012.02.135

John, N. M., Da Luz Reis, A. T., De Lima, M. A., & Dias Lay, M. C. (2016). Evaluation of a popular shopping mall built to accommodate previous street vendors in downtown Porto Alegre. *Archnet-IJAR: International Journal of Architectural Research, 10*(1), 75–95. doi:10.26687/archnet-ijar.v10i1.832

Kafafy, N. A. (2016). Right to Urban Space in Post-Revolution Cairo: A Study for Street Vending Phenomenon. In Dynamics and Resilience of Informal Areas: International perspectives (pp. 97-112). Switzerland: Springer International Publishing. doi:10.1007/978-3-319-29948-8_6

Kafafy, N. A. (2017). Street Traders in Post-Revolution Cairo: Victims or Villains? In A. Brown (Ed.), *Rebel Streets and the Informal Economy: Street Trade and the Law* (pp. 220–235). New York, NY: Routledge. doi:10.4324/9781315641461-14

Karimi, K. (2016). The Shifting Paradigm of Space Syntax Research and its Future Prospects: A Reflection on the 10th International Space Syntax Symposium in London, 2015. *Journal of Space Syntax, 6*(2), 275–278.

Karimi, K. (2018). Space Syntax: Consolidation and Transformation of an Urban Research Field. *Journal of Urban Design, 23*(1), 1–4. doi:10.1080/13574809.2018.1403177

Karimi, K., & Parham, E. (2012). An Evidence Informed Approach to Developing an Adaptable Regeneration Programme for Declining Informal Settlements. *8th International Space Syntax Symposium*.

Koohsari, M. J., Kaczynski, A., Mcormack, G. R., & Sugiyama, T. (2014). Using Space Syntax to Assess the Built Environment for Physical Activity: Applications to Research on Parks and Public Open Space. *Leisure Sciences: An Interdisciplinary Journal, 36*(2), 206–216. doi:10.1080/01490400.2013.856722

Lefebvre, H. (2003). *The Urban Revolution*. Minneapolis, MN: University of Minnesota Press.

Lerman, Y., & Lebendiger, Y. (2017). Visualizing Space Syntax Analyses for Decision Makers: Lessons from Israel. *11th International Space Syntax Symposium*.

Lerman, Y., Rofe, Y., & Omer, I. (2014). Using Space Syntax to Model Pedestrian Movement in Urban Transportation Planning. *Geographical Analysis, 46*(4), 392–410. doi:10.1111/gean.12063

Li, J., Ren, L., Hu, T., & Wang, F. (2018). A city's "urban crack" at 4 a.m.: A case study of morning market vendors in Beijing's Longfu Temple area. *Habitat International, 71*, 14–21. doi:10.1016/j.habitatint.2017.10.006

Low, S., & Smith, N. (Eds.). (2005). *The Politics of Public Space*. New York, NY: Routledge.

Lynch, K. (1960). *The Image of the City*. Cambridge, MA: MIT Press.

Lyons, M., & Snoxell, S. (2005). Sustainable Urban Livelihoods and Marketplace Social Capital: Crisis and Strategy in Petty Trade. *Urban Studies (Edinburgh, Scotland)*, *42*(8), 1301–1320. doi:10.1080/00420980500150631

Maneepong, C., & Walsh, J. C. (2013). A New Generation of Bangkok Street Vendors: Economic Crisis as Opportunity and Threat. *Cities (London, England)*, *34*, 37–43. doi:10.1016/j.cities.2012.11.002

Marcus, L., Giusti, M., & Barthel, S. (2016). Cognitive Affordances in Sustainable Urbanism: Contributions of Space Syntax and Spatial Cognition. *Journal of Urban Design*, *21*(4), 439–452. doi:10.1080/13574809.2016.1184565

Marcus, L., Stahle, A., & Dehlhielm, M. (2010). Architectural Knowledge and Complex Urban Space Analysis of Five Proposals for Slussen in Stockholm. *Journal of Space Syntax*, *1*(1), 177–198.

Maureira, V., & Karimi, K. (2017). The Everyday and the Post-Disaster Urban Systems as One Thing: A Configurational Approach to Enhance the Recovery and Resilience of Cities Affected by Tsunamis. *11th International Space Syntax Symposium*.

McMillian, J. (2002). *Reinventing the Bazaar: A natural history of markets*. New York, NY: Norton.

Mehta, V. (2013). *The Street: A Quintessential Social Public Space*. New York, NY: Routledge.

Morales, A. (2009). Public Markets as Community Development Tools. *Journal of Planning Education and Research*, *28*(4), 426–440. doi:10.1177/0739456X08329471

Morales, A. (2011). Marketplaces: Prospects for Social, Economic, and Political Development. *Journal of Planning Literature*, *26*(1), 3–17. doi:10.1177/0885412210388040

Mumford, L. (1961). *The City in History: Its Origins, Its Transformations, and its Prospects*. New York, NY: Harcourt, Brace, and World.

Nagati, O., & Stryker, B. (2013). *Archiving the City in Flux: Cairo Shifting Landscape since the January 25th Revolution*. Cairo: Cluster.

Onyango, J.O., W.L.A, O., & Onyango, L. (2012). Dynamics of Street Vending Phenomenon in the Kisumu Municipality, Kenya. *International Journal of Arts and Commerce*, *1*(4), 107–120.

Osman, K. M., & Suliman, M. (1994). The Space Syntax Methodology: Fits and Misfits. *Arch. & Behav.*, *10*(2), 189–204.

Oz, O., & Eder, M. (2012). Rendering Istanbul's Periodic Bazaars Invisible: Reflections on Urban Transformation and Contested Space. *International Journal of Urban and Regional Research*, *36*(2), 297–314. doi:10.1111/j.1468-2427.2011.01076.x

Penn, A. (2008). Architecture Research. In A. Knight & L. Ruddock (Eds.), *Advanced Research Methods in the Built Environment* (pp. 14–27). Chichester, UK: Wiley-Blackwell.

Penn, A., Hillier, B., Banister, D., & Xu, J. (1998). Configurational Modelling of Urban Movement Networks. *Environment and Planning. B, Planning & Design*, *25*(1), 59–84. doi:10.1068/b250059

Penn, A., & Turner, A. (2004). Movement-Generated Land-Use Agglomeration: Simulation Experiments on the Drivers of Fine-scale Land-use Patterning. *URBAN DESIGN International*, *9*(2), 81–96. doi:10.1057/palgrave.udi.9000120

PPS. (2003). *Public Markets as a Vehicle for Social Integration and Upward Mobility. Phase 1 Report: An Overview of Existing Programs and Assessment Opportunities*. New York, NY: Project for Public Spaces, Inc.

Rajagopal. (2010a). Street Markets Influencing Urban Consumer Behavior in Mexico. *Latin American Business Review, 11*(2), 77-110.

Rajagopal. (2010b). Coexistence and Conflicts between Shopping Malls and Street Markets in Growing Cities: Analysis of Shoppers' Behavior. *Journal of Retail and Leisure Property, 9*(4), 277-301.

Raymond, A. (2007). *City of History*. Cairo: The American University in Cairo Press.

Recio, R. B., & Gomez, L. E. A. Jr. (2013). Street Vendors, their Contested Spaces, and the Policy Environment: A View from Caloocan, Metro Manila. *Environment & Urbanization Asia*, *4*(1), 173–190. doi:10.1177/0975425313477760

Reid, D. M., Fram, E. H., & Guotai, C. (2010). A Study of Chinese Street Vendors: How They Operate. *Journal of Asia-Pacific Business*, *11*(4), 244–257. doi:10.1080/10599231.2010.520640

Reyes, R. M. (2013). Crime, Street Vendors and the Historical Downtown in Post-Giuliani Mexico City. *International Journal of Criminology and Sociology*, *2*, 186–198.

Roever, S., & Skinner, C. (2016). Street Vendors and Cities. *Environment and Urbanization*, *28*(2), 359–374. doi:10.1177/0956247816653898

Sarpong, S., & Nabubie, I. B. (2015). Nuisance or Discerning? The social construction of street hawkers in Ghana. *Society and Business Review*, *10*(2), 102–117. doi:10.1108/SBR-03-2015-0005

Schappo, P., & Van Melik, R. (2017). Meeting on the marketplace: On the integrative potential of The Hague Market. *Journal of Urbanism*, *10*(3), 318–332. doi:10.1080/17549175.2016.1223741

Serra, M., & Pinho, P. (2013). Tackling the structure of very large spatial systems: Space syntax and the analysis of metropolitan form. *Journal of Space Syntax*, *4*(2), 179–196.

Setsabi, S., & Leduka, R. C. (2008). The Politics of Street Trading in Maseru, Lesotho. *Urban Forum, 19*(3), 221-241.

Sims, D. (2012). Understanding Cairo: The Logic of a City Out of Control. Cairo: The American University in Cairo Press.

Stryker, B., & Nagati, O. (Eds.). (2016). Creative Cities: Re-Framing Downtown Cairo. Cairo: Cluster.

Tandon, M., & Sehgal, V. (2017). Traditional Indian Religious Streets: A Spatial Study of the Streets of Mathura. *Frontiers of Architectural Research*, *6*(4), 469–479. doi:10.1016/j.foar.2017.10.001

Tangires, H. (2003). *Public Markets*. New York, NY: Norton.

Taylor, J., & Song, L. (2016). Return to the Streets. *Cityscape: A Journal of Policy Development and Research, 18*(1), 71-88.

Turner, A. (2003). Analyzing the Visual Dynamics of Spatial Morphology. *Environment and Planning. B, Planning & Design, 30*(5), 657–676. doi:10.1068/b12962

Turner, A. (2011). *UCL Depthmap: Spatial Network Analysis Software*. Retrieved from https://www.ucl.ac.uk/bartlett/architecture/research/space-syntax/depthmapx

United Nations General Assembly. (2016). *New Urban Agenda. United Nations Conference on Housing and Sustainable Urban Development (Habitat III)*, Quito, Ecuador.

Unlu-Yucesoy, E. (2013). Constructing the Marketplace: A Socio-Spatial Analysis of Past Marketplaces of Istanbul. *Built Environment, 39*(2), 190–202. doi:10.2148/benv.39.2.190

Varoudis, T. (2012). *DepthmapX Multi-Platform Spatial Network Analysis Software*. Retrieved from https://varoudis.github.io/depthmapX/

Vaughan, L. (2007). The Spatial Syntax of Urban Segregation. *Progress in Planning, 67*(3), 199–294. doi:10.1016/j.progress.2007.03.001

Von Schonfeld, K. C., & Bertolini, L. (2017). Urban Streets: Epitomes of Planning Challenges and Opportunities at the Interface of Public Space and Mobility. *Cities (London, England), 68*, 48–55. doi:10.1016/j.cities.2017.04.012

Watson, S. (2009). The Magic of the Marketplace: Sociality in a Neglected Public Space. *Urban Studies (Edinburgh, Scotland), 46*(8), 1577–1591. doi:10.1177/0042098009105506

Whyte, W. H. (1980). *The Social Life of Small Urban Spaces*. New York, NY: Project for Public Spaces, Inc.

Wiet, G. (1964). *Cairo, City of Art and Commerce*. Norman, OK: University of Oklahoma Press.

Wongtada, N. (2014). Street Vending Phenomena: A Literature Review and Research Agenda. *Thunderbird International Business Review, 56*(1), 55–75. doi:10.1002/tie.21596

Yatmo, Y. A. (2008). Street Vendors as 'Out of Place' Urban Elements. *Journal of Urban Design, 13*(3), 387–402. doi:10.1080/13574800802320889

Yatmo, Y. A. (2009). Perception of Street Vendors as 'Out of Place' Urban Elements at Day-time and Night-time. *Journal of Environmental Psychology, 29*(4), 467–476. doi:10.1016/j.jenvp.2009.08.001

ADDITIONAL READING

Al-Sayed, K. (2013). *Space Syntax Methodology*. London, UK: UCL.

Cardoso, R. C. V., Companion, M., & Marras, S. R. (Eds.). (2014). *Street Food: Culture, Economy, Health, and Governance*. New York, NY: Routledge. doi:10.4324/9781315776255

Chase, J. L., Crawford, M., & Kaliski, J. (Eds.). (2008). Everyday Urbanism. (Expanded Ed.) New York, US: The Monacelli Press.

Gehl, J., & Svarre, B. (2013). *How to Study Public Life*. Washington, D.C.: Island Press. doi:10.5822/978-1-61091-525-0

Hillier, B. (1996). *Space is the Machine*. Cambridge, UK: Cambridge University Press.

Karimi, K. (2012). A Configurational Approach to Analytical Urban Design: 'Space Syntax' Methodology. *URBAN DESIGN International*, *17*(4), 297–318. doi:10.1057/udi.2012.19

Mathews, G., Ribeiro, G. L., & Vega, C. A. (Eds.). (2012). *Globalization from Below: The World's Other Economy*. London, UK: Routledge. doi:10.4324/9780203106006

Turner, A. (2004). *Depthmap 4: A Researcher's Handbook*. London, UK: UCL.

Section 2
Conventional Methods

This section includes four chapters that examine the effectiveness of introducing terms as challenges in urban planning and design. These terms describe the characteristics of traditional cities – for instance, cities of singularity and cities of hardship. The term cities of hardship expresses the idea that the quality of the city is unmatched. The cities of singularity was presented to distinguish cities. The last four chapters deal with these two challenges through an integrative review about urban morphology, energy-efficient mobility, urban policies, and economic diagnosis in historical areas.

Chapter 13
An Argumentative Review Between Urban Morphology and Energy-Efficient Mobility

Omar M. Galal
https://orcid.org/0000-0002-4165-4663
Arab Academy for Science, Technology and Maritime Transport, Egypt & Technical University of Berlin, Germany

ABSTRACT

Urban settlements are recognized in literature to be a significant energy consumer. Among the primary contributors of such consumption is the transportation system. Recent literature that discusses the relation between energy efficient mobility and urban form has been continually branching and growing. However, as it grew, it becomes more specialized, less self-conscious, and to some extent contradicting. After adopting Conzen's approach to urban morphology, this chapter discusses through an argumentative review the relationship between energy efficient mobility and buildings density, land use, and streets networks. In addition to providing a snapshot to critical studies in the last three decades, two arguments are refuted. First, the residents' behavior resulting from socio-economic factors or individuals' choice has more impact than urban morphology characteristics on energy-efficient mobility. Second, morphological characteristics of urban settlements that lead to energy efficient mobility do not vary with the variation of context.

INTRODUCTION

Throughout the last decades, urban form has been continuously centered in the debates of sustainability and energy efficiency. The fact that buildings produce 43 percent of the world's carbon dioxide emissions and consume 48 percent of the energy produced has raised calls for rethinking the methods by which the built environment is created (Newman and Kenworthy, 2009). Centre to these debates, the

DOI: 10.4018/978-1-5225-9238-9.ch013

compact, dense, mixed-use, pedestrian, and transit-oriented urban form was recognized to be sustainable and energy efficient in the literature (Newman and Kenworthy, 1989; Calthorpe, 1993; Beatley, 2000; Banister, 2011a).

Despite the presence of arguments for less compact urban forms, the adverse consequences of low density and separation of uses have caused a gradual reduction of the impact of such opinions. The inevitable dependence on personal vehicles for travel is perhaps the most negatively significant impact of such urban forms with both economic (infrastructure) and environmental (climate change) implications (Filion 2013). On the other hand, either active modes of commuting (like walking and cycling which are non-motorized) or commuting through means of public transportation (like buses, rapid bus transit, light rail transits and Metros) have been argued by many to be more energy efficient (Newman, 2013; Rogers, 2008; Farr, 2008). The questions remain; what are the characteristics of urban form that empowers Energy Efficient Mobility (EEM)? Is there any relation to begin with or is it a matter of socio-economic factors and individual preferences? Moreover, whether there is a single formula that works everywhere? In other words, what is fixed from context to context, and what is not?

Although it seems that currently the majority of opinions are oriented to more compact urban forms, there are many ambiguities. For example, the intangibility of some of the new vocabulary introduced by a movement like the New Urbanism, such as mixed-use, access, and diversity have been highlighted (Talen, 2003; Elshater, 2012; Elshater, 2015) or even doubts regarding whether they achieve their objectives (Handy, 2005). Indeed, when the concept of mixed-use is stated, it was not specified on which scale and by what manner a sustainable mix should be formed (Evans, Aiesha, Foord, 2009).

Moreover, Boarnet and Crane (2001) have argued that other than traffic calming, no other feature of urban design seems to have a clear and consistent impact on vehicles miles traveled (VMT) reduction. Lo (2016) has further argued that daily sustainability practice is more dependent on personal factors such as the socio-economic features of the residents than contextual factors that are related to urban form. Additionally, studies that relate urban form to sustainable mobility sometimes vary drastically with the variation of the context. As expected as that may be, it still counters all the arguments for the possibility that cities can be formed through a set of standards causing them to become more energy-efficient despite the variation of the context.

Although most previous studies on EEM focused urban form, this chapter focuses on Urban Morphology (UM) as it accounts for the factor of time and the manner of growth, transformation or change occurring in the form of urban settlements. Hence, advocating a specific urban form is advocacy of static conditions of form that may or may not be extremely different from or similar to the urban form of any other settlement in the world. Such advocacy also may imply that there is only a single set of procedures to adopt. On the other hand, in the classical theories of UM exist wisdom that may guide to a diverse range of changes. Consequently, provide a pathway to different future urban forms that are more energy efficient than current ones.

There is a continually growing body of literature investigating the relation between urban morphology/form and energy efficiency in terms of mobility and other disciplines. However, as it grew, it has changed to be more specialized and to some extent less self-conscious. Through an integrative review, this chapter aims to review the relationship between UM and EEM in a more inclusive manner. This chapter first provides a–necessarily low resolution–snapshot for the research on the topic and mostly throughout the last 30 years since the influential study of Newman and Kenworthy (1989). Afterward, this overview is discussed in the light of literature interested in the impact of residents' related features and different climates–as they both vary with the variation of context–on EEM. Hence determine whether

UM that supports EEM is context sensitive or not. The relevance of this review to the concept of cities of singularity springs from this argument.

Firstly, definitions for Urban Morphology (UM) are sought, and their relation to Energy Efficient Mobility (EEM) is discussed. Later on, an overview of UM characteristics that are relevant to EEM is conducted. Furthermore, literature discussing the impact of city size on EEM is investigated as well. The time frame of the focused studies was mostly in succession to Newman & Kenworthy (1989) as it is recognized in the literature as one of most influential studies in the debate that occurred around the topic. It should further be stated that no attempt is made to discuss all or most of the publications from these locations. Instead, some of the main focal points of the debate are augmented and represented to cover a broad range of debates in a variety of contexts.

Such shortened overview is then discussed while questioning the consistency of UM's impact across the different contexts on EEM. The contextual characteristics of different cities are perhaps too diverse to be completely discussed in this review. However, in this chapter, two context-related aspects were chosen. The first is the socio-economic characteristics of the residents; due to the presence of the well-established body of literature discussing these characteristics' impact on the choice of mode of mobility. The second is the climatic conditions; due to the presence of a well-established body of literature discussing the impact of different morphological attributes on the urban climate. The second characteristic may not seem to be directly related to mobility. However, it is definitely relevant to energy efficiency. Hence, through integrating literature, this chapter discusses cases where usually recommended UM characteristics that achieve EEM; may reinforce or offset energy efficiency in terms of heating/cooling or even active forms of mobility due to improving or worsening the urban climate conditions.

BACKGROUND

Urban Morphology (UM) was defined as 'the study of the form and shape of settlements' (Carmona, Heath, Tiesdell, & Oc, 2003). It is also "a complex field whose primary concern is the structure and growth of urban form" (Moreno and Steadman 2014). Kropf (2014) has stated that within urban morphology there is a general consensus on the core, fundamental elements of the physical built form: streets, plots, and buildings. Carmona et. al (2003) also stated that it is widely accepted amongst morphologists that urban settlements could be seen through key elements (based on Conzen's opinion) which are land uses, building structures, plot pattern and street pattern. On the other hand, in his critique of the modern urban fabric Levy (1999) has used the same elements but with the substitution of the land use with the open space.

Conzen (2004) argued that every urban settlement has three fundamental attributes; its functional attribute, its morphological attribute, and its historical, geographical attribute. It was also Conzen who first introduced the tripartite division of the urban landscape into the town plan (streets, plots and block plans of buildings), the building fabric and the land and building utilization (Whitehand 2001). Morphological analysis is based on three main principles. The first is the form which is defined by buildings and their relation to opens spaces, plots, and streets. The second is the resolution which represents the different scales of analyses. The third is 'time' which is essential to take into account as the form elements interact, evolve or even be replaced by time (Moudon, 1997; Elshater, 2018).

There is a wide range of literature regarding urban form outside the domain of scholarship of urban morphology. Lynch and Rodwin (1958) specified the "element types, quantity, density, grain, focal organization and generalized spatial distribution" as characteristics of the physical form of the urban settlement to be analyzed. However, they stressed the separation between physical pattern and the pattern of activities during the analysis. Lynch (1960) has also identified five elements of the built environment on a physical and perceptive level which are: paths, edges, districts, nodes, and landmarks. Alberti (1999) chooses four elements that can be examined against various environmental aspects. The first is the form that refers to the degree by which the urban structure is centralized/decentralized. The second is density that refers to the relationship between populations or jobs at a given area. the third is the grain which refers to the diversity of land uses. Finally, the connectivity comes to refers to the mode of circulation.

Urban form was also more recently defined as a 'spatial construct that is inseparably in relation with the human activities and its resulting development patterns (Nedovic-Budic et al., 2016). Nedovic-Budic et al. state that it is possible to define urban form in terms of 'connectivity (street network and transportation); land use types, intensity, mix and proximity', the density of population and employment in addition to the manner of its distribution. Despite the diversity in terminologies and date of all these definitions, they still, to a significant extent, share the same main elements specified by Conzen.

On the same vein, UM can be studied on scales varying from an entire city region to a single plot and building as shown in Figure 1. For example, Ou, Liu, & Li (2013) aimed to quantify the relationship between the urban form of four of the most growing Chinese cities and CO_2 emissions. The study monitored the growth of the city at the interval between 1990 and 2010. On the other hand, another study was conducted on the informal morphing process occurring at social housing in Cairo throughout the interval between the 1970s until 2018 (El Gabalawy and Galal, 2019). The study has qualitatively questioned the extent of usefulness or harm resulting from such a phenomenon in the light of literature of sustainable urbanism.

At last but not least, the significance of time in the field of urban morphology springs from the changes that occur in urban form at every era. Conzen (1960) has discussed how different time periods lead to the manifestation of different reactions between "land use, streets, plots, and buildings." It was indeed argued that the peak of Conzen's investigations of the development and change occurring in an urban settlement was the "morphological regions" (Whitehand, 2001). These are areas that have to some extent a sense of unity in their form characteristics in comparison to other surrounding areas. Conzen has further demonstrated that not all four previously stated elements change with the same speed. For instance, land use was argued to be the quickest to change with time as new functions not always necessarily drive new forms (Conzen, 1960). On the other hand, once constructed, streets patterns are the hardest to change. Initially, UM had three classical schools; the English, the Italian and the French (Moudon, 1997).

The Italian school had a pivotal approach to UM based on the concept of 'type' and the assumption of the existence of a strong link between UM and buildings' typology (Marzot, 2002). On the other hand, the French school was said to consider the city's physical form to be the result of socio-economic relations and usually approach urban layout through a bottom-up approach (Moreno and Steadman, 2014). However, it was also stated that the study of urban form in France is significantly fragmented and non-hierarchical (Darin, 1998). Later on, a wider variety of approaches to urban form study and analysis were introduced (Kropf, 2009). Despite such variety, the author chooses to follow the simple English perception of the key four elements affecting urban morphology; lead by the German geographer Conzen. The cause is that most of the investigated literature tended to focus studying the relation between UM and EEM based on the primary four elements he specified (building form, streets pattern, plot patterns, and land use) as visualized in Figure 2. Furthermore, Conzen's simple perception of the

Figure 1. The first four satellite images show the evolution and growth of the form of The Greater Cairo Region during the interval between 1989 and till 2019. The last three photos demonstrate the evolution of a single building since its construction and till 2018.
Source: Photos and illustrations adopted from Google Earth and El Gabalawy and Galal (2019).

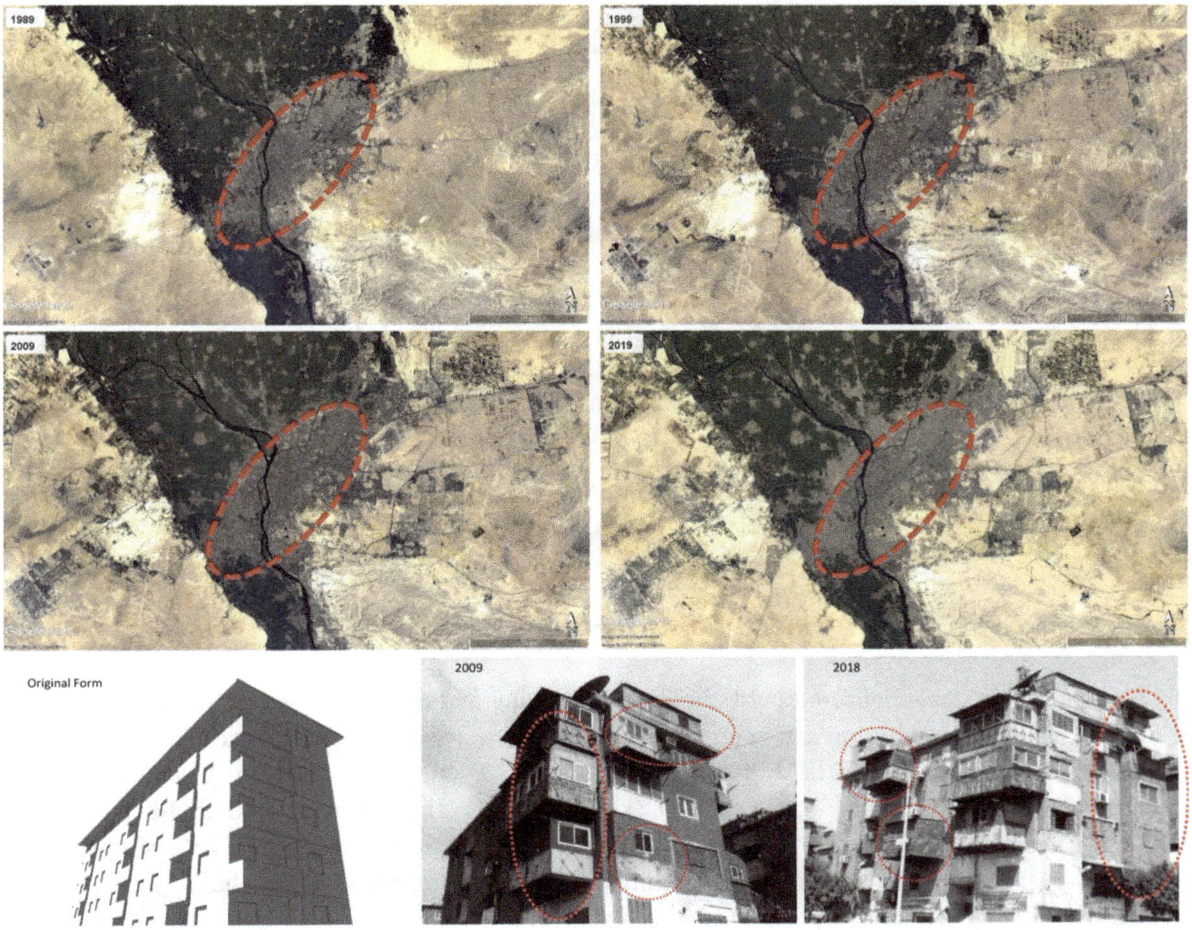

evolution/change occurring in urban form (UM) is perhaps more appropriate to formulate a common theoretical platform for integrating different disciplines of urban form studies.

Moreover, when studying the relationship between UM and EEM adopting definitions or classifications of UM that do not include land use, it would not make sense. From here on, this review adopts the term "morphology" - not form. The preference for the prior term is because it accounts for growth and changes with time. Although this is seldom done in studies related to urban mobility, the researcher believes that time is a critical dimension and demonstrate such importance in the discussion section. Furthermore, UM can guide the process of transformation from current inefficient urban forms to future forms that are more energy efficient.

Figure 2. Based on Conzen's approach, urban morphology is mainly focused on the study and different configurations of four main urban form elements which are buildings' form, plot patterns, street patterns, and land use
Source: Diagram developed by the author.

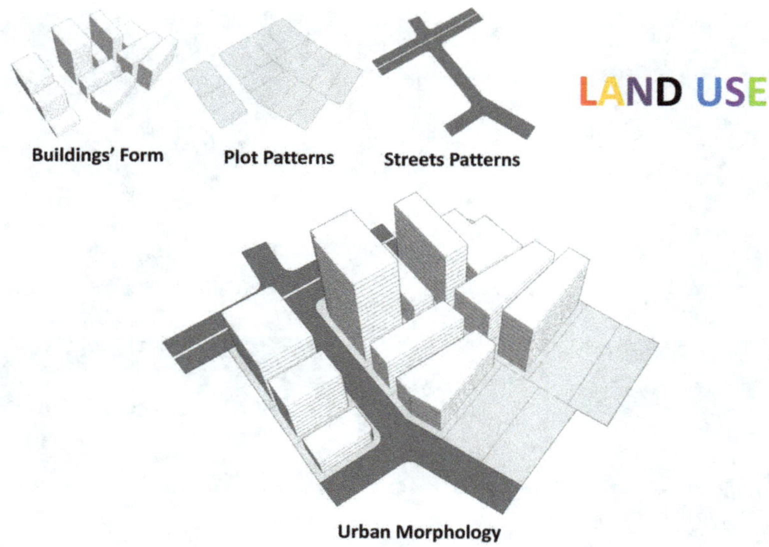

URBAN MORPHOLOGY: WAYS OF CLASSIFICATION

The gradual rise in dependence on public transport, walking and cycling is becoming more important as a mean of reducing pollution and energy consumption in the sector of transportation (Newman and Kenworthy, 1989; Newman, 2013; Rogers, 2008; Cervero, 2009). Although there are other emerging concepts of EEM such as light electric vehicles (i.e. electric bicycles, scooters, automobiles) (Mitchell, 2010). However, they do not require major changes in already existing urban forms. To the contrary they are usually praised for being possible to accommodate in the current settlements; hence they were not focused in this chapter. Some aspects of UM have a clear effect on facilitating EEM. However, not all aspects have a noticeable influence. For example, buildings' form is one of the key elements of interest in UM; yet, the outline of the plan of every single building has almost no influence on mobility. However, the overall density of the built environment affects mobility dramatically. Hence, based on what is found in literature, this section as shown in Figure 3 that focuses on overviewing the interrelation between:

- First: three main features of the UM which are: 1) urban population/ dwelling/ buildings' density as a sub-parameter of buildings' form. 2) Streets network pattern/shape. Although plot patterns are recognized among the main elements of UM, the single plot shape has a week effect on mobility. On the other hand, the block (group of plots) shape can be studied via studying the streets network. 3) Land use is focusing on the relationship between the different uses. Furthermore, a fourth aspect is investigated which is the city size. The goal is to investigate to what extent should towns be allowed to grow? Moreover, to what extent does settlements size affect mobility either positively or negatively?

Figure 3. Focusing mainly on books and peer-reviewed journal articles, the data gathered was structured based on Conzen's approach to Urban Morphology
Source: The author

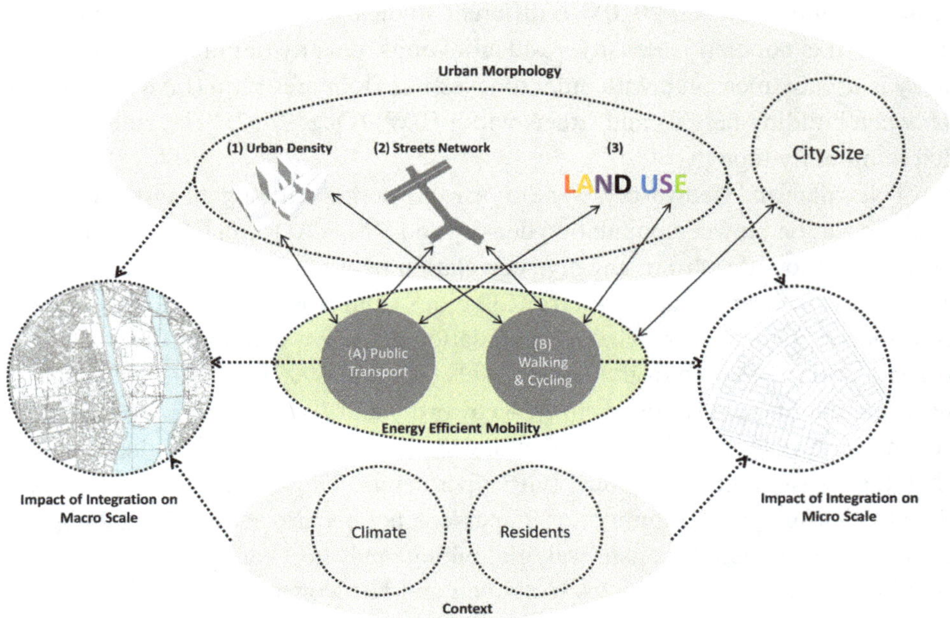

- Second: the main features of sustainable EEM which are: A) Public transportation as Motorized EEM. B) Walkability and cycling as non-motorized EEM.

The way of classification not only provides a more accurate understanding of the relation between UM and EEM but also, as a by-product that it studies UM on different scales. Consequently, the literature focusing on the relation between UM and public transport 1A, 2A, and 3A will mostly tend to concentrate on the macro scale of a city, town or districts. Meanwhile, 1B, 2B, and 3B which focus the relation between UM and non-motorized forms of mobility tends to concentrate on the micro scale of a neighborhood. At last but not least, the findings are then discussed in the light of literature interested in the relation between; a) socio-economic features of the residents and their impact on EEM; b) different climatic contexts and their impact on EEM.

URBAN MORPHOLOGY AND ENERGY EFFICIENT MOBILITY

Though it might be optimistically argued that the prevalence of the personal vehicle can continue under the conditions of technological improvements in fuel efficiency and alternative fuels, the recent progress of such advances is limited (Moriarty and Honnery, 2008). Hence, the gradual rise in dependence on public transport, walking and cycling become more important than ever before as a mean of reducing pollution and energy consumption in the sector of transportation. Consequently, morphing current cities to be more compatible with EEM becomes crucial.

Density and Motorized Mobility

Physical density is 'a numerical measure of the concentration of individuals or physical structures within a given geographical unit' (Cheng, 2010). In different studies, density is referred to in different manners. In some cases, it is population density, residential units' density or building density. However, the building density is usually measured with other indexes like floor area ratio (FAR) or the canyon aspect ratio (ratio between building heights and streets width H/W) (Oke, 2017). The latter is seldom used in studies related to mobility though.

The works of Newman and Kenworthy over the years are perhaps among the most debated in literature in studying the correlation between population density and VMT. At the end of their analysis, Newman and Kenworthy (1989) found a strong negative correlation between VMT and population density. The scope of the study involved aggregated data from 32 cities from North America, Australia, Europe, and Asia. However, their method was criticized as population density was found to be a too crude measure of urban form and hard to separate from income levels (Alberti, 1999).

In a more recent and more developed study, Kenworthy and Laube (1999) proved through an even more comprehensive sample of global cities that the correlation between VMT and density is far more significant than the Gross Regional Product (GRP) per capita. They have later argued that aggregated data from global cities suggest a minimum of 35 persons per hectare and additional approximately 18 jobs per hectare are needed before transit, walking, biking, and short car trips can combine to decrease the need to commute via personal vehicles (Newman and Kenworthy, 2013). However, Newman and Kenworthy were not the only researchers who specified preferable estimations for density.

The example is given, Calthorpe (1993) has argued that a Transit Oriented Development (TOD) would have an average of minimum densities varying between 10 to 25 dwelling units per net residential acre. Rogers and Power (2000) in the UK have recommended a minimum of 50 dwelling units per hectare. In the context of Athens, however, it was found that a residential density of a 200 persons/hectare is the most influential value by which at lowers values, the use of private vehicle increases (Milakis et al., 2008). Although this study, to some extent, further stresses on the significance population density for EEM, it is also a counter-argument for the concept of a universal standard for density advocated by Newman and Kenworthy or even in some of the international urban certificating systems (i.e., LEED-ND).

Population density – with its resulting dwellings density and buildings density - is perhaps the most influential aspect of urban form when it comes to successfully utilizing public transport. Vos and Witlox (2013) and Alonso et al. (2017) have demonstrated how sprawl and low density had a severe economic effect on public transport in both Belgian and Spanish contexts.

Although in some of the literature it seems that the denser, the better, which might imply that building high rise cities is what suits public transport the best, this is not always the case. For instance, Beatley (2000) argues that many European cities display compactness and density that supports public transports while preserving the human scale through low to mid-rise built environment. Furthermore, in a study focused on FAR, it was found that an urban sample taken from Manhattan had similar FAR to the one taken from Moscow, despite that the latter had more floors (Moreno and Steadman, 2014).

Density and Non-Motorized Mobility

Net residential density was found to be significantly correlated with increased walking (Lee and Moudon, 2006). It was additionally argued that a minimum density of 50 dwelling units/ hectare is needed if shops, buses, and services are to be situated within the range of walking distances from the residents (Rogers and Power 2000). Dwelling density higher values were also recognized to be one among of other attributes that reinforces walkability in a wide range of studies (Owen et al., 2007; Leslie et al., 2007; Sallis et al., 2009; Villanueva et al., 2014; Sundquist et al., 2011).

Furthermore, in several studies of these another density related feature was utilized which is FAR for the commercial uses. Such indicator was considered to be a sign of reduced areas dedicated for parking which is consequently expected to promote walkability as well. In a study conducted in Toronto, Canada, it was found that only a few areas contained higher residential density without many destinations or vice versa (Glazier et al., 2014). The study further demonstrated the presence of correlation between both characteristics and walking especially at locations where they coexist.

However, in Glasgow, a study based on questionnaires for four different neighborhoods was executed to test the relation between density and mixed uses on one hand and accessibility to city facilities and travel patterns on the other. It was found that residents of mixed-use settlements with either low or high density tended to find walking more convenient and enjoy more facilitated access to their destinations (Masnavi 2000). Such study could imply that that mix of uses is a more critical component for walkability than density; however, the densely mixed-use sample achieved best results.

Nonetheless, Townshend and Lake (2009) have highlighted how density alone might not lead to improved walking. Based on a small number of interviews and observation, the authors noticed how a dense development with only a drive-through fast food restaurant near the entrance and a wall around its boundary was car dominated. It can be concluded that the density of the built environment or dwelling density is essential for promoting non-motorized forms of mobility. However, density's importance to walkability stems from providing the preconditions necessary for the creation of destinations like transit stops and commercial uses within walking range. Hence, it was logically argued to increase the density of the built environment near transit stops either in the literature (Rogers, 2008; Farr, 2008) or LEED-ND (Gouda and Masoumi, 2017b).

Moreover, in Singapore, private developers were found to further reinforce this notion as a response to free market mechanisms. The results of a survey conducted by Zhu and Diao (2016) has demonstrated that newer housing schemes constructed near transit lines contained apartments that are smaller in size. Hence, it increased population density even though that FAR did not change much from other locations.

Land Use and Motorized Mobility

The traditional North-American method for zoning is currently being criticized for its frustration to urban density, limiting intensity and promoting segregation and isolation (Scheer, 2017). The social, economic, and environmental drawbacks of modern planning and its suburbs have led to the reconsideration of the mix of uses. The multifunctional land use can be defined as "the combination of different land use functions in the same area" (Dieleman and Wegener, 2004). Further, scholars state that in literature the allocation of these mixed-use locations near and around modes of public transport is often referred to as Transit Oriented Development. The mix of uses itself can occur at different scales rather than the city/town scale which is obviously the case in all cities and towns. The mixes could occur on the level of districts, neighborhoods, streets, street block and a single building (Rowley, 1996).

However, several studies have been skeptic towards the relation between mixed use and promoting EEM either on the level of applicability at existing communities, or reduction of traveling distance and promoting the use of public transport. Highlighting the slow rate of change of the built environment in

London, Williams (2000) has stated that the intensification land use planning policies are unable to keep up with the faster-growing rates of personal vehicle ownership. Simmonds and Coombe (2000) have referred to the relatively limited impact of more compact land uses to the residents' weak preference for visiting the nearest location.

In the Belgian context, an empirical study concluded that even at the event of applying strict location policies that concentrate housing and working near urban areas with extensive public transport, the resulting reduction in commuting via private cars could be reduced by 10-15% at most (Verhetsel and Vanelslander, 2010). Another empirical study at Lisbon showed that the allocation of a mixed-use employment center which is highly accessible for public and private transport in addition to sufficient parking space was not enough to create more sustainable commuting patterns (Vale, 2013). Additionally, the limited availability of private gardens or high-quality public gardens resulting from dense and intensely used urban environments may create other challenges. Holden and Norland (2005) for example have reported that such circumstances have led some residents to pursue cross-border leisure activities.

Meanwhile, despite the relatively slow rate of changing land use patterns stated above, it was argued that the manner of locating some of the new uses within the existing settlements could have a significant effect. For example, in the Netherlands, retail policy in 1973 prohibiting the establishment of out-of-town shopping malls and hypermarkets as a way of preserving town center's vitality and avoiding intensive car use, was said to be effective (Schwanen et al., 2004).

Furthermore, Banister and Hickman (2006) argued that the specified locations for new housing schemes have a substantial effect on travel patterns of their residents. Based on a study conducted at Copenhagen, it was found by Næss (2009) that residents living in central areas –which are denser and mixed in uses–tend to travel less by car. The same effect was also reported in the Chinese context by Ma et al. (2015) as they stated that residents living in inner city and traditional neighborhoods tend to have greener travel behavior. They further argued that among the main factors that increase CO_2 emissions resulting from travel demand is the mismatch of jobs and housing.

Land Use and Non-Motorized Mobility

Generally speaking and at the smallest scale, the mixed-use could occur either vertically (more than one use in the same building) or horizontally (more than one use in adjacent buildings) and is recognized for reducing distances between activities and the encouragement of walking (Pahl-Weber et al., 2012). Different studies have tried to address the mix of use in a quantitative manner. Older studies adopted an index named 'entropy' focused on the ratios between the areas of different uses in a given area and utilized indexes that score positively or negatively based on the diversity of uses (Frank and Pivo 1994). However, the entropy method was criticized, as it was argued that the presence of walkable land use is what matters regardless of the ratios between their areas (Brown et al., 2009).

More recent studies also articulated the significance of the extent to which the different uses are finely grained and how are they planned to integrate or complement each other, in comparison to just focusing the proportions of different uses (Manaugh and Kreider, 2013). However, there is no census or specific singular description or definition of how the right mix of uses would be (Leck 2006). That being said, several studies discussed several aspects and characteristics of the impacts of mixed use on making residents walk more.

Cervero (1996a) was among the older studies that demonstrated how proximity between dwelling and consumer services encourages transit use, walking and cycling even when controlling for residential density and car ownership. Handy (1996) has argued that the relationship between residential and commercial uses does matter. The shorter the distance between them, and the lesser the barriers such as arterial streets, the more the people would walk. However, according to her study, such pedestrian trips sometimes are in addition to the vehicular trips and do not substitute them.

Furthermore, the proximity between housing units and land uses that provide the daily food supply, and basic domestic provisions were found to raise walking rates (Lee and Moudon, 2006). The authors further demonstrated that uses with most extensive areas such as schools, large parcels housing and offices seemed to reduce walking. It was also argued that although the availability of destinations and density are both strongly related to walking and cycling, their coexistence has an even stronger effect (Glazier et al., 2014).

In the Netherlands, it was found that sometimes, the public transportation competed not only with private vehicles but with cycling as well. Hence it was advised to have them both more integrated through allowing cycling riders to use the metro for example (Schwanen et al., 2004). Furthermore, in another study, it was shown that the presence of green and recreational space – especially parks and sports fields- are associated with higher rates of cycling (WENDEL-VOS et al., 2004). It should be noted, however, that the previous study was not executed in a centralized, compact and crowded urban settlement; instead, it was executed at a relatively smaller town.

Streets Network and Motorized Mobility

Calthorpe (1993) stated that among the features of more transit-oriented development is that its street networks should be easy to recognize, inter-connected, connecting main transit stops, core commercial areas, schools, and parks. He also stated that the network itself needs to provide several alternatives to more intensely visited locations. Hence, smaller blocks achieve the latter objective. Moughtin (2003) has stated that a layout with acceptable permeability could be attained with street junctions occurring at 70 to 100 m. These numbers were purely based on qualitative debates though.

In this, Cervero (1996b) conducted a comparative study between several neighborhoods at the San Francisco Bay area. He found no significant impact for the transit-oriented neighborhoods with more interconnected grid-shaped street networks, in comparison to auto-oriented ones with curved shapes and cul-de-sacs on transit use. Though the use of the personal vehicle was less dominant in the first, dominance remained the same. Cervero then stated that neighborhood design should be incremented with other policies that discourage the use of the private vehicle and transit-oriented planning on the regional level. Moreover, it was even argued that the claim of many New Urbanists regarding the interconnected and small blocked street networks are baseless and could facilitate the use of the private vehicle instead (Leck, 2006).

In another comparative study executed at England between the new towns and historical towns, using Space Syntax, it was found that unlike the case in the new towns, in the historical towns, the most intensively visited uses were located in the most accessible streets (Karimi and Vaughan, 2014). Such alignment and correspondence of activities, density, and small blocks lead to a more intelligible structure for movement and wayfinding. However, despite the stated studies, there is not much empirical evidence found in the literature on the relation between streets networks' form and the use of public transport either positively or negatively.

Streets Network and Non-Motorized Mobility

In the USA, a study has shown that the increase of the streets intersection density (smaller blocks) can lead to increased rates of walking for non-work trips (Frank et al., 2007). Additionally, in Atlanta, it was found that intersection density and youth walking were positively related (Kerr et al., 2007).In the same context, the intersection density was recognized among others as one of the components of walkability index in several studies (Frank et al., 2010; Sallis et al., 2009; Villanueva et al., 2014; Sundquist et al., 2011). These authors' argument was based on the fact that higher intersection densities would lead to more direct paths between different destinations.

In another study conducted at the Netherlands, it was found that on the neighborhood level, loop networks are not preferred as they induce more motorized travel (Snellen et al., 2016). Furthermore, at Utah, USA, hierarchical streets pattern was found to compromise children safe while walking to school by funneling their walks from cul-de-sacs to arterials (Gallimore et al., 2011; Elshater, 2018). However, in another study at Adelaide in Australia, it was found at one of the neighborhoods that had curvilinear streets network with many cul-de-sacs was not affecting walkability on the perceptive level for the residents. The reason behind that was the presence of a pedestrian network that connected the main activities within it (Leslie et al., 2007).

Meanwhile, it was found at the Dutch, Danish and German cities that the provision of separate cycling lanes, parking, and rights of way along heavily traveled roads and intersection to be among some of the most important means by which cycling could become safe and convenient (Pucher and Buehler, 2008). More recently a more comprehensive concept named 'Complete Streets' is introduced that focus on expanding the interest of street design to include all modes of travel (Hui et al., 2017). However, not many quantitative studies were developed to define how can the concept be applied and assessed accurately.

In the same context, a study was conducted at Freiburg, Germany, with the aim to determine how the street is being used by the different modes of transportation. Even though Freiburg is known to have relatively more sustainable transportation, it was found that the mode with the most dominant ratio of the street was the private vehicle (Gössling et al., 2016). The study recommended assigning more space for cycling not only for being popular at Freiburg but also because it is cheap and energy efficient.

City Size Impact

Newman and Kenworth (1989) have stated that the correlation between city size and VMT is relatively weaker – compared to population density- and negative as well. In other words, smaller cities tend actually to have people using personal vehicles more. Naess (2012) also highlighted the moderate effect a settlement size could have. He argued that the smaller expected traveling distances resulting from smaller settlements may be offset by the lower densities of smaller towns and higher dependence on economic activities and services outside the settlement.

Klinger et al. (2013) argued that some smaller European cities like Graz in Austria and Freiburg in Germany are planned densely around non-auto modes. Furthermore, these scholars stated that the US cities of similar size have almost no public transport system at all which might have caused such an impression. In the Netherlands, it was further reported that cities with substantially larger than average population tended to have relatively lower bicycle use (Harms et al., 2015). Banister (2011b), however, has argued that traveling distance reduction and proximity between different activities can be one of the most influential aspects of an urban settlement and consequently should be one of the leading priorities of transport planning. Prior to that Banister (2008) has specified a city inhabited by 25000 residents (preferably 50000) to be among the key parameters of the sustainable city.

In conclusion, it can be stated that a higher population density and bigger population size may make investing in transportation infrastructure more economical. However, smaller towns also provide an advantage which is that all parts of the town can be accessible by bicycle and in a competitive manner time wise. Perhaps, seeking an answer for a preferable city size may not be found in arguments related to EEM alone and needs further research and investigation in a broader scope while taking other parameters in mind.

THE CONTEXTUAL DISCUSSION

Urban planning has a critical and significant role to play in promoting the use of public transport, walking and cycling (Hickman et al., 2013). Consequently, more than ever before, planners are being asked to implement changes in land use, transportation systems, and the built environment to decrease greenhouse gases emissions (Howard, 2009) significantly. However, it is the type of changes, and in which context that needs further interrogation. 'What is the optimal density for this city?' A question asked by Roaf (2010), and she stated that the answer varies for each city depending on the capacity and constraints of the city's social, economic and ecological aspects.

Perhaps no study validates the logic behind the previous answer more than (Gouda and Masoumi, 2017a) in their critique of urban sustainability certification (USC) systems. The authors have demonstrated how some of the LEED-ND indicators' scoring systems, tend to be in several ways context indifferent. This section of the chapter builds on these previous studies while reviewing the topic from two context related angels which are the characteristics of the residents and the climate.

The Behavioral Dimension

It is argued that settlements forms can play a significant role in consuming energy more efficiently. However, that very notion is questioned at several studies in the literature. Although the reasons might be diverse, eventually they can all be broken down into a single aspect, which is human behavior. It was found in some studies that the attitudes of the residents are more strongly associated with travel than land use characteristics (Kitamura et al., 1997; Rowley, 1996). Such studies suggested that land use policies that increase density and use mixtures cannot alter travel demands without the residents' attitudes changing. Even in a dense city like Tunisia, it was reported that increase in GDP combined with lack of investment in public transport has led to significant increase in usage of personal vehicles (Mraihi et al., 2015). Additionally, the proximity of different commercial uses, leisure activities, and services to residential units do not necessarily mean the residents always choose these facilities (Jacobs, 1961; Simmonds and Coombe, 2000). Concerning Dubai, Alawadi (2017) has discussed how some groups view that commuting via cycling is not culturally acceptable for the local females.

Sustainable mobility itself was argued to be only successful in the case of having the people understand and accept it (Banister, 2008). In previous reviews, the significance of concepts like self-selection as a major determinant of traveling behavior was highlighted (Cao et al. 2009). However, it should also be stated that UM combined with other policies can achieve changes to a certain extent on the residents' behavior. For instance, it was argued that policies which limit the number of parking lots or makes parking less convenient (Frank et al., 2007) or making traveling by car more expensive (Liddle, 2013; Lo, 2016) could cause people to shift into alternative modes.

In another comparative study between the USA and Germany, it was found that the Germans tend to walk, cycle and use public transports more. Nonetheless, the urban settlements in comparison were denser, mixed in use and enjoyed more proximity to public transports in favor of the USA (Buehler, 2011). The author argued that policies that promote alternative modes making them more convenient and faster, in addition to policies which are disadvantageous economically for the car users; caused such significant difference. These types of policies were also found to be one of the reasons behind the extensive use of bicycles in the Netherlands as well (Harms et al., 2015). All these studies demonstrate the significance of non-UM policies.

Furthermore, these studies make it clear that the impact of UM varies with the variation of residents. However, it must be understood that they complement and do not replace the impact of UM. Even if there were census that socioeconomic factors and concepts like self-selection had more impact on travel behavior, which is not the case (Cao et al., 2009), there are other dimensions that significantly add to the importance of UM. For example, it was found that although they may not cause a significant change in travel patterns, physical planning policies may make more choices for switching modes of mobility available (van de Coevering and Schwanen, 2006). In another comparative study in San Francisco between urban and suburban neighborhoods, it was found that in case of the later, the conditions of the environment prevailed over the preference of the residents, while the opposite was found in the first (Schwanen and Mokhtarian, 2005). Hence, residents who have the tendency to use public transportation do not do so if they are living in suburban and private vehicle oriented neighborhoods. On the other hand, more transit-oriented neighborhoods will not necessarily have all residents use public transport, but at least provide the option for those who chose to do so.

The other important dimension is time. Most of the previous studies are limited by their time. That would be time with a relative abundance of energy and resources. However, the price and availability of energy may not be the same after oil production peaks. For example, at some lowly dense fringe suburbs in Australia; commuting by car is too expensive because of the long distances that will need to be traveled. On the other hand, investing in public transportation system will not be feasible economically (Newman and Kenworthy, 2013). Furthermore, most of these studies are limited by context as well. Meaning that in countries with higher income and rates of car ownership, residents have the luxury of choosing their personal car even in the presence of an efficient transportation system and an urban form that supports it. Yet, the opposite is not true.

In a context where the car ownership rates are far lower if the built environment was not planned to promote. In that case, the present cheaper focuses on the modes of travel, serious complications related to mobility occurred. If the wealthier countries of the industrialized world face economic troubles in sustaining public transport system in some types of urban environments, then how would the countries with lower gross income per capita sustain one? Significant growth of informal settlements at the developing countries near major highways–where private group transportation methods flourish–despite the simultaneous existence of new ghost towns (Figure 6) planned based on the theories of the modern planning era, may be partially caused by this.

Despite all the different arguments made in literature, it is safe to say that both UM policies and non-UM policies do affect mobility. The matter in question is which is more effective and to what extent? The answer in the author's opinion is; there is no possible way to specify a consistent answer for such question in every context, and even for different time frames of the same context. However, the one consistent

Figure 4. diagram developed by the author to represent the impact of integrating UM characteristics with non-UM policies and strategies to reinforce EEM. Though all of them have the same degree of opacity here, in reality, the strength of impact of each element varies with the variation of context.

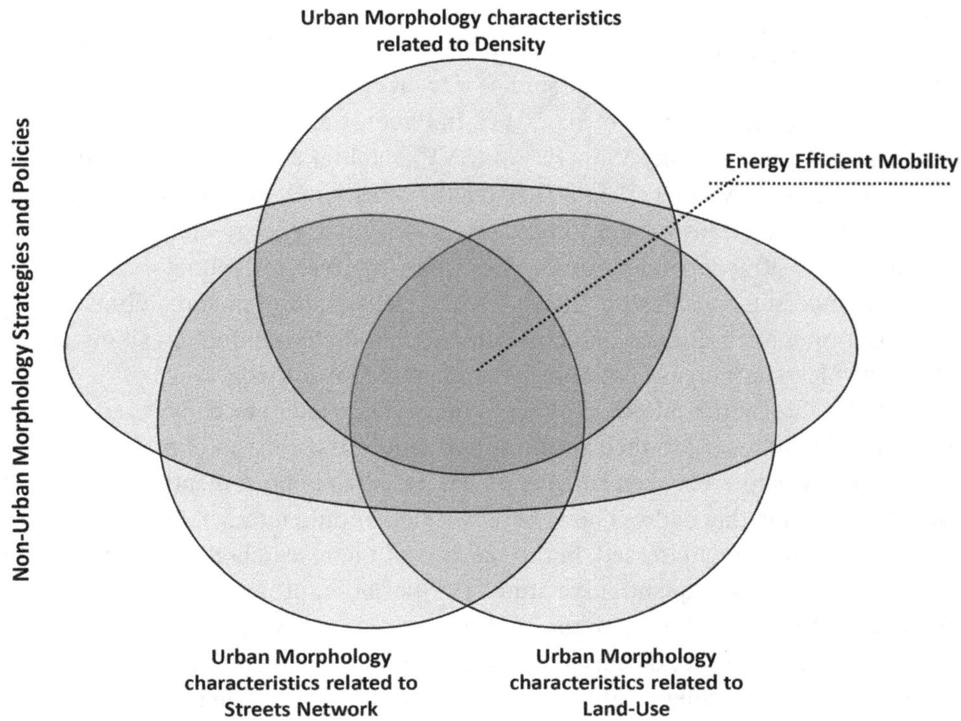

matter is that all the features related to UM introduced here perform the best when they integrate and complement each other as visualized in Figure 4. Not only regarding the three main elements – density, land use and streets network - but in terms of the scale of intervention as well.

The Climatic Dimension

For cities and towns to be energy efficient in the most real sense of the word or even just regarding EEM; focusing aspects related to only mobility is not be enough. For example, among the studies in this chapter, was one that discussed the relationship between population density and VMT at Athens where a density of 200 persons/hectare was found to be critical and by which at lower values, the use of private vehicle significantly increase (Milakis et al., 2008). Also, in Athens, it was found that the urban heat island (UHI) effect in summer at noon could make some of the central denser areas 6-7°C higher and some locations more than 10°C higher in comparison to rural areas (Littlefair, 2000). Such difference in temperature might offset the savings resulting from density in motorized mobility by two ways; a) causing a significant rise in cooling loads and b) causing a significant decline in walking and cycling.

There are other examples where specific features of the UM led to better results in terms of EEM and reduced UHI effect such as the availability of parks. The presence of parks was found to improve non-motorized forms of mobility (WENDEL-VOS et al., 2004; Brown et al., 2009) and also has a significant impact on cooling the urban environment and reducing the UHI (Wong and Chen, 2010; Oke, 2017). Such understanding of the role of the parks may promote its significance and priority even if led to slightly lower densities in a context like Athens. Non-UM strategies could then be implemented to make up for reductions that occurred in density.

Dwelling higher density and FAR values were also recognized to be one among of other attributes that reinforces walkability in a wide range of studies (Owen et al. 2007; Leslie et al., 2007; Sallis et al., 2009; Villanueva et al., 2014; Sundquist et al., 2011). In Phoenix at the USA, it was found that the more compact environments with lower Sky View Factor (SVF) and higher H/W tends to reduce temperature during peak time at Summer (Middel et al., 2014). On the other hand, in Copenhagen, it was found that if the built environment was transformed to be denser with time, energy consumption will rise. The increase may reach up to 30% depending on the function (office or residential) and orientation of the building(Strømann-Andersen and Sattrup, 2011). Hence, densification in some climatic contexts may have a useful impact on a microclimate and consequently indirectly reinforce walking and cycling. At other contexts, it may lead to increase other forms of energy consumption.

This chapter argues that more multi-disciplinary approaches need to be considered in future UM and EEM studies; especially, in regard to their resulting influence on the urban climate and vice-versa. Although the contradictory nature between EEM and UHI has already been mentioned before (Erell et al. 2011), no studies were found that address how these two fields could interact. For example, the minimal values of density are frequently addressed, but to the best of the researcher's knowledge, no maximum value of density is specified in a quantitative study. Furthermore, urban forms with similar FAR may lead to a wide variety of forms. Different urban configurations (even with similar FAR values) may lead to better or worse microclimatic conditions, consequently affects active forms of mobility.

On a conceptual level, the influence of EEM policies and UHI reduction policies on UM can be presented graphically as shown in Figure 5. Policies which achieve mutual benefit should be the ones most prioritized, along with policies which achieve one without negatively affecting the other. However, the third group – where the benefit in one is achieved on account of the other– should be carefully studied and calculated before applied. Either, in case of existing urban communities or new ones, the same

Figure 5. diagram developed by the author to represent how some planning policies targeting EEM may contradict with UHI. It also shows the manner by which design and planning decisions should be prioritized.

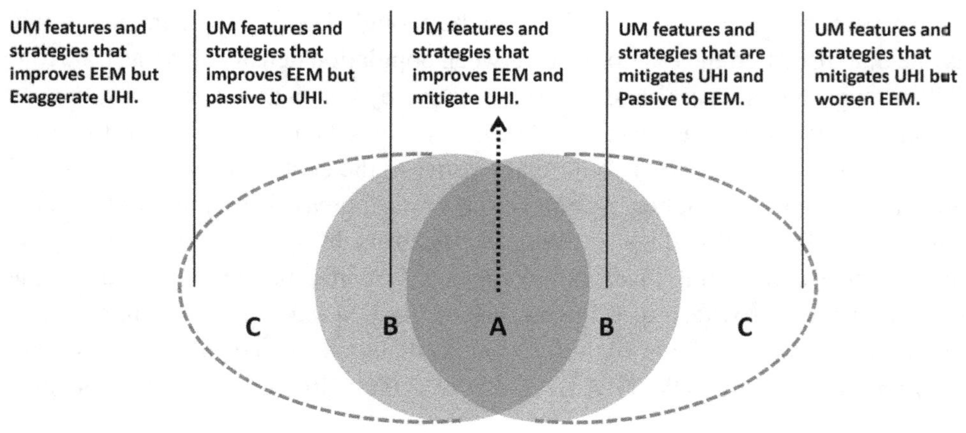

policy's priority might vary with variation of space and time. Both should be integrated with non-UM strategies and policies that reinforce them.

The importance of understanding how the weather influences travel behavior in the context of a warmer future climate has already been acknowledged (Liu et al., 2017). Besides, in warmer contexts like Qatar, the weather conditions were found to have a significant impact on the volumes of pedestrians within neighborhoods (Shaaban et al., 2017). That study has concurred that there were consistently more pedestrians observed in winter in comparison to the severely hot summers. Even before that, it was also acknowledged in most of the studies included in this chapter how travel behavior influences the change occurring in the climate globally. However, understanding how the policies adopted to promote energy efficient mobility may contradict or offset each other, influence the urban micro-climate and energy consumption in terms of cooling/heating, still need more research.

IMPLICATIONS FOR FUTURE RESEARCH AND POLICIES

It is neither possible nor required to deduce general guidelines from the previous overview of the assumption that these guidelines work anywhere. Actually, the main aim of this chapter was to formulate the counter-argument to such notions. Assuming that specific values of UM characteristics will solve all problems of mobility in a given urban settlement can be dangerously miss-leading. Contextual conditions relevant to the residents and climate can significantly reinforce or cripple EEM. Hence, future research should be also more integrative and holistic in approach. The first step towards such integration would be unifying the parameters of interest. For instance, parameters used to measure the built environment's density in the discipline of mobility are usually a number of dwellings per acre or population density. However, studies interested in the micro-climate usually focus on H/W or SVF. Perhaps more research should focus on unifying the morphological parameters focused on the different disciplines.

There is already a wide range of studies that aim to quantify the relationship between urban form and energy or micro-climate (Galal, 2017). Many computational tools were also developed to simulate energy consumption or micro-climatic condition for a given form. These computational tools have also the ability to test the performance of different scenarios. Among the widely used of these tools, there are Design-Builder (for energy simulations) and Envi-mat (for micro-climate simulations). Additionally, Geographic Information System (GIS) can play a significant role as a tool that can provide a common platform for the different disciplines. Hence, provide a common understanding of the different morphological parameters focused by different disciplines. Furthermore, GIS can be the optimum medium for integrating older research methods based on surveys and questionnaires with newer simulation methods.

Such an integrative approach to research might perhaps lead to more comprehensive research on the topic of urban form and energy efficiency in the future. Instead, many new urban communities were constructed during the last 50 years (especially in the MENA Region) based on building regulations that do not lead to energy efficient ends. In Upper Egypt for example, cities like New Aswan, New Assiut, and New Sohag were planned to be indifferent to EEM or their climatic conditions (Figure 6). Significant resources were invested and are planned to be invested in their construction. Hence, it is neither economically nor environmentally feasible to demolish them. Consequently, they need to be transformed to more energy efficient forms. Based on the previously stated studies of UM, land use is the easiest to change among the other aspects of urban form. Therefore, perhaps the priority of planning authorities in these cities should be injecting commercial uses and basic services within the residential wards in

Figure 6. Photos for the deserted New Aswan City in Upper Egypt. The author of the present work visited this city during the Summer of 2018. The city was planned based on separation of uses and with the goal of increasing the number of residential units at the desert. The urban forms characteristics are insensitive to the requirements of the residents, EEM, and climate
Source: The author.

these cities. Furthermore, the distribution of green infrastructure like parks, vertical green walls and vegetation have a significant effect on EEM and micro-climate and are relatively easy to execute as well.

The parameter to be manipulated will be building forms. Possible scenarios may include increasing densities through increasing height of existing buildings or the addition of new buildings through infill techniques. As for streets networks which are the hardest to edit, perhaps the addition of walking trails which are independent of the original streets network can reinforce walking. However, such decisions should be made after a thorough study of their impact in the light of their contextual conditions. This chapter sheds lights on a new dimension of the concept of cities of singularity (Abusaada, 2017). That dimension is based on transforming and tailoring urban forms specifically for their contextual conditions.

CONCLUSION

In times when research is becoming more and more specialized, this review aimed to discuss the relation between Urban Morphology (UN) and energy efficient forms of mobility through an integrative review. First, through adopting Conzen's approach, UM was broken down into three main components; building's form (urban density), land use and streets networks. The relation between these three components in addition to city size on the one hand and forms of EEM on the other were then overviewed. It was consistently found in both qualitative and quantitative studies, at a diverse range of contexts, that there is a relationship between EEM and UM.

Population/dwelling/building density where found to be strongly correlated with the use of public transportation, in addition to providing the preconditions for the availability of mixed-use developments and proximity with transit stops. Land use was more related to walking and cycling by creating proximity between residents and shops that provide their daily needs or transit stops. However, the allocation of specific uses at the center or fringe of the city proved to have an impact on private car use as well. As for the streets network, higher intersection densities were mostly found to have a significant impact on walkability by providing more direct access. The streets configuration was also found to be related to the promotion of cycling. Hence, indirectly reinforce the use of public transportation by providing easier access to transportation nodes.

The previous paragraph contains a brief description of the census in the literature regarding the relationship between EEM and elements of UM. However, there is no census on the impact strength or values for each element. For example, there might be a census regarding the avoidance of a single digit dwelling densities per acre. However, other than this, there is no evidence-based census of preferable maximum or minimum densities, even for a given context. There is also no census regarding land use mix. Till this moment, it is also neither clearly specified nor defined despite the several methods developed to quantify it. As for streets network, there is a census on the positive influence that intersection density has on walkability and cycling. However, that very census is based on the assumption of the exact alignment of streets networks with cycling and pedestrian lanes/paths.

It was also found in several other studies that socio-economic factors or individual behavioral aspects do affect the choice for mode of mobility. However, after a short review, it was argued that by all means, these factors should not lead to ignoring the impact that UM has on promoting means of EEM. This chapter has highlighted also another critical gap in already existing mobility literature. That would be the way some measures and policies may be contradictory concerning Urban Heat Island (UHI) and urban climate.

Future research should focus on quantifying and assessing how one policy or UM characteristic can save energy in the sector of mobility but create additional consumption in another. Regarding city size, buildings' form and density, land use configuration and street network and blocks and plots pattern, UM needs to be approached rigorously in a multi-disciplinary way and a context-sensitive manner. Indeed, the features of Urban Morphology that empower EEM are not fixed from context to context. This review highlights perhaps another perspective by which the concept of cities of singularity could be viewed.

ACKNOWLEDGMENT

The author would like to thank the reviewers and editors for their valuable reviews that have significantly contributed to improving the content of this chapter. The author also acknowledges the rich discussions he had with professor Elke Pahl-Weber which have greatly helped in the formulation and transformation of the ideas and main themes of this study.

REFERENCES

Abusaada, H. (2017). The Untold Story of the Urban Design Practice of New Cities in Egypt. In H. Elsharkawy, S. Zahiri, & J. Clough (Ed.), *International Conference for Sustainable Design of the Built Environment (SDBE 2017)* (pp. 787-798). London: SDBE.

Alawadi, K. (2017). Rethinking Dubai's urbanism: Generating sustainable form-based urban design strategies for an integrated neighborhood. *Cities (London, England)*, *60*, 353–366. doi:10.1016/j.cities.2016.10.012

Alberti, M. (1999). Urban Patterns and Environmental Performance: What Do We Know? *Journal of Planning Education and Research*, *19*(2), 151–163. doi:10.1177/0739456X9901900205

Alonso, A., Monzón, A., & Cascajo, R. (2017). Measuring Negative Synergies of Urban Sprawl and Economic Crisis over Public Transport Efficiency. *International Regional Science Review, 8*, 016001761668736. doi:10.1177/0160017616687361

Banister, D. (2008). The sustainable mobility paradigm. *Transport Policy, 15*(2), 73–80. doi:10.1016/j.tranpol.2007.10.005

Banister, D. (2011a). Cities, mobility and climate change. *Journal of Transport Geography, 19*(6), 1538–1546. doi:10.1016/j.jtrangeo.2011.03.009

Banister, D. (2011b). The trilogy of distance, speed and time. *Journal of Transport Geography, 19*(4), 950–959. doi:10.1016/j.jtrangeo.2010.12.004

Banister, D., & Hickman, R. (2006). How to design a more sustainable and fairer built environment: Transport and communications. *IEE Proceedings - Intelligent Transport Systems, 153*(4), 276. 10.1049/ip-its:20060009

Beatley, T. (2000). *Green urbanism: Learning from European cities / Timothy Beatley*. Washington, DC: Island Press.

Boarnet, M., & Crane, R. (2001). *Travel by design: The influence of urban form on travel. Spatial information systems*. New York: Oxford University Press.

Brown, B. B., Yamada, I., Smith, K. R., Zick, C. D., Kowaleski-Jones, L., & Fan, J. X. (2009). Mixed land use and walkability: Variations in land use measures and relationships with BMI, overweight, and obesity. *Health & Place, 15*(4), 1130–1141. doi:10.1016/j.healthplace.2009.06.008 PMID:19632875

Buehler, R. (2011). Determinants of transport mode choice: A comparison of Germany and the USA. *Journal of Transport Geography, 19*(4), 644–657. doi:10.1016/j.jtrangeo.2010.07.005

Calthorpe, P. (1993). *The next American metropolis: Ecology, communities, and the American dream / Peter Calthorpe*. New York: Princeton Architectural Press.

Cao, X., Mokhtarian, P. L., & Handy, S. L. (2009). Examining the Impacts of Residential Self-Selection on Travel Behaviour: A Focus on Empirical Findings. *Transport Reviews, 29*(3), 359–395. doi:10.1080/01441640802539195

Carmona, M., Heath, T., Tiesdell, S., & Oc, T. (2003). *Public Places-Urban Spaces: The Dimensions of Urban Design*. Oxford, UK: Architectural Press.

Cervero, R. (1996a). Traditional neighborhoods and commuting in the San Francisco Bay Area. *Transportation, 23*(4), 373–394. doi:10.1007/BF00223062

Cervero, R. (1996b). Mixed land-uses and commuting: Evidence from the American Housing Survey. *Transportation Research Part a: Policy and Practice, 30*(5), 361–377. doi:10.1016/0965-8564(95)00033-X

Cervero, R. (2009). Transit and the Metropolis: Finding Harmony. In S. Wheeler & T. Beatley (Eds.), *The sustainable urban development reader* (2nd ed.). London: Routledge.

Cheng, V. (2010). Understanding Density and High Density. In E. Ng (Ed.), Designing high-density cities for social and environmental sustainability (pp. 3–17). Sterling, VA: Earthscan.

Conzen, M. R. G. (1960). Alnwick, Northumberland: A Study in Town-Plan Analysis. *Transactions and Papers*, *3*(27), 3. doi:10.2307/621094

Conzen, M. R. G., & Conzen, M. P. (2004). *Thinking about urban form: Papers on urban morphology, 1932-1998 / M.R.G. Conzen* (M. P. Conzen, Ed.). Bern: Peter Lang.

Darin, M. (1998). The Study of Urban Form in France. *Urban Morphology*, *2*(2), 63–76.

de Vos, J., & Witlox, F. (2013). Transportation policy as spatial planning tool; reducing urban sprawl by increasing travel costs and clustering infrastructure and public transportation. *Journal of Transport Geography*, *33*, 117–125. doi:10.1016/j.jtrangeo.2013.09.014

Dieleman, F., & Wegener, M. (2004). Compact City and Urban Sprawl. *Built Environment*, *30*(4), 308–323. doi:10.2148/benv.30.4.308.57151

El Gabalawy, K., & Galal, O. (Eds.). (2019). The Unplanned Morphing of Social Housing in Cairo: A Positive or Negative Phenomena? Cairo: Academic Press.

Elshater, A. (2012). The New Urbanism Principles vs. Urban Design Dimensions towards Behavioral Performance in Egyptian Neighborhood. In *Future Communities: Socio-Cultural & Environmental Challenges* (pp. 826-843). Giza: AicE-Bs- ASIA Pacific International Conference on Environment-Behavior Studies. 10.1016/j.sbspro.2012.12.270

Elshater, A. (2015). Urban Design Paradigm: Working Manuscripts. London: Penguin Reader.

Elshater, A. (2018). What can the urban designer do for children? Normative principles of child–friendly communities for Responsive third places. *Journal of Urban Design*, *23*(3), 432–455. doi:10.1080/13574809.2017.1343086

Erell, E., Pearlmutter, D., & Williamson, T. J. (2011). *Urban microclimate: Designing the spaces between buildings / Evyatar Erell, David Pearlmutter, and Terry Williamson* (1st ed.). London: Earthscan.

Evans, G., Aiesha, R., & Foord, J. (2009). Mixed-Use or Mixed Messages? In R. Cooper, G. Evans, & C. Boyko (Eds.), *Designing sustainable cities* (pp. 189–217). Chichester, UK: Wiley-Blackwell Pub.

Farr, D. (2008). *Sustainable urbanism: Urban design with nature*. Chichester, UK: John Wiley. [distributor]

Filion, P. (2013). Recent Planning and Development in Toronto. In I. Vojnovic (Ed.), *Urban sustainability: A global perspective, edited by Igor Vojnovic* (pp. 509–529). East Lansing, MI: Michigan State University Press.

Frank, L., Bradley, M., Kavage, S., Chapman, J., & Lawton, T. K. (2007). Urban form, travel time, and cost relationships with tour complexity and mode choice. *Transportation*, *35*(1), 37–54. doi:10.100711116-007-9136-6

Frank, L. D., & Pivo, G. (1994). Impacts of Mixed Use and Density on Utilization of Three Modes of Travel: Single-Occupant Vehicle, Transit, and Walking. *Transportation Research Record: Journal of the Transportation Research Board*, (1466), 44–52.

Frank, L. D., Sallis, J. F., Saelens, B. E., Leary, L., Cain, K., Conway, T. L., & Hess, P. M. (2010). The development of a walkability index: Application to the Neighborhood Quality of Life Study. *British Journal of Sports Medicine*, *44*(13), 924–933. doi:10.1136/bjsm.2009.058701 PMID:19406732

Galal, O. (2017). The Impact of Digital Simulations on Planning Energy Efficient and Solar Friendly Urban Communities. *SSRN Electronic Journal.* Advance online publication. doi:10.2139srn.3163428

Gallimore, J. M., Brown, B. B., & Werner, C. M. (2011). Walking routes to school in new urban and suburban neighborhoods: An environmental walkability analysis of blocks and routes. *Journal of Environmental Psychology*, *31*(2), 184–191. doi:10.1016/j.jenvp.2011.01.001

Glazier, R. H., Creatore, M. I., Weyman, J. T., Fazli, G., Matheson, F. I., Gozdyra, P., ... Booth, G. L. (2014). Density, destinations or both? A comparison of measures of walkability in relation to transportation behaviors, obesity and diabetes in Toronto, Canada. *PLoS One*, *9*(1), e85295. doi:10.1371/journal.pone.0085295 PMID:24454837

Gössling, S., Schröder, M., Späth, P., & Freytag, T. (2016). Urban Space Distribution and Sustainable Transport. *Transport Reviews*, *36*(5), 659–679. doi:10.1080/01441647.2016.1147101

Gouda, A. A., & Masoumi, H. E. (2017a). Certifications systems as independent and rigorous tools for assessing urban sustainability. *International Journal of Urban Sciences*, *2*(1), 1–14. doi:10.1080/12265934.2017.1398103

Gouda, A. A., & Masoumi, H. E. (2017b). Sustainable transportation according to certification systems: A viability analysis based on neighborhood size and context relevance. *Environmental Impact Assessment Review*, *63*, 147–159. doi:10.1016/j.eiar.2016.10.005

Handy, S. (2005). Smart Growth and the Transportation-Land Use Connection: What Does the Research Tell Us? *International Regional Science Review*, *28*(2), 146–167. doi:10.1177/0160017604273626

Handy, S. L. (1996). Understanding the Link Between Urban Form and Nonwork Travel Behavior. *Journal of Planning Education and Research*, *15*(3), 183–198. doi:10.1177/0739456X9601500303

Harms, L., Bertolini, L., & Brömmelstroet, M. T. (2015). Performance of Municipal Cycling Policies in Medium-Sized Cities in the Netherlands since 2000. *Transport Reviews*, *36*(1), 134–162. doi:10.1080/01441647.2015.1059380

Hickman, R., Hall, P., & Banister, D. (2013). Planning more for sustainable mobility. *Journal of Transport Geography*, *33*, 210–219. doi:10.1016/j.jtrangeo.2013.07.004

Holden, E., & Norland, I. T. (2005). Three Challenges for the Compact City as a Sustainable Urban Form: Household Consumption of Energy and Transport in Eight Residential Areas in the Greater Oslo Region. *Urban Studies (Edinburgh, Scotland)*, *42*(12), 2145–2166. doi:10.1080/00420980500332064

Howard, J. (2009). Climate Change Mitigation and Adaptation in Developed Nations: A Critical Perspective on the Adaptation Turn in Urban Climate Planning. In S. Davoudi, J. Crawford, & A. Mehmood (Eds.), *Planning for climate change: Strategies for mitigation and adaptation for spatial planners* (pp. 19–32). London: Earthscan.

Hui, N., Saxe, S., Roorda, M., Hess, P., & Miller, E. J. (2017). Measuring the completeness of complete streets. *Transport Reviews*, *38*(1), 73–95. doi:10.1080/01441647.2017.1299815

Karimi, K., & Vaughan, L. (2014). An Evidence-based approach to designing new cities: The English New Towns revisited. In M. Carmona (Ed.), *Explorations in urban design: An urban design research primer* (pp. 261–274). Surrey, UK: Ashgate Publishing Limited.

Kenworthy, J. R., & Laube, F. B. (1999). Patterns of automobile dependence in cities: An international overview of key physical and economic dimensions with some implications for urban policy. *Transportation Research Part A, Policy and Practice*, *33*(7-8), 691–723. doi:10.1016/S0965-8564(99)00006-3

Kerr, J., Frank, L., Sallis, J. F., & Chapman, J. (2007). Urban form correlates of pedestrian travel in youth: Differences by gender, race-ethnicity and household attributes. *Transportation Research Part D, Transport and Environment*, *12*(3), 177–182. doi:10.1016/j.trd.2007.01.006

Kitamura, R., Mokhtarian, P. L., & Daidet, L. (1997). Article. *Transportation*, *24*(2), 125–158. doi:10.1023/A:1017959825565

Klinger, T., Kenworthy, J. R., & Lanzendorf, M. (2013). Dimensions of urban mobility cultures – a comparison of German cities. *Journal of Transport Geography*, *31*, 18–29. doi:10.1016/j.jtrangeo.2013.05.002

Kropf, K. (2009). Aspects of urban form. *Urban Morphology*, *13*(2).

Kropf, K. (2014). Ambiguity in the definition of built form. *Urban Morphology*, *18*(1), 41–57.

Leck, E. (2006). The Impact of Urban Form on Travel Behavior: A Meta-Analysis. *Berkeley Planning Journal*, *19*(1), 37–58.

Lee, C., & Moudon, A. V. (2006). The 3Ds+R: Quantifying land use and urban form correlates of walking. *Transportation Research Part D, Transport and Environment*, *11*(3), 204–215. doi:10.1016/j.trd.2006.02.003

Leslie, E., Coffee, N., Frank, L., Owen, N., Bauman, A., & Hugo, G. (2007). Walkability of local communities: Using geographic information systems to objectively assess relevant environmental attributes. *Health & Place*, *13*(1), 111–122. doi:10.1016/j.healthplace.2005.11.001 PMID:16387522

Levy, A. (1999). Urban morphology and the problem of the modern urban fabric: Some questions for research. *Urban Morphology*, *3*(2), 79–85.

Liddle, B. (2013). Urban density and climate change: A STIRPAT analysis using city-level data. *Journal of Transport Geography*, *28*, 22–29. doi:10.1016/j.jtrangeo.2012.10.010

Littlefair, P. J. (2000). *Environmental site layout planning: Solar access, microclimate and passive cooling in urban areas*. Building Research Establishment.

Liu, C., Susilo, Y. O., & Karlström, A. (2017). Weather variability and travel behaviour – what we know and what we do not know. *Transport Reviews*, *37*(6), 715–741. doi:10.1080/01441647.2017.1293188

Lo, A. Y. (2016). Small is green? Urban form and sustainable consumption in selected OECD metropolitan areas. *Land Use Policy*, *54*, 212–220. doi:10.1016/j.landusepol.2016.02.014

Lynch, K. (1960). *The image of the city. Publications of the Joint Center for Urban Studies*. Cambridge, MA: MIT Press.

Lynch, K., & Rodwin, L. (1958). A Theory of Urban Form. *Journal of the American Institute of Planners*, *24*(4), 201–214. doi:10.1080/01944365808978281

Ma, J., Liu, Z., & Chai, Y. (2015). The impact of urban form on CO2 emission from work and non-work trips: The case of Beijing, China. *Habitat International*, *47*, 1–10. doi:10.1016/j.habitatint.2014.12.007

Manaugh, K., & Kreider, T. (2013). What is mixed use? Presenting an interaction method for measuring land use mix. *Journal of Transport and Land Use*, *6*(1), 63. doi:10.5198/jtlu.v6i1.291

Marzot, N. (2002). The Study of Urban Form in Italy. *Urban Morphology*, *6*(2), 59–73.

Masnavi, M.-R. (2000). The New Millennium and the New Urban Paradigm: The Compact City in Practice. In K. Williams, M. Jenks, & E. Burton (Eds.), *Achieving sustainable urban form* (pp. 64–73). London: E & FN Spon.

Middel, A., Häb, K., Brazel, A. J., Martin, C. A., & Guhathakurta, S. (2014). Impact of urban form and design on mid-afternoon microclimate in Phoenix Local Climate Zones. *Landscape and Urban Planning*, *122*, 16–28. doi:10.1016/j.landurbplan.2013.11.004

Milakis, D., Vlastos, T., & Barbopoulos, N. (2008). Relationships between Urban Form and Travel Behaviour in Athens, Greece: A Comparison with Western European and North American Results. *Ejtir*, *8*(3), 201–215.

Mitchell, W. J. (2010). Sustainable Urban Mobility through Light Electric Vehicles. In M. Mostafavi & G. Doherty (Eds.), *Ecological urbanism* (pp. 382–397). Baden: Lars Müller.

Moreno, G., & Steadman, P. (2014). The relationship between density, built form and design. In M. Carmona (Ed.), *Explorations in urban design: An urban design research primer* (pp. 193–205). Surrey: Ashgate Publishing Limited.

Moriarty, P., & Honnery, D. (2008). Low-mobility: The future of transport. *Futures*, *40*(10), 865–872. doi:10.1016/j.futures.2008.07.021

Moudon, A. V. (1997). Urban morphology s an emerging interdisciplinary field. *Urban Morphology*, *1*, 3–10.

Moughtin, C. (2003). *Urban design: Method and techniques (* 2nd ed.). Oxford, UK: Architectural.

Mraihi, R., Harizi, R., Mraihi, T., & Bouzidi, M. T. (2015). Urban air pollution and urban daily mobility in large Tunisia's cities. *Renewable & Sustainable Energy Reviews*, *43*, 315–320. doi:10.1016/j.rser.2014.11.022

Næss, P. (2009). Residential Self-Selection and Appropriate Control Variables in Land Use: Travel Studies. *Transport Reviews*, *29*(3), 293–324. doi:10.1080/01441640802710812

Naess, P. (2012). Urban form and travel behavior: Experience from a Nordic context. *Journal of Transport and Land Use*, *5*(2). doi:10.5198/jtlu.v5i2.314

Nedovic-Budic, Z., Knaap, G. J., Shahumyan, H., Williams, B., & Slaev, A. (2016). Measuring urban form at community scale: Case study of Dublin, Ireland. *Cities (London, England)*, *55*, 148–164. doi:10.1016/j.cities.2016.02.014

Newman, P., Beatley, T., & Boyer, H. (2009). *Resilient cities: Responding to peak oil and climate change*. Washington, DC: Island Press.

Newman, P., & Kenworthy, J. (2013). Urban Sustainability and Automobile Dependence in an Australian Context. In I. Vojnovic (Ed.), *Urban sustainability: A global perspective, edited by Igor Vojnovic* (pp. 227–253). East Lansing, MI: Michigan State University Press.

Newman, P. W. G., & Kenworthy, J. R. (1989). Gasoline Consumption and Cities. *Journal of the American Planning Association*, *55*(1), 24–37. doi:10.1080/01944368908975398

Oke, T. R. (2017). *Urban climates*. Cambridge, UK: Cambridge University Press. doi:10.1017/9781139016476

Ou, J., Liu, X., Li, X., & Chen, Y. (2013). Quantifying the relationship between urban forms and carbon emissions using panel data analysis. *Landscape Ecology*, *28*(10), 1889–1907. doi:10.100710980-013-9943-4

Owen, N., Cerin, E., Leslie, E., duToit, L., Coffee, N., Frank, L. D., ... Sallis, J. F. (2007). Neighborhood walkability and the walking behavior of Australian adults. *American Journal of Preventive Medicine*, *33*(5), 387–395. doi:10.1016/j.amepre.2007.07.025 PMID:17950404

Pahl-Weber, E., Seelig, S., & Ohlenburg, H. (2012). The Shahre Javan community detailed plan: Planning for a climate responsive and sustainable Iranian urban quarter. Young cities research paper series: v. 3. Berlin: Universitätsverlag der TU Berlin; BHRC, Building and Housing Research Center.

Pucher, J., & Buehler, R. (2008). Making Cycling Irresistible: Lessons from The Netherlands, Denmark and Germany. *Transport Reviews*, *28*(4), 495–528. doi:10.1080/01441640701806612

Roaf, S. (2010). The Sustainability of High Density. In E. Ng (Ed.), Designing high-density cities for social and environmental sustainability (pp. 27–39). Sterling, VA: Earthscan.

Rogers, R. (2008). *Cities for A Small Planet*. Boulder, CO: Basic Books.

Rogers, R., & Power, A. (2000). *Cities for a small country*. London: Faber.

Rowley, A. (1996). Mixed-use Development: Ambiguous concept, simplistic analysis and wishful thinking? *Planning Practice and Research*, *11*(1), 85–98. doi:10.1080/02697459650036477

Sallis, J. F., Saelens, B. E., Frank, L. D., Conway, T. L., Slymen, D. J., Cain, K. L., . . . Kerr, J. (2009). Neighborhood built environment and income: Examining multiple health outcomes. *Social Science & Medicine (1982)*, *68*(7), 1285–1293. doi:10.1016/j.socscimed.2009.01.017

Scheer, B. C. (2017). *The evolution of urban form: Typology for planners and architects / Brenda Case Scheer, AICP*. Abingdon, UK: Routledge. doi:10.4324/9781351179751

Schwanen, T., Dijst, M., & Dieleman, F. M. (2004). Policies for Urban Form and their Impact on Travel: The Netherlands Experience. *Urban Studies (Edinburgh, Scotland)*, *41*(3), 579–603. doi:10.1080/0042098042000178690

Schwanen, T., & Mokhtarian, P. L. (2005). What affects commute mode choice: Neighborhood physical structure or preferences toward neighborhoods? *Journal of Transport Geography*, *13*(1), 83–99. doi:10.1016/j.jtrangeo.2004.11.001

Shaaban, K., Muley, D., & Elnashar, D. (2017). Temporal variation in walking behavior: An empirical study. *Case Studies on Transport Policy*, *5*(4), 671–680. doi:10.1016/j.cstp.2017.07.001

Simmonds, D., & Coombe, D. (2000). The Transport Implications of Alternative Urban Forms. In K. Williams, M. Jenks, & E. Burton (Eds.), *Achieving sustainable urban form* (pp. 121–130). London: E & FN Spon.

Snellen, D., Borgers, A., & Timmermans, H. (2016). Urban Form, Road Network Type, and Mode Choice for Frequently Conducted Activities: A Multilevel Analysis Using Quasi-Experimental Design Data. *Environment & Planning A*, *34*(7), 1207–1220. doi:10.1068/a349

Strømann-Andersen, J., & Sattrup, P. A. (2011). The urban canyon and building energy use: Urban density versus daylight and passive solar gains. *Energy and Building*, *43*(8), 2011–2020. doi:10.1016/j.enbuild.2011.04.007

Sundquist, K., Eriksson, U., Kawakami, N., Skog, L., Ohlsson, H., & Arvidsson, D. (2011). Neighborhood walkability, physical activity, and walking behavior: The Swedish Neighborhood and Physical Activity (SNAP) study. *Social Science & Medicine (1982)*, *72*(8), 1266–1273. doi:10.1016/j.socscimed.2011.03.004

Talen, E. (2003). Measuring urbanism: Issues in smart growth research. *Journal of Urban Design*, *8*(3), 195–215. doi:10.1080/1357480032000155141

Townshend, T., & Lake, A. A. (2009). Obesogenic urban form: Theory, policy and practice. *Health & Place*, *15*(4), 909–916. doi:10.1016/j.healthplace.2008.12.002 PMID:19201641

Vale, D. S. (2013). Does commuting time tolerance impede sustainable urban mobility? Analysing the impacts on commuting behaviour as a result of workplace relocation to a mixed-use centre in Lisbon. *Journal of Transport Geography*, *32*, 38–48. doi:10.1016/j.jtrangeo.2013.08.003

Van de Coevering, P., & Schwanen, T. (2006). Re-evaluating the impact of urban form on travel patterns in Europe and North-America. *Transport Policy*, *13*(3), 229–239. doi:10.1016/j.tranpol.2005.10.001

Verhetsel, A., & Vanelslander, T. (2010). What location policy can bring to sustainable commuting: An empirical study in Brussels and Flanders, Belgium. *Journal of Transport Geography*, *18*(6), 691–701. doi:10.1016/j.jtrangeo.2009.11.003

Villanueva, K., Knuiman, M., Nathan, A., Giles-Corti, B., Christian, H., Foster, S., & Bull, F. (2014). The impact of neighborhood walkability on walking: Does it differ across adult life stage and does neighborhood buffer size matter? *Health & Place*, *25*, 43–46. doi:10.1016/j.healthplace.2013.10.005 PMID:24239702

Wendel-Vos, G. C. W., Schuit, A. J., De Niet, R., Boshuizen, H. C., Saris, W. H. M., & Kromhout, D. (2004). Factors of the Physical Environment Associated with Walking and Bicycling. *Medicine and Science in Sports and Exercise*, *36*(4), 725–730. doi:10.1249/01.MSS.0000121955.03461.0A PMID:15064601

Whitehand, J. W. R. (2001). British urban morphology: The Conzenian tradition. *Urban Morphology*, *5*(2), 103–109.

Williams, K. (2000). Does Intensifying Cities Make them More Sustainable? In K. Williams, M. Jenks, & E. Burton (Eds.), *Achieving sustainable urban form* (pp. 30–45). London: E & FN Spon.

Wong, N.-H., & Chen, Y. (2010). The Role of Urban Greenery in High-Density Cities. In E. Ng (Ed.), Designing high-density cities for social and environmental sustainability (pp. 227–262). Sterling, VA: Earthscan

Zhu, Y., & Diao, M. (2016). The impacts of urban mass rapid transit lines on the density and mobility of high-income households: A case study of Singapore. *Transport Policy*, *51*, 70–80. doi:10.1016/j.tranpol.2016.03.013

ADDITIONAL READING

Carmona, M. (2014). *Explorations in urban design: An urban design research primer*. Farnham, Surrey, UK: Ashgate.

Davoudi, S., Crawford, J., & Mehmood, A. (Eds.). (2009). *Planning for climate change: Strategies for mitigation and adaptation for spatial planners*. London: Earthscan. doi:10.4324/9781849770156

Erell, E., Pearlmutter, D., & Williamson, T. J. (2011). *Urban microclimate: Designing the spaces between buildings / Evyatar Erell, David Pearlmutter, and Terry Williamson* (1st ed.). London: Earthscan.

Newman, P., & Kenworthy, J. (2009). Traffic Calming. In S. Wheeler & T. Beatley (Eds.), *The sustainable urban development reader* (2nd ed., pp. 123–129). London: Routledge.

Ng, E. (Ed.). (2010). *Designing high-density cities for social and environmental sustainability. London.* Sterling, VA: Earthscan.

Oke, T. R. (2017). *Urban Climates*. Cambridge: Cambridge University Press. doi:10.1017/9781139016476

Pahl-Weber, E., Seelig, S., & Ohlenburg, H. (2012). The Shahre Javan Community Detailed Plan: Planning for a Climate Responsive and Sustainable Iranian Urban Quarter. Young Cities Research Paper Series: v. 3. Berlin, Tehran: Universitätsverlag der TU Berlin; BHRC, Building and Housing Research Center.

Williams, K., Jenks, M., & Burton, E. (Eds.). (2000). *Achieving Sustainable Urban Form*. London: E & FN Spon.

Chapter 14
An Economic Diagnosis of Historical Vilas in Alexandria:
Reuse vs. Demolition

Heba Hatem Aggour
Independent Researcher, Egypt

ABSTRACT

Alexandria's old town center is like passing into an extensive gallery include various architectural forms of several ancient buildings. Since theses listed buildings in Egypt are not profitable from the financial perspective, the owners have plans to destroy them and locate an option for their monetary needs. The author, in the research objective of this chapter, tries to discover the best approach to raise the financial estimation of some listed villas in Alexandria until they have the same estimation of or more than the building if decimated and replaces by a tall structure. The results provide solutions for the owners who are ready to keep it and safeguard the cultural heritage of the city. The concluded remarks provide answers to the primary research question raises of how to achieve a win-win situation between listed buildings in Egypt and the economic needs of the users.

INTRODUCTION

Passing by Alexandria's old town center is like crossing through an open museum incorporate different architectural styles: Italian, British, Greek and French buildings stand there where it exhibits Alexandria's history. Alexandria used to attract famous creative writers such as EM Foster and Lawrence Durrell, architects and poets like Greek poet Constantine Cavafy. Unfortunately, through this trip a scene might be noticed, many buildings are found in a fragile situation or completely demolished. The decay situation provokes activists to protest against neglecting the heritage of their city, asking the government to act and save it. Desoki (2007) mentioned, since the end of the 19th century Alexandria was a cosmopolitan city including a high number of foreigners lived together and influenced the city to have a collection of an astonishing variety of architectural styles. The historical importance of the cosmopolitan Alexandria's

DOI: 10.4018/978-1-5225-9238-9.ch014

buildings; have a unique architectural value such as cosmopolitan, contemporary and Islamic style buildings; built by a famous international architect; buildings belong to well- known historical character; and buildings have Moral value, a special memory for the citizen or witnessed notable events.

One of the most pressing issues today confronting communities with historic preservation laws is restricted proprietors who don't have the well or the funds to keep up their historic properties. The view of the subjects nowadays towards the authentic structures is identified with how financially significant it is. Since the listed buildings in Egypt are not profitable from the financial perspective, the owners are willing to destroy it and locate a substitution option for their monetary needs. The primary objective of the chapter is to discover how much the owners of the listed villas lose or profit from restricting its demolition in Alexandria. The central research questions raised, do the owners of the listed villas lose financially from restricting its demolition? and how big are their losses?

RESEARCH METHODOLOGY

To reach the aim of the research, a balance between two existing scenarios is studied in the chapter. Scenario one is the demolition of the listed building for a real estate construction towards higher financial profits, while scenario two is the keeping of the building and investing in it. The research intends to analyze the local situation of Alexandria's city center and surrounding areas and creating scenarios of their predicted future situation, apply the previous knowledge on Alexandria's case and measuring the cost-benefit analysis of the two scenarios.

The monetary investigation is going to determine the cost-benefit analysis of some existing cases in the focus area of study which express whether the real estate high-rise building projects on lands of previously existed listed villas, restoration project, reuse project. It also minds considering in the analysis other listed villas which have an unknown future, where the research will assess the cost-Benefits of them in various hypothetical expected scenarios. The aim is to comprehend the measure of the financial misfortunes the proprietor is confronting because of the limitations put on the property he possesses and whether it is feasible to be solved according to the current approach of the city or not. In the economic diagnosis, the focus will be on the private owned listed villas as the worst scenario causing the financial losses for the owner. In this chapter, two new divisions of the registered villas are the reused villas and the renovated villas. Some of the previous classifications of the deterioration cases happening in Alexandria mentioned are chosen to occur in the second ranking; the demolished villa, the replaced with a high-rise project and the neglected villa as shown.

Mixed Methods Research

Using the mixed methods research to combine the quantitative and qualitative data together by a Question-driven perspective to accomplish the following:

1. Explanations, understandings realities and relationships, and evaluations on the previous decisions taken towards the phenomenon occurs in Alexandria, its dimensions, and situations in the form of scenarios and observations and then by focusing on the claim.

By: collecting numbers of the demolished buildings, why, when and how, collect statistics of the relation between the variables, financial data, including the rents, taxes and class level, focusing on the period after the revolution, and the development of the law and procedures related to the heritage of Alexandria, using journal articles, previous surveys, and observations.

2. Field studies, collecting data, and analysis of a similar case study on the focused level.

By: Interviews, surveys, and questionnaires about the actual outcomes and ways of evaluating the historical buildings economically.

3. Reflect the theoretical data and the case studies on the future scenarios for the situation, to reach a win-win situation between the preservation of the historical buildings and the economic viability of the restricted owner.

By: comparing future scenarios scenario planning process of the building when demolition and when preserved, Questionnaires with the focus group

Research Problem Analysis

Due to the development pressure going fast in the context, prompting the expansion of the area esteem and in like manner the rapid increment of the gap between the value of the land itself, being more than the land with the listed building (Licciardi, 2002). The owners' perception towards the listed buildings in Alexandria is devastating the listed building and having an empty land, implies a high financial worth that he could not even dream for half of it while the building kept and preserved especially in downtown. "It's pointless to put these villas on a heritage list that doesn't protect them or maintain their value. It just creates dead spaces." Shahed said (Morayef, 2012). The research assumes and tries in this economic diagnosis to know, how enormous are the financial losses? Could it be covered by investing in the building under the existing law, regulations, methods and knowledge of managing these listed structures? can the owner benefit from the conservation as the benefits from the real estate project? If not, is it achievable, if some changes in economic policies occur?

Problem Indicators

In Figure 1.a, the yellow line reflects the greatest height stature that is permitted by the building regulations in the area and nobody ought to surpass it. The tall structures developed during and after the revolution are the ones under the red arrows while the ones under the dark arrows are the structures that exceeded the extreme allowed tallness, before 2011. In the master thesis of Borg, two relative photographs of the horizon of the Bahary prior skyline before and after the revolution (See Figure 1.b, It also represents a pointer for the activism of Alexandria construction mafia, who had an incredible part in this quick development of illegal structures through and after the political disarray occurred in Egypt (Save Alex, 2012). Al-Raml Station and Al-Manshia areas are the most critical chronicled locale in Alexandria representing the European city. It is visible downtown of the sea waterfront of Alexandria, passed through evident pulverization activities. Why such a respectable place similar to the center of Alexandria can't survive of the decay actions?

Figure 1. a: Visual Indicators of the existence of a high number of illegal buildings constructed in the historical sites of Alexandria in 2011. b and c: before and after the constraction of high illegal buildings appear in the skyline of Bahary district.
Source: Save Alex. Facebook page, 2012. b. Borg, 2013

There are some indicators that ensure the active existence of the phenomenon of deterioration in Alexandria. One of these indicators is the decrease in the number of the recorded buildings in the main legacy list for Alexandria that was issued by the Prime Minister in 2008, the main complete list for the entire city in Egypt. According to Borg, this legacy list had 1135 recorded structures, as indicated by Borg (2013), 193 structures in Alexandria had encountered new circumstances from the complete decimation, whether the area left vacant or was supplanted by an elevated structure. The cases connected to the court for expelling from the legacy list officially dismissed from the legacy list and the neglected structures. 80 buildings in 2011 were unlisted, 40 proprietors applied for cases to expel their property from the legacy list. 35 structures were entirely torn down, an elevated structure supplanted ten from them, and the last seven structures were left deserted and in moderate disintegration circumstance, because of the absence of upkeep (Borg 2013).

While the initiative Save Alex posted a photo demonstrates the quantity of the elevated structures built soon after the revolution of 2011, which is regularly related to the pulverization of the recorded structures. Additionally, shown by the abnormal state of obliteration licenses taken by Wasat locale in 2011, this is in any case, the most widely recognized illicit devastation happens with a significant portion of properties particularly the villas (Morayef 2012).

FINANCIAL IMPORTANCE OF HERITAGE PRESERVATION

Numerous powers and inhabitants trivialize the essentialness of the legacy to the nation, or they do not have the knowledge of its positive effect. The impact of heritage protection accumulated in a few cases (Haspel, 2011). First, the monetary advancement impact, where a study was done in Germany, saw that incomes from the sponsorships and state subsidizing to the proprietors of the enlisted structures were higher by four to twelve times than the anticipated standard yields. In Haspel, as per the leibniz-Institut für regionalentwicklung und Strukturplanung (IRS), impact studies of the conservation and legacy activities demonstrated that it provoked a financial growth and social coherence to the capital. Also, it encourages private parties to spend in the safeguarding arrangements (2011).

Heritage protection is considered as an advertising system for the city and is the overwhelming apparatus to improve tourism and thus support the financial benefits over the long term. Haspel likewise said that after 1999, Germany and other European nations accomplished stable relation within the legacy protection and monetary determinants. The derivation in the execution of protection undertakings and legacy posting exercises had a number of advantages, one of them that it made a beeline for the enhanced number of guests in Germany and the quantity of the overnight stays, particularly Berlin. Berlin and Munich in Germany enlisted in 2010 to be from the most astounding ten visited towns in the world and recorded a more remarkable number of overnight stays than Rome.

Heritage Conservation Perception

Through the Roman era, the perception of the natives to their the culture heritage primarily concerned its lifetime, elegance, and authenticity (Licciardi, 2012). He said that, though in the 18th century and while the Italian Renaissance, the perception of the citizens shifted to how satisfying the structure mirror the culture, history, and uniqueness of the city. This opinion obtained from a large number of culture preservation schemes and the issuing of the initial charters like Venice charter, the United Nations Educational, Scientific, and Cultural Organization (UNESCO) and International Council on Monuments and Sites (ICOMOS) (Licciardi, 2012). Although since the modern era till now, the attitude of the locals towards the heritage evolved entirely, due to several reasons, one of those grounds powerfully presented in the evolution of accelerated economic transformations. The power of the private sector, the financial development, and the globalization competition are determinants forced the citizens to perceive the heritage from an economic point of view. The estimate of the historical building value enhanced according to how effective is the structure, if the financial profits are not sufficient enough, then it doesn't't deserve to be protected anymore (Licciardi, 2012). According to Licciardi, in this case, the attempts to keep the heritage became complicated, multiple values and stakeholders need attention and to be honestly managed to settle a successful sustainable conservation process (2012).

Consequently, citizens' altered perception motivated the researcher of the present wprk to concentrate on the financial gap and realize the fiscal position in Egypt for a possible conflict resolution between the heritage conservation and the owners' economic interests and to achieve the win-win situation between them. The intention of the study seeks to seize a bargain among the involved participants, and the financial condition of the town controls the conservation of the designated culture heritage.

ECONOMIC DIAGNOSIS IN ALEXANDRIA

Observing Deterioration Types

The first type of decay is the deserted structures that have been left with no systems for upkeeps whether it is possessed or not. The second shape is the vertical expansion happens over an authentic building, debilitate its structure and prompt the twisting of the significant unique state of the building. At that point comes the third country of crumbling, which shows up as visual hindrance of the recorded structures on account of the fast-vertical extension of the encompassing structures, prompted the vanishing of the authentic working as though it doesn't't exist. The last shape is the harming of the recorded structures, then the decimation move makes put, this is the most troublesome circumstance to be handled. For more elucidation of these events, a couple of cases.

Its economic losses occur strongly in the cases of the villa, because of its small area which if even invested in it might not gain high profits for the owner. Ibrahim El-Henawy, the operation administration of SIGMA organization, said that the remained villas in the downtown are in a crucial situation because there are now very few villas surrounding with a big number of high-rise buildings, which threatens the destruction of the listed villas. Accordingly, they are in a significant need for a quick new strategy to be saved (EL-Henawy, Personal interview, May 03, 2016).

The Focus Area

The focus area is located in Al-Raml district, Moharem Bek, and Al-Hadara district. The city center area contains a big number of villas for the economic diagnosis and the famous monuments; the Library of Alexandria, the Shalalat Park, Saad Zaghloul Square, the old stadium of Alexandria, Masr Station, and Manshia square. The research intended to analyses not only one concentrated urban space but also include Moharem Bek and Al-Hadara district to pick the comparative cases from it. Although the other two spots are near the city center, they are different in the real estate value. The aim of analyzing them is to have different results which lead to a more accurate conclusion of the economic situation of the two studied scenarios. It can also help to see how much the influence of being near the heritage can change the value of the property, and dependently it shifts in the way of solving the economic gap occur between the two scenarios.

The next step was to choose the cases in the focus area to apply the economic diagnosis on it. There are four cases in Al-Raml district; real estate project, invested in building, restored villa and the neutral villa. Another two cases are located in Moharem Bek area; demolished villa and another housing project. one case of the neglected villa Ambron was chosen and it is located in Al-Hadara region.

Case one is two real estate projects replaced the previously existed listed villas. In this case, the economic analysis will just assess the possible housing price of the project and how much earns the investor will own at the end of contracting all the units. The object of this case is to recognize how the benefit calculations are arranged to be similarly executed over the comparison of the two scenarios. A telephone interview with an employee of the real estate company "Seif" and another one from the property company "Zamzam" was produced to recognize the value of the meter in the two projects. These two projects are real estate projects which were built on the land of a demolished listed villas just after its destruction. Another semi-structured interview with an experienced contractor "Fathy El-Sharabasy", to estimate the price of construction, the manpower needed and the further expenses.

Case two is regarding the reuse of the historical building like L passage, executed a restaurant food court design on the ground floor of a historical multi-story building. The analysis of this example has not represented in a villa building like the rest of the cases since there was no successful and economically profitable example of reuse project in a villa in the focus area of study. The research prepared a semi-structured interview with the investor of L Passage "El-Leithy Mekawy." Another semi-structured interview with the architect "Maha Magdy" and with "Ibrahim El-Henawy", the operation administration of the company owns the historical building hosting L Passage project "SIGMA" to know the procedures, the revenues and the costs of investing in heritage.

Case three is about the restoration of the villa, represented in Sarsaa Pasha Pal- ace on Fouad street that is restored by the authority to lease it later to a private foreigner company who sustain its management. The objective of this case is to appreciate how much restoration project may require and to practice this on the subsequent comparative cost-benefit analysis. The unstructured interview then was fulfilled with the restoration expert working on the project "Engineer Ramadan", to identify the costs of restoring the elevation and the interior of an old building and to know the number of handicrafts-men needed in the project.

From the beginning of case four to case six, the comparative cost-benefit analysis of the two scenarios will be implemented and then measuring the economic profit gap. In case four, the destroyed villa Aghion, which either in the future will have another real estate project on its land (scenario one), or the government may reconstruct it again, as an international followed way of heritage conservation trend, and reuse it as a museum or a restaurant (scenario two). The cost-benefit analysis of the two scenarios will be measured in this case. Case number five, the neglected villa will calculate the cost-benefit price on scenario one and two and conclude the economic interest gap occurred among them. In case four and five, first data collection was done to obtain information on the villa since one villa is demolished and the other one is totally overlooked and falling apart; the owners are unreachable. The sixth case is a neutral occupied and well-maintained villa, which the corresponding estimations are done on it also to test the economic gaps that can occur in a different situation and land values in Alexandria. In this case, a structured interview was done with the proprietor of a listed villa Nehad Abo-ElFadl. The aim of the interview is to identify her opinion about the conservation of the villa, if she applied for a case if the authority is paying the compensations stated in the regulations and what she is looking forward to doing on the property. Extra semi-structured interview with an owner of a likewise situation villa, "Loatfy", in kafr Abdo district outside the focus area, prepared including identical questions directed to him in addition to open-ended questions.

A snowball formation ran through the interviews of the research, began with three recognized specialists in Architecture and urban planning field, two from Alexandria and one from Stuttgart. The first professional led to "L passage investor" - who later attached another interviewee "L passage owner" - the "Economic expert", the "real estate expert" and different professional who conducted the research to the questioned "Restoration expert." The second expert combined the research with a member of the Technical security of Heritage Committee of Alexandria, who alongside led to the interviewed "Member in Technical security of Heritage Committee of Alexandria and got her Ph.D. on the heritage conservation in Alexandria focusing on the legislative aspects". The second profession- al additionally attached "Save Alex member" and "NOUH member". The third professional guided to the "German conservation expert", who is aware of the German conservation policies and regulations.

Before each economic diagnosis, the chapter will record initial documentation of some important structures; this documentation depended principally on the original contracts of these properties. The documentation contains the proprietorship exchanges, description of the assets in this time, the sale value through years, the range of the area and the annually rent valuation of the property by the government. The past literature discussed these structures and the circumstances happened to it as indicated by the observation done in the research.

At that point, a few structures were being financially broke down according to their unique situations: demolished and supplanted by a tall structure, destroyed, and still empty land, investments occurred in it, restored the property and the dismissed structures, neglected villas, and neutral well-maintained listed villas. On account of the annihilated and supplanted by high construction cases, will be utilized to figure the price of the units and the returns originate from offering the new commercial and private properties to know the impact of the location on the returns of the project. The second case of investing in the building will be used to ascertain the interests originate from investment ventures occurring on the ground floor as the most popular approach to be used in case four, five and six to compare the two scenarios. The third case which is the restored building will be used to compute the expense of restoring an authentic building to be utilized later in the comparative cases.

Scenario one represented in case one of the listed buildings, which is the benefits comes from a real estate tower project. Scenario two is described in case two and three consolidated: the ventures money saving advantage investment and the restoration projects in the property. After knowing the critical financial costs and benefits in the territory, case four, five and six of the trapped, the neglected and the neutral listed properties will be used to compare the two scenarios that can occur on the same piece of land for a direct monetary correlation (See Figure 2).

The prices of the lands, the buildings, especially the listed buildings and the investment projects are regularly changing quickly and differ sharply from place to another. Accordingly, the research is not calculating the prices, costs, returns and profits to be implemented on the ground with the exact same numbers. The principal aim is to initiate a discussion of the demolition problem by discussing the significant financial gap occur within the two mentioned scenarios, the diverse causes for its existence and the ways to balance it. Because of the difficulty of having accurate information about the costs, returns and costs of land properties, several interviews with experts, professionals, and highly experienced actors was necessary, as mentioned before, to know the average prices.

Case Studies Economic Analysis

CASE 1- Demolished Villa and Replaced by High Rise Building: Villa number 1774 and 1772:

Villa number 1772 in the Heritage list is located in 11 Hussien Nouh Street, extended from the primary Fouad road, before being demolished and supplanted by an elevated structure. The villa was categorized

Figure 2. The diagram shows the method the economic diagnoses will be calculated to reflect on the value of each scenario.
Source: The author

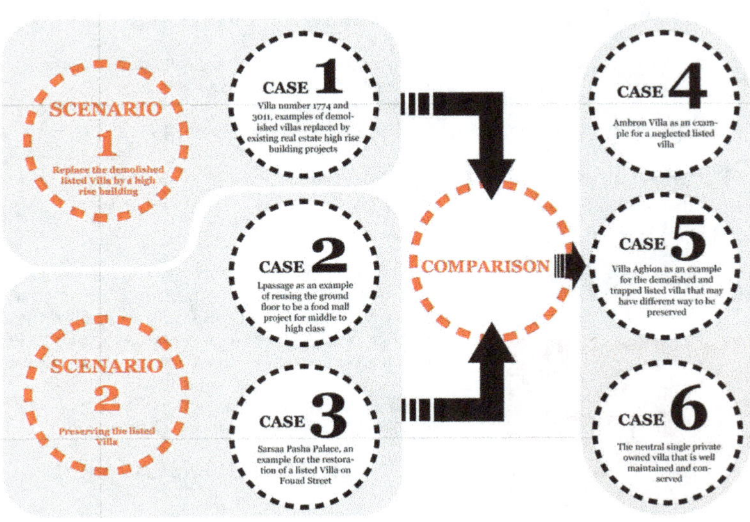

under level C (The local level). The range of the land area is 790 square meter, and private multiple proprietors have possessed it. According to the pronouncement number 222 for the year 2011 from Wasat district entity, the building had the pulverization permit ready. When the surrounded security men were asked about the date when the demolition action was done, they said that it started around the middle of 2012, and the real estate project started after half a year from the demolition in 2013 and was finished around the beginning of 2016.

For villa number 1774 in the list, it was situated on 13 Abdel Said road; now it is supplanted by the same tall structure of the villa number 1772 In the heritage list. Before 1968 it was possessed by two private foreigners, in 1968, the proprietorship was transferred to two individual Egyptians. In 1971, the owner sold it to another Egyptian owner for 12,000 EGP (/0.43≈ 64,899 USD, nowadays = 576,673 EGP) and in another contract it was for 10,000 EGP (/0.43 ≈ 23,256 USD, nowadays = 204,653 EGP) Egyptian pounds. Sadly, it had the same instance of annihilation and substitution like the previous case. The annually rent- al cost evaluated by the government was equal to 1,068.64 EGP (/0.43 ≈ 2,485 USD, nowadays = 21,870 EGP). In this case the meter square used to cost 26.46 EGP (/0.43 ≈ 61.5 USD, nowadays = 546 EGP) in 1971. Another demolition decree number 347, for the year 2012, offered the permit of a new building with a basement floor, ground floor and first floor for commercial shops, and five stories for residential units. Accordingly, the obligation of developing the new facility was given to the real estate company 'Seif'. A phone call with a worker in "Sief" organization was done keeping in mind the end goal to know the cost of every apartment (Sief, Telephone interview, February 20, 2016).

Figure 3 reveals that the final returns of the residential projects are very high, which explains the quick and great turnout of the owners, contractors, and investors to invest in this type of projects.

Figure 3. The calculation of the returns of Seif real estate project in Al-raml district.
Source: by Author

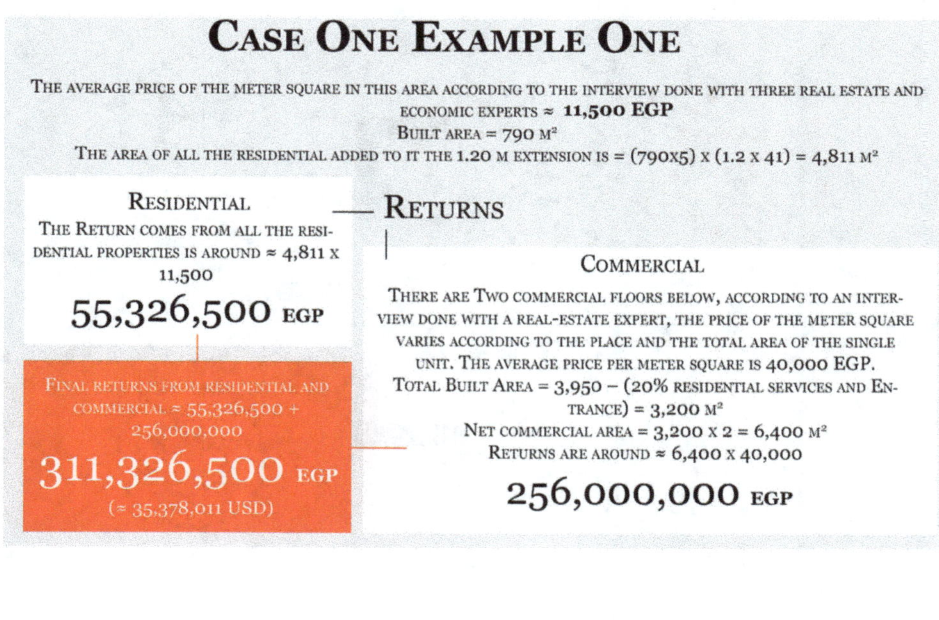

Villa Number 3011 in the Heritage List

Villa number 3011 in the list, situated in 4 Pierre, wabour El-maya region, with a land ranges for 1,244 square meter area. The category the villa is inserted under is level C (Local level). In 2003, the villa was acquired by 12 Egyptian proprietors who inherited the property after a few generations till they got to be numerous with the need to distribute the benefit of the land among each other, which drove them to think whether to offer it to an outside individual or to damage it totally. As mentioned by the proprietor of the listed villa number 589 in the list, situated on Sultan Hussein road, that it is a legacy that must be dispersed if it is not circulated when the villa is standing, they need to consider another way to get the profits in cash and divide it among themselves (Aboelfadl, N. Email Questions, February 23, 2016).

A buyer was found for the villa number 3011, who purchased the property for ten million EGP (/0.56 ≈ 17,857,143 USD, nowadays = 158,673,136 EGP) in 2007, which had yearly rental valuation for 840 EGP (/0.56 ≈ 1,500 USD, nowadays = 13,329 EGP).

In different cases, it was difficult to find a buyer who needs to purchase a restricted property, for this situation, the investment projects can bring income which can be dispersed between the inheritors. The result of the high demand for housing, the best and might be the primarily accessible option and quick answer for procuring cash and disseminate it for one time is to fabricate an elevated structure. In this arrangement, everybody will have the capacity to have his benefit interests after last two years as expressed by El-Lethy Mekawy and Mohamed El-Awady, investor and financial specialist (Mekawy, Personal interview, April 20, 2016), (El-Awady, Personal interview, Spring, 2016). In 2011, a destruction decree number 44 issued, and the building promptly decimated, which trailed by another decree number 899 for a new structure permit: basement, ground floor shops, first floor for administrative and commercial uses and another eight stories for residential units. For this case, the meter square used to cost 5,882 EPG (/0.56 ≈ 1,050 USD, nowadays = 9,330 EGP), in 2007.

After asking the security guards of the surrounding buildings, they said that the demolition, as the previous example, was done in 2012, and the real estate project started in 2013 and finished around the beginning of 2016. This can ensure that one of the important reasons for the high rate of demolition happened in Alexandria was due to the weak follow-ups from the authorities and the powerful existence of the construction mafia who aimed for the demolition of the listed buildings during the political chaos occurred in Alexandria at this time. Another telephone questionnaire prepared with an employee of Zamzam real estate company, the one in charge of the development of the new structures to know the prices of the apartments as shown in Figure 4 (Zamzam, Telephone interview, February 20, 2016).

CASE 2: Investing in a Listed Building: Building Number 33

Situated in 54 Gamal Abdel Nasser road, with a land of 1,800 square meters. In 2011, the ground floor was leased to a designer financial specialist. The building is under level B (city level) in the heritage list of Alexandria. The monthly rent in 2011 was around 4,000 EGP, and when it got to be celebrated, and numerous high-class individuals began to go there, the rent of the private lofts abruptly raised to be 12,000 EGP (/8.8 ≈ 1,364 USD).

As stated on the website of the owner company, SIGMA properties are one of the real estate investor groups, but with a different perspective, SIGMA is specialized in developing and re-utilizing heritage buildings instead of building a new high-rise residential building. The company started in 1959, but only lately it began to achieve great popularity, especially after completing the L passage project. Its

Figure 4. The calculations of the returns of Zamzam real estate project.
Source: by Author

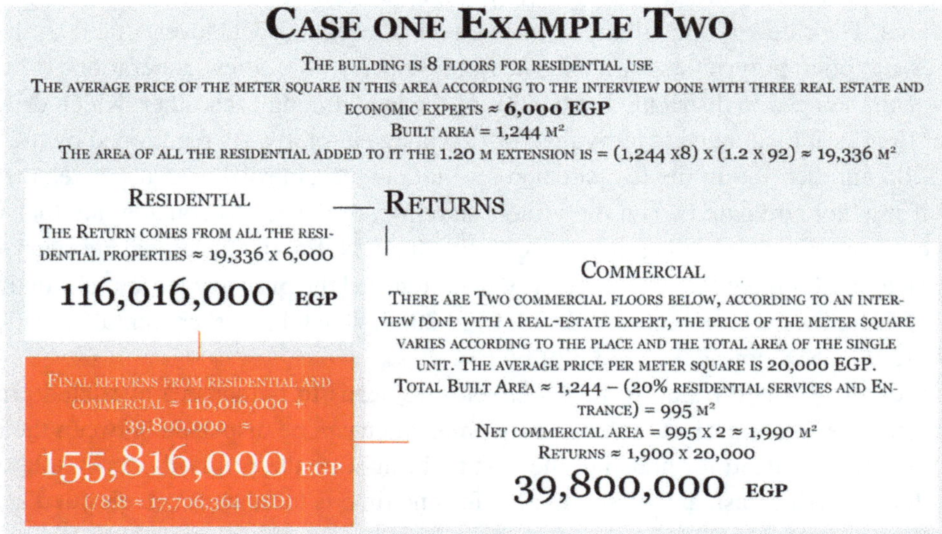

work is summarized in buying the old real estates, utilize and develop it, add investment projects. Since the recent years, they targeted Fouad Street as the oldest street in Alexandria, and they bought some properties there the most famous one from them is the prosperity listed with number 33 in the Heritage list of Alexandria. El-Nasr company for televisions owned the building, which started to stop working and to have some financial crisis. In this period SIGMA company bought the property and began to look for the owners of the apartments and the tenants if they want to sell it or stop renting it and buy it from them (Magdy, Personal interview, May 03, 2016).

According to an interview done in SIGMA company with Ibrahim El-Henanawy, he said that the firm usually faces the problem of the slow return of capital. This slow return is usually due to the occurrence of an average of 80% of the apartments under the old rent law, which makes it difficult for the company to have quick revenues from investing in the listed property. That's what drove them to depend on the investment projects as a way of fast earnings ensure and fast (El-Henawy, Personal interview, May 03, 2016). Ibrahim El-Mekahwy stated that the revenues of the company come from the direct rents from the tenant and by the increase the value of the prosperity throughout time by increasing the utilities, maintaining the infrastructure, restoring the elevation and raise the quality of the building. Then when they start to sell the building, they will earn the difference of the increased value of the property that was sharply higher in a short period (El-Henawy, Personal interview, May 03, 2016).

El-Henawy also said also that, this increase in the value of the property leads to their losses as well as their profit because the compensations of the tenant to leave the apartment for the company will be very high, as the compensation is calculated by half of the value of the normal property in the building. Also, the persuasion of the tenant to leave will be so difficult, due to not only the low rent but also the improved high quality of the building done by the company that he is living in now. Maha mentioned, the company tries to buy the apartments from their owner to rent it with a high value. Also, they try to compensate the tenants for half the price of the property to leave the house. The organization is focusing on a particular class of individuals which is the high level and need them to be incorporated in their ventures rather than the primary occupants, to be able to obtain enough revenues cover their expenses on the reuse of the recorded properties (Magdy, Personal meeting, May 03, 2016). The fear of gentrification is strongly attached to the reuse of the heritage that was commonly used by the low and middle class of citizens. When Maha was asked if all the apartments' owners were replaced by a high-class resident and the historical shop is sold to another class of people for a new investment targeting the high level. After 10- or 20-years, Fouad Street will be mostly used by completely different people and in an entirely different way, and she agreed that it should be because not only SIGMA is doing that but other companies as well. Maha and Ibrahim argued that SIGMA company owns several listed properties where they try to implement the same strategy of replacing the people with a high class to increase the value of the property and so increase the revenues throughout time (Magdy, Personal interview, May 03, 2016).

In an interview with Maha Magdy, she stated that the ground floor was decided to be rented to an investor and to be used as a food mall, where different restaurants and cafes take part in it. El-Lethy group office was the one responsible for the design of the project. As Maha Magdy, a senior architect in SIGMA said that the maintenance of the building is the responsibility of the company, with sometimes the occupant's union except for the ground floor, it is the responsibility of the investor renting it (Magdy, Personal interview, May 03, 2016). The second part of the ground floor was rented to an antique shop called "El- khawy," which they finally succeeded to compensate him and take the shop for implanting phase two of L passage project as shown in (Magdy, Personal interview, May 03, 2016).

"El-lethy Mekawy," an architect and investor who is looking forward to investing in and develop the heritage of Alexandria a solution for its deterioration. He utilized it as an aggregate spot for nourishment eateries as an investment. As per his expert studies in Turkey in the reuse of historic structures, he chose to make the same in Alexandria and picked Fouad road as the ultimate place to start with. In that time, it was odd to have another present-day venture focusing on the higher class, rather than the authentic shops in this street due to the unsafe situation of it. Amid the project, he confronted a few issues, first the lack of facilities to record the new use of the property and the withdrawal of his partner as a consequence of the fear of the indistinct circumstance of such a newly introduced type of project to the area. After the redesign of the place, no companies wanted to partake in the project because of the same reasons mentioned, and all declined to lease in the area and preferred to rent in the places already focusing on the targeted high class of customers. Before long, at last, a company acknowledged to do as such, and the work began. Now and after the situation changed gradually and another similar venture started to take place in the area, numerous organizations are asking for to rent a section of this project, and the waiting list is increasing (Mekawy, Personal interview, April 20, 2016).

The financial advantages of investing in a recorded building determined by preparing a meeting with the architect. El-Lethy expressed that around 2013, the destruction phenomenon diminished, due to the showed up of ventures like L passage and Leila eatery, putting resources into the legacy, as a non-ordinary option of the land investments. The proprietor was attempting to devastate his property, so as to plan an estate investment and benefit from it. once he had another option as the restaurant projects and understood that he could pick up a general interest, the owners began leasing their entire building or a portion of it for extended run investments. The authorities shut down, in 2015, a couple of reuse ventures, as Leila eatery in light of the limitation of the permitted utilizes that can be located in the neighborhood, which the eatery projects were not one of them. Then the proprietors returned considering demolition of the properties to have a benefit rather than starting investment and find it closed after a while (Mekawy, Personal interview, April 20, 2016). These are critical reasons to push the proprietor seeking the destruction of his property. This is one of the reasons driving the research attempting to locate a significant motivation to encourage the owner to seek for the keep up and the support for their properties.

L passage was the peak point, and its success prompted many investors to start focusing on the city center and its potentials (Mekawy, Personal interview, April 20, 2016). "The cost expected to make a task like L passage was in 2011 around 5.5 Million Egyptian pounds, now it may cost around 55 million Egyptian pounds (/8.8 ≈ 6,250,000 USD, nowadays =55,535,597 EGP)" (Mekawy, Personal interview, April 20, 2016). This is another confirmation of the backhanded financial effect of heritage investments on the surrounding context of the long term, which ought to be considered while figuring the cost-benefit analysis of the diverse situations. The benefits valuation stated by El-Lethy to be around 150,000 EGP (/8.8 ≈ 17,045 USD, nowadays = 151,457 EGP) monthly, which guarantees an organized return for the owner during his whole life.

The researcher of the current work decided to learn from the process of reuse and investments happened in the project of L passage and start to look for other general examples of the cost-benefit analysis of restaurant projects in Egypt. The research analyzed a financial study of a restaurant project in Alexandria to conclude the average of revenues the investor and the owner can gain. In the project, the area of the building is 366 square meter which is adjacent to the field of the studied villas in case four, five and six. The study took into consideration the detailed measuring of the sale cost of building, restaurant area, kitchen area, toilet area, administration area, the number of tables and persons. The research calculated the constituent expenses of the project including the costs of the equipment, furniture, storage materials and the organization with total costs equal to 510,820 Egyptian pounds. The continue expenses included the costs of the services and workers' salaries, the costs of the meals, bank interests, the transportations, insurance installments, advertisement, and electricity. The returns were calculated according to the average price of the meal the customer pay; the daily average predicted some the client and the number of working days in the year. According to this analysis in 2005, the average monthly benefits are around 40,800 EGP (/5.7 = 7158 USD, nowadays = 63,604 EGP). while case two is concerned with the procedures and challenges that appeared and the research should be aware of its presence for the final recommendations. This case ended with the average expected benefits of the reuse eatery venture that will be utilized as a part of case four, five and six to measure the benefits of the two scenarios and appraisal the financial crevice between them in every case.

CASE 3: Restoration Project: Sarsaa Pasha Palace

The palace situated in 58 Fouad street before, has the number 596 on the Heritage list. The category of the villa is level B (city level). It was occupied at the outset by the Lebanese Sarsaa Pasha, after years it got to be under the responsibility of the city, which then were reused to be the National Union Bank (Marine commercial Bank before). Presently, the palace is under restorations by a legislative entity; however, it is set up to be utilized as a Council for Saudi Arabia, as engineer Ramadan stated that the restoration expert is dealing with this project. This procedure of restoration and transference of the proprietorship to an outsider proprietor was to guarantee the constant of its support and significant circumstance. However, much as could be expected and diminish the obligations of the legislature towards its administration, as on account of the gems of jewelry museum, the government restored it, and now it is typically kept up by the proprietor. As indicated by the interview with specialist Ramadan, their primary standard was to convey the Palace to its unique original condition, which was extremely hard to perceive in light of the fragmented documentation accomplished for it and most of the structures located in Alexandria. For the procedures, as mentioned by Ramadan, the building has insurance for a long time. The insurance includes the obligation of the organization doing the restoration works to tackle any technical issues confronting the palace amid these five years, and the new owner ought to keep the palace up again after the end of the five years (Ramadan, Personal interview, April 12, 2016). According to the costs of the interior and the elevation, restoration works expressed by engineer Ramadan, the calculations of the expense of the whole restoration project of Sarsaa palace, as shown in Figure 5.

This case is concerned with finding the expenses of the restoration works of a recorded structure in the same zone, that can then be utilized as a part of computing scenario two of the reuse ventures in the listed properties of case four, five and six. The expenses of the restoration investment are not all that high which is assumed to be covered effortlessly by the profits of the eatery venture taking place in the listed villas in the subsequent cases of the economic diagnosis.

Case Four, Five and Six

There are laws addressed in the district guidelines to regulate the heights and built areas in the region, while in the heritage list of Alexandria, there are additional controls for the new constructions in a site full of historical buildings. In the following cases, the calculations of the fiscal gap within the rehabilitation and the reuse of the listed villa and the replacement of the villa with a real estate project will be estimated when the regulations stated in the heritage list of Alexandria is assumed to be performed and when regulations of the district are executed.

In case four and five, the rules in the region of wasat state that the maximum allowed height to be 24 meters and the maximum authorized built area is 100% from the entire land area. On the other side, the regulations in the heritage list of Alexandria state that the maximum approved height to be 12 meters and the maximum allowed built area is 60% of the whole area of land. This huge variation in regulations drive to a large variance in the costs, revenues and profits of the two scenarios, which commenced the

Figure 5. Costs of the restoration in Sarsaa Palace in Fouad street, Alexandria.
Source: The author

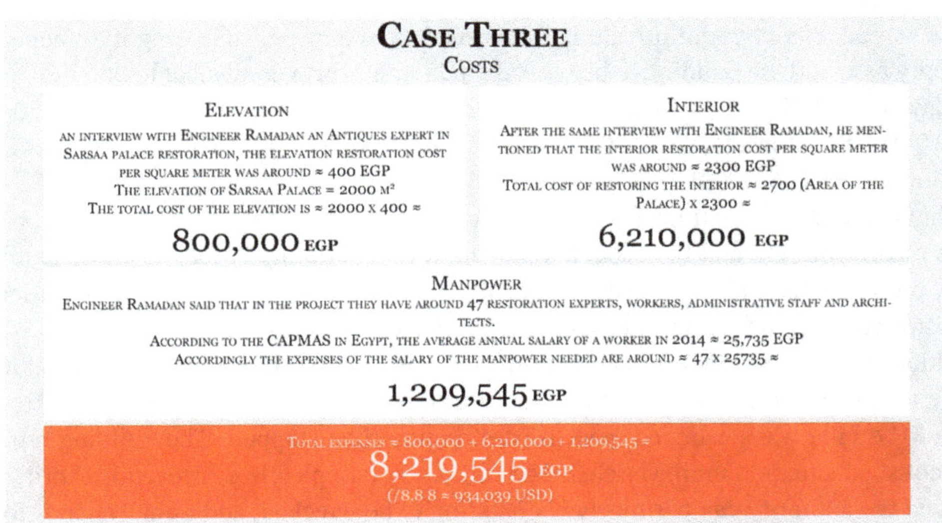

research to choose to calculate scenario one first when the regulations of the heritage list are executed and second when the statutes of the district are performed. In case four, the reconstruction of the devastated villa is regularly conducted, from the viewpoint of the locals, to be reused as a museum. The museum reuse project sometimes is not the ideal reuse model for the region, which leads the chapter to assess the reuse plan in case four when it is used as a museum use and when it is used for restaurant use. Although in case five the reuse of the villa will be only measured in the case of being reused as a restaurant. The ordinances of the district and the heritage list correlated to the height and built area restriction are the same on Al-Sultan Hussien Street, which is 100% built area and maximum height to be 24 meters. Accordingly, in case six the real estate project in scenario one will be expected once, depending on these numbers and will determine the reuse project of scenario two when only the villa is reused as a restaurant.

CASE 4: The trapped Demolished Building

During an interview with Loatfy, an owner of one of the listed villas placed in Kafr Abdo quarter, when he was asked if he stands with or against devastating the listed buildings? He replied with completely against it, and he indeed thought it is a violation that should have adequate punishments as declared in the regulations. He also quoted, "In the previous years, it was seldom feasible for an owner who requested his restricted villas to get out of the Heritage list, to win the case and destroy it for the sake of building another high-rise building. Instantly since the city is dressing a deficit in extending a suitable infrastructure for all the new real estates created rather of the old ones, an entire denying will face any other future demand of restricted villas destruction, indeed wherever it warrants to get removed" (Loatfy, Personal interview, March 13, 2016).

Villa Aghion Number 631

Villa Aghion was a standout amongst the most imperative recorded estates in Alexandria from level B (city level) villas in the heritage list of Alexandria. The villa was situated in 75 Minesha Street, with an area around 1830.25 square meters. It is famous for its Palladian style. In 1923, the Italian architect Gustave Aghion purchased the land empty from a foreigner proprietor in the mixed court. In 1958, after the passing of Gustav Aghion.

The estate was acquired by four private Italian owners, and they began leasing it for annual charges. In 1980, the city assessed the building to be with annual rental worth equivalent to 660 EGP (/0.7 ≈ 943 USD, nowadays = 8,379 EGP). In 1982, it was sold to 12 private Egyptian proprietors for 20,0000 EGP (/0.77 ≈ 259,740 USD, nowadays = 2,307,971 EGP), and kept leasing it annually for 400 EGP (/0.77 ≈ 52 USD, nowadays = 462 EGP).

Villa Aghion considered the trapped illustration of a villa, where the time is solidified, between the not presence of the building and the expected new high-rise building. After the entire annihilation of the villa, few propositions said to reconstruct the villa again with the same previous condition, because of its high historical esteem for Alexandria (Mekawy, Personal interview, April 20, 2016).

Villa Aghion passed through several stages of demolition, first, in 2006 it had partial destruction when many citizens objected to these actions and the authorities stopped sabotage activities. In 2012, a huge part of the villa was destroyed, and the destruction then stopped. The building remains in its partially demolished situation until March 2016 until it was completely destroyed in April 2016. How the citizens and the authority can permit such a building to be wrecked, additionally speak to a red line of the enormous need for awareness campaigns to the residents and owners, of how critical is this building and protecting it is productive to the city in the long term and in a backhanded way. “We wanted to preserve it and make it something for tourists, but there was no interest from anyone,” said Hamed the proprietor of villa Aghion to the Guardian (Cairo observer, 2012).

According to Kingsley (2014), “It might be the most valuable villa that we might ever have had here in Alexandria” Mohamed Aboulkheir clarified. This sort of Estates is at a level of significance that in the international case, it should be rebuilt again as it was before demolition. Along these lines of protection will be examined in the following part. For this situation, the meter square costs 109.27 Egyptian pounds (/0.77 $\approx$ 142 USD, nowadays = 1262 EGP), in 1982.

In the case of villa Aghion, the two scenarios will be compared the establishment of a real estate project on the land and the reconstruction of the same old building. The measurement of the returns of the housing project on the land of villa Aghion will be done in two different ways, first, when the stated built area and height regulations according to the Heritage list of Alexandria are implemented on the ground. These regulations are not usually executed, which drives the research to calculate secondly the practical scenario expected to happen, which is the implementation of the general requirements of heights and built area in the district guidelines instead of the one for the Heritage list of Alexandria. As shown in Figure 6, the costs of constructing a new real estate building are higher when the regulations of the district are implemented instead of the statutes of the heritage list. The reason is the large area that is allowed to build on it which consumes more material and money to be constructed.

In Figure 6, due to the little-permitted built area in the controls of the heritage list, the benefits came from the real estate project is with low interests, which implies then financial losses to the owner. while the returns of the housing project on the same land are still highly profitable when the standards of the built area stated in the district guidelines is executed.

Figure 7 demonstrates that the eatery venture is more profitable than the museum investment in the long term. The negative result does not imply that the task is not beneficial, but rather it relies on having benefits in the following two years of the execution.

In Figure 8, the gap between the two situations was computed from four calculations with different variables every time. Chart number A demonstrates the crevice between the two situations when the reuse venture sort is an eatery and the guidelines of the heritage list of Alexandria is executed. The resulted gap of benefits is a negative number, which implies that the interest of the reuse venture is higher than the real estate investment. In diagram B the crevice is likewise with the negative result despite the fact that the reuse of the property as a museum is less beneficial than the eatery use sort in diagram A. As indicated by this comparison, the decision of the reuse design type encourages the increase of the benefits in the short term and bit by bit to rise repeatedly in the long term. In diagram C and D, the gap is exceptionally wide between the two scenarios where situation one of the real estate investments is a great deal more efficient than scenario two with a difference hard to be adjusted smoothly which forms a major impediment towards fulfilling the purpose of the research and saving the registered property from decimation.

Comparing diagram, A and B with c and d reveals the vital enormous issued financial gap among the two scenarios resulted due to the not utilized statutes of the heritage list and the execution of the district issued guidelines instead. The implementation of the present necessary rules of the heritage list ultimately can influence the owner not to annihilate his recorded home since the option of the real estate venture is considered then as an economic loss. While not executing it directs to the existence of high

Figure 6. Costs of the construction of new real estate building on the land of villa Aghion for case Four.
Source: The author

COSTS OF CONSTRUCTING A NEW RESIDENTIAL BUILDING

CASE 4 SCENARIO ONE

ACCORDING TO THE REAL ESTATE EXPERT (WAEL GHONIEM) THE PRICE OF THE LAND IN THIS AREA VARY FROM 20,000 TO 30,000 G, SO THE AVERAGE IS = 25,000 PER METER SQUARE ≈ 1830 X 25,000 ≈ 45,750,000 EGP.

COSTS OF NEW CONSTRUCTION

REGULATIONS IN THE HERITAGE LIST IS EXECUTED	REGULATIONS IN THE HERITAGE LIST IS NOT EXECUTED
ACCORDING TO THE BUILDING REQUIREMENTS IN THE HERITAGE LIST, THE MAXIMUM HEIGHT IS 12M AND MAXIMUM ALLOWED BUILT AREA IS 60% OF THE WHOLE LAND. AREA OF THE LAND= 1830 M^2, THE BUILT AREA = 1098 M^2 ACCORDING TO AN INTERVIEW WITH TWO REAL ESTATE EXPERTS, THE COSTS OF CONSTRUCTION PER M^2 WITH A MIDDLE LEVEL OF FINISHES VARIES A LOT WITH AN AVERGAE ≈ 1000 EGP THE COST OF BUILDING A RESIDENTIAL TOWER ON A 1098 M^2 AND 12-METER HEIGHT ≈ 1000 X (1098X4) ≈ **4,392,000 EGP**	AREA OF THE LAND= 1830M2 ACCORDING TO AN INTERVIEW WITH TWO REAL ESTATE EXPERTS, THE COST OF CONSTRUCTION PER M^2 WITH A MIDDLE LEVEL OF FINISHES ≈ 1000 EGP THE COST OF BUILDING A RESIDENTIAL TOWER ON A 1830M2 AND 24-METER HEIGHT ≈ 1000 X (1830X8) ≈ **14,640,000 EGP**

COSTS OF MANPOWER

FROM THE INTERVIEW DONE WITH FATHY ELSHARABASY, A PROFESSIONAL CONTRACTOR, HE SAID THAT THIS PROJECT NEEDS FROM 50 TO 70 WORKERS (AVERAGE OF 60 WORKERS).

ACCORDING TO AN INTERVIEW WITH EL-LEITHY AND A PROFESSIONAL INVESTOR, THE PROJECT MAY TAKE FROM 8 MONTHS TO ONE YEAR AND HALF. BECAUSE THE TOTAL AREA OF THE PROJECT IS SMALL, THE RESEARCH WILL ASSUME, IT WILL TAKE AROUND A YEAR.

ACCORDING TO THE CAPMAS IN EGYPT, THE AVERAGE ANNUAL SALARY OF A WORKER IN 2014 = 25,735 EGP

COST OF SALARIES PAID TO THE WORKERS IN THE PROJECT THOUGH OUT THE YEAR IS ASSUMED TO BE AROUND ≈ 60X25,735 ≈

1,544,100 EGP

NOTE: ACCORDING TO THE INTERVIEWS DONE WITH THE ECONOMIC, REAL ESTATE AND INVESTMENT EXPERTS, THE PRICE OF THE LICENSES AND THE ILLEGAL PROCEDURES IS AROUND 5,000,000 EGYPTIAN POUNDS. HOWEVR, IT IS NOT CONSIDERED TO BE CALCULATED IN THE RESEARCH BECAUSE OF THE UNCERTAINTY OF THE NUMBER AND THE RESEARCH IS CALCULATING IN A LEGAL CALCULATIONS

THE SUM OF COSTS

REGULATIONS IN THE HERITAGE LIST IS EXECUTED	REGULATIONS IN THE HERITAGE LIST IS NOT EXECUTED
(NEW CONSTRUCTION AND MANPOWER SALARY) ≈ 45,750,000 + 4,392,000 + 1,544,100 ≈ **51,686,100 EGP** (/8.8 ≈ 5,873,420 USD)	(NEW CONSTRUCTION AND MANPOWER SALARY) ≈ 45,750,000 +14,640,000 + 1,544,100 ≈ **61,934,100 EGP** (/8.8 ≈ 7,055,216 USD)

benefits from the housing investment, which pushes the proprietor towards the demolition of the historic property to obtain returns from another alternative of scenario one.

CASE 5: The Neglected Building

Villa Ambron is the best example of a magnificent neglected villa. Located in 19 Al-Ma'amount Street, Moharem Bek district, with a total area of land 1800 meter square, and built area of 540-meter square. villa Ambron dragged in the Heritage list with number 1687 under the category of level B (City level), and was designed by the famous Italian architect, Aldo Ambron, in a Baroque style and with roman columns, as shown in (Borg, 2013).

Figure 7a. The Returns and Profits from the real estate project in the case the regulations in the Heritage list are implemented and, in the case, it is not implemented for case four.
Source: by Author

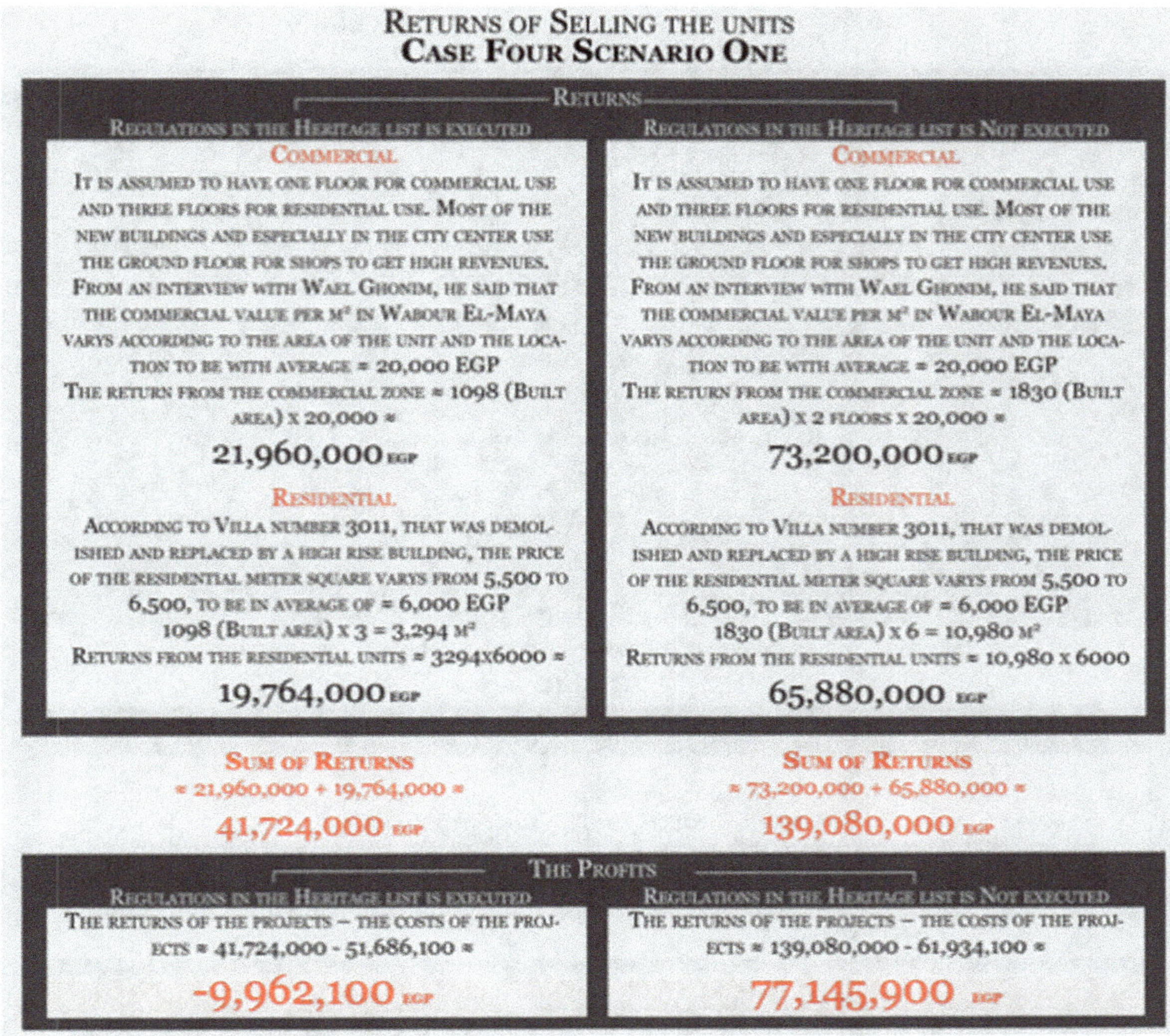

According to Borg, several famous characters used to live in the villa like the famous novelist Durrell Lawrance, and was leased by the famous Egyptian artist, Effat Nagui. After his demise, it was sold to a real estate company, when it attempted to expel it from the Heritage list which was met by refusal. The fence surrounding the villa was demolished and left open allowing anybody to enter and damage it, sitting tight for a few years till the villa is entirely destroyed, compelling the government to offer the authorization of its expulsion from the heritage list, regardless the horizontal extension of a studio on its land (2013). The worse the condition, the more must be repaired and replaced and the more expensive the renovation must be. In other words, there is a relationship between the level of costs of a construction project and the scope of the monument destruction. In the following figures, the cost-benefit analysis of the new constructed real estate project on Ambron villa land, the cost-benefit analysis of the reuse of

Figure 7b. The Cost-Benefit and Profits calculations of a Museum and restaurant reuse project, and the gap between them and the real estate project for case four.
Source: by Author

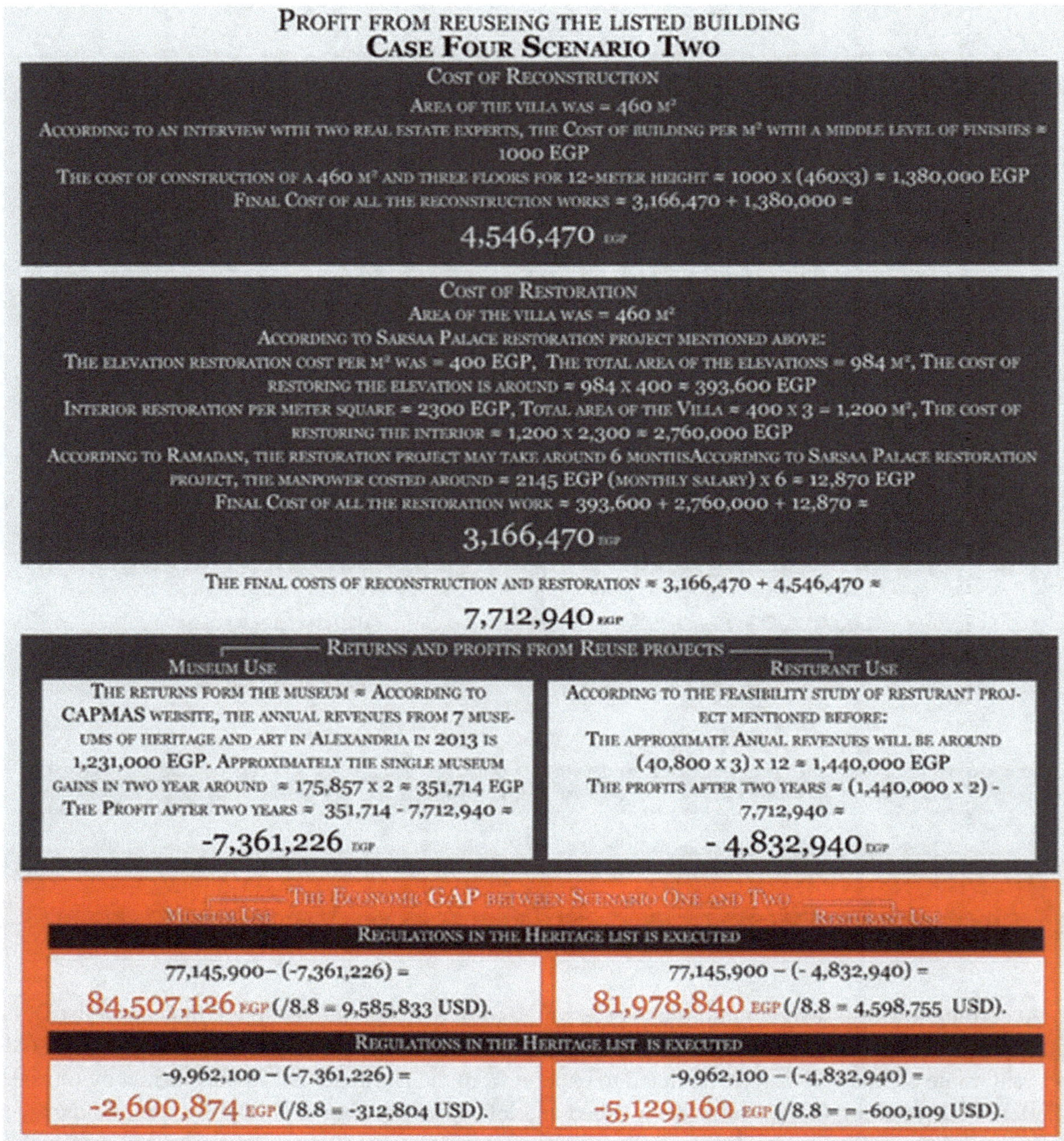

Figure 8. Collective diagram of the profits and gaps between the two scenarios with a comparison for case four.
Source: The author

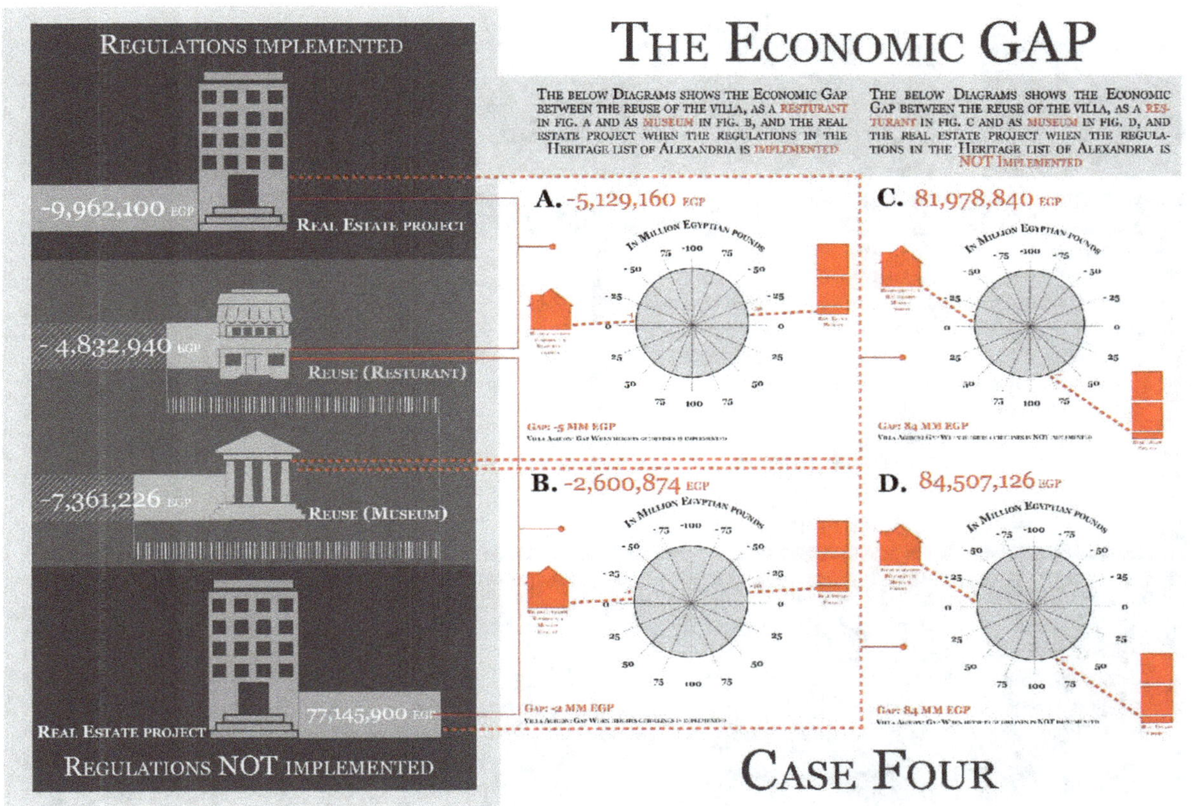

the resto- ration, reuse as a restaurant and the calculation of the gap occurred between the two scenarios will be considered (See Figure 9, 10 and 11).

Case five have the same circumstances of case four, but it differs in some variables. The location of case four is in Wabour Al-Maya, while Ambron villa in case five is located in Al-Hadara district. Ambron villa in case five exists, not demolished, so in case five the demolition costs are included in the calculation of scenario one, which were not included in case four. The expenses of the reconstruction of villa Aghion in case four was considered in measuring scenario two, while in case five it was not included. The last difference is that in villa Ambron case calculations of the reuse project is a use restaurant type only without estimating the museum use type like case four.

The aim of the previous case is to determine the resulted profits of the two scenarios with slightly different variables from case four. In the earlier calculations of case five, it ensured the result of case four. Although the profit of the real estate project was profitable opposite to the situation in case four, the benefit of the housing project was similar to case four with remarkably higher profits when the regulations of the district in implemented compared to the implementation of the guidelines stated in the heritage list. The gap between the two scenarios was shown in the two conditions if the regulations of the heritage list are implemented and the other condition of executing the guidelines of the district.

Figure 9. Costs of the construction of new real estate building on the land of Villa Ambron for case five.
Source: The author

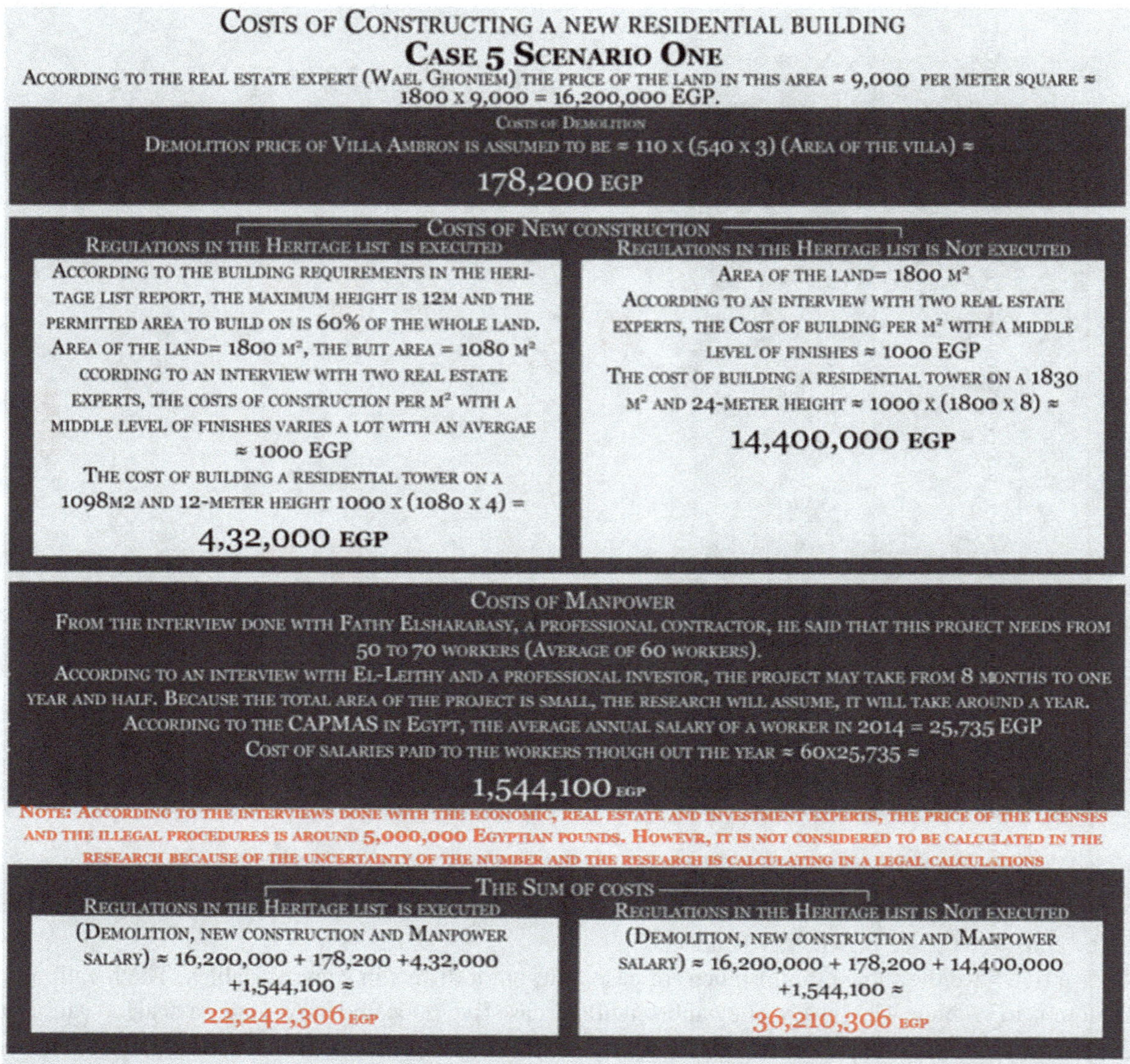

They indicate that when the controlled built area regulations stated in the heritage list of Alexandria is applied, the gap between the two scenarios are sharply less than the gap might occur when the district guidelines are applied (As shown in diagram B in Figure 11).

CASE 6: The Neutral Villa

Since Wabour El-Maya district and Al-Hadara districts, where Aghion villa and Ambron are located, are not highly costly zones, the research can't just rely on upon their computations for measuring the

Figure 10a. The Returns and Profits from the real estate project for case four.
Source: The author

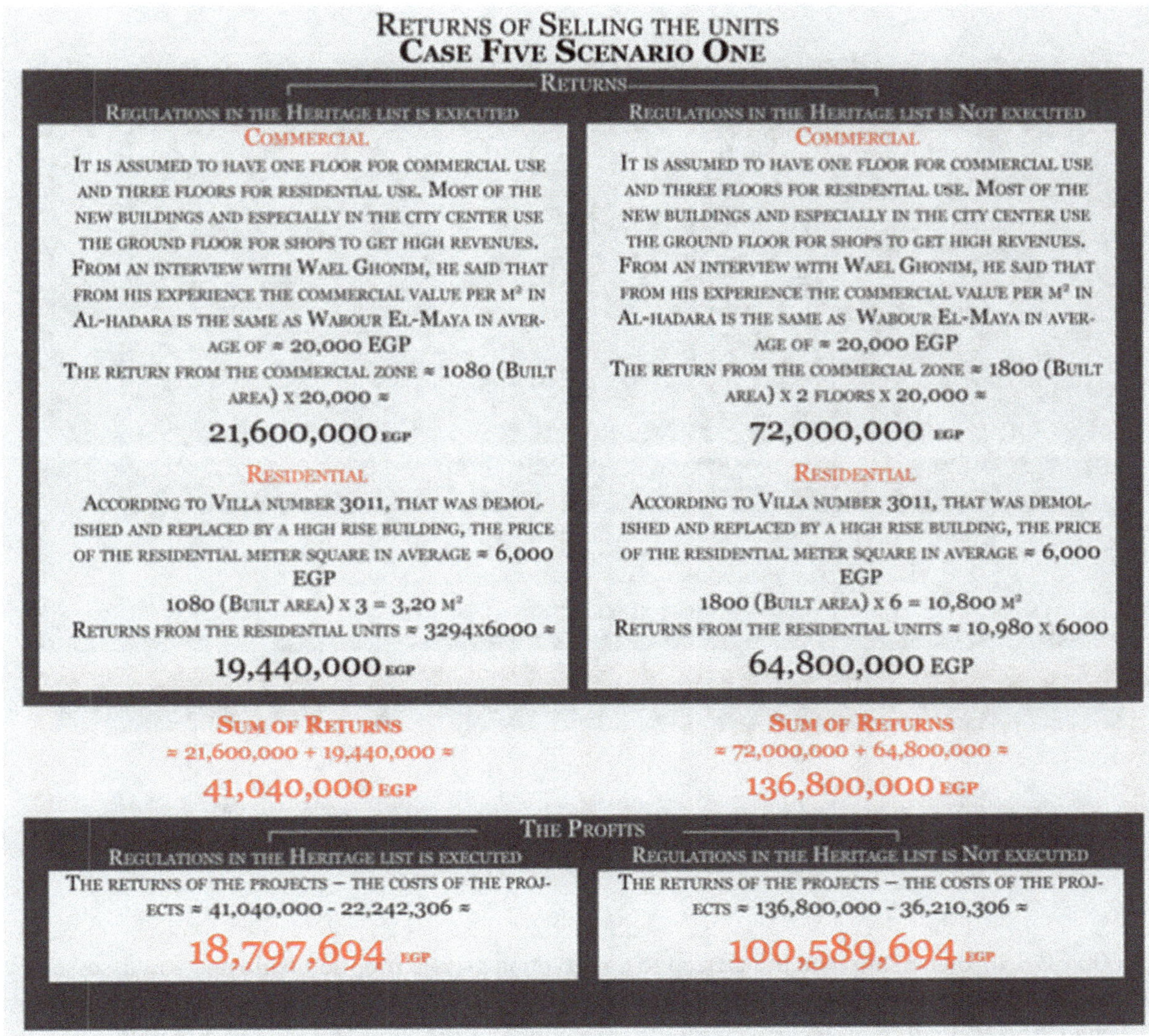

cost-benefit analysis investigation, that the subsequent sections ought to consider figuring of the two scenarios if they occurred in a costly urban range. In the interview done with Wael Ghonim, Real estate expert, the cost of the area and the land properties situated on Sultan Hussien road are the costliest while Wabour El-Maya is with lower price in the two instances of residential and commercial units (Ghoniem, Telephone interview, Spring, 2016, Spring).

Villa number 590 in the Heritage list, is situated on Sultan Hussien road, a standout amongst the costliest zones for residential and commercial units as mentioned recently, will be likewise ascertained in the current of the two scenarios. The villa placed in 127 Sultan Hussien road, the range of the land area, is around 1188 square meter, and the area of the villa is around 320 square meters. The villa is under category C (Local value). A British man owned it in 1976, and after inheritance, it was sold for

Figure 10b. The Cost-Benefit calculations of a restaurant reuse project, and the gap between it and the real estate project for case five.
Source: The author

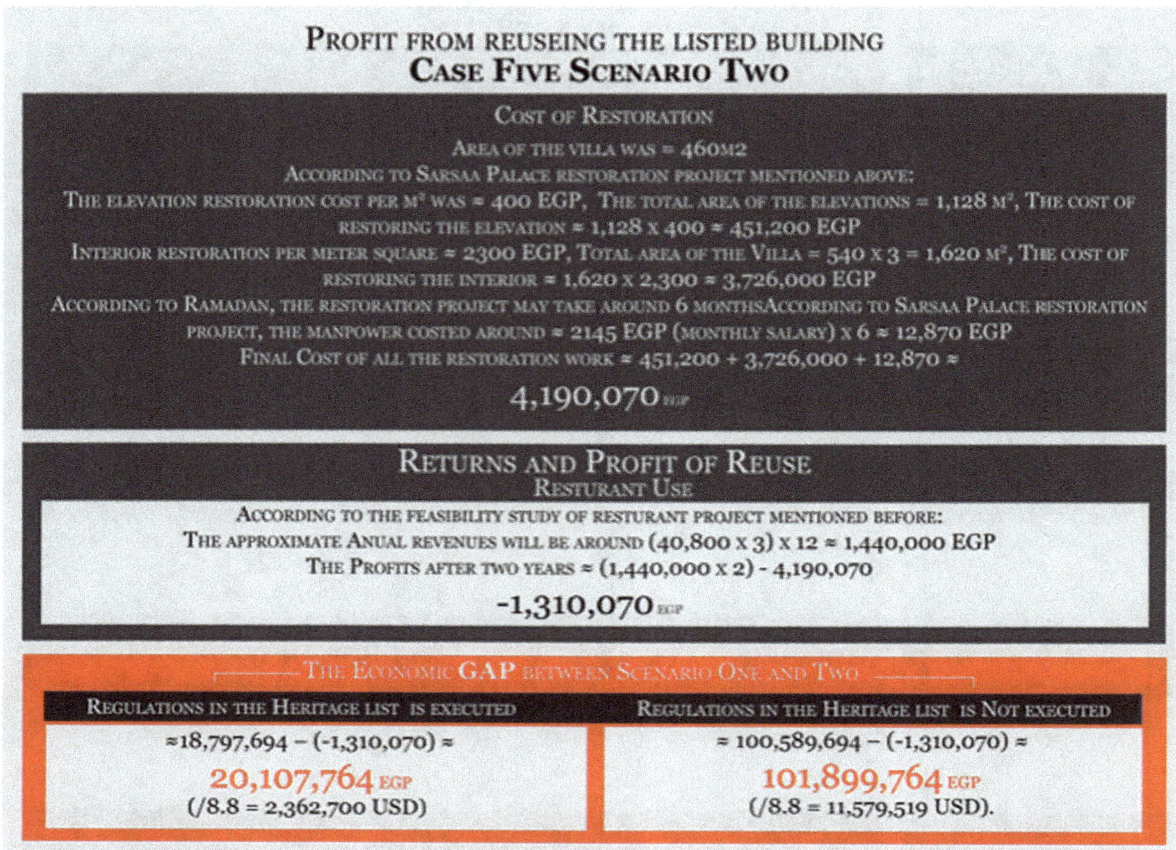

120,000 EGP. Its ownership then transferred to an Egyptian private proprietorship and was assessed by the city for an annual rent of 480 EGP.

Nehad Abo El-Fadl, the owner of the villa, responded to the topic concerning her willingness to keep and preserve the villa, with yes, she is, because she perceives the villa as a desirable architecture which can't be compensated with a different one and keeping it is essential. Although questioning her if there is any direct or indirect compensation given to her from the state, she said "No." She also replied that there is no support from the authority as pronounced in the law. The villa is a personal residential use with frequent use, and she preferred to convert the villa into a museum if it had undergone a reuse project. When she was requested to state the main reasons for demolition phenomenon, from her point of view, which drive the owner to tear the villa down, her answer was the owner's need to the money or his need to another residence for himself and his sons. For these purposes, she suggested that the city should reimburse the owner the price of the market and convert the villa into a museum or touristic destination, so it can be easily preserved after restoring it (Aboelfadl, Email Questions, February 23, 2016). The same previous calculations will be calculated on this villa also as shown in Figures 12 and 13)

Figure 11. Collective diagram of the profits and gaps between the two scenarios with a comparison for case five.
Source: The author

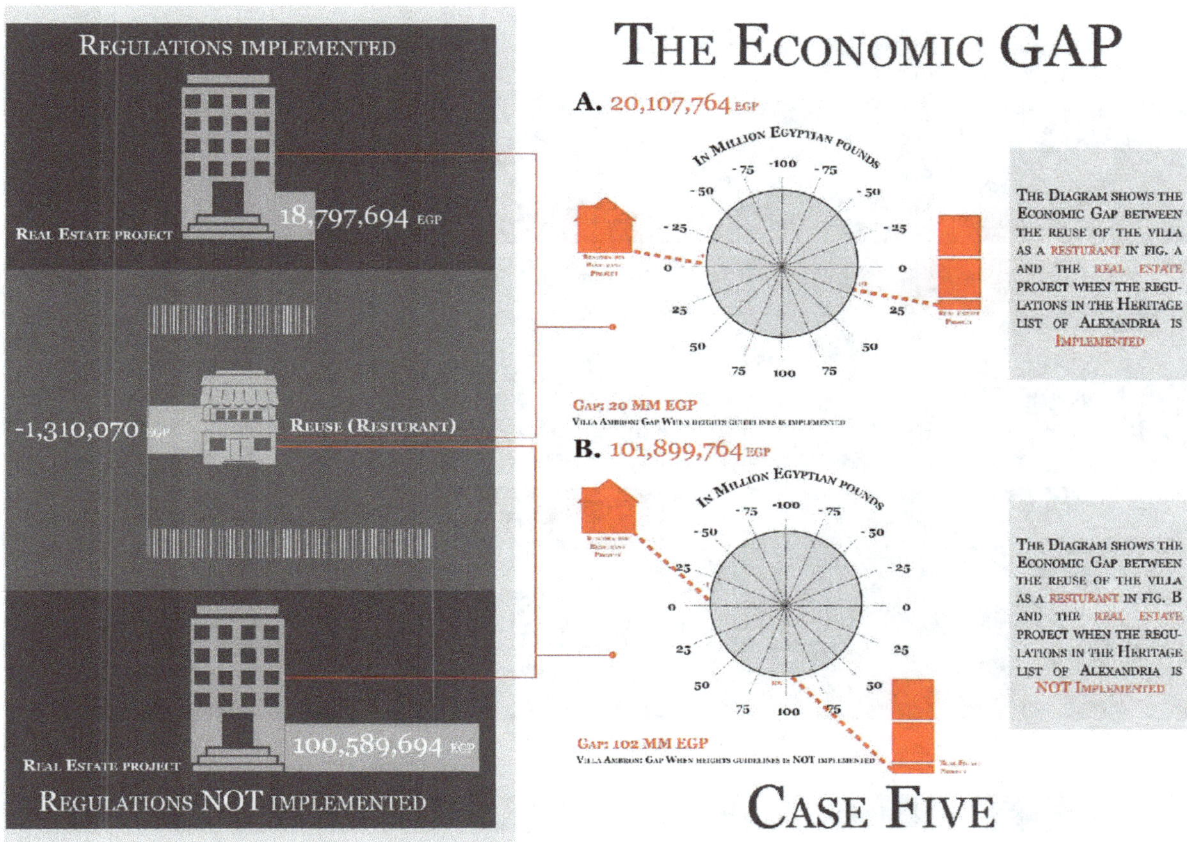

In cases like the one in case six, the long-term calculations must take place as shown in Figure 14 by considering the break-even period, break-even point and the profits during the years in both scenarios. The economic expert, Mohamed El-Awady, said that the investment projects in the listed buildings could accomplish the benefit of the real estate project, after around five to ten years. The proprietor went for the real estate investment, has the chance of investing again in a bigger and quicker venture afterward. Real estate projects don't only achieve high benefits, but it also has a fast stock cycle which prompts the significant increment of advantages contrasted with the eatery venture.

Additionally, there is some factors influence scenario one like the costs of the redevelopment of the infrastructure to be able to serve the new high density, must be put into consideration which will not be considered in scenario two. Another aspect should be considered which is the long-term positive indirect impact of the up keeping of the historical constructions on the city increases throughout time. These aspects should push the country to maintain the listed buildings as extraordinary indirect revenues instead of razing it for the execution of scenario one. However, the value of the heritage shouldn't be considered only by the financial method; the research is using the economic approach because the financial gap

Figure 12. Costs of the construction of new real estate building on the land of for case six.
Source: The Author

COSTS OF CONSTRUCTING A NEW RESIDENTIAL BUILDING

CASE 6 SCENARIO ONE

ACCORDING TO THE REAL ESTATE EXPERT (WAEL GHONIEM) THE PRICE OF THE LAND IN THIS AREA VARY FROM 50,000 TO 70,000 EGP , SO THE AVERAGE IS ≈ 60,000 PER METER SQUARE ≈ 1188 X 60,000 ≈ 71,280,000 EGYPTIAN POUNDS.

COSTS OF DEMOLITION

ACCORDING TO INTERVIEWS DONE WITH 3 REAL ESTATE INVESTORS AND EXPERTS, THE AVERAGE NUMBER SAID TO BE EQUAL TO 110 EGP PER M²

DEMOLITION PRICE IS ASSUMED TO BE ≈ 110 X (320 X3) (AREA OF THE VILLA) ≈

105,600 EGP

COSTS OF NEW CONSTRUCTION

ACCORDING TO THE BUILDING REQUIREMENTS IN WASAT DISTRICT OF ALEXANDRIA, THE SAME IN THE HERITAGE LIST REGULATIONS, THE MAXIMUM HEIGHT IS 24M AND THE PERMITTED AREA TO BUILD ON IS 100% OF THE WHOLE LAND.

AREA OF THE LAND= 1188 M²

FROM INTERVIEWING TWO REAL ESTATE EXPERTS, BUILDING PER M² WITH A HIGH LEVEL OF FINISHES ≈ 1,350 EGP

COST OF RESIDENTIAL TOWER ON A 1188 M² AND 24-M HEIGHT = 1,350 X (1188X8) ≈

12,830,400 EGP

COSTS OF MANPOWER

FROM THE INTERVIEW DONE WITH FATHY ELSHARABASY, A PROFESSIONAL CONTRACTOR, HE SAID THAT THIS PROJECT NEEDS FROM 50 TO 70 WORKERS (AVERAGE OF 60 WORKERS).

ACCORDING TO AN INTERVIEW WITH EL-LEITHY AND A PROFESSIONAL INVESTOR, THE PROJECT MAY TAKE FROM 8 MONTHS TO ONE YEAR AND HALF. BECAUSE THE TOTAL AREA OF THE PROJECT IS SMALL, THE RESEARCH WILL ASSUME, IT WILL TAKE AROUND A YEAR.

ACCORDING TO THE CAPMAS IN EGYPT, THE AVERAGE ANNUAL SALARY OF A WORKER IN 2014 = 25,735 EGP

COST OF SALARIES PAID TO THE WORKERS THOUGH OUT THE YEAR ≈ 60X25,735 ≈

1,544,100 EGP

NOTE: ACCORDING TO THE INTERVIEWS DONE WITH THE ECONOMIC, REAL ESTATE AND INVESTMENT EXPERTS, THE PRICE OF THE LICENSES AND THE ILLEGAL PROCEDURES IS AROUND 5,000,000 EGYPTIAN POUNDS. HOWEVR, IT IS NOT CONSIDERED TO BE CALCULATED IN THE RESEARCH BECAUSE OF THE UNCERTAINTY OF THE NUMBER AND THE RESEARCH IS CALCULATING IN A LEGAL CALCULATIONS

THE SUM OF COSTS

(DEMOLITION, NEW CONSTRUCTION AND MANPOWER SALARY) ≈ 105,600+12,830,400 +38,603 + 71,280,000 ≈

84,254,603 EGP

between the real value of the land and the property is one of the biggest problems facing the owners and causing an unfair situation for them.

WAYS OF RISING TECHNIQUES OF HERITAGE ECONOMIC VALUE

The government can issue different ways to compensate the owners for applying restrictions on their historic properties. There are direct, and indirect methods of incentives may be introduced to the national heritage conservation policies. The Direct way of Incentives is presented mainly in grants and loans, one of the common techniques applied by the government. The grants classified in Taskforce of Commonwealth Report (2004), into entitled grants, discretionary grants, and performance grants. Through the report, it was also mentioned that the government offers the Entitlement grants to any owner holds a listed building, whether he made efforts to preserve and maintain it or not. The disadvantages of this type are

Figure 13. a: The Returns and Profits from the real estate project for case six. Source: by Author; b: The Cost-Benefit and Profits calculations of a restaurant reuse project, and the gap between it and the real estate project for case six.
Source: The author

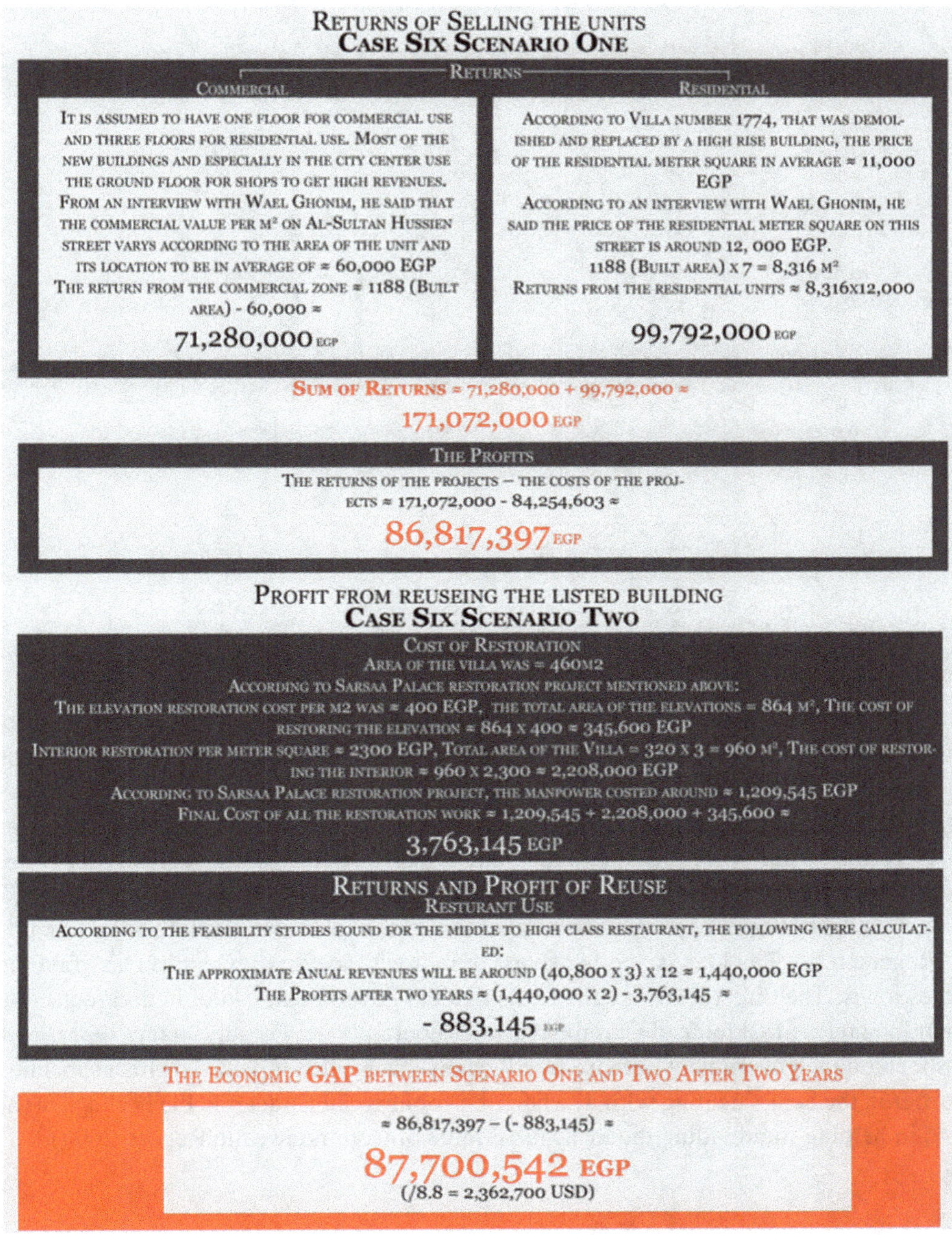

Figure 14. Graph illustrates the Break-even point, Break-even time and Profits on the long-term for a case like a case six.
Source: The author

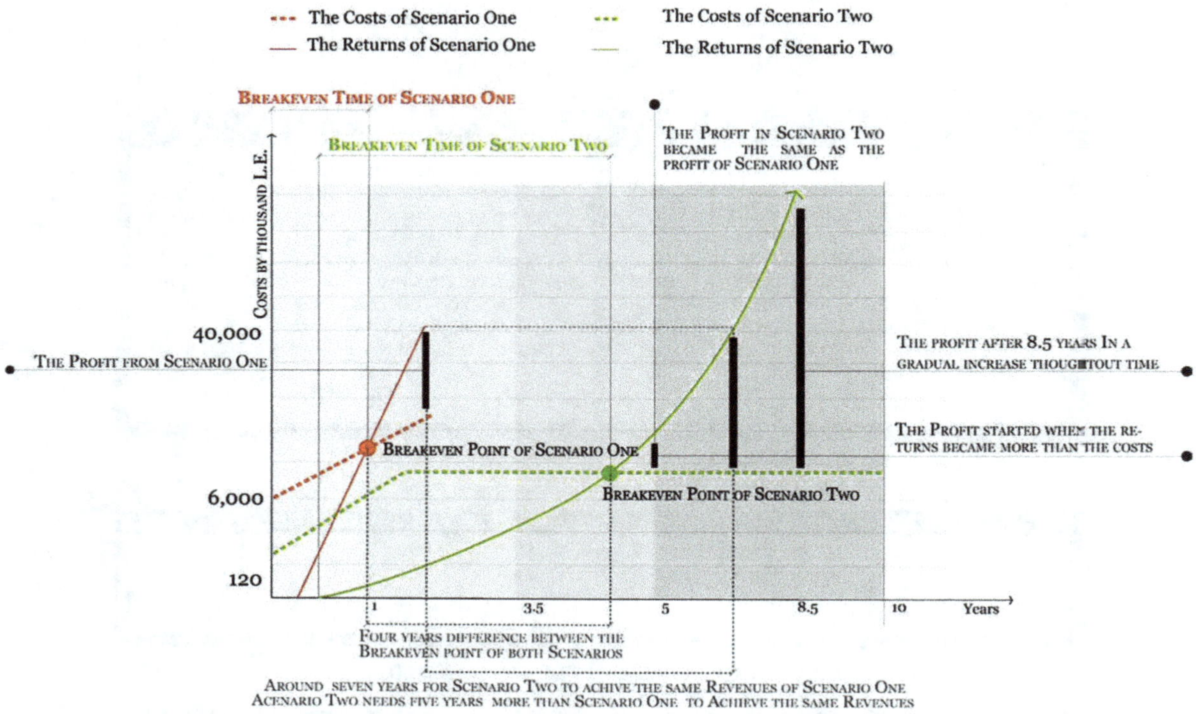

that it is not fair for the owners who spends significant amount of money on maintaining their property. It may encourage the owner to list his building but doesn't help him to improve it, and it causes a huge load on the government, which lead to unsustainable heritage conservation (2004). While in discretionary grants, the government set criteria for the selected properties that deserve to receive the awards. The discretionary grants are fairer for the community, and it motivates the owners to apply for conservation projects for their construction to obtain the grants (Rogers, n.d.). The government in the performance grants select a few national level projects to fund according to strict criteria with a significant amount of capital (Rogers, n.d.). The loan is another shape of indirect Incentives which is classified into direct and subsidies loans. The direct loans are the loans that the government can offer and request its return after short or long term (Taskforce of Commonwealth Report, 2004). The advantages of this process are having less load on the government because it will receive back what it paid. The loan subsidies are the same but differ in the actors dealing with it, where the commercial lender and the heritage organization are involved in helping in providing the loans (Taskforce of Commonwealth Report, 2004).

CONCLUSION

The economic diagnosis explained the gap between the two scenarios which varied according to the change in variables. In case four, the implementation of the regulation or not, turned the profit from the villa by reconstructing and reusing it to be higher than the benefits comes from a real estate project on the same land. The reason helped in the improvement of the advantages of scenario two in case four and five was the low value of the land in the real estate market, due to its location outside the city center. The other reason, the regulations in the heritage list of Alexandria have restricted controls on the new constructions in a historical area. According to an interview done with Nassar, a member of the Technical Securities Heritage Committee of Alexandria and a professor in Alexandria University, the committee tends to put guidelines in the regulations concerning the heights of the newly constructed constructions in a site containing many historical buildings. The execution of these instructions can spare the villa from demolition and be preserved to be more valuable than the real estate project or any other project requests the destruction of the building. The selection of the reuse project can influence a lot the revenues come from the villa. It is well-known in Egypt that the reuse project should be whether a museum or a cultural center, but nowadays the most profitable projects that can continuously manage the villa are mostly not the museums or the culture centers (See Figure 15).

In case five, it was ensured that the implementation of the regulation already states in the heritage list of Alexandria and the law can change the situation of the villa from being very lower than the other investment project to be near the same amount of revenues or even higher. In case six, the implementa-

Figure 15. The Reasons for the gaps between scenario one and Two from all the cases of the economic diagnosis.
Source: The author

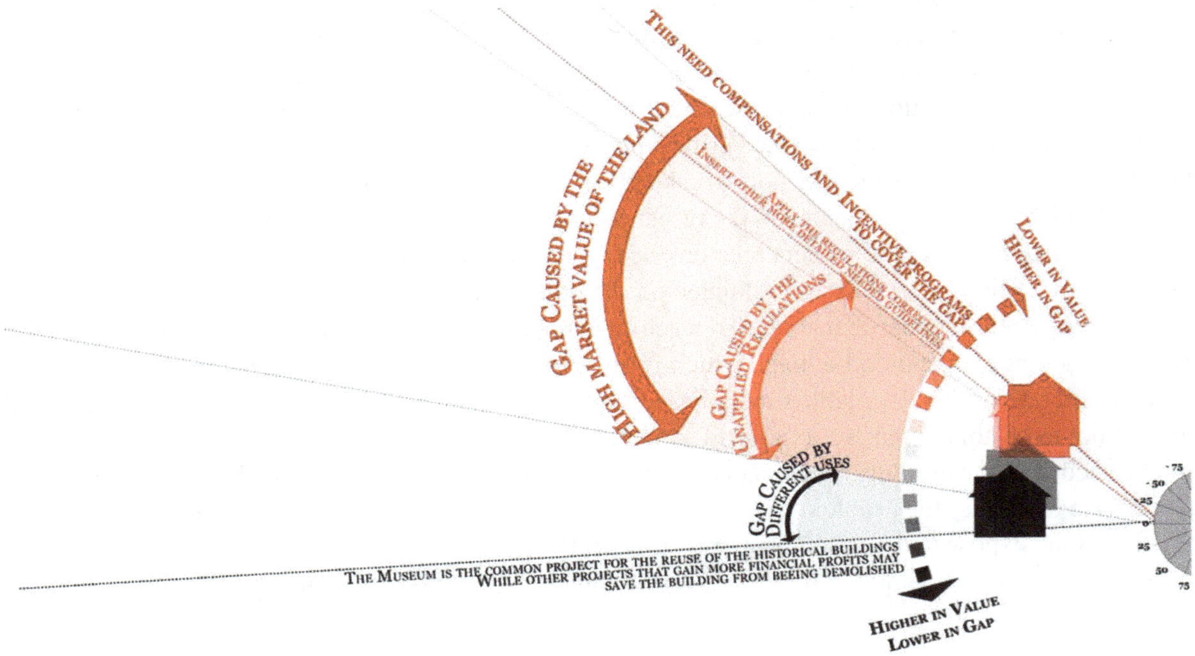

tion of the regulation can get no difference. The reason is the high value of the land is in the real estate market, due to the prominent location in the city center of Alexandria and the Sultan Hussien old street. The regulations allow the building on the whole plot area on this land because the street it is located on is wide enough to build on the entire area and go for a high height. In this case, the compensation of the owners is a necessity frequently performed by incentive programs asserted in the regulations on the country, state and city levels. According to the results came from the last case six, the gap between the two venture is clearly extraordinary.

likewise, scenario one can accomplish a high benefit in a brief time, while scenario two, is a long-haul nonstop interest investment. The gap essentially is not the gap of the capital as critical as the time difference between the benefits achievement. In the nowadays extraordinary pace of economic evolution, the demand for quick, high profits is the most aspired intention to survive and attain the life needs for him and his family. So, the main causes behind the economic gap are the not applied regulation of the heritage list, the wrong type of reuse and the not existed incentive programs to cover the difference in the real estate value in the market.

However, the owner still should have the advantage to obtain earnings from his held land in the short term, if he wants so. The research then commences finding a different indirect method to overcome this gap by first, lower the break-even time difference by addressing the possible economic incentives that can be granted to the landlord in an indirect form. The best kind of incentive to implement is the one that creates the lowest pressure on the governments and allows the owner sufficient financial interest to be satisfied, in a precise situation and with limited constraints, but indeed in an advantage to own a historical construction.

FUTURE RESEARCH DIRECTIONS

An action plan attempts to link the NGO like the initiative Save Alex, experts in economy, real estate, and social science. It should tend to include the active groups' minds for heritage like Wasf Iskandrya and Gudran in the awareness organizations and the training programs. It also involves the citizens' worries about heritage conservation to cooperate in the donations and culture activities. It attempts to include the real estate companies deal with the heritage preservation like SIGMA in Alexandria to take place in the NGO and to encourage the investors to ask for the participation in safeguarding the historical buildings. Behind utilizing this action plan, the research may have ended with an achievable intention of the win-win situation, where the owners are convinced with the incentives, they are granted to repay their restricted properties and the listed buildings, on the other side, can be well preserved in a sustainable decentralized superintendence manner. To reach an applicable on ground win-win situation a long-term study and action plan should be done, which is not studied in the current research because of several research limitations. The available time for the thesis was one of the challenges faced the research and drove it not to be able to reach a long-term action plan. Another challenge was the difficulty in accessing the needed data especially the financial estimations concerning the economic diagnosis the research depends on to discuss the problem of heritage demolition, main reasons and tackle them. However, the research didn't expose to the action plan, it is aware of the needed steps to be done in the further research to be able to achieve a workable framework to start applying in the context of Egypt.

ACKNOWLEDGMENT

I would like to thank Professor Abeer Elshater. The chapter wouldn't be written without her guidance, assistance, availability whenever and needed her advice and her persistent help when any difficulties confronted me amid the research. Furthermore, I want to thank Professor Josefine Fodkal from Stuttgart University and Professor Ayat Ismail from Ain Shams University for their help, critical notes, advice, and transference of their knowledge to this chapter.

REFERENCES

Alex, S. (2013). Workshop. Alexandria, Egypt SIGMA company. (2016). L. Passage re-use project (M. Magdy, Trans.). Alexandria, Egypt: Academic Press.

Borg, Y. (2013). *Social Contexts & Built Heritage. A case study of Alexandria, Egypt*. Cairo, Egypt: University Paris 1 Pantheon-Sorbonne.

Cairobserver. (n.d.). *11 Recent Cultural Disasters in Egypt*. Retrieved September 01, 2015, from Cairobserver: http://cairobserver.com/ post/75403717693/11-recent-cultural-disasters-in-egypt#.VL190oe7ggE

Dessouki, Y. (2007). *The Bibliotheca Alexandria and the preservation of Egypt's visual cultural heritage*. Academic Press.

Gudran Madina Motaaba blog. (2012). *Some of what was demolished in June*. Available online at http:// thewallsofalex.blogspot.com.eg/2012/06/2012. html

Heba, K. (2011). *Preserving architectural heritage in historical cities*. Study Of The Architectural Heritage With The End Of The Nineteenth Century And The Beginning Of The Twentieth Century In Alexandria, Egypt. Available online at http://www.cpasegypt.com/pdf/Khaled_Heba/Research/001%20 %20Preserving%20architectural%20heritage%20in%20historical%20cities.pdf

ICOMOS. (2001). *ICOMOS World Report on Monuments and Sites in Danger*. ICOMOS.

Licciardi, G., & Amirtahmasebi, R. (n.d.). The economics of uniqueness. Investing in historic city cores and cultural heritage assets for sustainable development. Academic Press. doi:10.1596/978-0-8213-9650-6

Morayef, S. (2012). *Demolition of disused Alexandria Greek Club highlights larger problems*. Available online at http://www.egyptindependent.com/news/demolition-disused-alexandria-greek-club-highlights-larger-problems

O'Brien, D. (2010). *Measuring the Value of Culture. Report to the Department of Culture and Sport*. Arts and Humanities Research Council.

Rogers, L. (n.d.). Incentives for Heritage Protection Handbook: A national guide for Local Government and the Community. Information Guide. *Commonwealth of Australia for the Heritage Chairs and of Cials of Australia and New Zealand*.

Throsby, D. (2006). The value of cultural heritage: What can economics tell us? In *Capturing the Public Value of Heritage: The Proceedings of the London Conference 25-26 January 2016*. London: English Heritage.

Tung, A. M. (2001). *Preserving the world's great cities. The destruction and renewal of the historic metropolis*. New York: Clarkson Potter.

UNESCO. (2013, August). *Warning: Looting of the Malawi National Museum in the Upper Egypt city of Minya.* Retrieved May 05, 2016, from http://www.unesco.org:http://www.unesco.org/new/en/culture/themes/illicit-traffic-of-cultural-property/emergency-actions/egypt/warning-looting-of-the-malawi-national-museum/

Wasft-Iskendrya. (2016). *Facebook page*. Retrieved on July 06, 2016, from https://www.facebook.com/groups/798916430159507/

Chapter 15
Rethinking Cairo's Informal Settlement in the Light of Hardships

Ahmed Hassan Abayazeed
The National Organization for Urban Harmony, Egypt

ABSTRACT

This chapter aims at understanding and interpreting the informal urban growth in Cairo as a physical objection act against the failing radical ways in managing the city. Accordingly, this chapter tries to analyze both informal physical actions and the formal policies and approaches adopted, and reach reasons for the successes and failures. The chapter first investigates briefly the evolution of Cairo's informal settlements and the hardship conditions behind. Then it examines these settlements through analyzing their two main types. Afterward, it discusses the reasons behind the success of these informal actions. Then it reviews policies and approaches. Consequently, it tries to discover the reasons behind the failure of formal policies and approaches. Thereafter, it discusses briefly using geospatial digital research methods in Cairo's informal settlements. Finally, the chapter ends with a concluded discussion tries to figure out how to reach the right path in dealing with Cairo's informal settlements.

INTRODUCTION

One of the most significant challenges during the last decades is the rapid urbanization which caused the emergence of new kind of informal settlements scattering in and around the growing cities in the developing world (UN-Habitat, 2003). This makes informality the often-primary mode of the production of 21st-century metropolitan space in the Global South cities (Roy, 2009). Cairo, as most of the Global South cities, suffers from the same phenomenon. Around 40% of the population in Cairo lives in informal settlements as estimated by the Ministry of Housing (Tadamun, 2014). This chapter argues that this phenomenon could be understood and interpreted as a physical objection act against the failing radical ways of managing the city and the futile urban context which made Cairo at the end, according to Sims (2010), a city out of control.

DOI: 10.4018/978-1-5225-9238-9.ch015

Ross (2011) argues that government policies are no longer reliable for solving society's most deep-rooted and hard problems. Instead of that, necessary actions should be taken by people themselves through a wholly new method of doing things. Informal urban growth performed by poor people is a real representation of such actions. Since the start of the literature on informal settlements around the 1960s, these settlements are seen as an unhealthy phenomenon which should be eradicated. Torres (2007) as cited in Hernández (2013) argues that the government should provide housing and facilities for all citizens not to leave them in the people's hands. By the time, new literature emerges as informality become more accepted as another way of doing things. This new literature argues that informality can be an opportunity, and it is not necessarily a problem as it can be considered as an alternative way of producing space (Roy, 2009). AlSayyad (1993) sees informal settlements as a creative and innovative way of taking the advantages of available resources and opportunities to enhance the daily life of the poor. This leads him to perceive informality as 'a way of life.' Some of the literature within this point of view sees that informal settlements are similar to old traditional cities which used to grow by people's hands without any intervention of architects (Egyptian Urban Action, 2012).

Accordingly, this chapter aims to reread the informal growth phenomenon in Cairo as a physical action responding to the inability of the policies and approaches adopted by the successive governments to provide shelters for the urban poor. This different point of view would lead to developing new policies, strategies, and plans when dealing with such settlements. Also, this kind of physical actions represented in the informal urban expansion could be seen as a practical response to a multi-layered complex failing context. This context should be discussed through understanding the reasons and conditions behind the emergence of Cairo's informal settlements.

THE EVOLUTION OF CAIRO INFORMAL SETTLEMENTS

There are some complex conditions behind the emergence of informal urban growth in Cairo. These conditions, in most cases, could not be interpreted apart from each other. They always interfere with mutual impacts. These conditions are investigated, within this chapter, through splitting them into three groups of conditions which are: the political-economic conditions, those related to laws and regulation, and the socio-cultural ones.

The Political-Economic Conditions

According to the 2016 census, around 10% of the Egyptian's population lives in Cairo (about a quarter of Egypt's urban population) (CAPMS, 2017). Cairo gains its importance as it is the center of the country's political and economic life as well as most of the industrial activities. This leads to the centralization of jobs which attract rural populations to work in Cairo and enhance their living situation. This was one of the main reasons for Cairo's Informality evolution. The informal growth phenomena in Cairo began after the Second World War, due to the housing pressure caused by the migration from Upper Egypt and the Delta. The Economic development took place in Cairo plus the massive industrialization policy launched by President Nasser[1] afterward attracted the internal migrants to Cairo (Séjourné, 2009). The first newcomer migrants, mostly young men, stayed in rented and shared rooms and flats in central and historic districts. Later, after collecting some savings, some of them were able to buy and build on land on the fringes of the villages attached to the city where the land prices were lower than in the central districts. In this period, the phenomenon of squatting on state-owned land started to appear mostly in the eastern part of Cairo. Some new informal districts appeared while others continued to grow. Low-income families-built houses on state-owned, which was almost desert land, through what is called "Wada' al-yad" (A phenomenon of squatting on a state-owned land through hand claim) as they could not afford to buy agricultural land. Thus, informal settlements, like Manshiet Nasser and Ezbet Al-Haggana, started to grow slowly on desert land (ibid).

A new phase of informal urbanization started at the beginning of the 1970s as 84% of the new units built during this period were considered informal (ABT Associates et al. 1982). People invested in informal constructions during this period due to some factors. First, the oil booms in 1973 and 1979 which enriched the near oil-producing countries like Libya, Saudi Arabia, and Iraq. This attracted the Egyptian manpower and thus, enables a big sector of the Egyptians to save money. Second, the insufficient and unaffordable housing supply proposed by the public sector. The third factor is the launching of the economic liberalization policy "Infitah" adopted by Sadat[2] in 1974 which enables Egyptians and goods to travel more freely. This policy helped people who belong to lower social classes to gain more money through trading and thus invest them in building houses informally. In 1977, a new town policy is introduced by Sadat. The goal was to solve the housing problem and the urban extension on agricultural land by relocating the demographic growth on the agricultural land into public housing on the desert fringes of the city. In the 1980s, the informal settlements in Greater Cairo3 slowed down in growth to some extent due to two reasons, the first one is the oil prices tumbling in 1983-1984 which caused decreasing the emigration of the Egyptian workers to oil countries, the second, Iran-Iraq War from 1980 till 1988 (Sims, n.d.). In the mid-1980s, the national demographic growth rates started to diminish and the migration from rural areas almost stopped (Bayat & Denis, 2000).

In the 1990s, although the growth of the informal settlements didn't slow down, no new informal settlements appeared. In spite of the Military decrees (1 and 7) which forbid building on the agricultural land, informal settlements in Cairo continued to grow and extend. From 1986 till 1996, the construction growth rate of informal settlements reached 3.2% per year compared to 1.1% of growth in formal housing areas (Denis & Séjourné, 2002). Till now, although horizontal expansion of informal settlements on desert land is nearly stopped, the vertical densification through demolishing old structures and building taller buildings with more stories is still taking place. On the other hand, sprawling on agricultural land by informal growth is still an ongoing phenomenon in spite of the governmental efforts made to face it. Losing agricultural land is so related to the negative effects of the issued laws of the agricultural land reform and regulations which is being discussed among other laws and regulations in the next section.

Conditions Related to Laws and Regulation

These conditions could be seen in the light of important laws released by President Nasser after the 1952 revolution to mobilize popular support for his policies (Al-Alaily, 2017). These laws are among the indirect reasons for the informal growth in Cairo. The first group of laws is used to limit and then reduce the rent of residential units. The implementation of these laws resulted in the reluctance of the private sector to invest in housing construction for several decades, which eventually led to a severe housing crisis on the long run and thus, encouraging informal growth.

The second group of laws is agricultural reform laws. The first law set the property per capita at 200 acres and was followed by a second law to determine the ownership of the family by 100 acres. Accordingly, the rest of the agricultural land was distributed among the farmers, five acres for each. With the succession of generations of the landowners, the land is split more and more. The resultant small land plots are not economically feasible at all especially, after the emergence of new techniques used in agricultural activities and the transformation of the agricultural economy into managing large holdings. This led to a significant reduction in the volume of food production for local consumption and accordingly, agricultural land has been gradually and severely converted into built areas (ibid).

The Socio-Cultural Conditions

The socio-cultural conditions in this section are used to describe the social and the cultural factors affecting people's preferences and attitudes leading, directly or indirectly, to the emergence of Cairo's informal settlements. The increasing of the population at high rates is one of these factors and considered as one of the challenges facing Egypt and plays a vital role in informal urban growth. According to CAPMS (2017), the Egyptian population growth rate is 2.4%. There are some social-related factors behind the existence of this phenomenon. According to rural Egyptian culture, families tend to have as many children as they could. This could be interpreted for two reasons. The first one is the harsh economic status, which pushes families to treat extra kids as additional sources for increasing income by introducing them to the labor market. The second one is related to feeling proud and ostentatious, in rural contexts especially in Upper Egypt, for belonging to an extended family of a large number of members. In the presence of tribal traditions, belonging to a big family or a tribe is seen as a source of protection and prestige.

Also, densification of informal settlements and their expansion is related to another social habit. According to Shehayeb & Eid (2007), people in rural areas prefer to move in groups and convincing relatives to live closer in the same area. After the enhancement of the socio-economic conditions within the settlement phase in Cairo's informal settlements, the settlers bring their families and invite their relatives, living in rural areas, to join them in the settlement (Eldefrawi, 2015). This leads to densification of the informal settlements through time and thus, their expansion.

The choice of poor people and families coming to Cairo searching for jobs to live in these settlements is not only related to their harsh economic situations and the unaffordability of formal housing, but also to their rural lifestyle. Most of the poor people coming to Cairo have a rural origin, and they prefer to live in areas look like their villages with their rural lifestyle and outdoor activities rather than staying in modern residential units in urban areas. El-Mouelhi (2014) in his research found clear similarities between living in Cairo's informal settlements and living in villages of origin in Upper Egypt. Accordingly, the first residents usually start building their new settlements the same way their villages were constructed. They usually choose locations near to formal services, infrastructure, and job opportunities. Some rural features like narrow alleys, raising poultry and sitting in front of the dwellings, and the spread of tiny economic activities are usually observed when going through these settlements.

All the previously mentioned conditions are the base to interpret the phenomenon of the growth of informal settlements in Cairo and to understand how these conditions are responsible for the informal physical act done by the poor to get a shelter.

THE HARDSHIPS WITHIN CAIRO'S INFORMAL SETTLEMENTS

The emergence of Cairo's informal settlements could be understood through the term 'Cities of Hardship.' This term started by referring to significance of the social and economic problems, besides the political isolation between cities and suburban (Nathan and Adams, 1976). In order to understand the term 'cities of hardship' the factors of hardship index created by Nathan and Adams (1976) should be cleared. The index is based on measuring six factors which are: unemployment, dependency, education, income level, crowded housing, and poverty. Other factors, according to Morse (2005), should be considered within the hardship index like access to health insurance, teenagers' pregnancy, homelessness, rates of housing issuance, the square acreage of the public space, violent crime rates, and business loans, and rates of volunteerism. Another group of factors related to visual and aesthetic aspects is still missed within this index (Abusaada and Elshater, 2019).

Most of these factors are shaping cities in the Global South in general and informal settlements in particular. When moving to Cairo's informal settlements, related hardships could be split into two categories. The first one 'The Causative Hardships' which are the direct cause of the internal migration or growth over agricultural land adjacent to formal Cairo. Lack of development, scarcity of job opportunities, and deterioration of services are examples of this category. The second one is 'The Resultant Hardships' which is split into two minor types. The first type is 'The Inner Hardships' which usually emerges inside the informal settlement itself due to lack of services, inadequate infrastructure and the dwellings poor conditions. The second type is 'The External Hardships' which begin to appear in formal Cairo due to the parasitizing of the informal settlements and the huge burden on the formal infrastructure and services (see Figure 1).

The main paradox is that the emergence of Cairo's informal settlements is directly related to social and economic hardships in villages and rural areas, and at the same time the informal physical act, represented by informal dwellings done by the poor to get a shelter, transfers these hardships or even create new ones in their new informal settlements in Cairo.

Accordingly, to Collier (2013), waves of migration today from Global South to Global North searching for better life could be seen as unconscious revenge of developing countries. This migration is accompanied by the fear of decreasing quality of life of Indigenous citizens. This process happen against their previous colonization by developed countries in the last decades which leaded to the postponement of

Figure 1. How Hardships are related to Cairo's informal settlements.
Source: The author

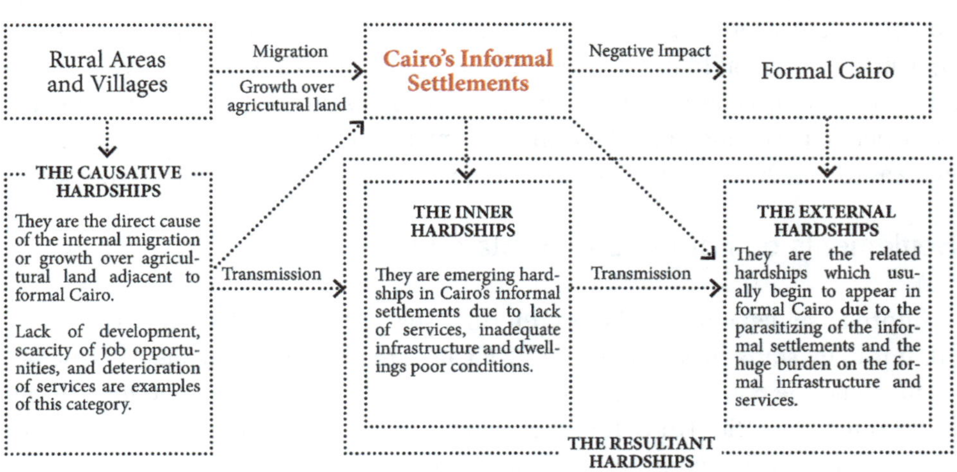

their development plans. Therefore, the increasing of hardships over the last decades was widely recognized in Cairo. The same concept could be used to understand Cairo's informal growth.

Migration from rural areas to Cairo could be seen as an unconscious revenge against the hardships resulting from Ignorance of rural development and investment, and lack of job opportunities in Upper Egypt and the countryside. This migration and therefore the resulting informal settlements are seen as a huge burden not only on the existing formal infrastructure and services in Cairo, but also in relation to lack of services, healthy dwellings, and suitable infrastructures in Cairo's informal settlements.

THE INFORMAL SETTLEMENTS: THE PHYSICAL ACT

The physical reaction against the failing context could be understood through the two main types[4] of Cairo's informal settlements. The first type is informal settlements on former agricultural land, and the second type is informal settlements on former desert state land.

Informal Settlements on Former Agricultural Land

Sims (2003) defines this type as residential buildings built on agricultural land bought from farmers without the existence of subdivision plans and building permits. Over half of the population of Greater Cairo and half the total residential area are contained within this type (ibid). This phenomenon started in the 1960s when farmers and brokers started to subdivide small agricultural lands on the fringes of Cairo to be sold to individual builders. This could interpret why the residents almost make use of the whole plot areas as the built area percentage per plot is almost 100% (Sims, 2003). It started to overgrow rapidly after launching "Infitah" policy in 1974. El Kadi (1987) as cited by Piffero (2009) states that, in such type, the owner may keep one of the housing units for himself and sell or rent the rest. It is important to mention that illegality, in this case, is not related to the ownership right, but it stems from the illegal conversion of the former agricultural land into building plots as well as not following the existing regulations concerning the plots sizes and the construction (ibid). Most of the buildings are made of a reinforced concrete skeleton with red bricks walls and designed for the height of five stories at least. A few older buildings were built of bearing walls and concrete slabs. The norm in this type is the incremental construction, room by room and story by story. Constructing taller buildings (up to 10 stories and more on larger plots in the fringes areas) adopted by small developers, started to appear3 (Sims, 2003). Currently, the buildings heights reach 10-12 stories (El-Mouelhi, 2014). As the residents are not threatened by eviction, as well as the quality of the housing constructions is right, an increasing process of densification is taking place over time accompanied by overcrowding especially in the more mature inner areas (Sims, 2003). Ezbet Al-Lamoon is an example of an informal settlement of that type.

Informal Settlements on Former Desert State Land

This type of informal settlements was built on former state land under the process of "wada' al-yad" (Sims, 2003). It usually starts by spreading out from an authorized initial nucleus. Usually, houses are gathered together, and the core settlement is built quickly to avoid risks of eviction. Spaces are given for streets and common facilities (El Batran & Arandel, 1997). These settlements started to expand due to illegal squatting of the surrounding vacant land (Piffero, 2009). They usually expand slowly with an apparent usual neglection of the government towards its property (Sims, 2003). The development process was completely informal within this typology.

It has no legal paperwork but relies on personal trust and mediation of the community when necessary. Although these settlements are illegal, the settlers have some customary rights originated from interpretations of parts of the civil code dealing with hand claims on desert land. Settlers keep receipts from paying "Tahkir" (a nominal rent imposed by a Governorate's Amlak (properties) Department), "Awayyid" (property tax) and receipts for electrical and water connections to have as much legality paper as possible (ibid). (Sims 2003; Piffero 2009) state that housing conditions in this type, compared to the previous one, are worse as there is a high percentage of dilapidated structures due to the risk of eviction. However, in older settlements, the building quality tends to become better as the residents invest more money in housing development as the risk of eviction decreases (Piffero, 2009). In this type, which is similar to the squatting, found in Latin America and the rest of the third world, each location has a particular history. For example; Manshiet Nasser started by relocating garbage collectors to the area in the 1960s (Sims, 2003).

Reasons Behind the Success

It is tough to discuss the complex reasons behind the partial success of such acts and ways of playing the role that the governments and real estate developers should play. Ross (2011) states nine principles to guide actions against failing systems and policies[5]. The first principle is excavating the convictions and identifying the target clearly. The second is the assessment of the financial condition and the landscape of power. The third is acting as if the means are the end. Referring to the Cosmopolitan Criterion is the forth. The fifth is addressing those suffering the most. The principle number six is keeping on consulting and negotiating. The seventh one is taking continuous small steps in the right direction of the plan. The eighth is depending on fully peaceful and non-violent acts. The ninth and the last principle is putting most of the efforts on achieving the results rather than focusing on the processes.

Reflecting on Cairo's informal growth through these principles could give hints on why this kind of growth succeeded. Through this reflection, it is found that most of these principles are applied by residents and the first builders of Cairo's informal settlements unconsciously as follows:

1. Building an informal dwelling is accompanied by a strong belief of the right to have a shelter even poor people couldn't afford it through the legal and formal ways.
2. Calculating risks related to exposure to punishment by the government is also put into consideration. A limited amount of money and a lot of courage is needed at the beginning of the process.
3. Informal growth, as a physical act, gains its strength from the simple idea mentioned by Ross (2011) where the means are, at the same time, the destination. To put it simply, building an informal dwelling is considered as resistance means against the failing housing policies, but at the same time, it achieves his final aim: to have a shelter.
4. The informal urban growth is backed with a real and direct representation of the individual's needs.
5. The building process is always accompanied by continuous and direct cooperation and participation of those who suffer the most from poverty and lack of houses.
6. Owners usually ask builders and small contractors to build their homes and deal with the technical issues related to the construction process.

7. People living in informal settlements negotiate with each other for public facilities and spaces and managing everyday activities.
8. One of the most characteristics of the informal settlements is its incremental growth and nature accumulative depending on the financial status of the residents and the growing sizes of families.
9. This kind of building acts is far away from violence as building communities survives and flourishes only in contexts of peace and calmness.
10. The strongest point within the whole process of informal urban growth initiated by poor people is that it focuses on the result which is, in our case, building shelters.

POLICIES AND APPROACHES USED AND APPLIED

In order to face the informality challenge, many efforts are made through policies and approaches which are adopted either by the government or civil society and other international cooperation actors. Accordingly, these efforts could be split into governmental and non-governmental efforts as follows:

Governmental Efforts

In the 1990s, the government started to pay attention to the growth of informal settlements as a result of security and humanitarian reasons (Khalifa, 2015). Accordingly, two successive programs for upgrading informal settlements were initiated. The first one was 'Informal Settlements Development Program' (ISDP) starting from 1994 till 2004. This program adopted an interventionist approach where the government intervenes through providing basic services and infrastructure like water supply, sewage systems and streets paving (ibid). Through this program, basic urban services were provided to about 325 informal areas and 13 areas of worse conditions were developed with a total cost of 3.2 Billion EGP (Algohary & El-faramawy, 2010).

According to Khalifa, 2015, the ISDP lacked the participation of the community and legalization through tenure security. It only focused on improving physical conditions while ignoring the social and economic aspects. The second program adopts a preventative approach, in which informal growth are limited through "Tahzeem Al-Ashwai'at" which means belting informal settlements by imposing UGB (urban growth boundaries), banning building on agricultural land, banning squatting on state-owned desert land, and using codes and regulations to prevent introducing new structures in these settlements (Khalifa, 2015; Tadamun, 2014; Algohary & El-faramawy, 2010). In 2008, Informal Settlements Development Facility (ISDF), following the Egyptian Cabinet, was established by presidential decree No. 305/2008 as a result of a rock fall over Duweiqa, one of Cairo's informal settlements located at the base of the Moqattam cliffs. Around 110 people were killed as a result of this catastrophe (Tadamun, 2014). The main target of ISDF is to contribute to ensuring safe housing and improve the quality of life of the residents in unsafe areas (Algohary & El-faramawy, 2010).

ISDF coordinates efforts and finance for the development of informal settlements. ISDF adopted new typology which categorized informal settlements into unplanned or unsafe areas. Grades from 1 to 4 are given to unsafe areas, where grade 1 includes areas threatened by the dangerous situations related to health and safety of its residents, and Grade 4 includes areas exposed to the least dangerous situations (Tadamun, 2014). This refers to the priority adopted by the ISDF and dealing with unsafe areas, which are in need to a quick intervention to improve the housing conditions, to prevent other catastrophes like

Duweiqa rock fall from retaking place again (ibid). The financial strategy adopted by ISDF is based on considering itself as a revolving fund seeking cost recovery and financial sources for focused grants providing a safe housing environment, market value compensation for properties, or their devaluation as a result of the development action (Khalifa, 2015). According to (Algohary & El-faramawy, 2010), ISDF's Resources of finance in 2010 were as follows:

1. A fund from The National Housing Program (NHP) for constructing of 25000 new units with a total cost of about 2.1 Billion EGP.
2. The land-based finance from the Development of Governmental Lands (DGL) with a total cost recovery of 17 Billion EGP.
3. The self-finance of the governmental authorities which own the lands with a total amount of 450 Million EGP.
4. Funds from international development agencies, donor countries, and financial institutions with a total amount of 300 Million EGP.

In the last years, the Cairo government adopts a plan to eliminate the dangerous informal settlements of grades 1 and 2 and resettle the residents to newly constructed housing projects. The plan will be implemented within two years at an estimated cost of LE 12 billion through constructing more than 62,000 housing units within 13 projects (Al-Sharkawy, 2017). The three-phase "Asmarat" project, which includes 18,300 housing units, the Shahba project with 7,720 housing units, Al Khayyala project with 3,200 housing units, and Al Mahrousa 1 and 2 projects in Al Salam City with 4776 units are examples for those projects (ibid). Most of these projects are being implemented in cooperation with the Ministry of Housing, ISDF and Tahia Masr Fund.

Non-Governmental Efforts

Besides the governmental efforts in developing Cairo's informal settlements, there were some attempts and interventions done by donor agencies, civil society organizations, and charities or religious institutions. According to (Abdelhalim, 2010), there are two main modes of these interventions. The first mode is 'upgrading' which include introducing services and enhancing sectors while maintaining most of the urban fabric and the physical structures. Upgrading of Hai Elsalam and Manshiet Nasser could be put under the umbrella of that mode. The second mode is 'redevelopment' which means demolishing the old physical settings and replacing it with new ones. In-Situ Zeinhum developed housing area, which depended on donations from businessmen under the follow-up of Cairo governorate, could be seen as an example of that mode.

One of the initiatives that attempt to apply the community participation concept in upgrading informal settlements is the Participatory Development Program in Urban Areas (PDP) which is initiated by the German Development Cooperation (GIZ) and the Egyptian Ministry of Planning and International Cooperation (MoPIC). The PDP aims at the upgrading of informal settlements within Greater Cairo Region through adopting participatory methodologies for urban upgrading with the cooperation of local authorities and civil society organizations to improve services and enhance living conditions. Two pilot projects, focusing on upgrading infrastructure and community services, were implemented by the PDP in Manshiet Nasser and Boulaq El Dakrour (ibid).

REASONS BEHIND THE FAILURE

It is important to mention that providing direct solutions for such a complex phenomenon is very hard, especially, because of the accumulation of the related faults and wrong strategies in the last 60 years. Dealing with such a phenomenon needs to deal directly with its main reasons and to direct all the efforts to address the logic behind its emergence. Most of the strategies and approaches are tackling the symptoms rather than the main disease.

By moving through the previously discussed policies and approaches, some main reasons behind their failure appear on the surface. They decrease their efficiency and inactivate reaching the development targets. These reasons could be stated as follows:

1. Adopting severe centralization policies: the government adopts a very intensive centralization approach instead of strengthening the local authorities especially in Upper Egypt, the main source of internal migration towards the capital. This approach is responsible for lack of internal and external investments in rural areas and Upper Egypt compared to Cairo and mega Egyptian cities and thus lack of job opportunities.
2. The weakness of the rural development: the rural and agricultural related development in Upper Egypt and the Delta is not among the priorities for the government compared to land reclamation projects outside the Nile Valley. Although many changes in agricultural technologies and the related industries took place in the last decades, which led to reduce employment, the government had not taken any concrete measures to mitigate the situation. Also, suitable alternatives for affordable housing are not enough to prevent people from building over agricultural land. In other words, most of the strategies and plans are approaching 'the Resultant Hardships' not 'the Causative' ones (see figure 1). A balanced strategy dealing with the resultant and the causative hardships would be a better alternative.
3. The existence of different authorities responsible for dealing with informal settlements, there are different bodies responsible for developing informal settlements in Cairo. Ministry of Agriculture is responsible for settlements built on agricultural land. Informal Settlements Development Facility (ISDF) is mainly responsible for settlements built on desert state-owned land. Deteriorated Historic Cores are in the hands of the local municipalities. This results in different policies and plans that sometimes conflict.
4. The lack of monitoring and control, the entity responsible for the continuous monitoring of informal growth is absent. Accordingly, actions are only taken after the situation became worse which means solutions of larger cost.
5. Dealing with financing development projects in unsustainable manner, the government and the policymakers See Cairo's Informal Settlements as just inadequate housing challenge. They don't develop policies based on making use of the inhabitants' abilities and their real partnership and participation. Also, they limit the economic component to just financing the development without any rewards.
6. As Cairo's informal settlements are urban contexts with complex social, economic and cultural dimensions, new research methods, techniques, and tools should be developed to process these huge data and information instead of depending only on old and traditional methods. Using geospatial digital research methods could be an ideal approach as discussed in the next section.

USING GEOSPATIAL METHODS: FUTURE RESEARCH DIRECTIONS

A huge challenge with the development of Cairo's informal settlements in addressing its growth and the large inflow of people in informal settlements is lack of data and the often-heavy methods used in enumerating. This usually delays suitable planning for informal settlement and leads to wasteful spending, poorly planned development. Great potential is shown by geospatial techniques like earth observation, mobile GIS, and geo-location-based services in providing data on informal settlements (Tao, 2013). Accordingly, researchers and urban planners tend to apply smart management techniques and intelligent planning approaches to deal with urbanization challenges (ibid).

As the dealing with the informal settlements in cities has become overwhelming for urban planners, using geospatial data could help and give support to urban planners and the decision makers in taking more accurate decisions (Pinfold, 2015). The physical and socio-economic data is necessary to guide and support informal settlement upgrading approaches (Kohli et al, 2012), and the traditional techniques are often time wasting and lack the wanted information to guide decision-making process on the informal settlements upgrading. With the rapid progress in geo-information tools, innovative and holistic techniques for collecting data on the physical and socio-economic attributes of informal settlements, there is a big potential in the collection of complex data on informal settlements (Ching and Ferreira, 2015). These techniques go beyond using information communication technology (ICT) to ensure that data is detailed and useful for decision-making process on informal settlements (ibid).

Emerging techniques and tools like mobile GIS, volunteered geographic information from social media, and earth observation enable planners to obtain up-to-date, real-time, and rich data that can be used to guide decision making (Li et al, 2013). Accordingly, geospatial technology transformed the extent to which data could be used for urban development across the globe (Chitekwe-Biti et al, 2012). There are a lot of studies and cases that used these geospatial digital research methods in informal settlements. For example, Pinfold (2015) used surveys and participatory mapping in enumerating informal settlements in Cape Town. He focused on the physical structure and layout with little socio-economic and demographic data collected. Also, Kakembo and van Niekerk (2014) applied remote sensing images and mobile GIS to calculate population and identify the growth of informal settlements in Nelson Mandela Bay South Africa. In conclusion, dealing with Cairo's informal settlements and their hardships should be developed using geospatial techniques and tools which will save time and provide the decision makers with more accurate information and thus, right decisions and approaches.

As the scale of this research area is extensive and broad, more investigations are needed to allow further understanding and to reach an appropriate and workable holistic approaches dealing with Cairo's informal settlements. The achievement of this goal could be facilitated by exploring the following as future research strategies:

1. Examining the physical settings of Cairo's informal settlements and exploring their dynamics of growth and expansion.
2. Investigating the real reasons for lack of development in rural areas in Egypt, and exploring new schemes for attracting more investments.
3. Shedding light on the dynamics of informal urbanization of agricultural land and reasons behind and searching for creative ways for getting shelters and exploring various densification methods.

4. Analysis of the existing housing rental policies and searching for more reforming solutions.
5. Going deep into geospatial digital research methods and developing more applications for addressing Cairo's informal settlements.

CONCLUSION

The growth of informal settlements in Cairo, like many cities in the Global South, is the most dominant way of urban growth. This is mainly because of local reasons related to the wrong development and housing policies adopted to face housing pressure, lack of development in poor cities and villages in the Delta and Upper Egypt compared to the centralized Cairo. The government sees informal settlements as an unwanted phenomenon and their physical settings are useless, worthless and should be totally demolished (Tadamun, 2014). This is because of the inability and the shortening related to lack of services and infrastructure, and the bad housing conditions. Although these shortcomings are totally admitted, it is important to adopt an objective position not only to admit the settled negatives but also to find out the real positives. However, evidence mentioned by several researchers (Roy 2009; Shehayeb 2009; AlSayyad 1993) points to the fact that these settlements have some positives and values and are considered as proof of the residents' creativity represented in using their limited resources to solve their housing problems and to achieve their real needs.

This research draws attention towards rethinking policies adopted by the government regarding developing informal settlements through demolishing these settlements and building new housing projects adopting imported inappropriate approaches. This could be done either through adopting approaches making use and maximizing the positives within the existing physical settings and addressing the related negatives and shortcomings or through introducing new physical settings comprising qualities already exist within informal settlements. This imposes on those in charge of development of informal settlements to examine and investigate these settlements not only in terms of socio-economic aspects but also in terms of their physical settings and patterns of outdoor activities in order to have a broader view of the whole settlement before starting the development process.

This chapter aims at paying attention to the core of the informal growth problem in Cairo which is so related to challenges far away from the capital city but have severe impacts on it in terms of migration and informality. Of course, building new houses and increasing the supply of new shelter for the urban poor would help, but this could go hand in hand with developing other strategies related to addressing the development of rural areas to decrease the rate of migration to urban areas.

In order to overcome the reasons behind the failure of the adopted strategies and policies, new strategies and policies should replace the old ones where some points should be considered as follows:

1. The severe centralization policies should be mitigated by giving greater power to local authorities in Upper Egypt to attract more investments and to adopt ambitious development plans.
2. Strict laws to protect the valuable agricultural plots and prevent the informal growth over it in the Delta and Upper Egypt should be established with the need to follow up on their implementation.
3. Rent-related laws should be reformed to encourage investors and developers to invest in housing for urban poor through Public- Private Partnership (PPP) approaches.
4. Suitable alternatives for affordable housing should be available.

5. The competent authorities responsible for dealing with informal settlements should be unified. This will lead to the integration and harmony of plans and policies.
6. A plan to monitor continuous random growth must be activated.
7. The residents should be integrated into the urban development process and treated as partners in the development process.

In spite of what is often mentioned about Cairo's informal settlements and their physical settings which are seen as fruitless, worthless and should totally be destroyed, these settings are the tangible evidence of their physical actions towards getting affordable shelter without relying on governmental aid and far away from the failed formal mainstream. For that reason, it is important to adopt new policies related to the main reasons behind the formation and growth of these settlements. Unfortunately, most of these challenges are far away from the city capital.

REFERENCES

Abdelhalim, K. (2010). *Participatory upgrading of informal areas: a decision-maker's guide of action.* Cairo: participatory development program in urban areas (PDP) in Egypt.

Abusaada, H., & Elshater, A. (2019). From Chaos to Order: Articulating the Urban Policies for Cities of Hardship. In U. G. Benna, Industrial and Urban Growth Policies at the Sub-National, National, and Global Levels. USA: IGI Global. DOI: 10.4018/978-1-5225-7625-9

Al-Alaily, S. A.-A. (2017). *The issue of old rents.* Retrieved April 29, 2017, from http://www.shorouknews.com/columns/view.aspx?cdate=15042017&id=25952aba-8542-4178-8514-0fae3cc85538

Algohary, S., & El-faramawy, A. (2010). *Egyptian Approach to Informal Settlements Development.* Retrieved October 13, 2016, from Slideshare: http://www.slideshare.net/GRFDavos/egypt-for-idrc-2-june-2010-finalpptx

Associates A. B. T. Dames and Moore & GOHBPR. (1982). Informal Housing in Egypt. Cairo: U.S. Agency for International Development.

CAPMS. (2017). *Central Agency for Public Mobilization and Statistics.* Available at: http://www.capmas.gov.eg/

Ching, T. Y., & Ferreira, J. Jr. (2015). *Planning Support Systems and Smart Cities: Smart Cities: Concepts, Perceptions and Lessons for Planners.* Springer.

Chitekwe-Biti, B., Mudimo, P., & Nyama, G. M. (2012). Developing an Informal Settlement Upgrading Protocolin Zimbabwe - the Epworth Story. *Environment and Urbanization, 24*(1), 131–148. doi:10.1177/0956247812437138

Collier, P. (2013). *Exodus: How Migration is Changing Our World.* New York: Oxford University Press.

Denis, E., & Séjourné, M. (2002). *ISIS: Information System for Informal Settlements.* Cairo: Ministry of Planning, GTZ, CEDEJ.

El-Mouelhi, H. (2014). *Culture and informal urban development - the case of Cairo's 'Ashwa'eyat (informal settlements).* Berlin: German academic exchange service (DAAD).

Kakembo, V., & van Niekerk, S. (2014). The Integration of GIS into Demographic Surveying of Informal Settlements: The Case of Nelson Mandela Bay Municipality, South Africa. *Habitat International*, *44*, 451–460. doi:10.1016/j.habitatint.2014.09.004

Khalifa, M. (2015). Evolution of informal settlements upgrading strategies in Egypt: From negligence to participatory development. *Ain Shams Engineering Journal*, *6*(4), 1151–1159. doi:10.1016/j.asej.2015.04.008

Kipper, R. (2009). Cairo: a broader view. In R. Kipper & M. Fischer (Eds.), *Cairo's informal areas between urban challenges and hidden potentials* (pp. 12–15). Cairo: GTZ Egypt.

Kohli, D., Sliuzas, R., Kerle, N., & Stein, A. (2012). An Ontology of Slums for Image Based Classification. *Computers, Environment and Urban Systems*, *36*(2), 154–163. doi:10.1016/j.compenvurbsys.2011.11.001

Li, D., Shan, J., Shao, Z., Zhou, X., & Yao, Y. (2013). Geomatics for Smart Cities-Concept, Key Techniques, and Applications. *Geo-Spatial Information Science*, *16*(1), 13–24. doi:10.1080/10095020.2013.772803

Morse, S. W. (2005, June 27). *An Update on Urban Hardship*. Retrieved from Smart Communities: http://smartcommunities.typepad.com/suzanne/2005/06/an_update_on_ur.html

Nathan, R. P., & Adams, C. (1976). Understanding Central City Hardship. *Political Science Quarterly*, *91*(1), 47–62. doi:10.2307/2149158

Piffero, E. (2009). Beyond rules and regulations: the growth of informal Cairo. In R. Kipper & M. Fischer (Eds.), *Cairo's informal areas between urban challenges and hidden potentials* (pp. 20–27). Cairo: GTZ Egypt.

Pinfold, N. (2015). *Community Mapping in Informal Settlements for Better Housing and Service Delivery, Cape Town, South Africa. Lisbon, Portugal.* INSPIRE Geo-spatial World Forum.

Ross, C. (2011). *The Leaderless Revolution: How Ordinary People Will Take Power and Change Politics in Twenty-First Century*. New York: Blue Rider Press.

Séjourné, M. (2009). The history of informal settlements. In R. Kipper & M. Fischer (Eds.), *Cairo's informal areas between urban challenges and hidden potentials* (pp. 16–19). Cairo: GTZ Egypt.

Shehayeb, D. (2009). Advantages of Living in Informal Areas. In R. Kipper & M. Fischer (Eds.), *Cairo's informal areas between urban challenges and hidden potentials* (pp. 34–43). Cairo: GTZ Egypt.

Shehayeb, D., & Eid, Y. (2007). Neighborhood Design and Community Building: A model of social interaction. In *The appropriate home: Can we design"appropriate" residential environments?* (pp. 129–144). Cairo: HBNRC.

Sims, D. (2003). *Cairo, Egypt - understanding slums: case studies for the global report 2003*. UN Habitat. Available from: http://www.ucl.ac.uk/dpu-projects/Global_Report/pdfs/ Cairo.pdf

Sims, D. (n.d.). *Library - analytical materials*. The Habitat International Coalition's specialized Member group: Housing and Land Rights Network (HIC-HLRN). Available at: http://www.landpedia.org/land-doc/Analytical_materials/Sims_Residential_Informality_in_Greater_Cairo. pdf

Tadamun. (2014). *Coming up short: Egyptian government approaches to informal areas*. The Tadamun Initiative. Available from: http://www.tadamun.info/

Tao, W. (2013). Interdisciplinary Urban GIS for Smart Cities: Advancements and Opportunities. *Geo-Spatial Information Science*, *16*(1), 25–34. doi:10.1080/10095020.2013.774108

ENDNOTES

1 Gamal Abdel Nasser (1918-1970) was one of the July revolution leaders and the second President of Egypt, serving from 1956 until his death. He is considered as an iconic figure in Egypt and the Arab world, particularly for his steps towards Arab unity and social justice, as well as his industrialization and modernization policies. Nationalization of the Suez Canal and constructing Aswan high dam were among his most important achievements. He adopted the neutralism movement during the cold war. Source: nasser.bibalex.org.

2 Mohammed Anwar El-Sadat (1918-1981) was the third President of Egypt after July revolution, serving from 1970 until his assassination. During his period, he adopted an economic liberalization policy *"Infitah"* which had negative social impacts on the Egyptian society. He launched October war against the Israeli forces occupying the Egyptian Sinai Peninsula in 1973. Afterwards, he signed the peace treaty with Israel in 1979. He was awarded the Nobel Peace Prize in 1978. Source: sadat.bibalex.org.

3 The Greater Cairo Area (GCA) is the largest metropolitan area in Egypt and the largest urban area in Africa and the Middle East, consisting of all cities in the Cairo, Giza, and Qalyubia Governorates. [44] According to Sims (2003), there are four main types of informal settlements in Cairo. Two of them are the most dominant which are: settlements built on former agricultural land and that built on former desert state land. The other two types are the deteriorated historic core and the deteriorated urban pockets.

5 According to Sims (2003), there are four main types of informal settlements in Cairo. Two of them are the most dominant which are: settlements built on former agricultural land and that built on former desert state land. The other two types are the deteriorated historic core and the deteriorated urban pockets.

Chapter 16
Cairo's Downtown in Narratives:
From Singularity to Hardship

Mougibelrahman Aboamer
Misr International University (MIU), Egypt & University Grenoble Alpes, France

Dalia Abdelfattah
Arab Academy for Science, Technology and Maritime Transport, Egypt

ABSTRACT

Cairo's downtown, through sociopolitical conditions, had been moved from a single city to a hardship one. The authors attempt using a method of multiple readings to provide a new comprehension for the city by using the historical review side by side with several examples from Egyptian literature that describes this dramatic evolution. For this neighborhood, it is considered an active part of Cairo. The period they suggest for scanning the literature begins from Cairo Great Fire in 1952 and its consequences. The year of 1952 and the constitution of the first republic until the dramatic fall of this last by the revolution of January 2011. This chapter aims to articulate the evolution of Downtown Cairo from the singularity to the hardship.

INTRODUCTION[1]

Cairo's Downtown evolution is the focus of the present research. The period was limited and selecting between since the 30th until the second decade of the third millennium. The review emphases on the turning points that influenced the area, deforming and its characteristics from a district of singularity hosted by the elites to a zone of hardship that lacks the essential levels of the quality of life (Abaza, 2006; AlSayyad, 2011; Ashour, 2003; El Kadi G., 2012; Mitchell, 2002). This analysis is considered an approach for re-tracing its character, comprehending its qualities, and understanding its overlapped layers of history. The current work goes in depth through looking for traces and mapping of its historical, political, social, and urban deformations.

Arguably, Cairo Downtown has suffered from substantial sociopolitical changes since the second half of the twentieth century (Ashour, 2003; Awatta, 2015; Taher, 2008). These changes are seen in the shift from its glorious birth of a unique urban setting and architecture character, combining European

DOI: 10.4018/978-1-5225-9238-9.ch016

styles with local influences and materials (Ashour, 2003; Abu-Lughod, 1971; AlSayyad, 2011; El Kadi & Kerdany, 2006; El Kadi & Kerdany, 2006; Hawas, 2001; Myntti, 1999). The district has a feature of hardship living (Elshahed, 2007; Haykal, 1973; Sims, 2012).

In this work, the characteristics of the cities of hardship are addressed to describe the lacking the essential quality of life principles. These are seen in the gradual replacement of famous palaces and villas by commercial activities on the ground floors of residential and mixed-use buildings.

Throughout this on-going evolution, shifting from a district of singularity, Cairo downtown was the platform of progressive dramatic political movements, leading to directional activities of visions and perceptions, influencing its qualities, and transforming it to a district of hardship (Abaza, 2006; Ashour, 2003; Assy, 2006; Elshahed, 2007; Essa, 2013; Mehrez, 2011; Raafat, 2003; Singerman, 2009).

The research problem is the transformation in political, social, cultural and economic changes, lead to deformations affected the main attributes of Cairo Downtown. The deformations are recognized in the change from a city of singularity to a city of hardship. The current chapter addresses several further questions on how Cairo Downtown is considered as a reflection of sociopolitical evolutions. Another interesting question in this context is: what are the impacts of those evolutions that led to make it a district of hardship after being a glorious one?

This work traces the continually shifting of the hardship multi-attributes throughout downtown's history that goes beyond reviewing its origin in the nineteenth century. Scanning the narratives from the Egyptian literature focus mainly on its transformation from the 40th and early 50th to the January revolution of the present state in the twenty-first century. In this reading, the current work begins with observations and readings of the Cairo Downtown transformation process and shifting manners from its origins as a city of singularity to the current tendency towards socio-political deviations.

BACKGROUND

Cities of Singularity

In Cambridge Dictionary, the word 'singular' is to describe a formal of an unusual quality or standard; noticeable. In English grammar, this word is used to talk about only one thing (Cambridge Dictionary, 2019). In Oxford living Dictionary the word ***singularity*** is more than the quality of being singular; it's *a hypothetical moment in time when artificial intelligence and other technologies have become so advanced that humanity undergoes a dramatic and irreversible change* (Ogilvie, 2012, p. 106).

According to Aldo Rossi, the city of singularity is "*beings in the event and in the sign, which marked the event*" (Rossi, 1982, p. 106). For Rossi, the history of the city and its "locus" make the singularity and presuppose the existence of the architecture artifacts. According to Rossi, the artifacts of the city pointed out by the geographer Maximilien Sorre presented to his work about the fractionation of space's theory. In Sorre's work, the locus appeared inside the undifferentiated spaces, the situations, and the qualities that are indispensable to understand urban artifact (p. 34). Besides, the efforts of Viollet Le Duc asserted that the in his efforts to architecture is interpreted *as a series of logical operations based on a few rational principles, he also admitted the difficulty of transposing a work of architecture from one place to another* (p. 103). So that, the sites are considered the locus in the city an essential agent in creating its singular unique and physical place. According to Rossi (1966), *each city is constituted of parts, and each one of these parts have characteristics; each city has some singular (Primary) elements around which buildings aggregate* (Rossi, 1982, p. 58). Those singular elements are not only its monuments nor its fixed activities, but the singular elements of the city are those who are capable of the urban process's acceleration, and the city's dynamics through the quality of those elements and their order, in addition of their placement, their individuality and their unfolding of a precise event. Rossi added, *Architecture is the ultimate moment in this process, also what emerges from this complex structure* (p. 62). In what could be added to the role of those elements, the sociopolitical contexts, and transformations, as a participatory agent, a witness and a source of inspiration.

The individual elements of the city create its singularity through a role that exceeds their functional identity or existence to a significant value (Lynch, 1960, p. 53). This significant value makes those singular elements of the city immeasurable and un-constructible (Rossi, 1966, p. 73). Rossi gave an example to better understand those "artifacts" who could be an event "which has" importance *that gives place to the spatial transformations of a site* (Bjorkdahl & Kappler, 2017).

Recently, Abusaada (2017) presents an attempt to revisits the criterion 'city of singularity' to be as a mechanism to differentiate between the cities, particularly in the urban design discipline. He coined the term 'Singular Urbanism' *to show the lifestyle of city citizens as exceptionally magnificent* (Abusaada, 2017, p. 3). This term is exploring the similarity in the new Arab cities; it is *an emphasis on the intangible quality of the city compared with other cities* (Abusaada, 2017, p. 5). It focuses *on the insubstantial ideas that create a better community* (p. 8). In his article "The untold story of the urban design practice of new cities in Egypt" wrote that singularity is not only exist *based on architectural form and formation but also in the differentiation of design ideas at the level of creating the city and the meaning stored in a hidden or inaccessible place in the public realm* (Abusaada, 2017, p. 789). Accordingly, the organization of a city's elements (physical and nonphysical) could be assisted by the recognition of its semantics and symbols of differentiation as follows (Abusaada, pp. 790-791; Rossi, 1982, pp. 29-32).

- The singularity of the place.
- The exceptionally magnificent lifestyle of city citizens as.
- The city as a work of art.
- Urban history imprinted the city with form.The event and the sign that has marked the event.
- Urban architecture arteficts.
- Landmarks are the essential physical characteristic.
- Figure-background clarity is necessary.
- Architecture, individual semantics, and symbols characterized by their history and form.
- The products of all contemporary paradigms.

Cities of Hardship

The ability of cities to capture growth has a clear and direct relationship to hardship conditions over time (Meller, 2001; Abusaada & Elshater, 2019). It explains the difficulties the city may confront and deal with its daily base. The "*Hardship allowance of a city*" refers to the significant deterioration in living conditions in the living city. Various references have investigated the deficiencies of the quality of life, the "*intercity hardship index.*" Richard Nathan and Charles Adams are the initiators of the *Urban Hardship Index in 1976* (Nathan & Adams, 1976), which examines key statistics that draws together six key factors into a composite index permitting comparing areas over time (Wright & Montiel, 2007). These factors include:

People

- Unemployment.
- Dependency.
- Education.
- Income level.
- Crowded Housing.

Poverty

Mercer's Quality of Living Methodology was established and developed by an international team of experts, working thoroughly with major multinational enterprises and other professionals in the field (Andersen, 2018). The Quality of Living Report is released annually, in early November". The total index is based on the following categories:

- Consumer goods.
- Economic environment.
- Housing.
- Medical and health considerations.
- Natural environment.
- Political and social environment.
- Public services and transport.
- Recreation.
- Schools and education.

Socio-Cultural Environment

The University of Melbourne, Health and Livability Research Program, has identified a large and diverse range of subjective and objective livability indicators in their report, which indicates their usefulness for the city (Lowe, et al., 2015). The identified indicators fit within 11 policy areas. These are:

- Crime and safety.
- Housing.
- Education.
- Employment and income.
- Health and social services.
- Transport.
- Public open space.
- Social cohesion and local democracy.
- Leisure and culture.
- Food and other local goods.

Natural Environment

Based on the previous background interpolation, concepts, and backgrounds, the authors propose the hardship multi-attribute index, merging between the shared variables, using them as a matrix for analyzing the evolution of the urban district (downtown) from singularity to hardship (Figure 1).

METHODOLOGY

The methodology adopted a descriptive method to trace the dramatic shifting in readings (Barthes, 1971; Barthes, 1994). This approach aims to reveal the relationship between Cairo Downtown and literary fiction. This reading relies on the perspective of overlapping, using the multi-media, literature of books and novels as a tool. The expected outcome of this review is to create the narrative of the transformation of the architectural and urban dramatic changes and its entanglement in different social, economic and political contexts within the city's history. The mutual instrumental relationship between architecture

Figure 1. The hardship multi-attribute index

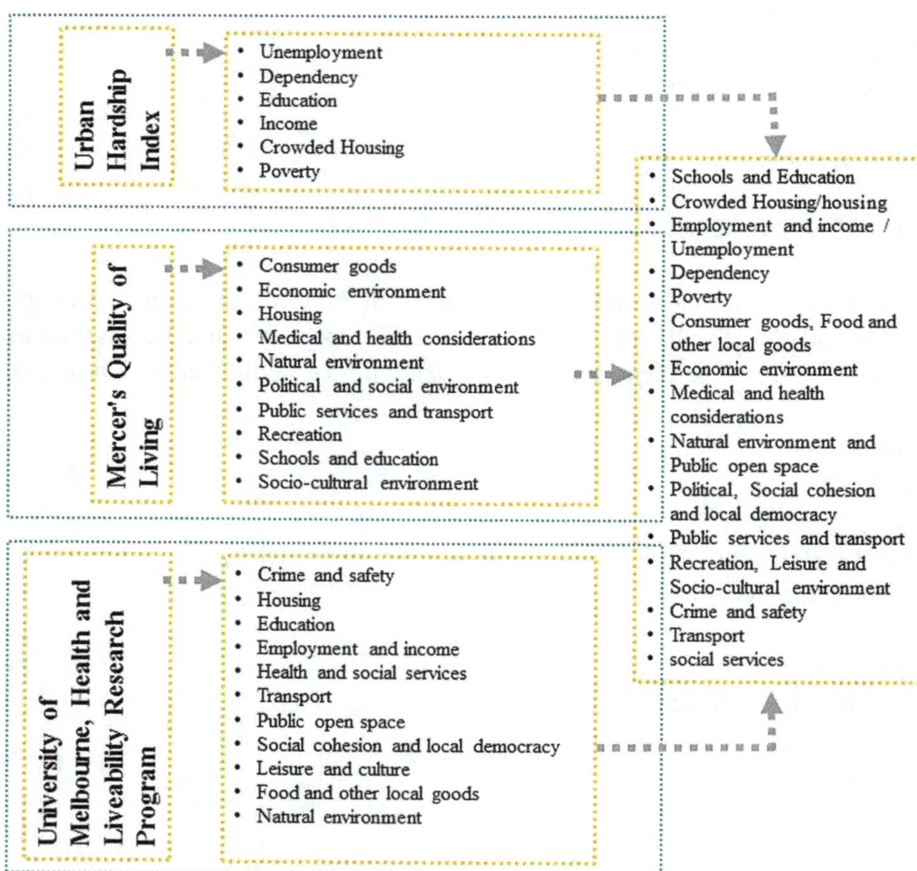

and urban spaces and literature can be far fruitful for attaining downtown evolutional regime, to consider a link between the significance of the singular elements/artifacts of the city of Rossi and the role of the signs through the multiple readings of Roland Barthes (Barthes, 1971; Barthes, 1994).

The next investigate the semantic charge and content of Cairo Downtown. This investigation undergoes too many transformations from the second half of the twentieth century, through the confrontation between the realities of the city and its literary representations to develop a language of this part of Cairo.

Yet, the multiple reading research approach allows making an improved cross tabulation that is provided through various types of data. The adopted database is comprised of a score of corpuses covering an account of Cairo Downtown history, the socio-political situation in Egypt, and its context transformations. Those fundamental texts are necessary for the theoretical framework, in order to provide an understanding of the relationship between the architecture of Cairo downtown, politics and society.

This reading reflects the transformations in the characteristics of Cairo downtown based on analysis selected novels. This analysis represents a spectrum of city readings, and that prompts the understanding of how the city was described and conceived by its writers within a specific sociopolitical context (Figure 2). The various images of the city that come from the fictional world could give some keys to understand the realities of the city. Those understanding keys answer the research questions. Notably, the main basic one is regarding the dramatic shift of Cairo Downtown from being a district of singularity into places of hardship living. The authors of the current work use this impact of the multiple readings to analyze the multi-attribute index.

The Deductive Logic

The results are expected to be conducted based on the reading of the overlapped layers of Cairo Downtown. The hardship is analyzed in the current work based on the hardship multi-attribute index, addressed on two manuscripts: the article of "Sémiotics and urbanism" and the book *the Semiotic Challenge (l'aventure sémiologique)* (Barthes, 1971; Barthes, 1994). These two works introduce a trial to discover the relationship between semiotics and urbanism. This trail he developed has three fundamental ideas which are as follows:

- **The first idea is the semantic content of the city**. There is a conflict between the functionalism of the city which he called "semiotic content" or its "semiotic power." The semiotic charge represents history. This is a conflict between significance and function provoked a sort of distress for the urbanists. The city has a rhythm between its functional elements and its significant elements (between importance and the non-significance). This rhythm might be based on the opposition, alternation or juxtaposition. Those are the forms of the relationship between the city's reality and its signification that physical maps can never explain.
- **The second idea is the language of the city.** Easily and metaphorically, this is the cinema's or Flowers's languages. However, it is difficult to pass from the metaphor to the analysis while discussing the language of the city. This is because of new scientific energy that is needed and would help with other data collected from other fields such as psychology, sociology, and geography to build a comprehensive city model. This new scientific energy is the semiotics, notably the role of the symbols which could transform the city's language from the metaphor to the description of meaning.

Figure 2. The methodology

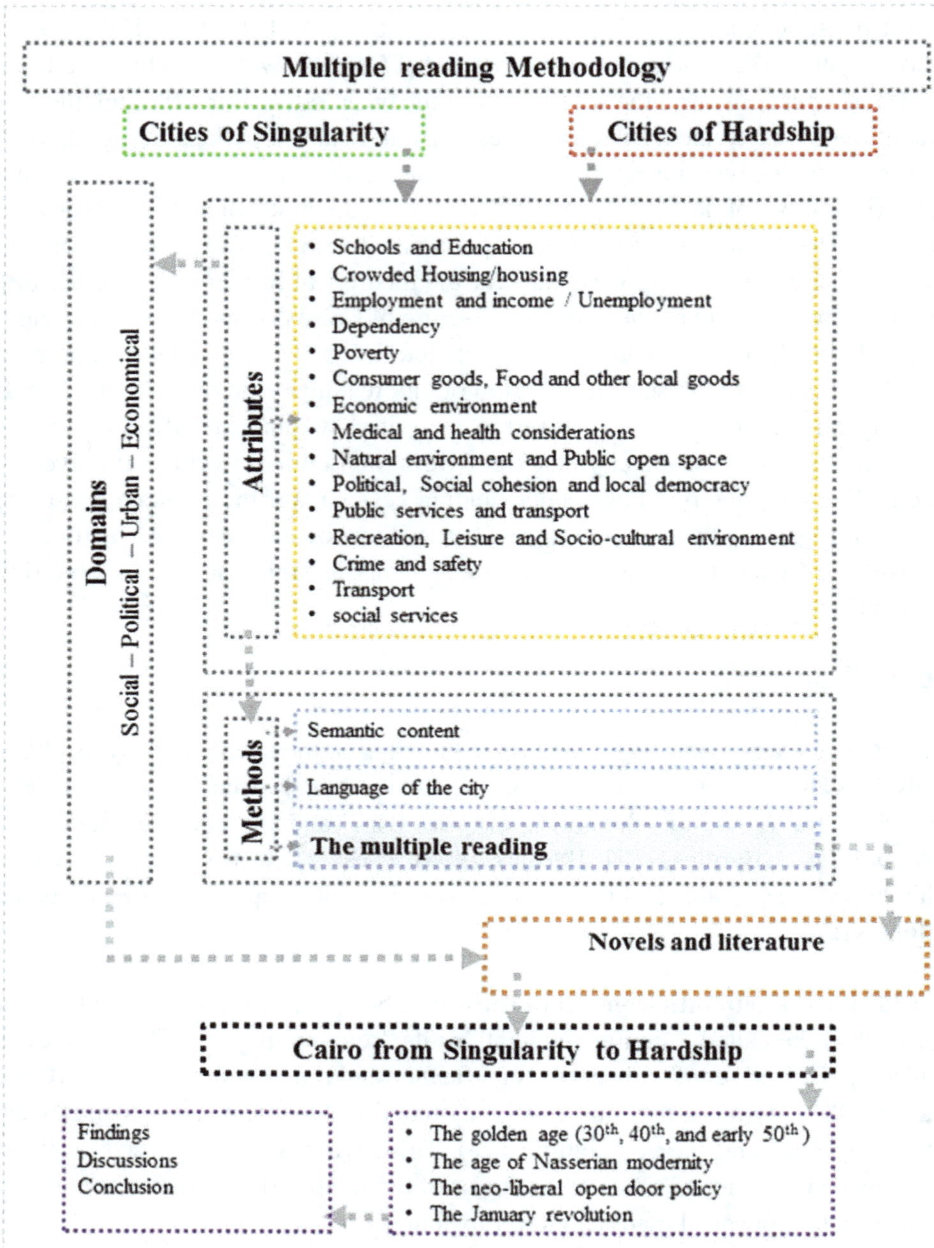

- **The third idea is according to the multiple readings**. It is a method that consists of readings coming from various categories of data. By dominating all these readings, we would develop the language of the city. In this matter, the most important thing is not so much to increase the investigations or the functional studies of the city than to multiply the readings of the city. These readings of the city are the elements that reconstitute the language or the code for the city.

CAIRO DOWNTOWN

In the Arabic language, Cairo is called emphatically *Al-Qāhira* (the victorious), and *Misr* (Egypt), *Oum El-Donia* (the mother of the world). These last two names signify the intimate relationship between Egypt and its capital Cairo. It is not only the capital of Egypt and the center of government but also considered the biggest metropolitan center in the Middle East. Cairo, as a city with over one thousand years of civilization since its establishment, has passed through several changes (Raymond, 2000, p. 307).

Cairo Downtown's breaking dawn started with the beginning of Ismail Pacha's era in the late nineteenth century (1863-1879). He wanted to turn Cairo to be Paris of the east (Influenced by Haussmann's re-planning of Paris). Ismail ambitiously attempted to build a 'Paris along the Nile' inspired by his French education and preparing the city to be with a European style to host the Suez Canal opening in 1869 (Abu-Lughod, 1971, p. 102). Downtown was the product of the great westernization of Cairo, adopted the idea of modernizing Egypt. He started a massive campaign to fulfill this objective, carried according to plans by Ali Mubarak–the Minister of Public Works – (p. 105), under the rule of Khedive Ismail. Besides, Downtown Cairo was developed to showcase a modern Cairo as *the symbol and showpiece of Egypt's progress* (Raymond, 2000, p. 311). Ali Mubarak was able to complete the infrastructure and physical development of the Ismailia quarter in record time. However, the area took several years to attract residents (AlSayyad, 2011, p. 126).

The exemplary architecture of value at the time meant that it followed contemporary European architectural styles. These buildings include residential units for the upper classes of society both Egyptian and European. Later, more businesses and commercial activity filled the apartments and ground floors of downtown buildings making the area a vibrant center for the city (Elshahed, 2007, p. 24).

Cairo Downtown has wide boulevards and great squares. Its neoclassical architecture was strongly influenced by French architecture in the 19th century. This style of architecture was known as the heart of the all political functions and activities, e.g., all the ministries' headquarters, the ruler's palace, the parliament and the embassies (Bierman, 2005, p. 37). This district was considered the economic, cultural hub. The luxurious stores and all the famous hotels were located and had witnessed many remarkable sociopolitical events.

During the rule of Ismail's son; Khedive Tawfiq (1879-1892), the area witnessed a significant boom and started evolving from a well-designed yet empty residential suburb to an urban center (Elshahed, 2007; Hawas, 2001; Raymond, 2000).

Tawfiq continued his father's development dream and created the adjacent Al-Tawfiqiyya with bustling commercial zones, and governmental entities, where hundreds of the common Egyptians acceded to the area increased the economic activities as well, expressed on their ground level, and upwards into higher floors as well (Hawas, 2001, p. 28). That development endures till it reached an uttermost point during World War II, with European immigrating to Egypt, avoidance from war, where Cairo in that time was a safe paradise with a stable, strong economy for their businesses. The rapid growth of foreigners who found the modern quarter a favorable place to live and work, turning it into the cosmopolitan center of the city (Raymond, 2000, p. 86). These factors plus innovation of cinema made the downtown center of entertainment with lots of shops, cinemas and became a vibrant center of culture and shopping destination for the elite.

Cairo as a city of singularity, the residential downtown had already transformed into a business district by the 1940s with many of its residents moving out to Cairo's new suburbs seeking quiet tree-shaded streets and lush gardens.

The new wave of high-rise apartment blocks in the new modern styles was built all over the city in districts that were once filled with villas and gardens (Arnaud, 1996, p. 57). Downtown grew in density with apartments already beginning to be divided into multiples as some of its old residents moved to other districts abandoning the increasingly busy streets of the central district.

Downtown Cairo remained the center for the city's elites and the hub for commercial and cultural activities up to the establishment of the Republic of Egypt (1952). The social fabric of downtown started to change in the due to the transformation of the political and economic ideology, which caused a general shift in the social fabric of downtown dwellers sparked by the state's ideological shift from a monarchy to a socialist state. This caused a transformation in the identity of downtown (Abu-Lughod, 1971; El Kadi, 2012).

Most foreigners left the city, which leads to a significant transformation in the socio-cultural make-up of the district (Haykal, 1973, p. 109). In this period, new and iconic buildings emerged. The example is given to Mogama Building (AlSayyad, 2011). The transformation of Downtown Cairo in general through an evolving consumer culture (Abaza, 2006; Awatta, 2015)

In the 1970s with Sadat's neoliberal open-door policy, downtown buildings witnessed their worst deterioration (a city of hardship). As the national urban policy aimed at developing desert-land and to establish new urban settlements instead of conserving the existing one. These new cities attracted most of the population, leaving Cairo downtown neglected due to its acquired low socioeconomic status (Elshahed, 2007, p. 36).

By the mid of 1970s development, the central district affected the number of households and type of jobs in the area. The commercial activities had a huge increase with more technical jobs. The evolution grew increasingly popular with bargain shoppers and professionals establishing more businesses in the upper floors of downtown buildings. This transforms the area into a fully commercial zone that is crowded with activity during the day (shop-owners, residents, street-vendors, tourists), and abandoned by night (Awatta, 2015, p. 49). The neoliberal open-door policy benefited a rising Egyptian bourgeoisie class interested in brand new lifestyles away from the old established centers. Downtown was out of the conversation shifting to a city of hardship.

Concerning downtown after the 1992 earthquake, some initiatives were taken by multiple groups and organization to maintain the surviving and rehabilitate the damaged buildings. This recent wave of studying, classifying, documenting downtown structures have resulted in the elite interest in reviving old structures for new functions. They prohibit the demolition of buildings of historical or architectural significance, to claim downtown as "national heritage" (Awatta, 2015, p. 51). This resulted in the addition of over sixty buildings into the national list of historical monuments (El Kadi & Kerdany, 2006, p. 348).

Experiencing deterioration of the heritage buildings is mainly caused by lack of maintenance, the absence of cohesive management, the spread of informal street urban-markets, traffic congestion pressure and the continuing crisis of the Egyptian economy (Elshahed, 2007, p. 20). Samir Raafat (2003) comments on the state of decay of downtown buildings as the result of *the lack of civic pride*. That generation is already being replaced by the nouveau riche, the new bourgeois of today who look at downtown as just another period in the past. These cause a shifting in the grounds by making the writing of the history of modem Cairo a tricky task (Abu-lughod, 1965).

MULTIPLE READINGS

The Golden Age

The first multiple readings are starting from the 30th to the early 50th, it includes three Egyptian. These novels are *I'm free 'Ana Horra'* by Ihsan Abdelkoudous (1954), *Midaq Alley 'Zuqaq al-Midaqq'* by Naguib Mahfouz (1972), and *Part of Europe by 'Qit'a Min Orouba' by* Radwa Ashour (2003) (Abdelkoudous, 2015; Ashour, 2003; Mahfouz, 1992). The three novels have been published in different timings while addressing downtown at the period of the 1930s. These novels have selected because they give an insightful vision in terms of social, cultural, economic, urban, and architecture views.

The First Novel: I'm free 'Ana Horra'

This novel revolves around a rebellious college girl, who was living with her aunt's family in the old Cairo, then she was transferred to live with her father in Cairo downtown as she chooses to pursue her studies at the American University of Cairo (AUC) there.

At the AUC Amina—which is a traditionally female name for former women in Egypt and means trustworthy or faithful—met a cosmopolite student's society which represents a societal shift that is contradicting with her old Cairene origins but ensures the personal liberty as she argued.

Although, Amina, the heroine of the novel, felt frustrated by the masculine dominance in society and had many experiences with men who were trying to limit her ambitions and rebellions. Throughout Amina's feelings, aspirations and fears, the novel shows a part of the feminist movement among the Egyptian society in the first half of the 20th-century.

From Amina's point of view, the feminist fight against the traditional mood lifestyle is considered the right way to gain societal freedom and valuable equality of rights. This fight between the rebellious girl and the conservative society was spatially represented between Cairo Downtown Street (Al-Bursa) and the AUC campus with the cosmopolitan culture and lifestyle as a symbol of a personal freedom in front of Al Khaleej Street in Old Cairo and its tramway seen by Amina as a giant snack flopping in the dirt. The wideness shows the apartment of Amina's father in Borsa Street with the characters of the cosmopolite society of Cairo Downtown. That was clearly described in this novel and the liberal educational atmosphere of the American University in Cairo that was in Tahrir Square.

The review, also, provides much more of "the personal liberty" against the "old fashioned oriental tradition" as well as the "Religious Intolerance". The goods consummation of Cairo Downtown through its great shops and the socio-cultural environment are provided by the constitution of the opera house and the coffee shops located in Al-Ismailiah district were observed by Al-Nazir (Ashour's narrator in this novel. So, many of the singular cities criteria were reflected.

The Second Novel: Midaq Alley 'Zuqaq al-Midaqq'

This novel describes a whole lifestyle in one of the alleys probably located in old downtown during the 1940s. The alley's houses and shops were detailed as well as the life of its people, whose Hamidah (the novel heroine) attributes. This novel represents a microcosm of all the specters of Egyptian society on the threshold of modern Egypt a while ago before the end of the monarchy.

Hamida is an ambitious young woman who found herself in the Midaq Alley with all its poverty, sadness and dirt. She decided to move from the alley. Hamida – whose name signifies "thankful" – is malcontent from her present and she wanted to be like women in the city with all their richness and elegance, surrounded by the luxury that she targeted as a goal for herself to achieve.

The Khedival Cairo or the European district was the incarnation of Hamida's dream, she followed "Ibrahim Farag" a middle-class draper and a pimp who recruit her as a prostitute, and latterly she changed her name to "Tity". The dual point of view between the city with all its temptations in front of the alley with all its ethical values are described in the novel, to recognize the wide gap between two worlds cohabited in the same great city.

The Third Novel: Part of Europe 'Qit'a Min Orouba' [2]

The novel shows through a detailed description for downtown's significant events that modified its physical and social morphology as well as it provides documentation for the evolution of some public spaces and buildings that were represented throughout the novel as silent heroes who witnessed those shifts along the city dramatic pathway from the singularity to the hardship.

Through the glasses of the Gazer or Al-Nazr, who is Ashour's narrator and main character, both singularity and hardship are signified, passing through Cairo Downtown since its foundation during Khedive Ismail's period till the downtown famous manifestations against the US war on Iraq (2003) which was the last sounding event before the novel's publication.

The main idea is a dialogue between a 65 years old Cairene, the Gazer who is born before WWII and is trying to write a book about the khedival Cairo, and his grandchild Shahrazad. The choice of the dialogue partner who inspired by Scheherazade the famous female character of One-Thousand and One Nights and storyteller in the Middle Eastern literature is assuring the story telling's mood of the novel.

Part of Europe is a nocturnal literature piece of work. The Gazer narrated downtown as remarkable personalities or families: Ismail Pacha, Sidnawi, Behler, Thomas cook, Shecorel, whom names remained in the collective national Egyptian identity, as well as the stories behind those names that are told through Downtown's great buildings nostalgically in a dual mood that is swinging between past and present.

By this dual treatment for the city's first transformations, this novel describes how the city moved from being Ismail's utopia in creating a European city into a dystopia of postcolonial underdevelopment and defeat. The scene witnesses several events, e.g., the Cairo Fire in 1952 to the opera big fire in 1971. The scene also witnessed the students' manifestations against the British force's occupation in February 1946 till the students' manifestations against the US invasion of Iraq in 2003.

For the reflections of this multiple readings approach (Figure 3), Cairo Downtown at that period before 1952 seemed to have had its golden age of singularity. Regarding the criteria (see Figure 1) and throughout the analysis of the Egyptian literature during this era, even from novels that were published many decades later as *Part of Europe*. This also documents the scene starting from the quality of life in sheriff Pacha's Street and the easiness of transportation with the clearness of urban public spaces and buildings qualities *Midaq Alley*.

Figure 3. Cairo downtown from the 30th to the early 50th.
Source: The authors

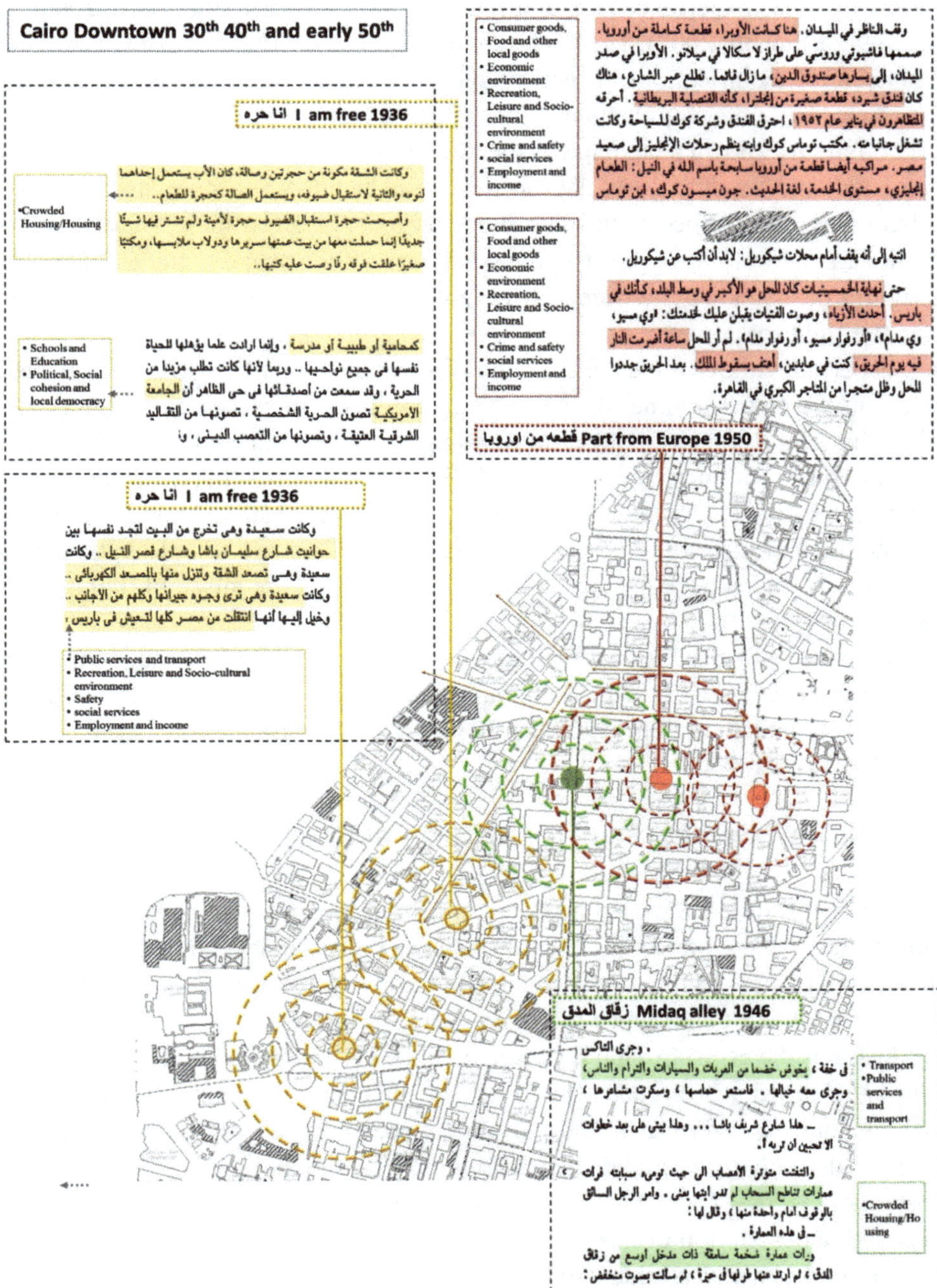

The Age of the Modernity

From the 1950s, Egypt had experienced a major political and societal shift (Abaza, 2006; Abu-lughod, 1965; Arnaud, 1996; Mitchell, 2002). The departure of the upper classes from Downtown had provoked social and architectural modifications. These neighborhoods subsequently opened to the middle and lower social classes. These developments had influenced the city visually and functionally during the 1960s. This is commonly known as the age of the modernity. The new era's architectural perspectives were based on economic construction. Flat facades, straight lines, and pure shapes were chosen for the concrete buildings embellished with only minor ornaments. Even these were a mass production technique. On the urban level, the authorities removed many of the squares' statues. The streets, named after famous figures of the monarchy were now re-labeled after ministers of the current government or former allies in the fight for power.

However, this period witnessed the construction of several landmarks in Downtown. New buildings started to appear in the Cairo skyline. Amongst them are large scale projects, such as the League of Arab States Building, the Arab Socialist Union Building, the Television and Radio tower, and the Cairo tower, the Ramses Hilton and more recently, the Ministry of Foreign Affairs. All these changes define the future role played by the city of Cairo as the heart of the Arabic national hub. This role is helped by displaying a contemporary modern image of a city. These buildings are what Egyptian engineer Milad Hanna refers to as "the golden peel that covers the surface, [hiding] the miserable conditions of popular areas" (Elshahed, 2007, p. 39).

The period between 1950 and 1960s is read socially and architecturally through three novels. These novels are *Don't Set the Sun Off 'La Tutf'e al-Shams'* by Ihsan AbdelKodous (1961), *That Smell and Notes from Prison* by Sonallah Ibrahim (1966), and *As Doha Said* by Bahaa Taher (1985) (AbdelKodous, 1960; Ibrahim, 2013; Taher, 2008).

The First Novel: Don't Set the Sun Off 'La Tutf'e al-Shams'

The novel describes life for a traditional aristocratic Egyptian family. It shows the differences in the way of thinking of members of one family regarding issues such as independence, the self-centrism and restricting social conventions. Behind the scenes, the family is a part of a society that represents the country during a critical phase in Egypt's modern history.

The novel illustrates the deep struggle between the old conservative mindset of the Egyptian society represented through Ahmed, who argued against the new liberal independent minds of his young brother Mamdouh and his younger sister Layla. The novel describes the inner fight that began inside Ahmed after the death of Mamdouh in an accident that becomes the turning point in his life. This coincides with the nationalization of the Suez Canal (1956) which was a critical moment in the recent history of Egypt. The novel ends with a description of how the triumphal new social independent mindset becomes that new sun alluded to in the title also related to the entire Nasserian Egypt.

For the luminous Groppi that it is "a humid, dark and silent place for the elderly." The protagonist's feeling is resumed seeming that "there [In Groppi] is the life's final stop." This emotional state of mind for the protagonist had transformed while describing the 'pleasant' crowd in front of the sandwich restaurant which – for him – is the top the livability and liberty! The socio-cultural environment was shown again through the strangers in Lappas café and bar; the clients were discriminated by the protagonist as (half strangers).

The Second Novel: That Smell and Notes from Prison

This novel describes a state of confusion and loss during the late 1960s, the chaotic downtown streets with the sewage crisis and the unsupportable smell. In this novel three image of Downtown during the late 1960s. A different picture of the transportation in downtown was illustrated in Figure 4 for a hazard meeting between the tramway – where the narrator was sitting – and a military train full of enthusiast soldiers who were coming back from Yemen's war. Nasser supported the Yemen revolution with 70,000 Egyptian troops and weapons which provoked an economic and military crisis in Egypt and paved the way for a resounding defeat in the six days war 1967. Between the enthusiasm shown by the returned soldiers from the train and the indifferent looks of the tramway passengers, it can be noticed the beginning of the societal crisis and the downtown began its hardship and difficulties.

The novel shows how Cairo and its Downtown as deeply absorbed in a very complicated situation by the end of the 1960s. The disappointing crisis totally consumes the city from Ibrahim's point of view. The city is diagnosed with symptoms of ruin, corruption, and collapse. A fascinating painting of loss is drawn in this novel further portraying the social malaise and social indifference for the -apparently-exploding political situation. A state of fear from the pervasive oppression is been merged with and is reflected in -the Egyptian realities of everyday life.

The Third Novel: As Doha Said

Bahaa Taher's novel is about the influences of July 1952 on Egyptian society. Through the personalities of the novel *As Doha Said*, the narrator himself and their loud or inner conversations. We can understand a part of the transformations that occurred in the Egyptian social morphology as well as the physical entity of the city of Cairo notably downtown. Our narrator is a late-thirty man who is working as a government employee. Doha is his attractive married friend as a rebellious woman who relative to him. She soon kicked him out. The novel time is the 1960s, which gave a clear image regarding the Egyptian society in the transitional phase between capitalism and socialism.

The qualities of public space and the recreational and leisure environmental were deeply changed according to the new social mobility. Those societal and spatial modifications were described through two public spaces in Cairo Downtown. Groppi (Tea salon at Talaat Harb Square) and a new sandwiches restaurant (Beans sandwich apparently). The first was café Groppi, since 1909, it was a cosmopolitan crossroads café, a meeting place for the upper class as said his owner Marco Groppi "Cairo at that time was at the center of the world, and Groppi was at the center of Cairo." Other words were provided by the same author concerning the socio-cultural environment of Groppi during the 1960s.

In this, the protagonist said that Groppi is "a country that doesn't belong to his country and there. A different living nation from his own that speaks a different language." Unlike this description, a feeling of a certain level of recreational and leisure experience from the author perspective, his protagonist was standing up in a long queue in front of a cinema. He began to describe the neck back of a dark-collar formal official who has already 6 children and an evil wife as the protagonist imagined and give an account about the new social class inhabiting Cairo Down during this era.

The author described Cairo Downtown notably in the last traces of its cosmopolitan society through the conversation between the narrator and Doha. The consumer goods, the socio-culture environment, and public spaces were shown through the duality of the "espresso" café where the owner in his shop abounded by the clients. The situation was looking sadly to Tala'at Harb new statue that replaced Soliman Pacha statue and unlikely people crowded the restaurant displays low prices bean sandwiches. Meanwhile, the radio programs are running some national songs that exalt Nasser and his era [!] the narrator commented "I don't understand Doha and myself…I understand nothing, and it is better to stop thinking about everything!"

Doha described a critical look for the new public buildings. The example is given to Hilton Hotel in Tahrir Square. The building looks like "a blue pie" that desecrate its neighbors that are recognized by the Egyptian Museum through its location and the hieroglyphic symbols drawn on its façades. The level of safety for women wearing newly the miniskirts were equally criticized in Kasr El Nile Street by mentioning "the ugly comments by the pedestrians and the people inside their cars." So, this showed the new class's behavior against women as well as personal liberty.

The hardship of Downtown was equally shown through the explosion of the infrastructure when the narrator described the smell "insupportable" in Soliman Pacha Square due to a sewage explosion that had been provoked by a social breakdown that could not be resisted by the current infrastructure system. The narrator reflected this urban explosion through the recreational and leisure activities notably the cinemas. The crowded cinema lobbies and in front of cinema halls where the majority presented "comedian movies" of Downtown. These long queues made the narrator wonder that "the current world is really – for him – "crazy"! These thoughts are reflecting a disappointing and absurd point of view for the current society living in a hardship city.

The Open-Door Policy

Another major shift that affected downtown was after Egypt's crossing Suez Canal in October 1973. The political and economic paradigms were restructured in the period of the seventies and eighties. One of the consequences was the infitah or the open-door policy which relied more on private investments, domestic and foreign, provoked new social mobility. The lower social class but high financial were affected by the behavioral setting of the neighborhood neglected already by both the landlords and the government. Lax enforcement of laws and regulations gave way to the entry of commercial establishment to invade the neighborhood urban and architectural form/ configuration into the neighborhood, mostly with no regard to maintaining aesthetic harmony or preserving the historic buildings, the community transformed totally to hardship after its glorious past.

This dramatic evolution was observed in two novels to show two images from Cairo's Downtown concerning societal and urban shifts. Ibrahim Aslân (1983) transferred the first image in *Sad Malek 'Malek Alhazaeen'* (Or the Sadness of the Heron) through the student's manifestations in 1972 (the dignity manifestations) in downtown against the state (no ware no peace). The second image offered by Ibrahim Issa in *Patriotic Ghosts 'Ashbah Wataneya'* (2013), (Aslan, 2009; Essa, 2013).

The First Novel: The Sadness of the Heron 'Malek Alhazaeen'

This novel discusses the influences of the open-door policy (Infitah era) on Egyptian society during the 1970s and the 1980s. It gives a discretion of the informal settlements in Cairo as well as the transformations in those neighborhoods physically and socially. Through the novel hero's (el Sheikh Hosny), Youssef el Naggar and others gave an image of the Egyptian society (the students' manifestations of dignity; 1972) and the beard riots manifestations (January 1977). Students' movements were an essential aspect of contentious politics on the streets. They were able to mobilize others to their cause and were

Figure 4. Cairo Downtown the 50th and 60th.
Source: The authors

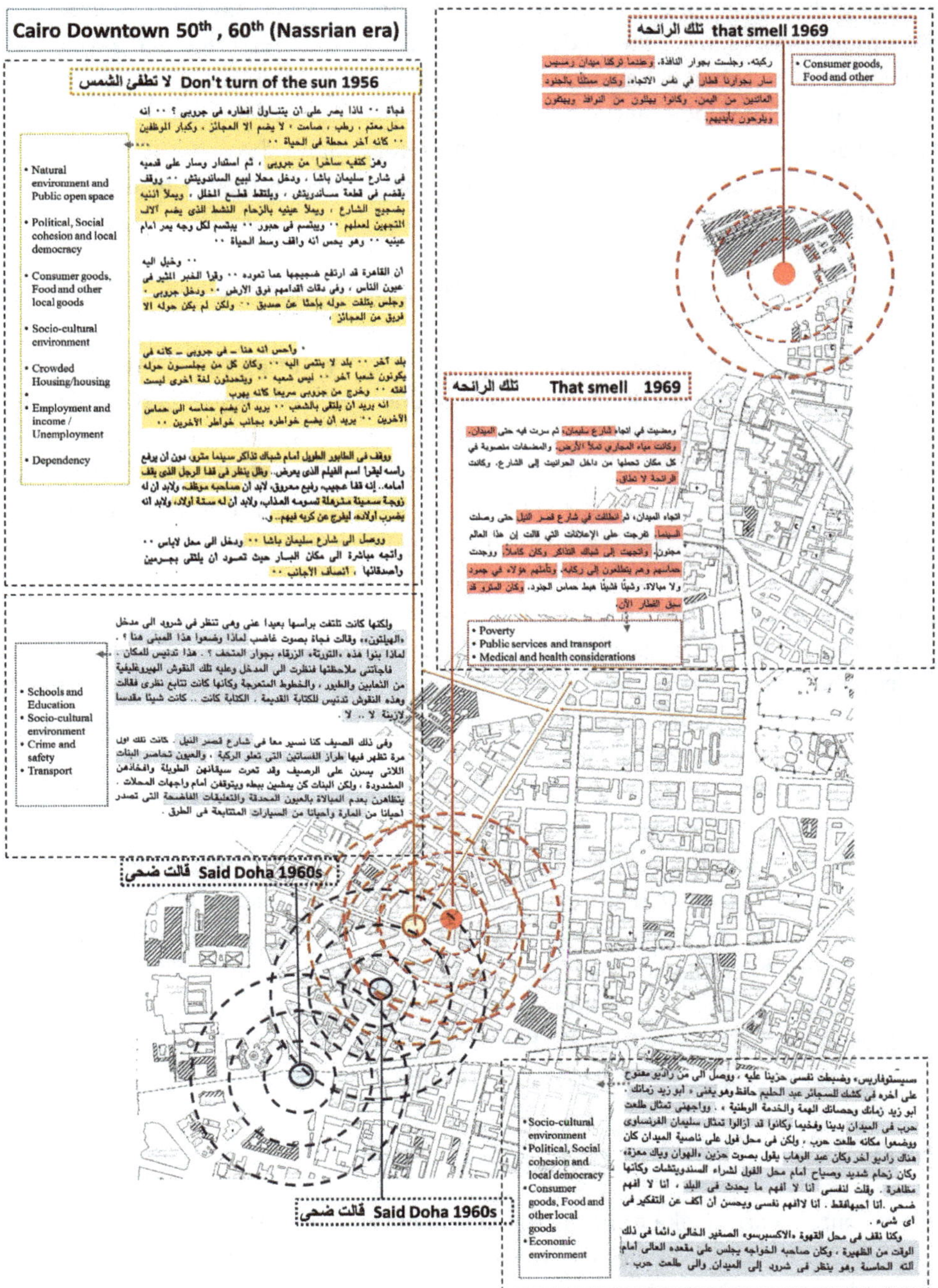

able to exert political pressure on the ruling regime (Hatem et al., 2016, p. 1-3)." Two of these manifestation movements were illustrated by Aslan and described a level of the socio-political environments and safety in Downtown streets and squares, i.e. Talaat Harb, Ahmed Orabi and Tahrir squares. The crowded streets by the demonstrators during the students riot 1972, and the bread riots 1977 against the poverty and corruption showed the social cohesion between the middle and low classes during the Sadatian era. The darkness, the ruins, smokiness and the fire illustrate the status of a country in full transformation as well as the rage and the political oppression. The narrator Aslan's describes Cairo Downtown streets during the most significant changes of the Egyptian society came in that as results of the open doors policies of al-Infitah (Figure 5).

The Second Novel: Patriotic Ghosts 'Ashbah Wataneya'

The novel describes the new parasitic class in Egyptian society during the 1980s and the 1990s. The appearance of this class was the resultant of the infitah era. Some of those businessmen gathered their wealth from suspicious sources as they tried to be engaged in politics to protect their wealth and their families. The novel heroes are represented as a real example of the Egyptian society such as businessmen, journalists, lawyers and opponents who participated in the system corruption. Those national ghosts draw an image of a hardship society. The pathway was chosen by them as full of suspicious deals and scandals. For Issa's point of view, it was recognized how corruption was inherited by a second generation from their parents. Through tracking the two generations, it is concluding how the Egyptian society as well as the physical aspects of their lives in the city (notably Cairo downtown), acted as the main events' theatre.

The novel describes the consumer good and the sociocultural environments under the new system (the open-door policies) illustrating Shawaraby Street. The street during this era was "*Al-Shawarby became a metaphor for infitah—quite literally a polluted landscape of dreams betrayed.*" The transformation of the street comes from a calm street till the mid of the 1970s to a shopping hub that contains all western goods that reflect the new characters of the downtown society. The new society aspirations are in the speed wealthy career as well as the power and the need to achieve a certain level of modernity by imitating western life modes. "*Al-Shawarby* Street" described by Issa is a place where "*people wanted to run away from an austere and coarse reality to a broad and consumption one, this world accomplishes the need (the thirstiness) for the showing graces and taking pride in the wealth*". The street shows cases full of dresses are dazzling the class who is aiming to the social promotion and the ascension careers. The shopping activities in "*Al-Shawarby* Street." This street became a vital characteristic of the "*inifitah*" era with the extremist cultures, the bribery, and corruption.

January Revolution

Since 1992 from Cairo's earthquake until the revolution erupted on 2011, the Egyptian government revived many initiatives for the conservation of the architectural heritage. Thus, many studies and projects had been launched to bring Cairo Downtown to its glory in accordance with many literary writings. These writings are *The Yacoubian Building 'Imarat Ya'qubyan'* by Alaa El Aswany (2002), *Saleh Heisa* by Khairi Shalabi (2000), and *Cairo 104* by Doha Assy (2006) (Aswany, 2004; Assy, 2006; Shalabi, 2008), as Figure 6 shows.

Figure 5. Cairo downtown, the 70th and 80th.
Source: The authors

The First Novel: The Yacoubian Building 'Imarat Ya'qubyan'

This novel describes the lifestyle in an old building in Downtown. The Yacoubian building was built in 1934. It was an architectural piece of art that witnessed, through the stories of its inhabitants which presented most of the remarkable moments of modern Egypt history. The building story is meanwhile the story of a country. The social mobility's phenomena were offered in the novel heroes and personalities:

Figure 6. Cairo downtown, the 90th, and 2000th.
Source: The authors

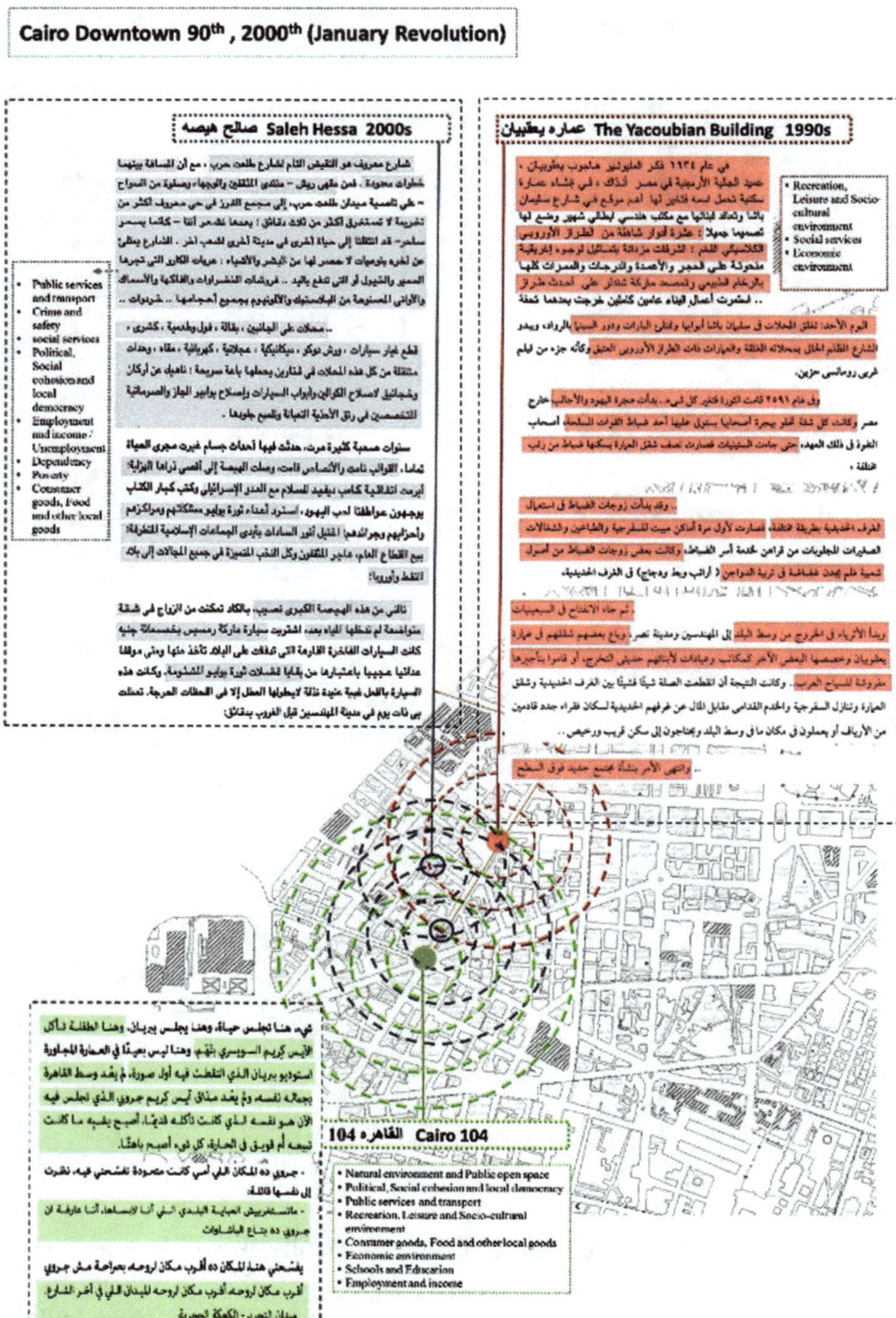

Zaki el Desouky, Taha, Boussayna, and Azzam. Every hero is a representative fora different Egyptian society spectrum. All the events that happened inside and outside the building are considered as metaphors for societal mobility as well as the spatial evolution during the second half of the 20th century.

Zaki ElDessouki, the protagonist of Yacoubian Building (Imârat Yacoubian), is telling to Bothayna the story of those old days in Cairo downtown. He resumed the current situation regarding downtown history. He describes these days as were different. In this, Cairo was like Europe. *It was clean and smart, and the people were well mannered and respectable, and everyone knew his/her place exactly.* Every night was filled with laughter and parties and drinking and singing. There were lots of foreigners in Cairo.

The Second Novel: Saleh Heisa

Khairy Shalaby's novel discusses the most significant periods in Modern Egypt after the July 1952 and the transformations that occurred among the Egyptian society between the Nasserian socialist politics and the infitah ones. The main subject is Saleh Heisa, a drug (hashish) dealer who represents the fall of the dream and the current confusion for the whole society. A silent hero was present in Saleh Heisa, the Ma'arouf neighborhood in Cairo downtown which is characterized as special "popular" spatial experiences unlike other spaces in Downtown. Although the drugs and the unconscious state of most of the heroes, this particular milieu witness the stimulating events in the country history. The British occupation, the July 1952 political revolution, then the defeat of the six days war, the partial victory of October 1973 and the infitah era passing by the initiation of President Sadat's 1977 trip to Jerusalem and the peace treaty.

Khairy Shalaby showed another perception of Cairo downtown during the decades 1980s and 1990s. Maarouf district is a popular district located in Cairo downtown, the district is unplanned and represents – according to Shalaby- the absolute opposite of the downtown environment because of its unique characteristics which can be found with new moods of life: "another life and another people!" The good consumption, as well as, the transportation was shown in the district description by the narrator "from the donkey/horse cars, handcars vegetables, fruit, and fish for sale; plastic and aluminum tableware in all shapes and sizes spread out over Blankets (Mehrez, 2011, p. 37). This image of Maarouf district shows a consumer society during the open-door policies era as well as a social cohesion but in a way that shows the middle and lower class's struggle against poverty and high prices. The narrator described the hardship through a series of political transformations. Camp David peace treaty, the pro-normalization writings, the return of the old bourgeoisie or the monarchist classes after Sadat's new policies. The extremist cultures and Sadat assassination by extremist military groups, societal transformations had affected Egyptian society.

The Third Novel: Cairo 104

This novel gives an account of Egyptians social life since the 1950s till the first decade of the twenty-one's century. Through the novel personalities, notably Inshraah Eweidah a poor none-educated woman, the novel represents the sense of a societal evolution that occurred during the second half of twentieth's century. An emotional, psychological struggle was described and given rise by the narrator, in order to present the current conflicts between the Egyptian's dreams or ambitions and their realities that block everything. Egyptian society as an accumulation of mixed cultures was equally shown through this societal painting with a web of interlocking relationships that are intersecting by Inshraah's circle.

This painting is considered as a microcosm of Egypt's spectra that combines between cultures and philosophies, fictions and realities. It mixes between dreams and facts. The nostalgia to the "belle époque," when Cairo Downtown was having its greatest days of glory. The city users discuss goods consumption in Groppi. Notably, the Suisse ice cream that the taste today has not the same as before. He added, "everything became pale"! For the narrator, Cairo Downtown of yesterday has the smell of old memories, but everything has been changed to the point that "the Cairo Downtown doesn't conserve the same beauty!"

FINDINGS AND DISCUSSION

The layers of Cairo Downtown's transformation are complex, create a debatable reflection of the multiple readings. "Once the rather exclusive domain of the 'colonial' foreigners and the Francophilic Egyptian elite (the district of singularity). This zone has undergone a dramatic transformation since the Revolution of 1952" (an area of hardship) (Abu-Lughod, 1971, p. 68).

The appraising of the methodology of multiple readings was an opportunity to bring new insights on how to tackle the evolution of Cairo downtown. The four phases that represented in the turning points for the history of downtown had their countless influence over the urban district including all its social, political and economic aspects. The mapping of the four phases gave an insight vision on specific critical locations and spots within the boundaries of the downtown district. There was always a drive that motivates the action, and the urban setting played the role of the platform that accommodated the event, and both affected and affected by its influence.

Political decisions usually had the upper hand in creating an image like Paris along the Nile. This vision a of Khedive Ismail was not a natural expansion of the old Cairo urban patterns, but an imposed squares and boulevards. There was nothing related to the existed previous fabric of Cairo. His vision has its ambitions and drawbacks. On the scale of tracing the hardship multi-attributes, it is quite apparent in the rise of individual variables. That reclaim downtown at that time as a district of singularity, such as education (shown by the multiple readings through the analysis of the novel *I am as free as liberty*). This novel describes the buildings of American University in Cairo and the importance of education as a tool for self-expression. This led the analytical study to focus on important social and urban spots that flourished in certain periods and faded in others.

The first phase is representing Cairo Downtown as being in the golden era (from the 30th to the early 50th): This era witnessed details narrated in books and novels simulating events. It is also describing the various locations and spots, freezing the moments, mixing realities with fantasies and fictions (Tahrir Square, Qasr El Nile Street, Soliman Pasha Street. The novel *I am as free as liberty*. The old opera house was also mentioned in this novel (el Opera Square; the novel "part of Europe"). The quality of transportation, with the new Highrise Buildings. The architectural styles of Sheriff Street also were mentioned in the novel *Midaq Alley*.

The second phase; the modern era (1952-1970), downtown starts gradually to lose her sense of place, and once again the new political vision of Abdel Nasser has another plan for Cairo, giving her back to downtown. Focal spots are demonstrated (Ramses square; the novel "that smell" the chaotic downtown streets with the sewage crisis and the unsupportable smell). For a very dispirit description for Groupie located in Talaat Harb Square, the novel "do not turn off the sun" gives a deep struggle. It was clearly showing the old conservative mindset in Egyptian society against the new liberal independent one. Once again, Tahrir Square was described in the novel by Doha. The change in the socio-cultural environment, the lyrics of AbdelHalim songs, the full groups at the local beans cuisine beneath the large statue of Talaat Harb replacing the old one of Soliman Pasha el Fransawy).

The third phase of the open-door policies (the new liberal era) (1970-1981), a composition of urban These socio-economic transformations emerged in the insurgency on every prototype product introduced in the previous era, for instance, reject the locally-made. Some disadvantages of this phase are the rise of tiny pathways for the international brands, like "*Al-Shawarby* Street"; as in the story patriotic ghosts. This approach based on the rejections of the standardization, where shopping in this Street become a sign for wealth and richness. In other words, this paradigm shift had a great deal of influence on downtown, particularly in commercial activities. Where the lower floors of the buildings turned into shops, the small pleasant walkways, and passages turned into street markets full of vendors, practicing all sorts of a district of hardship. The shift in the Egyptians' mentality changed the economic environment, the process of employment and income, the consumer goods, and the socio-culture in general.

The other example is "*Maarouf* Street", which described in the novel *Saleh Hessa*. This street becomes a very crowded street urban market, all kinds of goods and products, workshops for iron and metals within the elite district of downtown, a parallel universe in a couple of minutes walking distance from Talaat Harb Street and café rich. As if, all the basic original authenticity has collapsed, and another uncivilized layer is coming up to the surface.

The following graph (Figure 7) represents the significant shifts of the turning points. Their locations on the vertical level represent the hardship through the index of multi-attribute. On the horizontal level, the multiple readings tool is represented in novels and literature. Within each era, the reflections of the multiple readings elaborate a kind of impact that transformed downtown from a district of singularity to an area of hardship.

Figure 7. An analytical graph of the major evolutionary tuning points.
Source: The authors

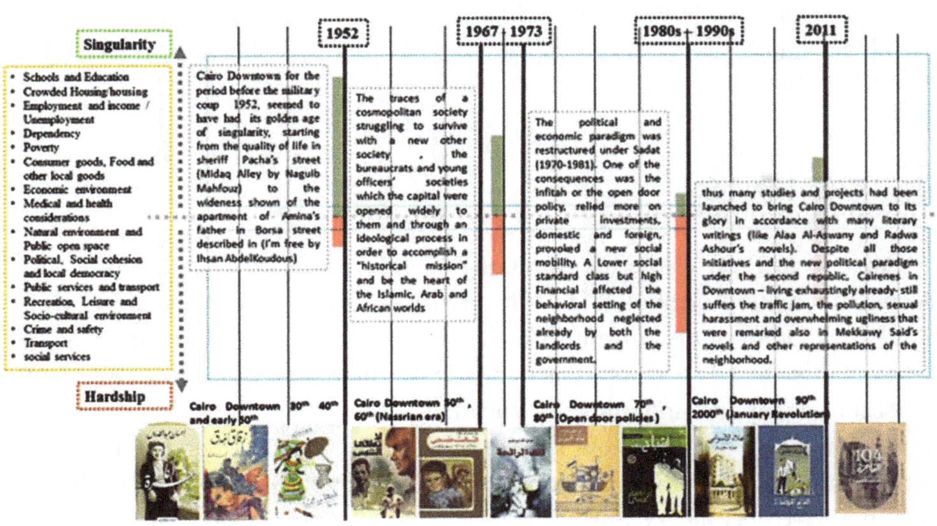

CONCLUSION

From the origins of the modem city district to its ongoing morphisms, downtown has been reflective of the city's social, cultural and economic shifts. These transformations came in many forms including the general urban expansion of highways through the city center. The settings have many forms of architectural buildings constructed in the early twentieth century. The moving in and out of institutions such as the Hilton Hotel, the American University in Cairo was built in later phases of development.

The multiple readings technique that has been adopted within this chapter, in the decoding of its characteristics. This has analyzed the moments of transition and their effect on the urban landscape of downtown Cairo and their cultural, social and historical significance. Throughout the political and economic evolutions. Downtown Cairo has moved from the ordered early modernist urbanism to redefine its experience, aesthetics, boundaries and scale as a city of singularity (where streets were visible in peoples' fashions, hairstyles, and choices of leisure activity, Café culture translated from its European context into downtown Cairo where urban dwellers went to see and be seen) to a city of hardship.

Cairo downtown, now, is the product of the fusion of diverse areas that were separate sectors in the past but have become central districts of today's metropolis. Attempts to decentralize superior tertiary activities towards the east or west have not however altered the position of the core, which has continued until now to retain most of its command/ administrative functions inherited from the monarchy and mandate periods.

What gave downtown the essence to become a district of singularity are qualities that had been embraced, and sometimes fade and should be revived. It is not merely a naive move to fulfill the lack of services and strive for maintenance (which is mandatory after all), but a vision to restore those characteristics. It is about public awareness that should come from a strong base of knowledge and education, scholarships that would give more exposure to a fruitful learning environment. True liberty of expressing and exchanging visions and dreams. A cosmopolitan district that hosts each and every nationality disregarding color or religious, encouraging investments, and creating a sense of belonging. In conclusion, the purpose of the present work is to focus on the core cause of the evolution and its influence on the social, urban, political and economic domains, in order to re-read Cairo Downtown morphisms through the multiple readings approach of novels and literature to de-code the transformation from a new perception of facts and fictions.

ACKNOWLEDGEMENTS

Our sincere thanks go to Prof. Catherine Maumi, Professor of History and Theories of Architecture and Urban Forms in Grenoble School of Architecture. Her support goes beyond her work as the PhD supervisor of the first author of the current work. The authors, here, would like to thank her for permitting them to extract this work from the thesis in progress under her supervision. We would also like to acknowledge her for her support and immense knowledge.

Foremost, we would love to express our sincere gratitude to the editors and anonymous reviewers of this book for their patience, motivation, encouragement and insightful comments. Without their support, this work would not have become a reality.

REFERENCES

Abaza, M. (2006). *The Changing Consumer Cultures of Modern Egypt Cairo's Urban Reshaping: Social, Economic and Political Studies of the Middle East and Asia* (Vol. 101). BRILL.

AbdelKodous, I. (1960). *Don't Set the Sun Off.* Cairo: Library of Egypt.

Abdelkoudous, I. (2015). *Ana Horra*. The Egyptian Lebanese Publishing House. (in Arabic)

Abu-lughod, J. (1965). Tale of Two Cities: The Origins of Modern Cairo. *Comparative Studies in Society and History*, *7*(4), 429–457. doi:10.1017/S0010417500003819

Abu-Lughod, J. (1971). *Cairo: 1001 Years of the City Victorious*. Princeton, NJ: Princeton University Press.

Abusaada, H. (2017). 'Singular Urbanism' as an Urban Design Trend to Create A 'Trans-Paradigms City' Approach. In *An International Conference on Urban Design, Cities of Good Intentions.* Rome: Sapienza Università di Roma, School of Engineering.

Abusaada, H. (2017). The untold story of the urban design practice of new cities in Egypt. In H. Elsharkawy, S. Zahiri, & J. Clough (Ed.), *International conference for sustainable design of the built environment* (pp. 787-798). London, UK: SDBE.

Abusaada, H., & Elshater, A. (2017). *Back to the Order: The Chaos in the Cities of Hardship*. Saarbrücken, Germany: Lambert Academic Publishing.

Abusaada, H., & Elshater, A. (2019). From Chaos to Order: Articulating the Urban Policies for Cities of Hardship. In Industrial and Urban Growth Policies at the Sub-National, National, and Global Levels (pp. 41-64). IGI Global. doi:10.4018/978-1-5225-7625-9

AlSayyad, N. (2011). *Cairo: Histories of a City*. Cambridge, MA: Harvard University Press. doi:10.4159/harvard.9780674060791

Andersen, M. (2018). *Vienna Tops Mercer's 20th Quality of Living Ranking*. Retrieved from https://www.mercer.com/newsroom/2018-quality-of-living-survey.html

Arnaud, J.-L. (1996). *Cairo: The Development of a Modern City (1867-1907).* Paris: CNRS.

Ashour, R. (2003). *Qit'a min Urubba: Riwaya [A Part of Europe: A Novel].* Cairo: Dar Al-Shorouk. (in Arabic)

Aslan, I. (2009). *The Heron* (E. Colla, Trans.). Cairo: The American University in Cairo Press.

Assy, D. (2006). *Cairo 104*. Amman, Jordan: Dar Al Yasmeen Publishing and Distribution.

Aswany, A. E. (2004). *The Yacoubian Building 'Imarat Ya'qubyan* (H. T. Davies, Trans.). Cairo: The American University of Cairo Press.

Awatta, H. (2015). *The Urban Transformation of Downtown Cairo between State and Non-State Stakeholders* (Master's thesis). Cairo: The Department of Sustainable Development, The American University in Cairo.

Barthes, R. (1971). Sémiotics and urbanism (in French). *L'Architecture d'Aujourd'hui*.

Barthes, R. (1994). The Semiotic Challenge First Edition Thus Edition (R. Howard, Trans.). University of California Press.

Bierman, I. (2005). Disciplining the Eye: Perceiving Medieval Cairo. In N. AlSayyad, I. A. Bierman, & N. Rabbat (Eds.), *Making Cairo Medieval*. New York: Lexington Books.

Bjorkdahl, A., & Kappler, S. (2017). *Peacebuilding and Spatial Transformation: Peace, Space and Place*. London, UK: Routledge. doi:10.4324/9781315684529

DictionaryC. (2019). *Singular*. Retrieved from dictionary.cambridge.org: https://dictionary.cambridge.org/dictionary/english/singular

El Kadi, G. (2012). *Le Caire - Centre en mouvement (in French)*. Collection Petit atlas urbain.

El Kadi, G. E., & Kerdany, D. (2006). *Belle-époque Cairo: The Politics of Refurbishing the Downtown Business District in Cairo Cosmopolitan*. Cairo, Egypt: The American University in Cairo Press.

Elshahed, M. (2007). *Facades of Modernity, Image, Performance, and Transformation in the Egyptian Metropolis* (Master's thesis). Massachusetts Institute of Technology.

Essa, I. (2013). *Patriotic Ghosts*. Bloomsbury Qatar Foundation Publishing.

Hawas, S. (2001). *Khedivian Cairo: Identification and documentation of urban architecture in Downtown Cairo*. Cairo: Arab Contractors Press.

Haykal, M. ò. (1973). *The Cairo documents: The Inside Story of Nasser and His Relationship with World Leaders, Rebels, and Statesmen*. Doubleday.

Ibrahim, S. (2013). *That Smell and Notes from Prison* (R. Creswell, Trans.). New York: New Directions.

Lowe, M., Whitzman, C., Badland, H., Davern, M., Aye, L., Hes, D., & Butterworth, I. (2015). Billie Giles-Corti, Planning Healthy, Liveable and Sustainable. *Urban Policy and Research*, *33*(2). doi:10.1080/08111146.2014.1002606

Lynch, K. (1960). *The Image of the City*. Cambridge, MA: The MIT Press.

Mahfouz, N. (1992). *Midaq Ally* (T. L. Gassick, Trans.). Anchor Books Edition.

Mehrez, S. (2011). *The Literary Life of Cairo: One-Hundred Years in the Heart of the City*. Cairo: The American University in Cairo Press.

Meller, H. (2001). *European Cities 1890-1930s: History, Culture and the Built Environment*. Cambridge University Press.

Nathan, R. P., & Adams, C. (1976). Understanding Central City Hardship. *Political Science Quarterly*, *1*(91), 47–62. doi:10.2307/2149158

Ogilvie, S. (2012). *Words of the World: A Global History of the Oxford English Dictionary*. Cambridge University Press. doi:10.1017/CBO9781139129046

Raafat, S. (2003). *Cairo, the Glory Years: Who built what, when, why and for whom*. Alexandria, Egypt: Harpocrates Publication.

Raymond, A. (2000). Cairo. Harvard University Press.

Rossi, A. (1982). *The Architecture of the City* (D. Ghirardo & J. Ockman, Trans.). Cambridge, MA: The MIT Press.

Shalabi, K. (2008). *Saleh Heisa*. Cairo: Dar Al Shorouk.

Taher, B. (2008). *As Doha Said* (P. Daniel, Trans.). Cairo: The American University in Cairo Press.

Wright, D. J., & Montiel, L. M. (2007). *Divided they fall: hardship in America's cities and suburbs*. Albany, NY: The Nelson A. Rockefeller Institute of Government.

ADDITIONAL READING

Armbrust, W. (1996). *Mass Culture and Modernism in Egypt*. Cambridge: Cambridge University Press.

United Nations Development Programme. (2016). *Human Development Report*. Human Development for Everyone.

Zayed, H., Sika, N., & Elnur, I. (2016). *The Student Movement in Egypt. A microcosm of contentious politics*. Power to youth working paper. AUC Press. Cairo. Egypt.

Zincke, F. (1871). *Barham. Vicar of Wherstead, Egypt of the Pharaohs and of the Khedive*. London: Smith, Elder and Company.

ENDNOTES

1 This work is extracted from a PhD thesis in process since 2016 by Mougibelrahman Aboamer under the entitle in French "Architecture et évolution sociopolitique- vers une lecture multiple du centre ville du Caire (1952-2015)", Supervised by Professor Catherine MAUMI, MHAevt research laboratory, Ecole Nationale Supérieure d'Architecture de Grenoble (ENSAG), University of Grenoble Alpes.

2 The choice of the novel's title (*part of Europe*) is based on the famous quote of Khedive Ismail "My country is no longer in Africa; we now form part of Europe" (Ashour, 2003). The spatial incarnation of this Ismailian part of Europe was indeed "Cairo Downtown".

Compilation of References

AbadaG. (2018). Retrieved from https://www.facebook.com/galal.abada/posts/10156270326245283?__xts__[0]=68.ARAVZlYtnWxJ5Ww3hK4sqpgrzDRFFCbA-4GGR4njWL9OTzcyuSGX02GlFd-i71-fW2hJH_JHacqV2kqmBcvSP-3WOUYwF_BPy_oSCD46RG5J1m8DJNwrKnQcXJ6889exAwanRsI40J41p&__tn__=-R

Abaza, M. (2006). *The Changing Consumer Cultures of Modern Egypt Cairo's Urban Reshaping: Social, Economic and Political Studies of the Middle East and Asia* (Vol. 101). BRILL.

Abaza, M. (2011). Critical Commentary. Cairo's Downtown Imagined: Dubaisation or Nostalgia? *Urban Studies (Edinburgh, Scotland), 48*(6), 1075–1087. doi:10.1177/0042098011399598

Abd El-Moneim, N. M., & Galal, O. M. (2017). Cairene Public Realm: An Analysis of the City Public Spaces and its Quality. *1st International Conference on Towards a Better Quality of Life*. 10.2139srn.3163037

Abd Elrahman, A. S. (2016). Tactical Urbanism: A Pop-Up Local Change for Cairo's Built Environment. *Prodecia-Social and Behavioral Science, 216*, 224–235. doi:10.1016/j.sbspro.2015.12.032

Abdelhalim, K. (2010). *Participatory upgrading of informal areas: a decision-maker's guide of action*. Cairo: participatory development program in urban areas (PDP) in Egypt.

AbdelKodous, I. (1960). *Don't Set the Sun Off*. Cairo: Library of Egypt.

Abdelkoudous, I. (2015). *Ana Horra*. The Egyptian Lebanese Publishing House. (in Arabic)

Abdelmonem, A., & Galán, S. (2017). Action-Oriented Responses to Sexual Harassment in Egypt: The Cases of HarassMap and WenDo. *Journal of Middle East Women's Studies, 13*(1), 154–167. doi:10.1215/15525864-3728767

Abu-lughod, J. (1965). Tale of Two Cities: The Origins of Modern Cairo. *Comparative Studies in Society and History, 7*(4), 429–457. doi:10.1017/S0010417500003819

Abu-Lughod, J. (1971). *Cairo: 1001 Years of the City Victorious*. Princeton, NJ: Princeton University Press.

Abusaada, H. (2016). An Approach of Making an Arabian City Branding: City Imprint (in Arabic). In 5th Jordan International Architectural Conference (pp. 1-13). Amman, Jordan: Jordanian Association of Engineers.

Abusaada, H. (2017). 'Singular Urbanism' as an Urban Design Trend to Create A 'Trans-Paradigms City' Approach. In *An International Conference on Urban Design, Cities of Good Intentions*. Rome: Sapienza Università di Roma, School of Engineering.

Abusaada, H. (2017). The untold story of the urban design practice of new cities in Egypt. In H. Elsharkawy, S. Zahiri, & J. Clough (Ed.), *International conference for sustainable design of the built environment* (pp. 787-798). London, UK: SDBE.

Abusaada, H. (2018). *Sympathizing Withthe the Footpaths (in Arabic).* Retrieved from www.facebook.com/sympathizingwiththefootpaths: https://www.facebook.com/sympathizingwiththefootpaths/?modal=admin_todo_tour

Abusaada, H., & Elshater, A. (2019). From Chaos to Order: Articulating the Urban Policies for Cities of Hardship. In Industrial and Urban Growth Policies at the Sub-National, National, and Global Levels (pp. 41-64). IGI Global. doi:10.4018/978-1-5225-7625-9

Abusaada, H., & Elshater, A. (2019). From Chaos to Order: Articulating the Urban Policies for Cities of Hardship. In U. G. Benna, Industrial and Urban Growth Policies at the Sub-National, National, and Global Levels. USA: IGI Global. DOI: 10.4018/978-1-5225-7625-9

Abusaada, H. (2016). Internal Displacement as a Silent Crisis: Managing Urban Chaos and Hardship—Back to the Order. In *RSA Annual Conference: New Pressures on Cities and Regions*. London, UK: Regional Studies Association.

Abusaada, H. (2017). The untold story of the urban design practice of new cities in Egypt. In H. Elsharkawy, S. Zahiri, & J. Clough (Ed.), *International conference for sustainable design of the built environment* (pp. 787-798). London: SDBE.

Abusaada, H. (2017). The Untold Story of the Urban Design Practice of New Cities in Egypt. In H. Elsharkawy, S. Zahiri, & J. Clough (Ed.), *International Conference for Sustainable Design of the Built Environment (SDBE 2017)* (pp. 787-798). London: SDBE.

Abusaada, H., & Elshater, A. (2017). *Back to the Order: The Chaos in the Cities of Hardship*. Saarbrücken, Germany: Lambert Academic Publishing.

Abusaada, H., & Elshater, A. (2019). From Chaos to Order: Articulating the Urban Policies for Cities of Hardship. In U. G. Benna (Ed.), *Industrial and Urban Growth Policies at the Sub-National, National, and Global Levels*. IGI Global. doi:10.4018/978-1-5225-7625-9.ch003

Acheampong, A. R., & Silva, E. A. (2015). Land use-transport interaction modeling: A review of the literature and future research directions. *Journal of Transport and Land Use*, *8*(3), 11–13.

Adams, D. M., Alig, R. J., Callaway, J. M., McCarl, B. A., & Winnett, S. M. (1996). *The Forest and Agricultural Sector Optimization Model (FASOM): Model structure and policy applications.* Research Paper PNW-RP-495. Portland, OR: U.S. Department of Agriculture, Pacific Northwest Research Station.

Adams, R. (1995). Assessing the performance of the NELUP hydrological models for river basin planning. *Journal of Environmental Planning and Management*, *38*(1), 53–76. doi:10.1080/09640569513110

Afifi, S. (2014). *Spatial Planning Support Systems in Egypt, PhD.* Ain Shams University.

Agarwal, C., Green, G. L., Grove, M., Evans, T., & Schweik, C. (2000). *A review and assessment of land-use change models: Dynamics of space, time, and human choice*. Bloomington, IN: Centre for the Study of Institutions, Population and Environmental Change, Indiana University.

Aji, A., Wang, F., Vo, H., Lee, R., Liu, Q., Zhang, X., & Saltz, J. (2013). Hadoop gis: a high performance spatial data warehousing system over map reduce. *Proceedings of the VLDB Endowment*, 1009-1020.

Al Sayed, K., Turner, A., Hillier, B., Iida, S., & Penn, A. (2014). *Space Syntax Methodology* (4th ed.). London: Bartlett School of Architecture, UCL.

Al Sayed, K., Turner, A., Hillier, B., & Iida, S. (2014). *Space Syntax Methodology*. London: University College London Press.

Al-Alaily, S. A.-A. (2017). *The issue of old rents.* Retrieved April 29, 2017, from http://www.shorouknews.ccm/columns/view.aspx?cdate=15042017&id=25952aba-8542-4178-8514-0fae3cc85538

Alawadi, K. (2017). Rethinking Dubai's urbanism: Generating sustainable form-based urban design strategies for an integrated neighborhood. *Cities (London, England), 60*, 353–366. doi:10.1016/j.cities.2016.10.012

Alberti, M. (1999). Urban Patterns and Environmental Performance: What Do We Know? *Journal of Planning Education and Research, 19*(2), 151–163. doi:10.1177/0739456X9901900205

Albrerchts, L. (2010). How to Enhance Creativity, Diversity, and Sustainability in Spatial Planning: Strategic Planning Revisited. In G. C. M. Cerreta (Eds.), Making Strategies in Spatial Planning, Knowledge and Values (pp. 3-25). Springer.

Alcamo, J., Kok, K., & Busch, G. (2006). Searching for the future of land: scenarios from the local to global scale. In E. F. Lambin & H. J. Geist (Eds.), *Land use and land-cover change: local processes and global impacts* (pp. 137–156). New York: Springer. doi:10.1007/3-540-32202-7_6

Alcamo, J., Leemans, R., & Kreileman, E. (1998). *Global change scenarios of the 21st Century. Results from the IMAGE 2.1 Model.* London: Elsevier.

Alex, S. (2013). Workshop. Alexandria, Egypt SIGMA company. (2016). L. Passage re-use project (M. Magdy, Trans.). Alexandria, Egypt: Academic Press.

Alexander, C. (1965). A City is Not a Tree. *The Architectural Forum, 122*(1), 58–62.

Alexander, C. (1966). A city is not a tree. *Design Magazine, 206*, 46–55.

Alexander, P. J., Fealy, R., & Mills, G. M. (2016). Simulating the impact of urban development pathways on the local climate: A scenario-based analysis in the greater Dublin region, Ireland. *Landscape and Urban Planning, 152*, 72–89. doi:10.1016/j.landurbplan.2016.02.006

Al-Ghatam, W. (2012). Cultural movement patterns and social implications in space of villages absorbed by cities in Bahrain. In Proceedings of the 8th International Space Syntax Symposium (8210:01-31). Santiago: Pontificia Universidad Católica.

Algohary, S., & El-faramawy, A. (2010). *Egyptian Approach to Informal Settlements Development.* Retrieved October 13, 2016, from Slideshare: http://www.slideshare.net/GRFDavos/egypt-for-idrc-2-june-2010-finalpptx

Ali-Toudert, F., & Mayer, H. (2006). Numerical study on the effects of aspect ratio and orientation of an urban street canyon on outdoor thermal comfort in hot and dry climate. *Building and Environment, 41*(2), 94–108. doi:10.1016/j.buildenv.2005.01.013

Aliyah, I., Setioko, B., & Pradoto, W. (2017). Spatial Flexibility in Cultural Mapping of Traditional Market Area in Surakarta: A Case Study of Pasar Gede in Surakarta. *City, Cultura e Scuola, 10*, 41–51.

Allen, J., & Browne, M. (2010). Sustainability Strategies for City Logistics. In *Green Logistics: Improving the Environmental Sustainability of Logistics* (pp. 282–305). London: Kogan Page Limited.

Al-Maimani, A., Salama, A. M., & Fadli, F. (2014). Exploring Socio-Spatial Aspects of Traditional Souqs: The Case of Souq Mutrah, Oman. *Archnet-IJAR: International Journal of Architectural Research, 8*(1), 50–65. doi:10.26687/archnet-ijar.v8i1.356

Al-Maqrizi, T.A. (1908). *Histoire d'Egypte de Makrizi* (E. Blochet, Trans.). Paris: E. Leroux.

Almeida, M. C., Batty, M., Monteiro, A. M. V., Câmara, G., & ... . (2003). Stochastic cellular automata modeling of urban land use dynamics: Empirical development and estimation. *Computers, Environment and Urban Systems, 27*(5), 481–509. doi:10.1016/S0198-9715(02)00042-X

Alonso, A., Monzón, A., & Cascajo, R. (2017). Measuring Negative Synergies of Urban Sprawl and Economic Crisis over Public Transport Efficiency. *International Regional Science Review, 8*, 016001761668736. doi:10.1177/0160017616687361

Alonso, W. (1964). *Location and land use*. Cambridge, MA: Harvard University Press. doi:10.4159/harvard.9780674730854

Al-Sayed, K. (2012). A Systematic Approach Towards Creative Urban Design. *Design Computing, and Cognition, 12*, 1–20.

AlSayyad, N. (2011). *Cairo: Histories of a City*. Cambridge, MA: Harvard University Press. doi:10.4159/harvard.9780674060791

Alsharif, A. A. A., & Pradhan, B. (2014). Urban sprawl analysis of Tripoli Metropolitan City (Libya) using remote sensing data and multivariate logistic regression model. *Photonirvachak (Dehra Dun), 42*(1), 149–163. doi:10.100712524-013-0299-7

Altshuler, A. A. (1965). *The City Planning Process: A Political Analysis*. Ithaca, NY: Cornell University Press.

Anas, A., Arnott, R., & Small, K. A. (1998). Urban spatial structure. *Journal of Economic Literature, 36*, 1426–1464.

Andersen, M. (2018). *Vienna Tops Mercer's 20th Quality of Living Ranking*. Retrieved from https://www.mercer.com/newsroom/2018-quality-of-living-survey.html

Andersen, L. E., Groom, B., Killick, E., Ledezma, J. C., Palmer, C., & Weinhold, D. (2017). Modeling land use, deforestation, and policy: A hybrid optimisation-heterogeneous agent model with application to the Bolivian Amazon. *Ecological Economics, 135*, 76–90. doi:10.1016/j.ecolecon.2016.12.033

Anderson, J. R. (1976). *A land use and land cover classification system for use with remote sensor data* (Vol. 964). US Government Printing Office. doi:10.3133/pp964

Anderson, N. B., & Bogart, W. T. (2001). The structure of sprawl: Identifying and characterizing employment centres in polycentric metropolitan areas. *American Journal of Economics and Sociology, 60*(1), 147–169. doi:10.1111/1536-7150.00058

Anholt, S. (2010). Definitions of place branding –Working towards a resolution. *Place Branding and Public Diplomacy, 6*(1), 1–10. doi:10.1057/pb.2010.3

Anjaria, J. S. (2006). Street Hawkers and Public Space in Mumbai. *Economic and Political Weekly, 41*(21), 2140–2146.

An, L., & Brown, D. G. (2008). Survival analysis in land change science: Integrating with GIScience to address temporal complexities. *Annals of the Association of American Geographers, 98*(2), 323–344. doi:10.1080/00045600701879045

Appleyard, D. (1981). *Livable Streets*. Berkeley, CA: University of California Press.

Arayici, Y. (2008). Towards building information modelling for existing structures. *Structural Survey, 26*(3), 210–222. doi:10.1108/02630800810887108

Arnaud, J.-L. (1996). *Cairo: The Development of a Modern City (1867-1907)*. Paris: CNRS.

Arthur, W. B. (2005). Out-of-equilibrium economics and agent-based modeling. In K. Judd, & L. Tesfatsion (Eds.), Handbook of Computational Economics, Vol. 2: Agent-Based Computational Economics. North-Holland: Elsevier.

Artz, M. (2010). GeoDesign. Changing Geography by Design. *GeoDesign. Changing Geography by Design. Selected Readings in GeoDesign*. ESRI Press. Retrieved from www.esri.com/library/ebooks/GeoDesign.pdf

Asawa, T., Hoyano, A., & Nakaohkubo, K. (2004, September). Thermal design tool for outdoor space based on numerical simulation system using 3D-CAD. In *Proc. 21th Int. Conf. on PLEA* (Vol. 2, pp. 1013-1018). Academic Press.

Ashour, R. (2003). *Qit'a min Urubba: Riwaya [A Part of Europe: A Novel]*. Cairo: Dar Al-Shorouk. (in Arabic)

Ashworth, G. (2009). The Instrument of Place Branding: How is it done? *European Spatial Research and Policy*, *16*(1), 9–22. doi:10.2478/v10105-009-0001-9

Aslan, I. (2009). *The Heron* (E. Colla, Trans.). Cairo: The American University in Cairo Press.

Aspinall, R. (2004). Modeling land use change with generalized linear models-A multi-model analysis of change between 1860 and 2000 in Gallatin Valley, Montana. *Journal of Environmental Management*, *72*(1-2), 91–103. doi:10.1016/j.jenvman.2004.02.009 PMID:15246576

Asplet, M. (2013). *Internal Displacement: Responsibility and Action*. Geneva: Inter-Parliamentary Union and UNHCR No 20.

Associates A. B. T. Dames and Moore & GOHBPR. (1982). Informal Housing in Egypt. Cairo: U.S. Agency for International Development.

Assy, D. (2006). *Cairo 104*. Amman, Jordan: Dar Al Yasmeen Publishing and Distribution.

Aswany, A. E. (2004). *The Yacoubian Building 'Imarat Ya'qubyan* (H. T. Davies, Trans.). Cairo: The American University of Cairo Press.

ATKearney. (2012). *2012 Global Index and Emerging Cities Outlook*. Retrieved 7 10, 2017, from www.atkearney.com: https://www.atkearney.com/research-studies/global-cities-index/2012

Atzori, L., Iera, A., & Morabito, G. (2010). The Internet of Things: A survey. *Computer Networks*, *54*(15), 2787–2805. doi:10.1016/j.comnet.2010.05.010

Augoyard, J.-F., & Torgue, H. (1995). *A l'écoute de l'environnement. Répertoire des effets sonores*. Marseille: Parenthèses.

Awatta, H. (2015). *The Urban Transformation of Downtown Cairo between State and Non-State Stakeholders* (Master's thesis). Cairo: The Department of Sustainable Development, The American University in Cairo.

Ayeni, B. (2005). The design of spatial decision support systems in urban and regional planning. In H. Timmermans (Ed.), *Decision Support Systems in Urban Planning* (pp. 2–15). E & FN SPON.

Azadi, H., Barati, A. A., Rafiaani, P., Taheri, F., Gebrehiwot, K., Witlox, F., & Lebailly, P. (2017). Evolution of land-use change modeling: Routes of different schools of knowledge. *Landscape and Ecological Engineering*, *13*(2), 319–332. doi:10.100711355-016-0311-9

Azhar, S., Khalfan, M., & Maqsood, T. (2012). Building information modeling (BIM): Now and beyond. *Australasian Journal of Construction Economics and Building, 12*, 15. Retrieved from http:// doi:10.5130/ajceb.v12i4.3032

Azhar, S. (2011). Building Information Modelling (BIM): Trends, benefits, risks and challenges for the AEC industry. *Leadership and Management in Engineering*, *11*(3), 241–252. doi:10.1061/(ASCE)LM.1943-5630.0000127

Bafna, S. (2003). Space syntax: A brief introduction to its logic and analytical techniques. *Environment and Behavior*, *35*(1), 17–29. doi:10.1177/0013916502238863

Baik, A., Yaagoubi, R., & Boehm, J. (2015). Integration of Jeddah historical BIM and 3D GIS [data] for documentation and restoration of historical monument [structures]. *The International Archives of the Photogrammetry, Remote Sensing and Spatial Information Sciences*, *XL-5*(W7), 29–34. doi:10.5194/isprsarchives-XL-5-W7-29-2015

Bai, L., & Nasu, S. (2018). Common Space Spatial Layout Transition in Japanese Nursing Home - By Space Syntax point of view. *Asian Journal of Environment-Behaviour Studies*, *3*(7), 19–30. doi:10.21834/aje-bs.v3i7.256

Baker, W. L. (1989). A review of models of landscape change. *Landscape Ecology*, *2*(2), 111–133. doi:10.1007/BF00137155

Balzter, H., Braun, P. W., & Kohler, W. (1998). Cellular automata models for vegetation dynamics. *Ecological Modelling*, *107*(2/3), 113–125. doi:10.1016/S0304-3800(97)00202-0

Banister, D., & Hickman, R. (2006). How to design a more sustainable and fairer built environment: Transport and communications. *IEE Proceedings - Intelligent Transport Systems, 153*(4), 276. 10.1049/ip-its:20060009

Banister, D. (2008). The sustainable mobility paradigm. *Transport Policy*, *15*(2), 73–80. doi:10.1016/j.tranpol.2007.10.005

Banister, D. (2011a). Cities, mobility and climate change. *Journal of Transport Geography*, *19*(6), 1538–1546. doi:10.1016/j.jtrangeo.2011.03.009

Banister, D. (2011b). The trilogy of distance, speed and time. *Journal of Transport Geography*, *19*(4), 950–959. doi:10.1016/j.jtrangeo.2010.12.004

Barredo, J. I., Kasanko, M., McCormick, N., & Lavalle, C. (2003). Modeling dynamic spatial processes: Simulation of urban future scenarios through cellular automata. *Landscape and Urban Planning*, *64*(3), 145–160. doi:10.1016/S0169-2046(02)00218-9

Barreteau, O., & Bousquet, F. (2000). SHADOC: A multi-agent model to tackle viability of irrigated systems. *Annals of Operations Research*, *94*(1/4), 139–162. doi:10.1023/A:1018908931155

Barthes, R. (1971). Sémiotics and urbanism (in French). *L'Architecture d'Aujourd'hui.*

Barthes, R. (1994). The Semiotic Challenge First Edition Thus Edition (R. Howard, Trans.). University of California Press.

Bartlett, S., Dodman, D., Hardoy, J., Satterthwaite, D., & Tacoli, C. (2011). Social Aspects of Climate Change in Urban Areas in Low and Middle-Income Nations. In *Cities and Climate Change: Responding to an Urgent Agenda: Main report (English)* (pp. 670–726). Washington, DC: World Bank, Urban development series. Retrieved from http://documents.worldbank.org/curated/en/613201468149671438/Main-report

Basiago, A. (1999). Economic, social, and environmental sustainability in development theory and urban planning practice. *The Environmentalist*, *19*(2), 145–161. doi:10.1023/A:1006697118620

Basse, M. R., Omrani, H., Charif, O., Gerber, P., & Bodis, K. (2014). Land use changes modeling using advanced methods: Cellular automata and artificial neural networks. The spatial and explicit representation of land cover dynamics at the cross-border region scale. *Applied Geography (Sevenoaks, England)*, *53*, 160–171. doi:10.1016/j.apgeog.2014.06.016

Batten, D. (2001). Complex landscapes of spatial interaction. *The Annals of Regional Science*, *35*(1), 81–111. doi:10.1007001680000032

Batty, M. (1998). Urban evolution on the desktop: Simulation with the use of extended cellular automata. *Environment & Planning A*, *30*(11), 1943–1967. doi:10.1068/a301943

Batty, M. (2001). Agent-based pedestrian modeling. *Environment & Planning B*, *28*(3), 321–326. doi:10.1068/b2803ed

Batty, M. (2005). *Cities and complexity: Understanding cities with cellular automata, agent-based models and fractals.* Cambridge, MA: The MIT Press.

Baudrillard, J., & Nouvel, J. (2002). *The Singular Objects of Architecture.* Minneapolis, MN: Minnesota Press.

Beard, V. A., Miraftab, F., & Silver, C. (2008). *Planning and Decentralization: Contested Spaces for Public Action in the Global South*. Routledge.

Beatley, T. (2000). *Green urbanism: Learning from European cities / Timothy Beatley*. Washington, DC: Island Press.

Beaufoy, H. (2010). Order out of chaos': The London Society and the planning of London 1912–1920. *Planning Perspectives*, *12*(2), 135–164. doi:10.1080/026654397364708

Beauregard, R. (2013). The chaos and complexity of gentrification. In N. Smith & P. Williams (Eds.), Gentrification of the City (pp. 35-55). Routledge.

Beaussier, T., Caurla, S., Bellon-Maurel, V., & Loiseau, E. (2019). Coupling economic models and environmental assessment methods to support regional policies: A critical review. *Journal of Cleaner Production*, *216*, 408–421. doi:10.1016/j.jclepro.2019.01.020

Begg, I. (1999). Cities and Competitiveness. *Urban Studies (Edinburgh, Scotland)*, *36*(5-6), 795–809. doi:10.1080/0042098993222

Behrens, K., Gaigne, C., & Thisse, J. F. (2009). Industry location and welfare when transport costs are endogenous. *Journal of Urban Economics*, *65*(2), 195–208. doi:10.1016/j.jue.2008.11.003

Ben Slama, H. (2007). *Parcours urbains quotidiens, L'habitude dans la perception des ambiances*. Grenoble: PhD, ENSAG, CRESSON.

Benedikt, M. L. (1979). To take hold of space: Isovists and isovist fields. *Environment & Planning B*, *6*(1), 47–65. doi:10.1068/b060047

Benediktsson, M. O., Lambert, B., & Larsen, E. (2015). Taming a "chaotic concept": Gentrification and segmented consumption in Brooklyn, 2002–2012. *Urban Geography*, 1–21. doi:10.1080/02723638.2015.1096113

Berger, T. (2001). Agent-based spatial models applied to agriculture: A simulation tool for technology diffusion, resource use changes and policy analysis. *Agricultural Economics*, *25*(2-3), 245–260. doi:10.1111/j.1574-0862.2001.tb00205.x

Berliant, M., & Wang, P. (2008). Urban growth and subcenter formation: A trolley ride from the Staples Centre to Disneyland and the Rose Bowl. *Journal of Urban Economics*, *63*(2), 679–693. doi:10.1016/j.jue.2007.04.009

Berry, J. K. (1987). Fundamental operations in computer assisted map analysis. *International Journal of Geographical Information Systems*, *1*(2), 119–136. doi:10.1080/02693798708927799

Berry, M., Hazen, B., MacIntyre, R. L., & Flamm, R. (1996). LUCAS: A system for modeling land-use change. *IEEE Computational Science & Engineering*, *3*(1), 24–35. doi:10.1109/99.486758

Bertolini, L. (2006). *Why and how can complexity theories help spatial planners?* Amsterdam: De Universiteit van Amsterdam.

Betsill, M. (2001). Mitigating Climate Change in US Cities: Opportunities and obstacles. *Local Environment*, *6*(4), 393–406. doi:10.1080/13549830120091699

Beurden, J. B., Bakema, A., & Tijbosch, H. (2007). A land use modeling system for environmental impact assessment: recent applications of the LUMOS toolbox. In E. Koomen, J. Stillwell, A. Bakema, & H. J. Scholten (Eds.), *Modeling land use change progress and applications* (pp. 281–296). Dordrecht: Springer. doi:10.1007/1-4020-5648-6_16

Bhowmik, S. (2003). National Policy for Street Vendors. *Economic and Political Weekly*, *38*(16), 1543–1546.

Bhowmik, S. (2005). Street Vendors in Asia: A Review. *Economic and Political Weekly*, *40*(22/23), 2256–2264.

Bierman, I. (2005). Disciplining the Eye: Perceiving Medieval Cairo. In N. AlSayyad, I. A. Bierman, & N. Rabbat (Eds.), *Making Cairo Medieval*. New York: Lexington Books.

Bilsborrow, R. E., & Okoth Ogondo, H. W. O. (1992). Population-driven changes in land use in developing countries. *Ambio*, *21*, 37–45.

Bishop, C. M. (2006). *Pattern recognition and machine learning*. New York: Springer.

Bjorkdahl, A., & Kappler, S. (2017). *Peacebuilding and Spatial Transformation: Peace, Space and Place*. London, UK: Routledge. doi:10.4324/9781315684529

Bloomfield, J., & Bianchini, F. (2002). *Planning for the Cosmopolitan City: A Research Report for Birmingham City Council*. Leicester, UK: Comedia, International Cultural Planning and Policy Unit.

Blumler, M., MacDonald, G., & Millington, A. (2011). Biocomplexity. In M. Blumler, G. MacDonald, A. Millington, & U. Schickhoff (Eds.), *Handbook of Biogeography* (pp. 469–487). London: Sage Publications.

Boarnet, M., & Crane, R. (2001). *Travel by design: The influence of urban form on travel. Spatial information systems*. New York: Oxford University Press.

Bockstael, N. E. (1996). Modeling economics and ecology: The importance of a spatial perspective. *American Journal of Agricultural Economics*, *78*(5), 1168–1180. doi:10.2307/1243487

Bockstael, N. E., & Irwin, E. G. (2000). Economics and the land use-environment link. In H. Folmer & T. Tietenberg (Eds.), *The international yearbook of environmental and resource economics 1999/2000* (pp. 1–54). Northampton, MA: Edward Elgar Publishing.

Boileau, I. (1959). La Ciudad Lineal: A Critical Study of the Linear Suburb of Madrid. *The Town Planning Review*, *30*(3), 230–238. doi:10.3828/tpr.30.3.377v40725m3m0771

Bolton, T., Francis, N., & Froy, F. (2018). The Impact of Space Syntax on Urban Policy Making: Linking Research into UK Policy. *11th International Space Syntax Symposium*.

Bommel, P., Becu, N., Le Page, C., & Bousquet, F. (2016). Cormas: An Agent-Based simulation platform for coupling human decisions with computerized dynamics. In T. Kaneda, H. Kanegae, Y. Toyoda, & P. Rizzi (Eds.), *Simulation and gaming in the network society. Translational Systems Sciences* (Vol. 9). Singapore: Springer. doi:10.1007/978-981-10-0575-6_27

Boonjubun, C. (2017). Conflicts over streets: The eviction of Bangkok street vendors. *Cities (London, England)*, *70*, 22–31. doi:10.1016/j.cities.2017.06.007

Borg, Y. (2013). *Social Contexts & Built Heritage. A case study of Alexandria, Egypt*. Cairo, Egypt: University Paris 1 Pantheon-Sorbonne.

Borning, A., Waddell, P., & Förster, R. (2008). Urbanism: Using Simulation to Inform Public Deliberation and Decision-Making. In Digital Government: E-Government Research, Case Studies, and Implementation (pp. 439-464). Academic Press. doi:10.1007/978-0-387-71611-4_22

Boswell, C. (2002). *New Issues in Refugee Research, Addressing the causes of migratory and refugee movements: the role of the European Union*. Institute for Peace Research and Security Policy University of Hamburg.

Boucsein, W. (2012). *Electrodermal Activity* (Vol. 2). Allemagne: Université de Wupperta.

Bouhali, A. (2018). Negotiating Streets and Space in Transnational Trade Marketplaces in Oran (Algeria) and Cairo (Egypt): "Place Struggle" in the Commercial City. *Articulo-Journal of Urban Research,* (17/18), 1-21.

Bousquet, F., Bakam, I., Proton, H., & Le Page, C. (1998). Cormas: Common-pool resources and multiagent systems. *Lecture Notes in Artificial Intelligence*, *1416*, 826–837.

Bouton, S., Cis, D., Mendonca, L., Pohl, H., Remes, J., Ritchie, H., & Woetzel, J. (2013). *How to Make a City Great?* McKinsey & Company.

Brabham, D. C. (2008). Crowdsourcing as a Model for Problem Solving. *The International Journal of Research into New Media Technologies*, *14*(1), 75–90. doi:10.1177/1354856507084420

Brandsma, A., Kancs, A., Monfort, P., & Rillaers, A. (2015). RHOMOLO: A dynamic spatial general equilibrium model for assessing the impact of cohesion policy. *Papers in Regional Science*, *94*(S1), S197–S221. doi:10.1111/pirs.12162

Braudel, F. (1979). *The Wheels of Commerce: Civilization and Capitalism, 15th-18th Century* (Vol. 2). New York, NY: Harper & Row.

Brecard, D., Fougeyrollas, A., Le Mouel, P., Lemiale, L., & Zagamé, P. (2006). Macroeconomic consequences of European research policy: Prospects of the NEMESIS model in the year 2030. *Research Policy*, *35*(7), 910–924. doi:10.1016/j.respol.2006.03.001

Briassoulis, H. (2000). Analysis of land use change: theoretical and modeling approaches. In S. Loveridge (Ed.), *The web book of regional science*. Morgantown, WV: West Virginia University.

Britz, W. (Ed.). (2004). *CAPRI modeling system documentation, final report of the FP5 shared cost project CAP-STRAT: Common Agricultural Policy strategy for regions, agriculture and trade, QLTR-2000-00394*. Bonn University.

Bromley, R. (2000). Street Vending and Public Policy: A Global Review. *The International Journal of Sociology and Social Policy*, *20*(1/2), 1–29. doi:10.1108/01443330010789052

Bromley, R. D. F. (1998). Market-place Trading and the Transformation of Retail Space in the Expanding Latin American City. *Urban Studies (Edinburgh, Scotland)*, *35*(8), 1311–1333. doi:10.1080/0042098984376

Bromley, R. D. F., & Mackie, P. K. (2009). Displacement and the New Spaces for Informal Trade in the Latin American City Centre. *Urban Studies (Edinburgh, Scotland)*, *46*(7), 1485–1506. doi:10.1177/0042098009104577

Brown, F. E. (1990). Comment on Chapman: Some Cautionary Notes on the Application of Spatial Measures to Prehistoric Settlements. In The Social Archaeology of Houses. Edinburgh, UK: Edinburgh University Press.

Brown, A. (Ed.). (2006). *Contested Space: Street Trading, Public Space, and Livelihoods in Developing Countries*. Rugby, UK: ITDG Publications. doi:10.3362/9781780444703

Brown, A., Kafafy, N., & Hayder, A. (2017). Street Trading in the Shadows of the Arab Spring. *Environment and Urbanization*, *29*(1), 283–298. doi:10.1177/0956247816673559

Brown, A., Lyons, M., & Dankoco, I. (2010). Street Traders and the Emerging Spaces for Urban Voice and Citizenship in African Cities. *Urban Studies (Edinburgh, Scotland)*, *47*(3), 666–683. doi:10.1177/0042098009351187

Brown, B. B., Yamada, I., Smith, K. R., Zick, C. D., Kowaleski-Jones, L., & Fan, J. X. (2009). Mixed land use and walkability: Variations in land use measures and relationships with BMI, overweight, and obesity. *Health & Place*, *15*(4), 1130–1141. doi:10.1016/j.healthplace.2009.06.008 PMID:19632875

Brown, D. G., Page, S. E., Riolo, R. L., Zellner, M., & Rand, W. (2005). Path dependence and the validation of agent-based spatial models of land use. *International Journal of Geographical Information Systems*, *19*(2), 153–174. doi:10.1080/13658810410001713399

Brown, D. G., Verburg, P. H., Pontius, R. G. Jr, & Lange, M. D. (2013). Opportunities to improve impact, integration, and evaluation of land change models. *Current Opinion in Environmental Sustainability*, *5*(5), 452–457. doi:10.1016/j.cosust.2013.07.012

Brown, D. G., Walker, R., Manson, S., & Seto, K. (2004). Modeling land use and land cover change. In G. Gutman, A. C. Janetos, C. O. Justice, & ... (Eds.), *Land Change Science* (pp. 395–409). Kluwer Academic Publishers.

Brown, D., Boano, C., Vivekananda, J., & Walker, J. (2015). *Urban Crises and Humanitarian Responses: A Literature Review. The DFID Humanitarian Policy and Partnerships Group*. CHASE.

Bruse, M. (2009, June). Analysing human outdoor thermal comfort and open space usage with the Multi-Agent System BOTworld. In *Seventh International Conference on Urban Climate ICUC-7* (*Vol. 29*). Academic Press.

Bryant, D., Nielsen, D., & Tangley, L. (1997). *The last frontier forests: Ecosystems and economies on the edge*. Washington, DC: World Resources Institute.

BSI Standards Publication. (2014). *Smart city framework – Guide customer service to establishing strategies for smart cities and communities*. BSI Standards Publication.

Buehler, R. (2011). Determinants of transport mode choice: A comparison of Germany and the USA. *Journal of Transport Geography*, *19*(4), 644–657. doi:10.1016/j.jtrangeo.2010.07.005

Bullock, N., & Dickens, P. (1975). Use of models in planning. In L. Martine & L. March (Eds.), *Urban space and structures* (pp. 97–112). Cambridge, UK: Cambridge University Press.

Bunio-Mroczek, P. (2017). Gentrification, revitalization and children raising. Family gentrifiers in a post-socialist city. *Przegląd Socjologii Jakościowej, 13*(4), 71-91. Doi:10.18778/1733-8069.13.4.05

Burnett, P. (2012). Urban industrial composition and the spatial expansion of cities. *Land Economics*, *88*(4), 764–781. doi:10.3368/le.88.4.764

Byrne, J. P. (2003). Two Cheers for Gentrification. *Georgetown Law Faculty Publications and Other Works*, 405-432. Retrieved from https://scholarship.law.georgetown.edu/facpub/930

Cairobserver. (n.d.). *11 Recent Cultural Disasters in Egypt*. Retrieved September 01, 2015, from Cairobserver: http://cairobserver.com/ post/75403717693/11-recent-cultural-disasters-in-egypt#.VL l90oe7ggE

Çakar, U., & Alakavuklar, O. N. (2011). Inherent Chaos in the Organizational Order: An Epistemological Approach. *International Journal of Business and Management Studies*, *3*(1), 391–405.

Calthorpe, P. (1993). *The next American metropolis: Ecology, communities, and the American dream / Peter Calthorpe*. New York: Princeton Architectural Press.

Cambridge Dictionary. (2018). *Meaning of "characteristic" in the English Dictionary*. Retrieved from https://dictionary.cambridge.org/dictionary/english/characteristic

Cambridge Dictionary. (n.d.). *Cambridge Dictionary | English Dictionary, Translations & Thesaurus*. Retrieved September 4, 2018, from https://dictionary.cambridge.org/

Can, I. (2012). *In-between space and social interaction: A case study of three neighborhoods in Izmir* (Doctoral thesis). University of Nottingham.

Cao, X., Mokhtarian, P. L., & Handy, S. L. (2009). Examining the Impacts of Residential Self-Selection on Travel Behaviour: A Focus on Empirical Findings. *Transport Reviews*, *29*(3), 359–395. doi:10.1080/01441640802539195

CAPMAS. (2018). *Egypt in Figures: 2018 (Statistical Yearbook, Ref. No: 71-01112-2018).* Arab Republic of Egypt.

CAPMS. (2017). *Central Agency for Public Mobilization and Statistics*. Available at: http://www.capmas.gov.eg/

Capros, P., Van Regemorter, D., Paroussos, L., & Karkatsoulis, P. (2013). *GEM-E3 Model documentation. No. EUR26034 EN*. European Commission Joint Research Centre. Retrieved from ftp://139.191.159.82/pub/EURdoc/JRC83177.pdf

Carmona, M., Heath, T., Oc, T., & Tiesdell, S. (2003). *Public Places- Urban Spaces: The Dimensions of Urban Design*. New York, NY: Architectural Press.

Carmona, M., Heath, T., Tiesdell, S., & Oc, T. (2003). *Public Places-Urban Spaces: The Dimensions of Urban Design*. Oxford, UK: Architectural Press.

Carrillo, A. C. (2009). Internal displacement in Colombia: Humanitarian, economic and social consequences in urban settings and current challenges. *International Review of the Red Cross*, *91*(875), 527–546. doi:10.1017/S1816383109990427

Caruso, G., Rounsevell, M., & Cojocaru, G. (2005). Exploring a spatio-dynamic neighbourhood-based model of residential behavior in the Brussels peri urban area. *International Journal of Geographical Information Science*, *19*(2), 103–123. doi:10.1080/13658810410001713371

Casselati, A. (1997). The Nature of Liveability. In *Making cities liveable. International Making Cities Liveable Conferences*. Gondolier Press.

Castells, M. (2010). End of Millennium: The Information Age: Economy, Society, and Culture, Volume III (2nd ed.). Wiley-Blackwell.

Castells, M. (1996). *The Rise of the Network Society*. London: Blackwell.

Castells, M. (1997). *The Power of Identity, The Information Age: Economy, Society and Culture* (Vol. 2). Cambridge, MA: Blackwell.

Castillo-Villar, F. R. (2018). City branding and the theory of social representation. *Bitácora Urbano Territorial*, *28*(1), 33–38. doi:10.15446/bitacora.v28n1.52939

Cavailhes, J., Gaigne, C., Tabuchi, T., & Thisse, J. F. (2007). Trade and the structure of cities. *Journal of Urban Economics*, *62*(3), 383–404. doi:10.1016/j.jue.2006.12.002

Certoma, C., Corsini, F., & Rizzi, F. (2014). Crowdsourcing urban sustainability. Data, people and technologies in participatory governance. *Futures*, *74*, 93–106. doi:10.1016/j.futures.2014.11.006

Cervero, R. (1996a). Traditional neighborhoods and commuting in the San Francisco Bay Area. *Transportation*, *23*(4), 373–394. doi:10.1007/BF00223062

Cervero, R. (1996b). Mixed land-uses and commuting: Evidence from the American Housing Survey. *Transportation Research Part a: Policy and Practice*, *30*(5), 361–377. doi:10.1016/0965-8564(95)00033-X

Cervero, R. (2009). Transit and the Metropolis: Finding Harmony. In S. Wheeler & T. Beatley (Eds.), *The sustainable urban development reader* (2nd ed.). London: Routledge.

Ceuclelis, H. (1985). Cellular worlds: A framework for modeling micro-macro dynamics. *Environment & Planning A*, *17*(5), 585–596. doi:10.1068/a170585

Cheng, V. (2010). Understanding Density and High Density. In E. Ng (Ed.), Designing high-density cities for social and environmental sustainability (pp. 3–17). Sterling, VA: Earthscan.

Cheng, J., Masser, I., & Ottens, H. (2003). Understanding Urban Growth System: Theories and Methods. *Sendai: 8th International Conference on Computer Techniques for Urban Planning and Management.*

Chen, H., Ooka, R., Huang, H., & Tsuchiya, T. (2009). Study on mitigation measures for outdoor thermal environment on present urban blocks in Tokyo using coupled simulation. *Building and Environment*, *44*(11), 2290–2299. doi:10.1016/j.buildenv.2009.03.012

Chen, L., & Ng, E. (2012). Outdoor thermal comfort and outdoor activities: A review of research in the past decade. *Cities (London, England)*, *29*(2), 118–125. doi:10.1016/j.cities.2011.08.006

Chen, X. (2017). Research on the Application of Space Syntax in Urban Planning, *World. Journal of Engineering Technology*, *5*, 29–35.

Cheong, S., Brown, D. G., Lopez-Carr, D., & Kok, K. (2012). Mixed methods in land-change research: Towards integration. *Transactions of the Institute of British Geographers*, *37*(1), 8–12. doi:10.1111/j.1475-5661.2011.00482.x

Cheung, S. (2012). How Companies Can Leverage Crowdsourcing. Cambridge, UK: Composite Information Systems Laboratory (CISL).

Ching, T. Y., & Ferreira, J. Jr. (2015). *Planning Support Systems and Smart Cities: Smart Cities: Concepts, Perceptions and Lessons for Planners*. Springer.

Chitekwe-Biti, B., Mudimo, P., & Nyama, G. M. (2012). Developing an Informal Settlement Upgrading Protocolin Zimbabwe - the Epworth Story. *Environment and Urbanization*, *24*(1), 131–148. doi:10.1177/0956247812437138

Chomitz, K., & Gray, D. (1996). Roads, land use and deforestation: A spatial model applied to Belize. *The World Bank Economic Review*, *10*(3), 487–512. doi:10.1093/wber/10.3.487

Cho, W., & Choi, E. (2015). A GPS trajectory map-matching mechanism with DTG big data on the HBase system. *Proceedings of the 2015 International Conference on Big Data Applications and Services*, 22-29. 10.1145/2837060.2837062

Clarke, K. C., & Gaydos, L. J. (1998). Loose-coupling a cellular automaton model and GIS: Long-term urban growth prediction for San Fransisco and Washington/Baltimore. *International Journal of Geographical Information Science*, *12*(7), 699–714. doi:10.1080/136588198241617 PMID:12294536

Clark, G., Couturier, J., Kelly, J., & Moonen, T. (2015). *Globalisation and Competition: The New World of Cities*. Cities Research Center.

Clifton, K., Ewing, R., Knaap, G., & Song, Y. (2008). Quantitative analysis of urban form: A multidisciplinary review. *Journal of Urbanism*, *1*(1), 17–45. doi:10.1080/17549170801903496

Cocina, G. Giulia (2017). Responding to User Cultural Needs in Hospitals with the Support of SPACE Syntax Analysis. In *Proceedings of the 11th Space Syntax Symposium* (pp. 64.1-64.10). Lisbon, Portugal: Instituto Superior Técnico, Departamento de Engenharia Civil, Arquitetura e Georrecursos.

Cohen, J., & Arato, A. (1994). *Civil society and political theory*. Cambridge, MA: The MIT Press.

Collier, P. (2013). *Exodus: How Migration is Changing Our World*. New York: Oxford University Press.

Collins. (2018). *Definition of 'characteristic'*. Retrieved from www.collinsdictionary.com: https://www.collinsdictionary.com/dictionary/english/characteristic

Collins, G. R., Sitte, C., & Collins, C. C. (2006). *Camillo Sitte: The Birth of Modern City Planning*. Meniola, NY: Courier Corporation.

Conroy-Ddalton, R. (2003). The Secret is to follow your nose: Route path selection and angularity. *Environment and Behavior*, *35*(1), 107–131. doi:10.1177/0013916502238867

Conroy, M. M., & Gordon, S. I. (2004). Utility of interactive computer-based materials for enhancing public participation. *Environmental. Planning. Management*, *47*, 19–33.

Convery, S., & Shahumyan, H. (2009). *Applications of GIS and Urban Modeling for Environmental Policy.* Presentation in UEP Events. Retrieved from: www.uep.ie/downloads/

Conzen, M. R. G. (1960). Alnwick, Northumberland: A Study in Town-Plan Analysis. *Transactions and Papers*, *3*(27), 3. doi:10.2307/621094

Conzen, M. R. G., & Conzen, M. P. (2004). *Thinking about urban form: Papers on urban morphology, 1932-1998 / M.R.G. Conzen* (M. P. Conzen, Ed.). Bern: Peter Lang.

Corong, E., Hertel, T., McDougall, R., Tsigas, M., & van der Mensbrugghe, D. (2017). The Standard GTAP Model, Version 7. *Journal of Global Economic Analysis*, *2*(1), 1–119.

Costanza, R., Voinov, A., Boumans, R., Maxwell, T., Villa, F., Wainger, L., & Voinov, H. (2002). Integrated ecological economic modeling of the Patuxent River watershed, Maryland. *Ecological Monographs*, *72*(2), 203–231. doi:10.1890/0012-9615(2002)072[0203:IEEMOT]2.0.CO;2

Costanza, R., Wainger, L. A., & Folke, C. (1993). Modeling complex ecological economic systems. *Bioscience*, *43*(8), 545–556. doi:10.2307/1311949

Council of Europe. (2009). *Resolution No. 2 on The European regional/spatial planning charter (Torremolinos Charter).* Retrieved 11 30, 2012, from European Conference of Ministers Responsible for Regional/Spatial Planning http://www.coe.int/t/dg4/cultureheritage/heritage/cemat/confminist1-15/6eresolution2_EN.asp

Craib, I. (1997). *Classical Social Theory: An Introduction to the Thought of Marx, Weber, Durkheim, and Simmel.* Oxford, UK: Oxford University Press.

Crooks, A., & Heppenstall, A. (2012). Introduction to Agent-Based Modelling. In Agent-Based Models of Geographical Systems (pp. 85-105). Academic Press. doi:10.1007/978-90-481-8927-4_5

Crooks, A. (2013). Reviews: A Framework for geodesign: changing geography by design. *Environment and Planning. B, Planning & Design*, *40*, 1122–1123. doi:10.1068/b4006rvw

Crossa, V. (2009). Resisting the Entrepreneurial City: Street Vendors' Struggle in Mexico City's Historic Center. *International Journal of Urban and Regional Research*, *33*(1), 43–63. doi:10.1111/j.1468-2427.2008.00823.x

Cross, J. (2000). Street Vendors, Modernity and Postmodernity: Conflict and Compromise in the Global Economy. *The International Journal of Sociology and Social Policy*, *20*(1/2), 29–51. doi:10.1108/01443330010789061

Croxford, B., Penn, A., & Hillier, B. (1996). Spatial Distribution of Urban Pollution: Civilizing Urban Traffic. *The Science of the Total Environment*, *189-190*, 3–9. doi:10.1016/0048-9697(96)05184-4

Dalton, N. (2001). Fractional configurational analysis and a solution to the Manhattan problem. *Proceedings 3rd International Space Syntax Symposium.*

Dalton, N., Peponis, J., & Dalton Conroy, R. (2003). To tame a TIGER, one has to know its nature 1: Extending weighted angular integration analysis to the description of GIS road-centerline data for large scale urban analysis. In *Proceedings of the 4th International Space Syntax Symposium.* London: University College London.

Dalton, R. C., & Bafna, S. (2003). The Syntactical Image of the City: A Reciprocal Definition of Spatial Elements and Spatial Syntaxes. *Proceedings of the 4th International Space Syntax Symposium.*

Darin, M. (1998). The Study of Urban Form in France. *Urban Morphology*, *2*(2), 63–76.

Davidson, M. D. (2017). Equaty and the conservation of the global ecosystem services. *Sustainability*, *9*(3), 339. doi:10.3390u9030339

Davies, A. (2012). IDPs in Host Families and Host Communities: Assistance for hosting arrangements. United Nations High Commissioner for Refugees (UNHCR).

Dawes, M., & Ostwald, M. J. (2013). Precise Locations in Space: An Alternative Approach to Space Syntax Analysis using Intersection Points. *Architecture Research*, *3*(1), 1–11.

Dawson, P. C. (2003). Analysing the Effects of Spatial Configuration on Human Movement and Social Interaction in Canadian Arctic Communities. *Proceedings of the 4th international space syntax symposium.*

De Groot, R. S., Alkemade, R., Braat, L., Hein, L., & Willemen, L. (2010). Challenges in integrating the concept of ecosystem services and values in landscape planning, management and decision making. *Ecological Complexity*, *7*(3), 260–272. doi:10.1016/j.ecocom.2009.10.006

De Groot, R. S., Wilson, M. A., & Boumans, R. M. J. (2002). A typology for the classification, description and valuation of ecosystem functions, goods and services. *Ecological Economics*, *41*(3), 393–408. doi:10.1016/S0921-8009(02)00089-7

De Koning, R. E., Van Nes, A., Ye, Y., & Roald, H. (2017). Strategies for Integrated Densification with Urban Qualities: Combining Space Syntax with Building Density, Land Usage, Public Transport and Property Rights in Bergen City. *11th International Space Syntax Symposium.*

De Nijs, T. C. M., de Niet, R., & Crommentuijin, L. (2004). Constructing land use maps of the Netherlands in 2030. *Journal of Environmental Management*, *72*(1-2), 35–42. doi:10.1016/j.jenvman.2004.03.015 PMID:15246572

De Rosa, M., Trydeman Knudsen, M., & Hermansen, J. E. (2016). A comparison of land use change models: Challenges and future developments. *Journal of Cleaner Production*, *113*, 183–193. doi:10.1016/j.jclepro.2015.11.097

De Soto, H. (1989). *The Other Path: The Invisible Revolution in the Third World.* New York, NY: Harper and Row.

de Vos, J., & Witlox, F. (2013). Transportation policy as spatial planning tool; reducing urban sprawl by increasing travel costs and clustering infrastructure and public transportation. *Journal of Transport Geography*, *33*, 117–125. doi:10.1016/j.jtrangeo.2013.09.014

deChazal, J. (2010). A systems approach to livability and sustainability: Defining terms and mapping relationships to link desires with ecological opportunities and constraints. *Systems Research and Behavioral Science*, *27*(5), 585–597. doi:10.1002res.1058

Dendoncker, N., Rounsevell, M., & Bogaert, P. (2007). Spatial analysis and modeling land use distributions in Belgium. *Computers, Environment and Urban Systems*, *31*(2), 188–205. doi:10.1016/j.compenvurbsys.2006.06.004

Denis, E., & Séjourné, M. (2002). *ISIS: Information System for Informal Settlements.* Cairo: Ministry of Planning, GTZ, CEDEJ.

Densham, P. (1991). Spatial decision support systems. In M. S. J. Maguire (Ed.), *In Geographical information systems: Principles and applications* (pp. 403–412). London: Longman Publishing Group.

Dessouki, Y. (2007). *The Bibliotheca Alexandria and the preservation of Egypt's visual cultural heritage.* Academic Press.

Desyllas, J., & Duxbury, E. (2001). Axial maps and visibility graph analysis. In *Proceedings of the 3rd International Symposium on Space Syntax* (pp. 27.1–27.13). Atlanta, GA: Georgia Institute of Technology.

Dezecache, C., Salles, J. M., Vieilledent, G., & Hérault, B. (2017). Moving forward socio-economically focused models of deforestation. *Global Change Biology*, *23*(9), 3484–3500. doi:10.1111/gcb.13611 PMID:28055134

DictionaryC. (2019). *Singular*. Retrieved from dictionary.cambridge.org: https://dictionary.cambridge.org/dictionary/english/singular

Dieleman, F., & Wegener, M. (2004). Compact City and Urban Sprawl. *Built Environment*, *30*(4), 308–323. doi:10.2148/benv.30.4.308.57151

Dingley, J. (2008). *Nationalism, Social Theory and Durkheim*. Hampshire, UK: Palgrave. doi:10.1057/9780230593107

Dobronogov, A., & Iqbal, F. I. (2005). Economic Growth in Egypt: Constraints and Determinants. *SSRN Electronic Journal*, 1-29. doi:10.2139srn.872328

Donovan, M. G. (2008). Informal Cities and the Contestation of Public Space: The Case of Bogota's Street Vendors, 1988-2003. *Urban Studies (Edinburgh, Scotland)*, *45*(1), 29–51. doi:10.1177/0042098007085100

Dovey, K. (1996). Social Theory and Space Syntax: Exploring Power and Spatial Programming. In *Evolving Environmental Ideals - Changing Way of Life, Values and Design Practices: IAPS 14 Conference Proceedings* (pp. 149-158). Stockholm, Sweden: Royal Institute of Technology.

Downing, T. E. (2002). *Avoiding New Poverty: Mining-Induced Displacement and Resettlement*. International Institute for Environment and Development.

Drummond, L. B. W. (2000). Street Scenes: Practices of Public and Private Space in Urban Vietnam. *Urban Studies (Edinburgh, Scotland)*, *37*(12), 2377–2391. doi:10.1080/00420980020002850

Duneier, M. (1999). *Sidewalk*. New York, NY: Farrar, Status, and Giroux.

Durkheim, E. (1893). The Division of Labor in Society. New York: Macmillan Publishing Company.

Echeverria, C., Coomes, D. A., Hall, M., & Newton, A. C. (2008). Spatially explicit models to analyze forest loss and fragmentation between 1976 and 2020 in southern Chile. *Ecological Modelling*, *212*(3-4), 439–449. doi:10.1016/j.ecolmodel.2007.10.045

Economic and Social Affairs. (2001). *Decentralization and Poverty Reduction: Africa and Asia Experience*. Dakar, Senegal: United Nations.

EEA. (2008). *Modeling environmental change in Europe: Towards a model inventory* (SEIS/Forward). EEA Technical Report No. 11/2008. Copenhagen: European Environmental Agency.

EEA. (2009). *Ensuring quality of life in Europe's cities and towns, Tackling the environmental challenges driven by European and global change*. Copenhagen: European Environmental Agency.

Efroymson, D. (2015). *Urban Menace or Valuable Asset? The Social and Economic Role of Street Vendors in Cities*. Health Bridge.

Eid, Y. Y., Salheen, M. A., & Toimah, A. M. (2012, July). Deriving Balance Indicators for Egyptian Urban-Rural Continuum (As a tool for Balanced Regional Development). *Al Azhar University Engineering Sector*, *9*(24), 1673-1694.

Eid, Y., Attia, A., El-Refaei, M., & Afifi, S. (2014). Spatial Planning and need to Spatial Planning Support Systems – UrbanSim as example. *Al Azhar University Engineering Sector, JAUES*, *9*(32), 1063–1074.

Eisenman, P. (1982). Editor's Introduction by Peter Eisenman. In A. Rossi, *The Architecture of the city* (D. Ghirardo & J. Ockman, Trans.). Cambridge, MA: The MIT Press.

El Fayoumi, M. A. (2017). Street Vendors' Role in Main Squares Utilization as a Type of Tactical Urbanism Application in G.C.R.: A Criticism Review on Ramses Square. *Procedia Environmental Sciences*, *37*, 153–163. doi:10.1016/j.proenv.2017.03.031

El Gabalawy, K., & Galal, O. (Eds.). (2019). The Unplanned Morphing of Social Housing in Cairo: A Positive or Negative Phenomena? Cairo: Academic Press.

El Kadi, G. (2012). *Le Caire - Centre en mouvement (in French).* Collection Petit atlas urbain.

El Kadi, G. E., & Kerdany, D. (2006). *Belle-époque Cairo: The Politics of Refurbishing the Downtown Business District in Cairo Cosmopolitan*. Cairo, Egypt: The American University in Cairo Press.

El-Mouelhi, H. (2014). *Culture and informal urban development - the case of Cairo's 'Ashwa'eyat (informal settlements).* Berlin: German academic exchange service (DAAD).

Elnabawi, M. H., Hamza, N., & Dudek, S. (2013, August). *Use and evaluation of the ENVI-met model for two different urban forms in Cairo, Egypt: measurements and model simulations*. In *13th Conference of International Building Performance Simulation Association*, Chambéry, France.

Elshahed, M. (2007). *Facades of Modernity, Image, Performance, and Transformation in the Egyptian Metropolis* (Master's thesis). Massachusetts Institute of Technology.

Elshater, A. (2012). The New Urbanism Principles vs. Urban Design Dimensions towards Behavioral Performance in Egyptian Neighborhood. In *Future Communities: Socio-Cultural & Environmental Challenges* (pp. 826-843). Giza: AicE-Bs- ASIA Pacific International Conference on Environment-Behavior Studies. 10.1016/j.sbspro.2012.12.270

Elshater, A. (2014). Tactical Urbanism: A Method of Community Empowerment in Cairo Neighborhoods. World Barcelona SB14.

Elshater, A. (2015). Urban Design Paradigm: Working Manuscripts. London: Penguin Reader.

Elshater, A., & Abusaada, H. G. (2016). What is Urban Design? Teaching and Learning. London: Penguin Reader.

Elshater, A. (2015). Urban Design Redux: Redefining a Professional Practice of Specialization. *Ain Shams Engineering Journal*, *6*(1), 25–39. doi:10.1016/j.asej.2014.08.004

Elshater, A. (2017). Widen the Scale of Urban Design to the Level of City Planning: Argument Beyond a Case of Two Cities. UPLanD – Journal of Urban Planning. *Landscape & Environmental Design*, *2*(2), 207–221.

Elshater, A. (2018). What can the urban designer do for children? Normative principles of child–friendly communities for Responsive third places. *Journal of Urban Design*, *23*(3), 432–455. doi:10.1080/13574809.2017.1343086

Elsner, W. (2010). The process and a simple logic of 'Meso'. Emergence and the co-evolution of institutions and group size. *Journal of Evolutionary Economics*, *20*(3), 445–477. doi:10.100700191-009-0158-4

Engelen, G., Lavalle, C., Barredo, J. I., Meulen van der, M., & White, R. (2007). The Moland modeling framework for urban and regional land-use dynamics. In E. Koomen, J. Stillwell, A. Bakema, & H. J. Scholten (Eds.), Modeling Land-Use Change. Springer.

Engelen, G., White, R., & Uljee, I. (2005). Integrating constrained cellular automata models, GIS and decision support tools for urban planning and policy-making. In H. Timmermans (Ed.), *Decision Support Systems in Urban Planning* (pp. 95–117). E & FN SPON.

Engelen, G., White, R., Uljee, I., & Drazan, P. (1995). Using cellular automata for integrated modeling of socio-environmental system. *Environmental Monitoring and Assessment*, *34*(2), 203–214. doi:10.1007/BF00546036 PMID:24202047

English Oxford Living Dictionaries. (2018). *Definition of characteristic in English*. Retrieved from https://en.oxforddictionaries.com/definition/characteristic

Engwicht, D. (1992). *Towards an Eco-City: Calming the Traffic*. Sydney, Australia: Envirobook.

Epple, D., Gordon, B., & Sieg, H. (2010). 'Drs. Muth and Mills meet Dr. Tiebout: Integrating location specific amenities into multi-community equilibrium models. *Journal of Regional Science*, *50*(1), 381–400. doi:10.1111/j.1467-9787.2009.00649.x

Erell, E., Pearlmutter, D., & Williamson, T. J. (2011). *Urban microclimate: Designing the spaces between buildings / Evyatar Erell, David Pearlmutter, and Terry Williamson* (1st ed.). London: Earthscan.

Ericson, B., & Lloyd-Jones, T. (2001). Design Problem. In M. Roberts, & G. Harlow (Eds.), Appreciating Urban Design: The Design Process. Longman.

Eriksson, K., Johnson, C., & Estep, D. (2004). Lorenz and the Essence of Chaos. In K. E. Authors (Ed.), *Eriksson, Applied Mathematics: Body and Soul* (pp. 849–858). Berlin: Springer. doi:10.1007/978-3-662-05800-8_5

Ertur, C., & Koch, W. (2007). Growth, technological interdependence and spatial externalities: Theory and evidence. *Journal of Applied Econometrics*, *22*(6), 1033–1062. doi:10.1002/jae.963

Essa, I. (2013). *Patriotic Ghosts*. Bloomsbury Qatar Foundation Publishing.

Etter, A., McAlpine, C., Wilson, K., Phinn, S., & Possingham, H. (2006). Regional patterns of agricultural land use and deforestation in Colombia. *Agriculture, Ecosystems & Environment*, *114*(2-4), 369–386. doi:10.1016/j.agee.2005.11.013

European Environmental Agency (EEA). (2009). *Ensuring quality of life in Europe's cities and towns, Tackling the environmental challenges driven by European and global change*. Copenhagen: European Environmental Agency.

Evans, G., Aiesha, R., & Foord, J. (2009). Mixed-Use or Mixed Messages? In R. Cooper, G. Evans, & C. Boyko (Eds.), *Designing sustainable cities* (pp. 189–217). Chichester, UK: Wiley-Blackwell Pub.

Evans, P. (2002). *Livable Cities? Urban Struggles for Livelihood and Sustainability* (Vol. 4). Berkeley, CA: University of California Press. doi:10.1525/california/9780520230248.001.0001

Evans, T. P., & Manson, S. (2007). Guest editorial: Space, complexity, and agent-based modeling. *Environment & Planning B*, *34*(2), 196–199. doi:10.1068/b3402ed

Fabbri, K., Di Nunzio, A., Antonini, E., & Boeri, A. (2015). *Outdoor Comfort: the ENVI-BUG tool to evaluate PMV values point by point*. Building Simulation Applications BSA.

Faguet, J.-P., Fox, A. M., & Poeschl, C. (2014). *Does decentralization strengthen or weaken the state? Authority and social learning in a supple state*. London, UK: Department of International Development, London School of Economics and Political Science.

Fahrig, L. L. (2003). Effects of Habitat Fragmentation on Biodiversity. *Annual Review of Ecology Evolution and Systematics*, *34*(1), 487–515. doi:10.1146/annurev.ecolsys.34.011802.132419

Fainstein, S. S. (2005). Cities and Diversity: Should we want it? Can we Plan for it? *Urban Affairs Review*, *41*(1), 3–19. doi:10.1177/1078087405278968

Fan, C. (1999). Migration in a Socialist Transitional Economy: Heterogeneity, Socioeconomic and Spatial Characteristics of Migrants in China and Guangdong Province. *The International Migration Review*, *33*(4), 954–987. doi:10.1177/019791839903300406 PMID:12349707

FAO. (2016). *State of the World's forests 2016. Forests and agriculture: Land-use challenges and opportunities*. Online Report. Rome: FAO. Retrieved from http://www.fao.org/3/a-i5588e.pdf

Farazmand, A. (2003). Chaos and Transformation Theories: A Theoretical Analysis with Implications for Organization Theory and Public Management. *Public Organization Review: A Global Journal*, *3*, 339–372.

Farooq, A. (2007). Social Malice by Housing Type. In *Proceedings of the 6th International Space Syntax Symposium*. Istanbul: ITU Faculty of Architecture.

Farquhar, M. (1995). Definitions of Quality of Life: A Taxonomy. *Journal of Advanced Nursing*, *22*(3), 502–509. doi:10.1046/j.1365-2648.1995.22030502.x PMID:7499618

Farr, D. (2008). *Sustainable urbanism: Urban design with nature*. Chichester, UK: John Wiley. [distributor]

Fielden, A. (2008). *Ignored Displaced Persons: the plight of IDPs in urban areas*. UNHCR, Policy Development and Evaluation Service, United Nations High Commissioner for Refugees.

Filion, P. (2013). Recent Planning and Development in Toronto. In I. Vojnovic (Ed.), *Urban sustainability: A global perspective, edited by Igor Vojnovic* (pp. 509–529). East Lansing, MI: Michigan State University Press.

Fischer, G., & Sun, L. X. (2001). Model based analysis of future land-use development in China. *Agriculture, Ecosystems & Environment*, *85*(1-3), 163–176. doi:10.1016/S0167-8809(01)00182-7

Fitz, H. C., De Bellevue, E. B., Costanza, R., Boumans, R., Maxwell, T., Wainger, L., & Sklar, F. H. (1996). Development of a general ecosystem model for a range of scales and ecosystems. *Ecological Modelling*, *88*(1-3), 263–295. doi:10.1016/0304-3800(95)00112-3

Flaxman, M. (2010). Fundamentals of Geodesign. In *Digital Landscape Architecture 2010* (pp. 28–41). Dessau: Wichmann. Retrieved from http://www.kolleg.loel.hs-anhalt.de/landschaftsinformatik/fileadmin/user_upload/_temp_/2010/Proceedings/Buhmann_28-41.pdf

Fontaine, C. M., & Rounsevell, M. D. A. (2009). An agent-based approach to model future residential pressure on a regional landscape. *Landscape Ecology*, *24*(9), 1237–1254. doi:10.100710980-009-9378-0

Fock, L. L. (2014). Towards Eco-cities in Europe and Asia—Sharing of Best Practices and Experiences: An Introduction. In W. Hofmeister, P. Rueppel, & L. L. Fook (Eds.), Eco-Cities: Sharing European and Asian Best Practices and Experiences (pp. 7-18). Singapore: Konrad-Adenauer Stiftung and European Union.

Forester, J. (1999). *The Deliberative Practitioner: Encouraging Participatory Planning Processes*. Cambridge, MA: MIT Press.

Forkuor, J. B., Akuoko, K. O., & Yeboah, E. H. (2017). Negotiation and Management Strategies of Street Vendors in Developing Countries: A Narrative Review. *SAGE Open*, *7*(1), 1–13. doi:10.1177/2158244017691563

Forman, R. T. T. (1995). Land Mosaics. *Networks*.

Fosu, R., Suprabhas, K., Rathore, Z., & Cory, C. (2015). Integration of Building Information Modeling (BIM) and Geographic Information Systems (GIS) – A literature review and future needs. *Proceedings of the 32nd CIB W78 Conference*, 196–204.

Frances Agyapong, T. K. (2018). Managing traffic congestion in the Accra Central Market, Ghana. *Journal of Urban Management*, *7*(2), 85–96. doi:10.1016/j.jum.2018.04.002

Frank, L. D., & Pivo, G. (1994). Impacts of Mixed Use and Density on Utilization of Three Modes of Travel: Single-Occupant Vehicle, Transit, and Walking. *Transportation Research Record: Journal of the Transportation Research Board*, (1466), 44–52.

Frank, L. D., Sallis, J. F., Saelens, B. E., Leary, L., Cain, K., Conway, T. L., & Hess, P. M. (2010). The development of a walkability index: Application to the Neighborhood Quality of Life Study. *British Journal of Sports Medicine*, *44*(13), 924–933. doi:10.1136/bjsm.2009.058701 PMID:19406732

Frank, L., Bradley, M., Kavage, S., Chapman, J., & Lawton, T. K. (2007). Urban form, travel time, and cost relationships with tour complexity and mode choice. *Transportation*, *35*(1), 37–54. doi:10.100711116-007-9136-6

Franz, G., & Wiener, J. M. (2005). Exploring isovist-based correlates of spatial behavior and experience. In A. van Nes (Ed.), *Proceedings of the 5th International Space Syntax Symposium.* Delft, The Netherlands: TU Delft Press.

Franz, G., Von der Heyde, M., & Bülthoff, H. H. (2005). Predicting experiential qualities of architecture by its spatial properties. In B. Martens & A. G. Keul (Eds.), *Designing Social Innovation: Planning, Building, Evaluating* (pp. 157–166). Cambridge, MA: Hogrefe and Huber.

Frenchman, D. (2000). Planning shapes urban growth and development. In L. B. Sanyal (Ed.), *The Profession of City Planning: Changes, Images and Challenges 1950-2000* (pp. 27–30). Center for Urban Policy Research, Rutgers, The State University of New Jersey.

Fujita, M., Thisse, J. F., & Zenou, Y. (1997). On the endogenous formation of secondary employment centres in a city. *Journal of Urban Economics*, *41*(3), 337–357. doi:10.1006/juec.1996.2002

Furundzic, D., & Furundzic, B. (2012). Infrastructure Corridor as Linear City. *1st International Conference on Architecture & Urban Design* (pp. 721-728). EPOKA University, Department of Architecture.

Galal, O. (2017). The Impact of Digital Simulations on Planning Energy Efficient and Solar Friendly Urban Communities. *SSRN Electronic Journal.* Advance online publication. doi:10.2139srn.3163428

Gallimore, J. M., Brown, B. B., & Werner, C. M. (2011). Walking routes to school in new urban and suburban neighborhoods: An environmental walkability analysis of blocks and routes. *Journal of Environmental Psychology*, *31*(2), 184–191. doi:10.1016/j.jenvp.2011.01.001

Gao, P. C., Liu, Z., Xie, M. H., & Tian, K. (2016). Low-cost cloud computing solution for geo-information processing. *Journal of Central South University*, *23*(12), 3217–3224. doi:10.100711771-016-3387-3

Garvin, A. (2016). *What Makes a Great City?* Washington, DC: Island Press. doi:10.5822/978-1-61091-759-9

Gaunaridis, D., Chorianopoulos, I., & Koukoulas, S. (2018). Exploring prospective urban growth trends under different economic outlooks and land-use planning scenarios: The case of Athens. *Applied Geography (Sevenoaks, England)*, *90*, 134–144. doi:10.1016/j.apgeog.2017.12.001

Geddah, R. (2017, January 26). The costs of marriage. *Al-Ahram Weekly,* (1329).

Gehl Architects. (2007). *Public Spaces, Public Life*. Sydney: City of Sydney.

Gehl, J. (1987). Life between buildings: Using public space (J. Koch, Trans.). New York: Van Nostrand Reinhold.

Gehlawat, M. (2011). Space, the Self, and Singularity in Le Corbusier and Christopher Isherwood. *Literary Imagination*, *15*(3), 349–364. doi:10.1093/litimag/imr140

Gehl, J. (2010). *Cities for People*. Washington, DC: Island Press.

Gehl, J. (2011). *Life between buildings: Using public space*. Island Press.

Gehl, J. (2011). *Life Between Buildings: Using Public Spaces*. Washington, DC: Island Press.

Gehl, J., & Svarre, B. (2013). *How to study public life*. Washington, DC: Island Press. doi:10.5822/978-1-61091-525-0

Geist, H. J., & Lambin, E. F. (2002). Proximate causes and underlying driving forces of tropical deforestation. *Bioscience*, *52*(2), 143–150. doi:10.1641/0006-3568(2002)052[0143:PCAUDF]2.0.CO;2

Geoghegan, J., Wainger, L. A., & Bockstael, N. E. (1997). Spatial landscape indices in a hedonic framework: An ecological economics analysis using GIS. *Ecological Economics*, *23*(3), 251–264. doi:10.1016/S0921-8009(97)00583-1

Gertsev, V., & Gertseva, V. (2004). Classification of mathematical models in ecology. *Ecological Modelling*, *178*(3-4), 329–334. doi:10.1016/j.ecolmodel.2004.03.009

Ghandour, S. E. (2017). Towards More pedestrian-friendly streets in Cairo. *PPAD Working Paper Series,* (3), 1-19. Retrieved from https://documents.aucegypt.edu/Docs/GAPP/PPAD-Working-Paper-Series-3.pdf

Ghoraba, H. (2018). Defusing the population bomb. *Al-Ahram Weekly,* (1404).

Ghozi, R. (2015). Occupancy-based Analysis and Interpretation of Soundscape Auditory Complexity: Case of a Campus Restaurant. *J. Audio Eng. Sco*, *1*(1).

Giachetta, R. (2015). A framework for processing large scale geospatial and remote sensing data in Map Reduce environment. *Computers & Graphics*, *49*, 37–46. doi:10.1016/j.cag.2015.03.003

Giap, T. K., Thye, W. W., & Aw, G. (2014). A new approach to measuring the liveability of cities: The Global Liveable Cities Index. *World Review of Science, Technology and Sustainable Development*, *11*(2), 176–196. doi:10.1504/WRSTSD.2014.065677

Gibon, A., Sheeren, D., Monteil, C., Ladet, S., & Balent, G. (2010). Modeling and simulating change in reforesting mountain landscapes using a social-ecological framework. *Landscape Ecology*, *25*(2), 267–285. doi:10.100710980-009-9438-5

Giddens, A. (2009). *Sociology*. Oxford, UK: Wiley.

Giusto, D., Iera, A., Morabito, G., & Atzori, L. (Eds.). (2010). *The Internet of Things*. Springer. doi:10.1007/978-1-4419-1674-7

Glazier, R. H., Creatore, M. I., Weyman, J. T., Fazli, G., Matheson, F. I., Gozdyra, P., ... Booth, G. L. (2014). Density, destinations or both? A comparison of measures of walkability in relation to transportation behaviors, obesity and diabetes in Toronto, Canada. *PLoS One*, *9*(1), e85295. doi:10.1371/journal.pone.0085295 PMID:24454837

Gleick, J. (1987). *Chaos: Making a New Science*. New York: Knopf.

Golafshani, N. (2003). Understanding reliability and validity in qualitative research. *Qualitative Report*, *8*(4), 597–606.

Goodchild, M. F., & Haining, R. P. (2004). GIS and spatial data analysis: Converging perspectives. *Papers in Regional Science*, *83*(1), 363–385. doi:10.100710110-003-0190-y

Gössling, S., Schröder, M., Späth, P., & Freytag, T. (2016). Urban Space Distribution and Sustainable Transport. *Transport Reviews*, *36*(5), 659–679. doi:10.1080/01441647.2016.1147101

Gouda, A. A., & Masoumi, H. E. (2017a). Certifications systems as independent and rigorous tools for assessing urban sustainability. *International Journal of Urban Sciences*, *2*(1), 1–14. doi:10.1080/12265934.2017.1398103

Gouda, A. A., & Masoumi, H. E. (2017b). Sustainable transportation according to certification systems: A viability analysis based on neighborhood size and context relevance. *Environmental Impact Assessment Review*, *63*, 147–159. doi:10.1016/j.eiar.2016.10.005

Graaff, K., & Ha, N. (Eds.). (2015). *Street Vending in the Neoliberal City: A Global Perspective on the Practices and Policies of a Marginalized Economy*. New York, NY: Berghahn Books.

Greenberg, M. (2000). Branding Cities: A Social History of the Urban Lifestyle Magazine. *Urban Affairs Review*, *36*(2), 228–263. doi:10.1177/10780870022184840

Greene, M. (2002). *From Informal Settlements to Sustainable Neighbourhoods: The role of spatial configuration in the urban system* (Unpublished doctoral dissertation). University of London.

Griffiths, S. (2014). Space Syntax as Interdisciplinary Urban Design Pedagogy. In M. Carmona (Ed.), *Explorations in Urban Design: An Urban Design Research Primer* (pp. 157–167). London, UK: Ashgate Publishing.

Grinberger, A. Y., Lichter, M., & Felsenstein, D. (2015). Simulating urban resilience: Disasters, dynamics and (synthetic) data. In S. Geertman, J. Ferreira Jr, R. Goodspeed, & J. Stillwell (Eds.), *Planning Support Systems and Smart Cities. Lecture Notes in Geoinformation and Cartography*. Cham: Springer. doi:10.1007/978-3-319-18368-8_6

Groat, L. N., & Wang, D. (2013). *Architectural research methods*. John Wiley & Sons.

Groeneveld, J., Müller, B., Buchmann, C. M., Dressler, G., Guo, C., Hase, N., ... Schwarz, N. (2017). Theoretical foundations of human decision-making in agent-based land use models-A review. *Environmental Modelling & Software*, *87*, 39–48. doi:10.1016/j.envsoft.2016.10.008

Grosjean, M., & Thibaud, J. P. (2001). *L'espace urbain en méthode*. Marseille: Parenthèses Editions.

Guarcello, L., & Koseleci, N. (2009). *A profile of Cairo street children*. ILO, World Bank, and UNICEF programme.

Guattari, C., Evangelisti, L., & Balaras, C. A. (2018). On the assessment of urban heat island phenomenon and its effects on building energy performance: A case study of Rome (Italy). *Energy and Building*, *158*, 605–615. doi:10.1016/j.enbuild.2017.10.050

Gudran Madina Motaaba blog. (2012). *Some of what was demolished in June*. Available online at http://thewallsofalex.blogspot.com.eg/2012/06/2012. html

Guthrie, R. A. (2014). *The Causes and Consequences of Migration: How Displacement and Change Affect the Well-Being of Humli Tibetans in Katmandu*. Missoula, MT: Graduate Student Theses, Dissertations, & Professional Papers.

Haddad, M. A. (2015). Review: Steinitz, Carl. 2012. A Framework for Geodesign: Changing Geography by Design. *Journal of Planning Education and Research*, *35*(2), 228–230. doi:10.1177/0739456X15581606

Hadjicostandi, J. (2007). Urbanism/Urban Culture. In G. Ritzer & G. Ritzer (Eds.), *The Blackwell Encyclopedia of Sociology* (pp. 5154–5159). Blackwell Publishing. doi:10.1002/9781405165518.wbeosu022

Hahlweg, D. (1997). The City as a Family. In S. H. Lennard, S. v. Ungern-Sternberg, & H. L. Lennard (Ed.), *Making cities liveable. International Making Cities Liveable Conferences*. Gondolier Press.

Handy, S. (2005). Smart Growth and the Transportation-Land Use Connection: What Does the Research Tell Us? *International Regional Science Review*, *28*(2), 146–167. doi:10.1177/0160017604273626

Handy, S. L. (1996). Understanding the Link Between Urban Form and Nonwork Travel Behavior. *Journal of Planning Education and Research*, *15*(3), 183–198. doi:10.1177/0739456X9601500303

Han, H., Yang, C., & Song, J. (2015). Scenario simulation and the prediction of land use and land cover change in Beijing, China. *Sustainability*, *7*(4), 4260–4279. doi:10.3390u7044260

Han, J., Kamber, M., & Pei, J. (2012). *Data mining: concepts and techniques* (3rd ed.). Waltham, MA: Morgan Kaufmann, Elsevier Inc.

Hankinson, G., & Cowking, P. (1993). *Branding in Action*. London, UK: McGraw-Hill.

Hankinson, G. (2007). The management of destination brands: Five guiding principles based on recent developments in corporate branding theory. *Journal of Brand Management*, *14*(3), 240–254. doi:10.1057/palgrave.bm.2550065

Hannerz, U. (1996). *Transnational Connections: Culture, People, Places*. London: Routledge.

Han, S. Y., & Kim, T. J. (1989). Can expert systems help with planning? *Journal of the American Planning Association*, *55*(3), 296–308. doi:10.1080/01944368908975417

Hanski, I. (1999). Habitat Connectivity, Habitat Continuity, and Metapopulations in Dynamic Landscapes. *Oikos*, *87*(2), 209. doi:10.2307/3546736

Hanson, J. & Hillier, B. (1987). The architecture of community: some new proposals on the social consequences of architectural and planning decisions. *Architecture et Comportment/Architecture and Behaviour, 3*(3), 251 - 273.

Hanson, J. (1989). *Order and Structure in Urban Space: A Morphological History of the City of London*. London, UK: UCL.

Harary, F., & Rockey, J. (1976). A city is not a semilattice either. *Environment & Planning A*, *8*(4), 375–384. doi:10.1068/a080375

Harms, L., Bertolini, L., & Brömmelstroet, M. T. (2015). Performance of Municipal Cycling Policies in Medium-Sized Cities in the Netherlands since 2000. *Transport Reviews*, *36*(1), 134–162. doi:10.1080/01441647.2015.1059380

Harvey, D. (2008). The Right to the City. *New Left Review*, *53*, 23–40.

Harvey, D. (2012). *Rebel Cities: From the Right to the City to the Urban Revolution*. London, UK: Verso.

Harvey, E. P., Cardwell, R. C., McDonald, G. W., Van Delden, H., Vanhout, R., Smith, N. J., ... van den Belt, M. (2019). Developing integrated models by coupling together existing models; land use, economics, demographics and transport in Wellington, New Zealand. *Computers, Environment and Urban Systems*, *74*, 100–113. doi:10.1016/j.compenvurbsys.2018.07.004

Hasbani, J.-G. (2008). *Semi-automated calibration of a cellular automata model to simulate land-use changes in the Calgary region* (Ms thesis). University of Calgary, Department of Geomatics Engineering, Canada.

Hawas, S. (2001). *Khedivian Cairo: Identification and documentation of urban architecture in Downtown Cairo*. Cairo: Arab Contractors Press.

Haykal, M. ḥ. (1973). *The Cairo documents: The Inside Story of Nasser and His Relationship with World Leaders, Rebels, and Statesmen*. Doubleday.

Hays-Mitchell, M. (1994). Street Vending in Peruvian Cities: The Spatio-Temporal Behavior of Ambulantes. *The Professional Geographer*, *46*(4), 425–438. doi:10.1111/j.0033-0124.1994.00425.x

Hazen, B. C., & Berry, M. W. (1997). The simulation of land-cover change using a distributed computing environment. *Simulation Practice and Theory*, *5*(6), 489–514. doi:10.1016/S0928-4869(96)00026-2

Heba, K. (2011). *Preserving architectural heritage in historical cities*. Study Of The Architectural Heritage With The End Of The Nineteenth Century And The Beginning Of The Twentieth Century In Alexandria, Egypt. Available online at http://www.cpasegypt.com/pdf/Khaled_Heba/Research/001%20%20Preserving%20architectural%20heritage%20in%20historical%20cities.pdf

Hedden, W. P. (1929). *How great cities are fed*. New York, NY: Heath and Company.

Heistermann, M., Muller, C., & Ronneberger, K. (2006). Land in sight? Achievements, deficits and potentials of continental to global scale land use modeling. *Agriculture, Ecosystems & Environment*, *114*(2-4), 141–158. doi:10.1016/j.agee.2005.11.015

He, J., Hoyano, A., & Asawa, T. (2009). A numerical simulation tool for predicting the impact of outdoor thermal environment on building energy performance. *Applied Energy*, *86*(9), 1596–1605. doi:10.1016/j.apenergy.2008.12.034

Helmi, A. M., Farhan, M. S., & Nasr, M. M. (2018). A framework for integrating geospatial information systems and hybrid cloud computing. *Computers & Electrical Engineering*, *67*, 145–158. doi:10.1016/j.compeleceng.2018.03.027

Herd, M. (2014). *What makes your city so special?* Retrieved 7 23, 2015, from http://www.theguardian.com/cities/2014/jan/27/guardian-cities-website-welcome

Hernandez, B., Hidalgo, M. C., Salazar-Laplace, M. E., & Hess, S. (2007). Place Attachment and Place Identity in Natives and Non-natives. *Journal of Environmental Psychology*, *27*(4), 310–319. doi:10.1016/j.jenvp.2007.06.003

Herold, M., Couclelis, H., & Clarke, K. C. (2005). The role of spatial metrics in the analysis and modeling of urban land use change. *Computers, Environment and Urban Systems*, *29*(4), 369–399. doi:10.1016/j.compenvurbsys.2003.12.001

Hersperger, A. M., Gennaio, M. P., Verburg, P. H., & Burgi, M. (2011). Feedback loops in conceptual models of land change: Lost in complexity? *Ecology and Society*, *16*(2), r1. doi:10.5751/ES-04204-1602r01

Hickman, R., Hall, P., & Banister, D. (2013). Planning more for sustainable mobility. *Journal of Transport Geography*, *33*, 210–219. doi:10.1016/j.jtrangeo.2013.07.004

Hiebert, D., Rath, J., & Vertovec, S. (2015). Urban Markets and Diversity: Towards a Research Agenda. *Ethnic and Racial Studies*, *38*(1), 5–21. doi:10.1080/01419870.2014.953969

Hilferink, M., & Rietveld, P. (1999). Land use scanner: An integrated GIS based model for long term projections of land use in urban and rural areas. *Journal of Geographical Systems*, *1*(2), 155–177. doi:10.1007101090050010

Hillier, B. & Stonor, T. (2010) Space Syntax- Strategic Urban Design: City Planning Review. *The City Planning Institute of Japan*, *59*(3), 7-11.

Hillier, B. (1999). Centrality as a process: accounting for attraction inequalities in deformed grids. In *Proceedings of the 2nd International Space Syntax Symposium* (vol. 1, pp. 06.1: 06.20). Brasilia.

Hillier, B. (2009a). Spatial sustainability in cities: organic patterns and sustainable forms. In *Proceedings of the 7th International Space Syntax Symposium* (pp. k01.1-k01.20). Stockholm: KTH.

Hillier, B., & Sahbaz, O. (2008). *An Evidence-Based Approach to Crime and Urban Design or, Can We Have Vitality, Sustainability and Security All at Once?* Retrieved from https://spacesyntax.com/wp-content/uploads/2011/11/Hillier-Sahbaz_An-evidence-based-approach_010408.pdf

Hillier, B. (1989). The Architecture of the Urban Object. *Ekistics*, *56*, 5–21.

Hillier, B. (1996a). *Space Is the Machine: A Configurational Theory of Architecture*. Cambridge, UK: Cambridge University Press.

Hillier, B. (1996b). Cities as Movement Economy. *URBAN DESIGN International*, *1*(1), 41–60. doi:10.1057/udi.1996.5

Hillier, B. (1996b). *Space is the Machine*. Cambridge, UK: Cambridge University Press.

Hillier, B. (1997). Cities as movement economies. In P. Droege (Ed.), *Intelligent Environments: Spatial Aspects of the information revolution* (pp. 295–344). Elsevier B.V. doi:10.1016/B978-044482332-8/50020-2

Hillier, B. (1999). The Common Language of Space: A Way of Looking at the Social, Economic and Environmental Functioning of Cities on a Common Basis. *Journal of Environmental Sciences (China)*, *11*, 344–349.

Hillier, B. (1999b). Centrality as a Process: Accounting for Attraction Inequalities in Deformed Grid. *URBAN DESIGN International*, *4*(3-4), 107–127. doi:10.1057/udi.1999.19

Hillier, B. (2001). Centrality as a Process: Accounting for Attraction Inequalities in Deformed Grids. *URBAN DESIGN International*, *4*(3), 107–127.

Hillier, B. (2002). *Can Streets Be Made Safe?* Oxford, UK: University of Oxford.

Hillier, B. (2004). Can streets be made safe? *URBAN DESIGN International*, *9*(1), 31–45. doi:10.1057/palgrave.udi.9000079

Hillier, B. (2008). Space and Spatiality: What the Built Environment Needs from Social Theory. *Building Research and Information*, *36*(3), 216–230. doi:10.1080/09613210801928073

Hillier, B. (2009b) The genetic code for cities – is it simpler than we thought? In *Proceedings of Complexity in Cities conference*. Delft: TU Delft Faculty of Architecture.

Hillier, B. (2009c) What do we need to add to a social network to get a society? Answer: something like what we have to add to a spatial network to get a city. In: D. Koch, L. Marcus, & J. Steen (Eds.) *Proceedings of the 7th International Space Syntax Symposium* (pp. p. 42). Stockholm: Royal Institute of Technology (KTH).

Hillier, B., & Hanson, J. (1984). *The Social Logic of Space*. Cambridge, UK: Cambridge University Press. doi:10.1017/CBO9780511597237

Hillier, B., & Hanson, J. (1998). Space Syntax as a Research Programme. *Urban Morphology*, *2*(2), 108–111.

Hillier, B., Hanson, J., & Graham, H. (1987). Ideas Are in Things: An Application of Space Syntax Method to Discovering House Genotypes. *Environment and Planning. B, Planning & Design*, *14*(4), 363–385. doi:10.1068/b140363

Hillier, B., Hanson, J., Peponis, J., Hudson, J., & Burdett, R. (1983). Space Syntax: A Different Urban Perspective. *Architectural Journal*, *30*, 47–63.

Hillier, B., Hanson, J., Peponis, J., Hudson, J., & Burdett, R. (1983). Space Syntax: A different urban perspective. *The Architects Journal*, *178*, 47–63.

Hillier, B., & Iida, S. (2005). Network and Psychological Effects in Urban Movement. In A. G. Cohn & D. M. Mark (Eds.), Lecture Notes in Computer Science: Vol. 3693. *Spatial Information Theory. COSIT 2005*. Berlin: Springer. doi:10.1007/11556114_30

Hillier, B., Penn, A., Hanson, J., Grajewski, T., & Xu, J. (1993). Natural Movement: Or, Configuration and Attraction in Urban Pedestrian Movement. *Environment and Planning. B, Planning & Design*, *20*(1), 29–66. doi:10.1068/b200029

Hillier, B., & Shu, S. (1999). Design for secure space. *Planning in London*, *29*, 36–38.

Hillier, B., Turner, A., Yang, T., & Park, H. T. (2007). Metric and Topo-Geometric Properties of Urban Street Network: Some Convergences, Divergences and New Results. In *Proceedings of the 6th International Space Syntax Symposium*. Istanbul: ITU Faculty of Architecture.

Hillier, B., Yang, T., & Turner, A. (2012). Normalising least angle choice in Depthmap and how it opens up new perspectives on the global and local analysis of city space. *Journal of Space Syntax*, *3*(2), 155–193.

Holden, E., & Norland, I. T. (2005). Three Challenges for the Compact City as a Sustainable Urban Form: Household Consumption of Energy and Transport in Eight Residential Areas in the Greater Oslo Region. *Urban Studies (Edinburgh, Scotland)*, *42*(12), 2145–2166. doi:10.1080/00420980500332064

Holland, J., & Miller, J. H. (1991). Artificial adaptive agents in economic theory. *The American Economic Review*, *81*(2), 65–71.

Hollands, R. G. (2008). Will the real smart city please stand up? Intelligent, progressive or entrepreneurial? *City*, *12*(3), 303–320. doi:10.1080/13604810802479126

Hölscher, C., Dalton, R. C., & Turner, A. (2007). *Proceedings of the Workshop on Spatial Cognition: Space Syntax and Spatial Cognition.* Bremen: Universität Bremen.

Houssein, F. (2012). *Construction et expérimentation d'un protocole pour le traitementdes ambiances sonores urbaines adapté aux personnes âgées presbyacousiques.* Nantes: thèse de doctorat en cotutelle, ENSA nantes.

Howard, J. (2009). Climate Change Mitigation and Adaptation in Developed Nations: A Critical Perspective on the Adaptation Turn in Urban Climate Planning. In S. Davoudi, J. Crawford, & A. Mehmood (Eds.), *Planning for climate change: Strategies for mitigation and adaptation for spatial planners* (pp. 19–32). London: Earthscan.

Howe, J. (2008). *Crowdsourcing: Why the Power of the Crowd Is Driving the Future of Business.* Crown Business.

Hox, J. J., & Boeije, H. R. (2005). Data Collection, Primary vs. Secondary. In *Encyclopedia of Social Measurement* (pp. 593–599). Elsevier. doi:10.1016/B0-12-369398-5/00041-4

Huang, H., Ooka, R., & Kato, S. (2005). Urban thermal environment measurements and numerical simulation for an actual complex urban area covering a large district heating and cooling system in summer. *Atmospheric Environment*, *39*(34), 6362–6375. doi:10.1016/j.atmosenv.2005.07.018

Huigen, M. G. A. (2003). *Agent-based modeling in land use and land cover change studies.* Interim Report IR-03-044, International Institute for Applied Systems Analysis-IIASA, Austria.

Hui, N., Saxe, S., Roorda, M., Hess, P., & Miller, E. J. (2017). Measuring the completeness of complete streets. *Transport Reviews*, *38*(1), 73–95. doi:10.1080/01441647.2017.1299815

Humanitarian Fact Sheet. (2016). *Forced Displacement: Poverty and Financing, Seven facts you need to know.* Global Humanitarian Assistance and development initiatives. Retrieved from http://devinit.org/wp-content/uploads/2016/09/Forced_Displacement_Poverty_and_Financing_DI_Sept_2016.pdf

Ibn Khaldun, A. b. (1377). *The Muqaddimah "Prolegomenon".*

Ibn-Khaldun, A. (1967). The Muqaddimah: An Introduction to History (Rosenthal, F. Trans., and Dawood, N.J. Ed.). London, UK: Routledge & Kegan Paul Publisher.

Ibrahim, S. (2013). *That Smell and Notes from Prison* (R. Creswell, Trans.). New York: New Directions.

ICOMOS. (2001). *ICOMOS World Report on Monuments and Sites in Danger.* ICOMOS.

IDRC, CDDI, & UN-HABITAT. (2008). *Decentralization, local power, and women's rights.* Ottawa, ON, Canada: International Development Research Centre. Retrieved from https://www.idrc.ca/sites/default/files/sp/Documents%20EN/policy-recommendations-on-decentralization.pdf

Internal Displacement Monitoring Centre. (2018). *2018 Global Report on Internal Displacement (GRID 2018)*. Retrieved from reliefweb: https://reliefweb.int/report/world/2018-global-report-internal-displacement-grid-2018

Irwin, G. E., & Geoghegan, J. (2001). Theory, data and methods: Developing spatially explicit economic models of land use change. *Agriculture, Ecosystems & Environment*, *85*(1-3), 7–23. doi:10.1016/S0167-8809(01)00200-6

Isikdag, U., & Zlatanova, S. (2009a). A SWOT analysis on the implementation of BIM within [the] geospatial environment. In Urban and Regional Data Management, UDMS Annuals 2009. CRC Press.

Jacobs, J. (1961). The Death and Life of Great American Cities. Random House Inc.

Jacobs, A. (1993). *Great Streets*. Cambridge, MA: MIT Press.

Jacobs, A., & Appleyard, D. (1987). Toward an Urban Design Manifesto. *Journal of the American Planning Association*, *53*(1), 112–120. doi:10.1080/01944368708976642

Jacobs, J. (1961). *The Death and Life of Great American Cities*. New York: Random House.

Jalaladdini, S., & Oktay, D. (2012). Urban Public Spaces and Vitality: A Socio-Spatial Analysis in the Streets of Cypriot Towns. *Procedia: Social and Behavioral Sciences*, *35*, 664–674. doi:10.1016/j.sbspro.2012.02.135

Jankowski, P., & Richard, L. (1994). Integration of GIS-based suitability analysis and multicriteria evaluation in a spatial decision support system for route selection. *Environment & Planning B*, *21*(3), 323–340. doi:10.1068/b210323

Jansen, H., & Stoorvogel, J. J. (1998). Quantification of aggregation bias in regional agricultural land use models: Application to Giocimo County, Costa Rica. *Agricultural Systems*, *58*(3), 417–437. doi:10.1016/S0308-521X(98)00039-0

Jenerette, D. G., & Wu, J. (2001). Analysis and simulation of land-use change in the central Arizona-Phoenix region, USA. *Landscape Ecology*, *16*(7), 611–626. doi:10.1023/A:1013170528551

Jin, Y., & Wilson, A. (1993). Generation of integrated multispatial input-output models of cities (GIMIMoC): Initial stage. *Papers in Regional Science*, *72*(4), 351–367. doi:10.1007/BF01576848

John, N. M., Da Luz Reis, A. T., De Lima, M. A., & Dias Lay, M. C. (2016). Evaluation of a popular shopping mall built to accommodate previous street vendors in downtown Porto Alegre. *Archnet-IJAR: International Journal of Architectural Research*, *10*(1), 75–95. doi:10.26687/archnet-ijar.v10i1.832

Johnson, P.-A. (1994). *The Theory of Architecture: Concepts Themes & Practices*. New York: John Wiley & Sons.

Johnson, P., & Sieber, R. (2009). Agent-Based Modelling: A Dynamic Scenario Planning Approach to Tourism PSS. In *Planning Support Systems Best Practice and New Methods. The GeoJournal Library* (Vol. 95). Dordrecht: Springer. doi:10.1007/978-1-4020-8952-7_11

Jorgensen, S. (2008). Overview of the model types available for development of ecological models. *Ecological Modelling*, *215*(1-3), 3–9. doi:10.1016/j.ecolmodel.2008.02.041

Junfeng, J., & Boerboom, L. (2006). Transition rule elicitation methods for urban cellular automata models. In J. van Leeuwen & H. Timmermans (Eds.), *Innovation in design & decision support systems in architecture and urban planning*. Springer.

Jun, M.-J. (2004). A metropolitan input-output model: Multisectoral and multispatial relations of production, income formation, and consumption. *The Annals of Regional Science*, *38*(1), 131–147. doi:10.100700168-003-0138-9

Jun, M.-J. (2005). Forecasting urban land-use demand using a metropolitan input-output model. *Environment & Planning A*, *37*(7), 1311–1328. doi:10.1068/a3723

Kafafy, N. A. (2016). Right to Urban Space in Post-Revolution Cairo: A Study for Street Vending Phenomenon. In Dynamics and Resilience of Informal Areas: International perspectives (pp. 97-112). Switzerland: Springer International Publishing. doi:10.1007/978-3-319-29948-8_6

Kafafy, N. A. (2017). Street Traders in Post-Revolution Cairo: Victims or Villains? In A. Brown (Ed.), *Rebel Streets and the Informal Economy: Street Trade and the Law* (pp. 220–235). New York, NY: Routledge. doi:10.4324/9781315641461-14

Kakembo, V., & van Niekerk, S. (2014). The Integration of GIS into Demographic Surveying of Informal Settlements: The Case of Nelson Mandela Bay Municipality, South Africa. *Habitat International*, *44*, 451–460. doi:10.1016/j.habitatint.2014.09.004

Kanaroglou, P., & Scott, D. (2001). Integrated urban transportation and land-use models for policy analysis. In M. Dijst, W. Schenkel, & I. Thomas (Eds.), *Governing cities on the move*. Aldershot, UK: Avebury.

Karimi, K. (2016). The Shifting Paradigm of Space Syntax Research and its Future Prospects: A Reflection on the 10th International Space Syntax Symposium in London, 2015. *Journal of Space Syntax*, *6*(2), 275–278.

Karimi, K. (2018). Space Syntax: Consolidation and Transformation of an Urban Research Field. *Journal of Urban Design*, *23*(1), 1–4. doi:10.1080/13574809.2018.1403177

Karimi, K., & Parham, E. (2012). An Evidence Informed Approach to Developing an Adaptable Regeneration Programme for Declining Informal Settlements. *8th International Space Syntax Symposium.*

Karimi, K., & Vaughan, L. (2014). An Evidence-based approach to designing new cities: The English New Towns revisited. In M. Carmona (Ed.), *Explorations in urban design: An urban design research primer* (pp. 261–274). Surrey, UK: Ashgate Publishing Limited.

Katzschner, L., Bosch, U., & Röttgen, M. (2006, September). Behaviour of people in open spaces in dependence of thermal comfort conditions. *Proceedings of 23rd Conference on Passive and Low Energy Architecture.*

Kenworthy, J. R., & Laube, F. B. (1999). Patterns of automobile dependence in cities: An international overview of key physical and economic dimensions with some implications for urban policy. *Transportation Research Part A, Policy and Practice*, *33*(7-8), 691–723. doi:10.1016/S0965-8564(99)00006-3

Kerr, J., Frank, L., Sallis, J. F., & Chapman, J. (2007). Urban form correlates of pedestrian travel in youth: Differences by gender, race-ethnicity and household attributes. *Transportation Research Part D, Transport and Environment*, *12*(3), 177–182. doi:10.1016/j.trd.2007.01.006

Khalifa, M. (2015). Evolution of informal settlements upgrading strategies in Egypt: From negligence to participatory development. *Ain Shams Engineering Journal*, *6*(4), 1151–1159. doi:10.1016/j.asej.2015.04.008

Kim, Y. O. (2001). The Role of Spatial Configuration in Spatial Cognition. *Proceedings of the 3rd International Space Syntax Symposium.*

Kim, Y. O., & Penn, A. (2004). Linking the Spatial Syntax of Cognitive Maps to the Spatial Syntax of the Environment. *Environment and Behavior*, *36*(4), 483–504. doi:10.1177/0013916503261384

Kipper, R. (2009). Cairo: a broader view. In R. Kipper & M. Fischer (Eds.), *Cairo's informal areas between urban challenges and hidden potentials* (pp. 12–15). Cairo: GTZ Egypt.

Kitamura, R., Mokhtarian, P. L., & Daidet, L. (1997). Article. *Transportation*, *24*(2), 125–158. doi:10.1023/A:1017959825565

Kiunsi, R. (2013). A review of traffic congestion in Dar es Salaam city from the physical planning perspective. *Journal of Sustainable Development*, *6*(2), 94–103. doi:10.5539/jsd.v6n2p94

Klaasen, I. T. (2003). *Knowledge-based design: developing urban & regional design into a science*. Delft: Delft University Press/Techne Press.

Klarqvist, B. (1993). A space syntax glossary. *Nordic Journal of Architectural Research*, *2*, 11–12.

Klarqvist, B. (1993). Space Syntax Glossary. *Nordisk Arkitekturforskning*, *2*, 11–12.

Klein, L. J., Marianno, F. J., Albrecht, C. M., Freitag, M., Lu, S., Hinds, N., & Hamann, H. F. (2015). A scalable geospatial data analytics platform. In *International Conference In Big Data (Big Data)* (pp. 1290-1298). IEEE. 10.1109/BigData.2015.7363884

Klinger, T., Kenworthy, J. R., & Lanzendorf, M. (2013). Dimensions of urban mobility cultures – a comparison of German cities. *Journal of Transport Geography*, *31*, 18–29. doi:10.1016/j.jtrangeo.2013.05.002

Knez, I., & Thorsson, S. (2006). Influences of culture and environmental attitude on thermal, emotional and perceptual evaluations of a public square. *International Journal of Biometeorology*, *50*(5), 258–268. doi:10.100700484-006-0024-0 PMID:16541241

Koerniawan, M. (2015). *The Simulation Study of Thermal Comfort in Urban Open Spaces of Commercial Area Using EnviMet Software*. Academic Press.

Kohler, T. A., & Gumerman, G. J. (2000). *Dynamics in human and primate societies: Agent based modeling of social and spatial processes. Sante Fe Institute Studies in the Sciences of Complexity*. Oxford, UK: Oxford University Press.

Kohli, D., Sliuzas, R., Kerle, N., & Stein, A. (2012). An Ontology of Slums for Image Based Classification. *Computers, Environment and Urban Systems*, *36*(2), 154–163. doi:10.1016/j.compenvurbsys.2011.11.001

Kollman, K., Miller, J. H., & Page, S. E. (1997). Political institutions and sorting in a Tiebout Model. *The American Economic Review*, *87*, 977–922.

Koohsari, M. J., Kaczynski, A., Mcormack, G. R., & Sugiyama, T. (2014). Using Space Syntax to Assess the Built Environment for Physical Activity: Applications to Research on Parks and Public Open Space. *Leisure Sciences: An Interdisciplinary Journal*, *36*(2), 206–216. doi:10.1080/01490400.2013.856722

Koomen, E. (2008). *Spatial analysis in support of physical planning* (PhD thesis). Amsterdam: Vrije Universiteit Amsterdam.

Koomen, E., Stillwell, J., Bakema, A., & Scholten, H. J. (Eds.). (2007). *Modeling land use change: Progress and applications*. Springer.

Koresawa, A., & Konvitz, J. (2001). Towards a new role for spatial planning. In O. Secretariat (Ed.), Towards a New Role for Spatial Planning (pp. 11-32). Paris: Organisation for Economic Co-Operation.

Kraftl, P. (2014). Liveability and urban architectures: Mol(ecul)ar biopower and the 'becoming lively' of sustainable communities. *Environment and Planning. D, Society & Space*, *32*(2), 274–292. doi:10.1068/d21012

Krishnamurthy, K., & Arasan, V. T. (2012). Effect of road width and traffic volume on vehicular interactions in heterogeneous traffic. *Journal of Advanced Transportation*. doi:10.1002/atr.196

Krohe, J. (2000). Return to Broadacre City. *Illinois Issues.*, *26*(4), 27–29.

Kropf, K. (2009). Aspects of urban form. *Urban Morphology*, *13*(2).

Kropf, K. (2014). Ambiguity in the definition of built form. *Urban Morphology*, *18*(1), 41–57.

Krüger, M. J. T. (1979). An approach to build-form connectivity. *Environment and Planning. B, Planning & Design*, *6*(1), 67–88.

Kubat, A. S., Istek, C., & Asami, Y. (2001). Characterization of the Street Networks in the Traditional Turkish Urban Form. *Environment and Planning. B, Planning & Design*, *28*(5), 777–795. doi:10.1068/b2718

Kunzmann, K. R. (1993). Geodesign: Chance oder Gefahr? In Bundesforschungsanstalt für Landeskunde und Raumordnung (Ed.), Planungskartographie und Geodesign (pp. 389–396). Informationen zur Raumentwicklung.

L'Heureux, A., Grolinger, K., Higashino, W. A., & Capretz, M. A. (2017). A Gamification Framework for Sensor Data Analytics. In *IEEE International Congress on Internet of Things (ICIOT 2018)* (pp. 74-81). Honolulu, HI: IEEE. 10.1109/IEEE.ICIOT.2017.18

Lambin, E. F. (1997). Modeling and monitoring land-cover change processes in tropical regions. *Progress in Physical Geography*, *21*(3), 375–393. doi:10.1177/030913339702100303

Lambin, E. F., Rounsevell, M. D. A., & Geist, H. J. (2000). Are agricultural land-use models able to predict changes in land-use intensity? *Agriculture, Ecosystems & Environment*, *82*(1-3), 321–331. doi:10.1016/S0167-8809(00)00235-8

Landis, J. D., Monzon, J. P., Reilly, M., & Cogan, C. (1998). *Development and pilot application of the California Urban and Biodiversity Analysis (CURBA) Model.* Paper presented at the 1998 ESRI International User Conference.

Landis, J. (1995). Imagining land use futures: Applying the California urban futures model. *Journal of the American Planning Association*, *61*(4), 438–457. doi:10.1080/01944369508975656

Landis, J., & Zhang, M. (1998a). The second generation of the California urban futures model. Part 1: Model logic and theory. *Environment & Planning A*, *25*(5), 657–666. doi:10.1068/b250657

Landis, J., & Zhang, M. (1998b). The second generation of the California urban futures model. Part 2: Specification and calibration results of the land-use change submodel. *Environment & Planning A*, *25*(6), 795–824. doi:10.1068/b250795

Landry, C., & Wood, P. (2008). *The Intercultural City: Planning for Diversity Advantage.* Earthscan.

Lash, H. (1976). *Planning in a human way: personal reflections on the regional planning experience in Greater Vancouver.* Ottawa: Ministry of State for Urban Affairs.

Lavalle, C., Baranzelli, C., & Batista e Silva, F. (2011). A high resolution land use/cover modeling framework for Europe: Introducing the EU-ClueScanner100 Model. In B. Murgante et al. (Eds.), ICCSA-Part I (pp. 60-65). Springer.

Layeb, S. (2008). *Intervention urbaine dans un tissu urbain ancien: cas de la petite Sicile de Tunis* (Unpublished doctoral dissertation). Tunis: ENAU.

Layeb, S. (2012). *Parcours commenté, parcours segmenté: recherche exploratoire sur les liens entre configurations urbaines et stress sonore en milieu urbain* (Unpublished doctoral dissertation). Tunis: Academic Press.

Layeb, S. (2016). Caractérisation ambiantale des espaces d'accueil pour enfants autistes:Vers une projectuelles des ambiances. *Séminaire de Conception Architecturale Numérique, SCAN'16*, 163-172.

Leck, E. (2006). The Impact of Urban Form on Travel Behavior: A Meta-Analysis. *Berkeley Planning Journal*, *19*(1), 37–58.

Lee, D. B., Jr. (1973, May). Requiem for large-scale models. *AIP Journal*, 163–177.

Lee, B. (2007). Edge or edgeless cities? Urban spatial structure in US metropolitan areas, 1980 to 2000. *Journal of Regional Science*, *47*(3), 479–515. doi:10.1111/j.1467-9787.2007.00517.x

Lee, C., & Moudon, A. V. (2006). The 3Ds+R: Quantifying land use and urban form correlates of walking. *Transportation Research Part D, Transport and Environment*, *11*(3), 204–215. doi:10.1016/j.trd.2006.02.003

Lefebver, H. (1991). *The Production of Space* (D. Nicholson-Smith, Trans.). Malden, MN: Blackwell Publishers.

Lefebvre, H. (1984). *Everyday Life in the Modern World.* New Brunswick, NJ: Transaction Publishers.

Lefebvre, H. (1991). Critique of Everyday Life: Vol. I. *Introduction.* London: Verso.

Lefebvre, H. (2003). *The Urban Revolution.* Minneapolis, MN: University of Minnesota Press.

Leh, M., Bajwa, S., & Chaubey, I. (2013). Impact of land use change on erosion risk: An integrated remote sensing, geographic information system and modeling methodology. *Land Degradation & Development*, *24*, 409–421.

Lennard, S. H., & Lennard, H. L. (1995). *Livable cities observed: a source book of images and ideas for city officials, community leaders, architects, planners and all other committed to making their cities livable.* Southampton, UK: Gondolier Press.

Lerman, Y., & Lebendiger, Y. (2017). Visualizing Space Syntax Analyses for Decision Makers: Lessons from Israel. *11th International Space Syntax Symposium.*

Lerman, Y., Rofe, Y., & Omer, I. (2014). Using Space Syntax to Model Pedestrian Movement in Urban Transportation Planning. *Geographical Analysis*, *46*(4), 392–410. doi:10.1111/gean.12063

Leslie, E., Coffee, N., Frank, L., Owen, N., Bauman, A., & Hugo, G. (2007). Walkability of local communities: Using geographic information systems to objectively assess relevant environmental attributes. *Health & Place*, *13*(1), 111–122. doi:10.1016/j.healthplace.2005.11.001 PMID:16387522

Levins, R. (1969). Some demographic and genetic consequences of environmental heterogeneity for biological control. *Bulletin of the Entomological Society of America*, *15*, 237–240. doi:10.1093/besa/15.3.237

Levitas, S. (1978). Anthropology and the sociology of the streets. In S. Anderson (Ed.), *On streets* (pp. 225–240). Cambridge, MA: The MIT Press.

Levy, A. (1999). Urban morphology and the problem of the modern urban fabric: Some questions for research. *Urban Morphology*, *3*(2), 79–85.

Ley, A., & Newton, P. (2010). Creating and sustaining liveable cities. In S. Kallidaikurichi & B. Yuen (Eds.), Developing Living Cities: From Analysis to Action (pp. 191-230). Singapore: World Scientific. doi:10.1142/9789814304504_0008

Licciardi, G., & Amirtahmasebi, R. (n.d.). The economics of uniqueness. Investing in historic city cores and cultural heritage assets for sustainable development. Academic Press. doi:10.1596/978-0-8213-9650-6

Li, D., Shan, J., Shao, Z., Zhou, X., & Yao, Y. (2013). Geomatics for Smart Cities-Concept, Key Techniques, and Applications. *Geo-Spatial Information Science*, *16*(1), 13–24. doi:10.1080/10095020.2013.772803

Liddle, B. (2013). Urban density and climate change: A STIRPAT analysis using city-level data. *Journal of Transport Geography*, *28*, 22–29. doi:10.1016/j.jtrangeo.2012.10.010

Ligtenberg, A., Bregt, A. K., & van Lammeren, R. (2001). Multi-actor-based land use modeling: Spatial planning using agents. *Landscape and Urban Planning*, *56*(1-2), 21–33. doi:10.1016/S0169-2046(01)00162-1

Li, J., Ren, L., Hu, T., & Wang, F. (2018). A city's "urban crack" at 4 a.m.: A case study of morning market vendors in Beijing's Longfu Temple area. *Habitat International*, *71*, 14–21. doi:10.1016/j.habitatint.2017.10.006

Lin, B., Li, X., Zhu, Y., & Qin, Y. (2008). Numerical simulation studies of the different vegetation patterns' effects on outdoor pedestrian thermal comfort. *Journal of Wind Engineering and Industrial Aerodynamics*, *96*(10-11), 1707–1718. doi:10.1016/j.jweia.2008.02.006

Lin, T. P. (2009). Thermal perception, adaptation, and attendance in a public square in hot and humid regions. *Building and Environment*, *44*(10), 2017–2026. doi:10.1016/j.buildenv.2009.02.004

Li, S., Verburg, P. H., Shihai, L., Wu, J., & Xiaobing, L. (2012). Spatial analysis of the driving factors of grassland degredation under conditions of climate change and intensive use in Inner-Mongolia, China. *Regional Environmental Change*, *12*(3), 461–474. doi:10.100710113-011-0264-3

Littlefair, P. J. (2000). *Environmental site layout planning: Solar access, microclimate and passive cooling in urban areas*. Building Research Establishment.

Liu, C., Susilo, Y. O., & Karlström, A. (2017). Weather variability and travel behaviour – what we know and what we do not know. *Transport Reviews*, *37*(6), 715–741. doi:10.1080/01441647.2017.1293188

Liu, X., & Anderson, C. (2004). Assessing the impact of temporal dynamics on land-use change modeling. *Computers, Environment and Urban Systems*, *28*(1-2), 107–124. doi:10.1016/S0198-9715(02)00045-5

Li, W., & Milburn, L. A. (2016). The evolution of geodesign as a design and planning tool. *Landscape and Urban Planning*, *156*, 5–8. doi:10.1016/j.landurbplan.2016.09.009

Li, X., Hijazi, I., Reinhard Koenig, Z. L., Zhong, C., & Schmitt, G. (2016). Assessing Essential Qualities of Urban Space with Emotional and Visual Data Based on GIS Technique. *International Journal of Geo-Information*, *5*(11), 1–18. doi:10.3390/ijgi5110218

Li, X., & Yeh, A. G. (2000). Modeling sustainable urban development by the integration of constrained cellular automata and GIS. *International Journal of Geographical Information Science*, *14*(2), 131–152. doi:10.1080/136588100240886

Lo, A. Y. (2016). Small is green? Urban form and sustainable consumption in selected OECD metropolitan areas. *Land Use Policy*, *54*, 212–220. doi:10.1016/j.landusepol.2016.02.014

Loibl, W., Tötzer, T., Köstl, M., & Steinnocher, K. (2007). Simulation of polycentric urban growth dynamics through agents: model concept, application, results and validation. In E. Koomen, J. Stillwell, A. Bakema, & H. J. Scholten (Eds.), *Modeling land use change progress and applications* (pp. 219–235). Dordrecht: Springer.

Longley, P., Higgs, G., & Martin, D. (1994). The predictive use of GIS to model property valuations. *International Journal of Geographical Information Systems*, *8*(2), 217–235. doi:10.1080/02693799408901995

Long, Y., Baran, P. K., & Moore, R. (2007). The Role of Space Syntax in Spatial Cognition: Evidence from Urban China. In *Proceedings of the 6th International Space Syntax Symposium*. Istanbul: ITU Faculty of Architecture.

Lopez, M. J. J., & Nes, A. V. (2007). Space and Crime in Dutch Built Environments: Macro and Micro Scale Spatial Conditions for Residential Burglaries and Thefts from Cars. In *Proceedings of the 6th International Space Syntax Symposium*. Istanbul: ITU Faculty of Architecture.

López, R. C., Arredondo, C. I., & Salcedo, J. (2011). *The Effects of Internal Displacement on Host Communities: A Case Study of Suba and Ciudad Bolívar Localities in Bogotá, Colomb*. Bogotá: Brookings Institution-London School of Economics, Project on Internal Displacement.

Lorenz, E. (1993). *The Essence of Chaos*. Seattle, WA: University of Washington Press. doi:10.4324/9780203214589

Lor, J., & Britz, J. J. (2007). Is a knowledge society possible without freedom of access to information? *Journal of Information Science*, *33*(4), 387–397. doi:10.1177/0165551506075327

Losch, A. (1954). *The economics of location* (W. H. Woglom & W. F. Stolper, Trans.). New Haven, CT: Yale University Press.

Lowe, M., Whitzman, C., Badland, H., Davern, M., Hes, D., Aye, L., . . . Giles-Corti, B. (2013). *What Are the Key Indicators for Melbourne Neighbourhoods?: Place, Health and Liveability Research Program.* Melbourne, Australia: Melbourne Sustainable Society Institute.

Lowe, M., Whitzman, C., Badland, H., Davern, M., Aye, L., Hes, D., & Butterworth, I. (2015). Billie Giles-Corti, Planning Healthy, Liveable and Sustainable. *Urban Policy and Research, 33*(2). doi:10.1080/08111146.2014.1002606

Low, S., & Smith, N. (Eds.). (2005). *The Politics of Public Space*. New York, NY: Routledge.

LUBW (Landesanstalst für Umwelt Baden-Württemberg). (2010). Bewertung von Böden nach ihrer Leistungsfähigkeit. Leitfaden für Planungen und Gestattungsverfahren. *Reihe Bodenschutz, 23*.

LUBW. (2012). *Landesanstalst für Umwelt Baden-Württemberg*. Fachplan Landesweiter Biotopverbund. Arbeitshilfe.

Lucas, E. R., & Rossi-Hansberg, E. (2002). On the internal structure of cities. *Econometrica, 70*(4), 1445–1476. doi:10.1111/1468-0262.00338

Lynch, K. (1960). *The Image of the City*. Cambridge, MA: The MIT Press.

Lynch, K. (1960). *The image of the city. Publications of the Joint Center for Urban Studies*. Cambridge, MA: MIT Press.

Lynch, K. (1976). *L'image de la cité*. Paris: Bordas.

Lynch, K. (1984). The Immature Arts of City Design. *Places, 1*(3), 10–21.

Lynch, K., & Rodwin, L. (1958). A Theory of Urban Form. *Journal of the American Institute of Planners, 24*(4), 201–214. doi:10.1080/01944365808978281

Lyons, M., & Snoxell, S. (2005). Sustainable Urban Livelihoods and Marketplace Social Capital: Crisis and Strategy in Petty Trade. *Urban Studies (Edinburgh, Scotland), 42*(8), 1301–1320. doi:10.1080/00420980500150631

Magda, M. (2003). *An investigation of the relation of space to society: a discussion on Giddens, H. Lefebvre and space syntax* (M.Sc. thesis). University College London.

Mahfouz, N. (1992). *Midaq Ally* (T. L. Gassick, Trans.). Anchor Books Edition.

Ma, J., Liu, Z., & Chai, Y. (2015). The impact of urban form on CO2 emission from work and non-work trips: The case of Beijing, China. *Habitat International, 47*, 1–10. doi:10.1016/j.habitatint.2014.12.007

Malczewski, J. (1997). *Spatial Decision Support Systems, NCGIA Core Curriculum in GIScience*. Retrieved 06 30, 2011, from http://www.ncgia.ucsb.edu/giscc/units/u127/u127.html

Malczewski, J., & Keller, C. P. (1997). *Unit 127 - Spatial Decision Support Systems*. Retrieved 11 20, 2018, from NCGIA Core Curriculum in Geographic Information Science: http://www.geo.upm.es/postgrado/CarlosLopez/materiales/cursos/www.ncgia.ucsb.edu/giscc/units/u127/index.html

Manaugh, K., & Kreider, T. (2013). What is mixed use? Presenting an interaction method for measuring land use mix. *Journal of Transport and Land Use, 6*(1), 63. doi:10.5198/jtlu.v6i1.291

Maneepong, C., & Walsh, J. C. (2013). A New Generation of Bangkok Street Vendors: Economic Crisis as Opportunity and Threat. *Cities (London, England), 34*, 37–43. doi:10.1016/j.cities.2012.11.002

Manski, C. F. (2000). Economic analysis of social interactions. *The Journal of Economic Perspectives, 14*(3), 115–136. doi:10.1257/jep.14.3.115

Manzo, L. C. (2011). 'We are the Fruit Bowl': Place, Cultural Identity and Social Ties among Immigrant Residents in Public Housing. In J. Hou, & J. L. Roberts (Ed.), *Transcultural Cities: Symposium Proceedings* (pp. 14-23). Washington, DC: Department of Landscape Architecture, University of Washington.

Marcus, L., Giusti, M., & Barthel, S. (2016). Cognitive Affordances in Sustainable Urbanism: Contributions of Space Syntax and Spatial Cognition. *Journal of Urban Design*, *21*(4), 439–452. doi:10.1080/13574809.2016.1184565

Marcus, L., Stahle, A., & Dehlhielm, M. (2010). Architectural Knowledge and Complex Urban Space Analysis of Five Proposals for Slussen in Stockholm. *Journal of Space Syntax*, *1*(1), 177–198.

Marshall, A. (1920). *Principles of economics*. London: McMillan.

Martin, L. (1975). The grid as generator in urban space and structures. In M. Carmona & S. Tiesdell (Eds.), *Urban design reader* (pp. 70–82). Architectural Press.

Marzot, N. (2002). The Study of Urban Form in Italy. *Urban Morphology*, *6*(2), 59–73.

Mas, J.-F., Kolb, M., Paegelow, M., Camacho Olmedo, M. T., & Houet, T. (2014). Inductive pattern-based land use/cover change models: A comparison of four software packages. *Environmental Modelling & Software*, *51*, 94–111. doi:10.1016/j.envsoft.2013.09.010

Masnavi, M.-R. (2000). The New Millennium and the New Urban Paradigm: The Compact City in Practice. In K. Williams, M. Jenks, & E. Burton (Eds.), *Achieving sustainable urban form* (pp. 64–73). London: E & FN Spon.

Matthews, R., Gilbert, N., Roach, A., Polhill, J., & Gotts, N. (2007). Agent-based land-use models: A review of applications. *Landscape Ecology*, *22*(10), 1447–1459. doi:10.100710980-007-9135-1

Maureira, V., & Karimi, K. (2017). The Everyday and the Post-Disaster Urban Systems as One Thing: A Configurational Approach to Enhance the Recovery and Resilience of Cities Affected by Tsunamis. *11th International Space Syntax Symposium*.

Mawhood, P., & Davey, K. (1980). Anglophone Africa. In D. C. Rowat (Ed.), *Anglophone Africa International Handbook on Local Government Reorganization* (pp. 404–414). London: Aldwych Press.

Maxwell, S. L., Fuller, R. A., Brooks, T. M., & Watson, J. E. M. (2016). Biodiversity: The ravages of guns, nets and bulldozers. *Nature*, *536*(7615), 143–145. doi:10.1038/536143a PMID:27510207

McCallin, B., Scherer, I., & Duyne, J. (2015). Urban informal settlers displaced by disasters: challenges to housing responses. Geneva, Switzerland: The Internal Displacement Monitoring Centre (IDMC), Norwegian Refugee Council.

McCormick, B., & Wahba, J. (2003). *Migration and Mobility in the Egyptian Labour Market*. Cairo, Egypt: Economic Research Forum.

McHarg, I. L. (1969). *Design with nature*. New York: Natural History Press.

McMillen, D. P. (2004). Employment densities, spatial autocorrelation, and subcenters in large metropolitan areas. *Journal of Regional Science*, *44*(2), 225–243. doi:10.1111/j.0022-4146.2004.00335.x

McMillen, D. P., & Smith, S. C. (2003). The number of subcenters in large urban areas. *Journal of Urban Economics*, *53*(3), 321–338. doi:10.1016/S0094-1190(03)00026-3

McMillian, J. (2002). *Reinventing the Bazaar: A natural history of markets*. New York, NY: Norton.

Mehrez, S. (2011). *The Literary Life of Cairo: One-Hundred Years in the Heart of the City*. Cairo: The American University in Cairo Press.

Mehta, V. (2013). *The Street: A Quintessential Social Public Space*. New York, NY: Routledge.

Meller, H. (1990). *Patrick Geddes: Social Evolutionist and City Planner* (N. Wrigley, Ed.). London: Taylor & Francis e-Library.

Meller, H. (2001). *European Cities 1890-1930s: History, Culture and the Built Environment*. Cambridge University Press.

Merriam Webster. (2018). *Characteristic*. Retrieved from www.merriam-webster.com: https://www.merriam-webster.com/dictionary/characteristic

Merriam Webster. (2018, September 30). *Chaos*. Retrieved from www.merriam-webster.com: https://www.merriam-webster.com/dictionary/chaos

Merriam-webster. (2018). *livability*. Retrieved from www.merriam-webster.com: https://www.merriam-webster.com/dictionary/livability

Mertens, B., & Lambin, E. (1997). Spatial modeling of deforestation in southern Cameroon. *Applied Geography (Sevenoaks, England)*, *17*(2), 143–162. doi:10.1016/S0143-6228(97)00032-5

Metcalf, G. (2002). *The path to a livable city*. San Francisco: Transportation for a Livable City.

Meyfroidt, P., Lambin, E. F., Erb, K.-H., & Hertel, T. W. (2013). Globalization of land use: Distant drivers of land change and geographic displacement of land use. *Current Opinion in Environmental Sustainability*, *5*(5), 438–444. doi:10.1016/j.cosust.2013.04.003

Michelson, W. (1970). *Man, and his urban environment: A sociological approach*. Reading, MA: Addison-Wesley.

Michener, W. K., Baerward, T. J., & Firth, P. (2001). Defining and unraveling biocomplexity. *Bioscience*, *51*(12), 1018–1024. doi:10.1641/0006-3568(2001)051[1018:DAUB]2.0.CO;2

Middel, A., Häb, K., Brazel, A. J., Martin, C. A., & Guhathakurta, S. (2014). Impact of urban form and design on mid-afternoon microclimate in Phoenix Local Climate Zones. *Landscape and Urban Planning*, *122*, 16–28. doi:10.1016/j.landurbplan.2013.11.004

Mijic, N., Sestic, M., & Koljancic, M. (2017). CAD—GIS BIM integration—Case study of Banja Luka city center. In M. Hadžikadić & S. Avdaković (Eds.), *Advanced technologies, systems, and applications, Lecture Notes in Networks and Systems, 3*. Cham: Springer. doi:10.1007/978-3-319-47295-9_22

Milakis, D., Vlastos, T., & Barbopoulos, N. (2008). Relationships between Urban Form and Travel Behaviour in Athens, Greece: A Comparison with Western European and North American Results. *Ejtir*, *8*(3), 201–215.

Miller, W. R. (2012). *Introducing Geodesign : The Concept Director of GeoDesign Services*. Redlands: Esri Press. Retrieved from http://www.esri.com/~/media/Files/Pdfs/technology-topics/introducing-geodesign.pdf

Mills, E. S. (1967). An aggregative model of resource allocation in a metropolitan area. *The American Economic Review*, *57*, 197–210.

Mitchell, W. J. (2010). Sustainable Urban Mobility through Light Electric Vehicles. In M. Mostafavi & G. Doherty (Eds.), *Ecological urbanism* (pp. 382–397). Baden: Lars Müller.

Mohamed, A. (2010). *Understanding urban navigation: way-finding within the city* (Unpublished M.Sc. thesis). Ain Shams University.

Mohamed, A. M. R. M., & Brown, F. (2009). The Urban Grid and its Consequences Assiut: An Egyptian Case-Study. In *Proceedings of the 7th International Space Syntax Symposium*. Stockholm: KTH.

Mohamed, A. M. R. M. (2014). *The Spatial Configuration of Formal Housing Types in New Assiut City, Egypt. In Environment and Design*. Istanbul, Turkey: Re-Thinking Residential Environments.

Mohit, M. A., & Iyanda, S. A. (2016). Liveability and Low-income Housing in Nigeria. *Social and Behavioral Sciences,* (222), 863 – 871.

Montello, D. R. (2007). The Contribution of Space Syntax to a Comprehensive Theory of Environmental Psychology. In *Proceedings of the 6th International Space Syntax Symposium*. Istanbul: ITU Faculty of Architecture.

Montiel, L. M., Nathan, R. P., & Wright, D. J. (2007). *Divided they Fall: Hardship in America's Cities and Suburbs*. Albany, NY: The Nelson A. Rockefeller Institute of Government.

Morales, A. (2009). Public Markets as Community Development Tools. *Journal of Planning Education and Research, 28*(4), 426–440. doi:10.1177/0739456X08329471

Morales, A. (2011). Marketplaces: Prospects for Social, Economic, and Political Development. *Journal of Planning Literature, 26*(1), 3–17. doi:10.1177/0885412210388040

Morayef, S. (2012). *Demolition of disused Alexandria Greek Club highlights larger problems*. Available online at http://www.egyptindependent.com/news/demolition-disused-alexandria-greek-club-highlights-larger-problems

Moreno, G., & Steadman, P. (2014). The relationship between density, built form and design. In M. Carmona (Ed.), *Explorations in urban design: An urban design research primer* (pp. 193–205). Surrey: Ashgate Publishing Limited.

Moreno, N., Quintero, R., Ablan, M., Barros, R., Dávila, J., Ramírez, H., ... Acevedo, M. F. (2007). Biocomplexity of deforestation in the Caparo tropical forest reserve in Venezuela. An integrated multi-agent and cellular automata model. *Environmental Modelling & Software, 22*(5), 664–673. doi:10.1016/j.envsoft.2005.12.022

Moriarty, P., & Honnery, D. (2008). Low-mobility: The future of transport. *Futures, 40*(10), 865–872. doi:10.1016/j.futures.2008.07.021

Morse, S. W. (2005, June 27). *An Update on Urban Hardship*. Retrieved from Smart Communities: http://smartcommunities.typepad.com/suzanne/2005/06/an_update_on_ur.html

Mosel, I., & Jackson, A. (2013). *Sanctuary in the city? Urban displacement and Urban displacement and Pakistan*. Humanitarian Policy Group, Overseas Development Institute.

Moser, G. (1997). *Les stress urbains*. Paris: Armand Colin.

Moser, G. (2009). *Psychologie environnementale - Les relations homme-environnement*. Bruxelles: De Boeck Editions.

Mosko, M. S., & Damon, F. H. (2005). *On the Order of Chaos: Social Anthropology and the Science of Chaos*. Berghahn Books.

Mossberger, K., Tolbert, C. J., & Franko, W. W. (2013). *Digital Cities: The Internet and the Geography of Opportunity*. Oxford, UK: Oxford University Press.

Moudon, A. V. (1997). Urban morphology s an emerging interdisciplinary field. *Urban Morphology, 1*, 3–10.

Moughtin, C. (2003). *Urban design: Method and techniques (* 2nd ed.). Oxford, UK: Architectural.

Mraihi, R., Harizi, R., Mraihi, T., & Bouzidi, M. T. (2015). Urban air pollution and urban daily mobility in large Tunisia's cities. *Renewable & Sustainable Energy Reviews, 43*, 315–320. doi:10.1016/j.rser.2014.11.022

Muggah, R. (2000). Through the Developmentalist's Looking Glass: Conflict-Induced Displacement and Involuntary Resettlement in Colombia. *Journal of Refugee Studies, 13*(2), 133–164. doi:10.1093/jrs/13.2.133

Müller, D., & Zeller, M. (2002). Land use dynamics in the central highlands of Vietnam: A spatial model combining village survey data with satellite imagery interpretation. *Agricultural Economics*, *27*(3), 333–354. doi:10.1111/j.1574-0862.2002.tb00124.x

Mumford, L. (1938). *The Culture of Cities*. Orlando, FL: Harcourt Brace Jovanovich.

Mumford, L. (1961). *The City in History: Its Origins, Its Transformations, and its Prospects*. New York, NY: Harcourt, Brace, and World.

Murray, A. T. (2010). Advances in location modeling: GIS linkages and contributions. *Journal of Geographical Systems*, *12*(3), 335–354. doi:10.100710109-009-0105-9

Murray-Rust, D., Brown, C., van Vliet, J., Alam, S. J., Robinson, D. T., Verburg, P. H., & Rounsevell, M. (2014). Combining agent functional types, capitals and services to model land use dynamics. *Environmental Modelling & Software*, *59*, 187–201. doi:10.1016/j.envsoft.2014.05.019

Musgrave, R. A. (1958). *Multi-Level Finance*. New York: McGraw-Hill.

Muth, R. (1969). *Cities and Housing*. Chicago: University of Chicago Press.

Næss, P. (2009). Residential Self-Selection and Appropriate Control Variables in Land Use: Travel Studies. *Transport Reviews*, *29*(3), 293–324. doi:10.1080/01441640802710812

Naess, P. (2012). Urban form and travel behavior: Experience from a Nordic context. *Journal of Transport and Land Use*, *5*(2). doi:10.5198/jtlu.v5i2.314

Nagati, O., & Stryker, B. (2013). *Archiving the City in Flux: Cairo Shifting Landscape since the January 25th Revolution*. Cairo: Cluster.

Nam, T., & Pardo, T. A. (2011). Conceptualizing Smart City with Dimensions of Technology, People, and Institutions. In *Proceedings of the 12th Annual International Conference on Digital Government Research* (pp. 282-291). Academic Press.

Nathan, R. P., & Adams, C. (1976). Understanding Central City Hardship. *Political Science Quarterly*, *91*(1), 47–62. doi:10.2307/2149158

Nedovic-Budic, Z., Knaap, G. J., Shahumyan, H., Williams, B., & Slaev, A. (2016). Measuring urban form at community scale: Case study of Dublin, Ireland. *Cities (London, England)*, *55*, 148–164. doi:10.1016/j.cities.2016.02.014

Nelson, G. (2011). Measuring Poverty: The Official U.S. Measure and Material Hardship. *Poverty & Public Policy*, *3*(3), 1–35. doi:10.2202/1944-2858.1077

Netto, V. M. (2007). *Practice, communication and space: A reflection on the materiality of social structures* (Unpublished doctoral dissertation). University of London.

Netto, V. (2016). 'What is space syntax not?' Reflections on space syntax as sociospatial theory. *Urban Des Int.*, *21*(1), 25–40. doi:10.1057/udi.2015.21

New & Improved Website. Coming Soon. (2018, August 15). *misrinternationalfilms*. Retrieved from Cairo as Told: http://www.misrinternationalfilms.com/ar/%D8%A7%D9%84%D9%82%D8%A7%D9%87%D8%B1%D8%A9-%D9%85%D9%86%D9%88%D8%B1%D8%A9-%D8%A8%D8%A3%D9%87%D9%84%D9%87%D8%A7

New Cities Foundation. (2015). *Building New Cities: Challenges, Opportunities and Recommendations: Summary and analysis of themes emerging from Cityquest*. KAEC Forum 2014. Retrieved from http://bit.ly/Cityquest2014

Newman, O. (1972). *Defensible space: crime prevention through urban design*. New York: Macmillan.

Newman, O. (1996). *Creating defensible space*. Washington, DC: U.S. Dept. of Housing and Urban Development, Office of Policy Development and Research.

Newman, P. W. G., & Kenworthy, J. R. (1989). Gasoline Consumption and Cities. *Journal of the American Planning Association*, *55*(1), 24–37. doi:10.1080/01944368908975398

Newman, P., Beatley, T., & Boyer, H. (2009). *Resilient cities: Responding to peak oil and climate change*. Washington, DC: Island Press.

Newman, P., & Kenworthy, J. (2013). Urban Sustainability and Automobile Dependence in an Australian Context. In I. Vojnovic (Ed.), *Urban sustainability: A global perspective, edited by Igor Vojnovic* (pp. 227–253). East Lansing, MI: Michigan State University Press.

Nikolopoulou, M.-H., Baker, N., & Steemers, K. (2001). Thermal Comfort in Outdoor Urban Spaces: Understanding the Human Parameter. *Solar Energy*, *70*(3), 227–235. doi:10.1016/S0038-092X(00)00093-1

Nikolopoulou, M., & Lykoudis, S. (2007). Use of outdoor spaces and microclimate in a Mediterranean urban area. *Building and Environment*, *42*(10), 3691–3707. doi:10.1016/j.buildenv.2006.09.008

O'Brien, D. (2010). *Measuring the Value of Culture. Report to the Department of Culture and Sport*. Arts and Humanities Research Council.

O'Callaghan, J. R. (1995). NELUP: An Introduction. *Journal of Environmental Planning and Management*, *38*(1), 5–20. doi:10.1080/09640569513084

ODPM. (2003). *Sustainable Communities*. London: Building for the Future Office of the Deputy Prime Minister.

Ogilvie, S. (2012). *Words of the World: A Global History of the Oxford English Dictionary*. Cambridge University Press. doi:10.1017/CBO9781139129046

Oke, T. R. (2017). *Urban climates*. Cambridge, UK: Cambridge University Press. doi:10.1017/9781139016476

Okulicz-Kozaryn, A. (2013). City Life: Rankings (Livability) Versus Perceptions (Satisfaction). *Social Indicators Research. An International and Interdisciplinary Journal for Quality-of-Life Measurement*, *110*(2), 433–451.

Olsen, D. J. (1986). *The City as a Work of Art: London, Paris, Vienna*. New Haven, CT: Yale University Press.

Onyango, J.O., W.L.A, O., & Onyango, L. (2012). Dynamics of Street Vending Phenomenon in the Kisumu Municipality, Kenya. *International Journal of Arts and Commerce*, *1*(4), 107–120.

Osman, K. M., & Suliman, M. (1994). The Space Syntax Methodology: Fits and Misfits. *Arch. & Behav.*, *10*(2), 189–204.

Osterman, A. (2015). The Empowerment of 'Self Power'. *Self-building City, 43*, 104-107. Retrieved 8 2, 2017, from http://volumeproject.org/the-empowerment-of-self-power/

Ou, J., Liu, X., Li, X., & Chen, Y. (2013). Quantifying the relationship between urban forms and carbon emissions using panel data analysis. *Landscape Ecology*, *28*(10), 1889–1907. doi:10.100710980-013-9943-4

Overmars, K. P., Verburg, P. H., & Veldkamp, T. (2007). Comparison of a deductive and an inductive approach to specify land suitability in a spatially explicit land use model. *Land Use Policy*, *24*(3), 584–599. doi:10.1016/j.landusepol.2005.09.008

Owen, N., Cerin, E., Leslie, E., duToit, L., Coffee, N., Frank, L. D., ... Sallis, J. F. (2007). Neighborhood walkability and the walking behavior of Australian adults. *American Journal of Preventive Medicine*, *33*(5), 387–395. doi:10.1016/j.amepre.2007.07.025 PMID:17950404

Oz, O., & Eder, M. (2012). Rendering Istanbul's Periodic Bazaars Invisible: Reflections on Urban Transformation and Contested Space. *International Journal of Urban and Regional Research*, *36*(2), 297–314. doi:10.1111/j.1468-2427.2011.01076.x

Pacione, M. (1990). Urban liveability: A review. *Urban Geography*, *11*(1), 1–30. doi:10.2747/0272-3638.11.1.1

Pahl-Weber, E., Seelig, S., & Ohlenburg, H. (2012). The Shahre Javan community detailed plan: Planning for a climate responsive and sustainable Iranian urban quarter. Young cities research paper series: v. 3. Berlin: Universitätsverlag der TU Berlin; BHRC, Building and Housing Research Center.

Palacios, J. (2007). *The Catholic Social Imagination: Activism and the Just Society in Mexico and the United States*. Chicago: The University of Chicago Press. doi:10.7208/chicago/9780226645025.001.0001

Papadopoulou, C.-A., & Giaoutzi, M. (2014). Crowdsourcing as a Tool for Knowledge Acquisition in Spatial Planning. *Future Internet*, *6*(1), 109–125. doi:10.3390/fi6010109

Paquot, T. (2015). *Désastres urbains*. Paris, France: Éditions La Découverte.

Parker, D. C., Manson, S. M., Janssen, M. A., Hoffmann, M. J., & Deadman, P. (2003). Multi-agent systems for the simulation of land-use and land-cover change: A review. *Annals of the Association of American Geographers*, *93*(2), 314–337. doi:10.1111/1467-8306.9302004

Park, M., Hagishima, A., Tanimoto, J., & Narita, K. I. (2012). Effect of urban vegetation on outdoor thermal environment: Field measurement at a scale model site. *Building and Environment*, *56*, 38–46. doi:10.1016/j.buildenv.2012.02.015

Parsons, S., Gmytrasiewicz, P., & Wooldridge, M. (2002). *Game theory and decision theory in agent-based systems*. Kluwer Academic Publishers. doi:10.1007/978-1-4615-1107-6

Parsons, T. (1985). *Talcott Parsons on Institutions and Social Evolution: Selected Writings (Heritage of Sociology Series)*. Chicago: The University of Chicago Press.

Pasha, M. R. (2016). *Data Warehousing and the Unstructured Data*. Academic Press.

Pastore, L., Corrao, R., & Heiselberg, P. K. (2017). The effects of vegetation on indoor thermal comfort: The application of a multi-scale simulation methodology on a residential neighborhood renovation case study. *Energy and Building*, *146*, 1–11. doi:10.1016/j.enbuild.2017.04.022

Patil, S. D., Gu, Y., Dias, F. S. A., Stieglitz, M., & Turk, G. (2017). Predicting the spectral information of future land cover using machine learning. *International Journal of Remote Sensing*, *38*(20), 5592–5607. doi:10.1080/01431161.2017.1343512

Paul, A. (2012). Unit-segment Analysis: A Space Syntax Approach to Capturing Vehicular Travel Behavior Emulating Configurational Properties of Roadway Structures. *Journal of Transport & Infrastructure Research*, *12*(3), 275–290.

Pazur, R., & Bolliger, J. (2017). Land changes in Slovakia: Past processes and future directions. *Applied Geography (Sevenoaks, England)*, *85*, 163–175. doi:10.1016/j.apgeog.2017.05.009

Peersman, G. (2014). *Overview: Data Collection and Analysis Methods in Impact Evaluation, Methodological Briefs: Impact Evaluation 10*. Florence: UNICEF Office of Research. Retrieved from https://www.unicef-irc.org/publications/pdf/brief_10_data_collection_analysis_eng.pdf

Penn, A., & Dalton, N. (1994). The Architecture of Society: Stochastic Simulation of Urban Movement. In Simulating Societies: The Computer Simulation of Social Phenomena. London: UCL Press.

Penn, A. (2003). Space Syntax and Spatial Cognition or, Why the Axial Line? *Environment and Behavior*, *35*(1), 30–65. doi:10.1177/0013916502238864

Penn, A. (2008). Architecture Research. In A. Knight & L. Ruddock (Eds.), *Advanced Research Methods in the Built Environment* (pp. 14–27). Chichester, UK: Wiley-Blackwell.

Penn, A., Hillier, B., Banister, D., & Xu, J. (1998). Configurational Modelling of Urban Movement Networks. *Environment and Planning. B, Planning & Design*, *25*(1), 59–84. doi:10.1068/b250059

Penn, A., & Turner, A. (2004). Movement-Generated Land-Use Agglomeration: Simulation Experiments on the Drivers of Fine-scale Land-use Patterning. *URBAN DESIGN International*, *9*(2), 81–96. doi:10.1057/palgrave.udi.9000120

Peponis, J. (1989). Space, culture and urban design in late modernism and after. Ekistics, 56(334/335), 93-108.

Petrov, L., Lavalle, C., & Kasankob, M. (2009). Urban land use scenarios for a tourist region in Europe: Applying the MOLAND model to Algarve, Portugal. *Landscape and Urban Planning*, *92*(1), 10–23. doi:10.1016/j.landurbplan.2009.01.011

Pettit, C., Keysers, J., Bishop, I., & Klosterman, R. (2008). Applying the What if? Planning Support System for Better Planning at the Urban Fringe. In C. C. Pettit (Ed.), *Landscape Analysis and Visualisation: Spatial Models for Natural Resource Management and Planning* (pp. 435–454). Berlin: Springer. doi:10.1007/978-3-540-69168-6_21

Picard, R. (2009). Future affective technology for autism and emotion communication. *Philosophy Transactions of the Royal Society, Biological Sciences*, *364*(1535), 3575-3584.

Picard, R., Vyzas, E., & Healey, J. (2001). Toward machine emotional intelligence: Analysis of affective physiological state. *IEEE Transactions on Pattern Analysis and Machine Intelligence*, *23*(10), 1175–1191. doi:10.1109/34.954607

Piffero, E. (2009). Beyond rules and regulations: the growth of informal Cairo. In R. Kipper & M. Fischer (Eds.), *Cairo's informal areas between urban challenges and hidden potentials* (pp. 20–27). Cairo: GTZ Egypt.

Pijanowski, P. C., Brown, D. G., Shellito, B. A., & Manik, G. A. (2002). Using neural nets and GIS to forecast land use changes: A land transformation model. *Computers, Environment and Urban Systems*, *26*(6), 553–575. doi:10.1016/S0198-9715(01)00015-1

Pilkauska, N., Currie, J., & Garfinkel, I. (2012). The Great Recession, Public Transfers, and Material Hardship. *The Social Service Review*, *36*(3), 401–427. doi:10.1086/667993 PMID:24379487

Pinfold, N. (2015). *Community Mapping in Informal Settlements for Better Housing and Service Delivery Cape Town, South Africa. Lisbon, Portugal.* INSPIRE Geo-spatial World Forum.

Pongsmas, N. (2004). *Configuration of Urban Space and Social Sustainability of Urban Neighborhood: A Case Study on the City of San Diego at the Dawn of the Twentieth-First Century* (Unpublished doctoral thesis). Texas Tech University.

Pons, X., & Masó, J. (2016). A comprehensive open package format for preservation and distribution of geospatial data and metadata. *Computers & Geosciences*, *97*, 89–97. doi:10.1016/j.cageo.2016.09.001

Pontius, R. G. (2000). Quantification error versus location error in comparison of categorical maps. *Photogrammetric Engineering and Remote Sensing*, *66*, 1011–1016.

Pontius, R. G. Jr, Cornell, J. D., & Hall, C. A. S. (2001). Modeling the spatial pattern of land-use change with GEOMOD2: Application and validation for Costa Rica. *Agriculture, Ecosystems & Environment*, *85*(1-3), 191–203. doi:10.1016/S0167-8809(01)00183-9

Pontius, R. G. Jr, & Spencer, J. (2005). Uncertainty in extrapolations of predictive and-change models. *Environment & Planning B*, *32*(2), 211–230. doi:10.1068/b31152

Power, D. (2008). Decision Support Systems: A Historical Overview. In F. B. Holsapple (Ed.), *Handbook on Decision Support Systems 1 - Basic Themes* (pp. 121–140). Springer. doi:10.1007/978-3-540-48713-5_7

Pozoukidou, G., Gavanas, N., & Verani, E. (2017). Land use models and sustainable urban mobility plans: An integrative approach for strategic planning. *International Journal of Sustainable Development and Planning*, *12*(5), 867–882. doi:10.2495/SDP-V12-N5-867-882

PPS. (2003). *Public Markets as a Vehicle for Social Integration and Upward Mobility. Phase 1 Report: An Overview of Existing Programs and Assessment Opportunities*. New York, NY: Project for Public Spaces, Inc.

Project for Public Spaces, Inc. (2000). *How to Turn a Place Around: A Handbook for Creating Successful Public Spaces*. New York: Project for Public Spaces, Inc.

Psarra, S. (2009). The ghost of conceived space – what kind of work does or should space syntax perform for architecture. In *Proceedings of the 7th international space syntax symposium* (089:01-089:10). Stockholm: KTH.

Psarra, S. (2010). The Ghost of Conceived Space – What Kind of Work Does or Should Space Syntax Perform for Architecture. *The Journal of Space Syntax*, *1*(1), 17–29.

Pucher, J., & Buehler, R. (2008). Making Cycling Irresistible: Lessons from The Netherlands, Denmark and Germany. *Transport Reviews*, *28*(4), 495–528. doi:10.1080/01441640701806612

Puertas, O. L., Henriquez, C., & Meza, F. J. (2014). Assessing spatial dynamics of urban growth using an integrated land use model: Application in Santiago Metropolitan Area, 2010-2045. *Land Use Policy*, *38*, 415–425. doi:10.1016/j.landusepol.2013.11.024

PULSA. (1972). The city is an artwork. In G. Kepes (Ed.), *Arts of the Environment* (pp. 208–232). New York: George Braziller.

Pulsellic, R., & Tiezzi, E. (2011). City out of Chaos: Social Patterns and Organization in Urban Systems. In C. Brebbia (Ed.), Ecodynamics: The Prigogine Legacy (pp. 193-202). WIT Press.

Pyka, A., & Fagiolo, G. (2007). Agent-based modeling: A methodology for neo-schumpeterian economics. In H. Hanusch & A. Pyka (Eds.), *The Elgar companion to neo-schumpeterian economics*. Cheltenham, UK: Edward Elgar. doi:10.4337/9781847207012.00037

Qualitest20. (2018). *Advantages and Disadvantages of Crowdsourcing*. Retrieved from https://www.qualitestgroup.com/white-papers/advantages-disadvantages-crowdsourcing/

Raafat, S. (2003). *Cairo, the Glory Years: Who built what, when, why and for whom*. Alexandria, Egypt: Harpocrates Publication.

Rajagopal. (2010a). Street Markets Influencing Urban Consumer Behavior in Mexico. *Latin American Business Review*, *11*(2), 77-110.

Rajagopal. (2010b). Coexistence and Conflicts between Shopping Malls and Street Markets in Growing Cities: Analysis of Shoppers' Behavior. *Journal of Retail and Leisure Property*, *9*(4), 277-301.

Ramankutty, N., & Foley, J. A. (1999). Estimating historical changes in global land cover: Croplands from 1700 to 1992. *Global Biogeochemical Cycles*, *13*(4), 997–1027. doi:10.1029/1999GB900046

Randolph, J. (2004). *Environmental land use planning and management*. Washington, DC: Island Press.

Raymond, A. (2000). Cairo. Harvard University Press.

Raymond, A. (2007). *City of History.* Cairo: The American University in Cairo Press.

Recio, R. B., & Gomez, L. E. A. Jr. (2013). Street Vendors, their Contested Spaces, and the Policy Environment: A View from Caloocan, Metro Manila. *Environment & Urbanization Asia*, *4*(1), 173–190. doi:10.1177/0975425313477760

Reid, D. M., Fram, E. H., & Guotai, C. (2010). A Study of Chinese Street Vendors: How They Operate. *Journal of Asia-Pacific Business*, *11*(4), 244–257. doi:10.1080/10599231.2010.520640

Reinhart, C. F., Dogan, T., Jakubiec, J. A., Rakha, T., & Sang, A. (2013, August). *Umi-an urban simulation environment for building energy use, daylighting and walkability.* In *13th Conference of International Building Performance Simulation Association*, Chambery, France.

Reyes, R. M. (2013). Crime, Street Vendors and the Historical Downtown in Post-Giuliani Mexico City *International Journal of Criminology and Sociology*, *2*, 186–198.

Rindfuss, R. R., Entwisle, B., Walsh, S. J., An, L., Badenoch, N., Brown, D. G., ... Verburg, P. H. (2008). Land use change: Complexity and comparisons. *Journal of Land Use Science*, *3*(1), 1–10. doi:10.1080/17474230802047955 PMID:20354590

Roaf, S. (2010). The Sustainability of High Density. In E. Ng (Ed.), Designing high-density cities for social and environmental sustainability (pp. 27–39). Sterling, VA: Earthscan.

Robinson, D. T., Brown, D. G., Parker, D. C., Schreinemachers, P., Janssen, M. A., Huigen, M., ... Barnaud, C. (2007). Comparison of empirical methods for building agent-based models in land-use science. *Journal of Land Use Science*, *2*(1), 31–55. doi:10.1080/17474230701201349

Robinson, J. (2000). Cities on the Move: Urban chaos and global change–East Asian art, architecture and film now. *Ecumene*, *7*(1), 105–109. doi:10.1177/096746080000700105

Robinson, W. C. (2003). *Risks and Rights: The Causes, Consequences, and Challenges of Development-Induced Displacement.* Washington, DC: The Brookings Institution-SAIS.

Rocha, J., Ferreira, J. C., Simoes, J., & Tenedorio, J. A. (2007). Modeling coastal and land use evolution patterns through neural network and cellular automata integration. *Journal of Coastal Research*, *50*, 827–831.

Roever, S., & Skinner, C. (2016). Street Vendors and Cities. *Environment and Urbanization*, *28*(2), 359–374. doi:10.1177/0956247816653898

Rogan, J., Franklin, J., Stow, D., Miller, J., Woodcock, C., & Roberts, D. (2008). Mapping land-cover modifications over large areas: A comparison of Machine Learning Algorithms. *Remote Sensing of Environment*, *112*(5), 2272–2283. doi:10.1016/j.rse.2007.10.004

Rogers, L. (n.d.). Incentives for Heritage Protection Handbook: A national guide for Local Government and the Community. Information Guide. *Commonwealth of Australia for the Heritage Chairs and of Cials of Australia and New Zealand.*

Rogers, R. (2008). *Cities for A Small Planet.* Boulder, CO: Basic Books.

Rogers, R., & Power, A. (2000). *Cities for a small country.* London: Faber.

Ronald, N., Sterling, L., & Kirley, M. (2007). An agent-based approach to modeling pedestrian behavior. *International Journal of Simulation*, *8*(1), 25–38.

Rosa, I. M. D., Purves, D., Souza, C. Jr, & Ewers, R. M. (2013). Predictive modeling of contagious deforestation in the Brazilian Amazon. *PLoS One*, *8*(10), e77231. doi:10.1371/journal.pone.0077231 PMID:24204776

Rose-Ackerman, S. (1999). *Corruption and Government: Causes, Consequences, and Reform.* Cambridge, UK: Cambridge University Press. doi:10.1017/CBO9781139175098

Roseland, M. (1977). *Eco-city Dimensions: Healthy Communities, Healthy Planet.* New Society Publishers.

Roser, F. (2011). *Entwicklung einer methode zur großflächigen rechnergestützten Analyse des landschaftsästhetischen Potenzials.* Berlin: Weißensee.

Ross, C. (2011). *The Leaderless Revolution: How Ordinary People Will Take Power and Change Politics in Twenty-First Century.* New York: Blue Rider Press.

Rossi, A. (1981). *L'architecture de la ville.* Paris: L'Equerre Editions.

Rossi, A. (1982). *The Architecture of the City* (D. Ghirardo & J. Ockman, Trans.). Cambridge, MA: The MIT Press.

Rouchier, J., Bousquet, F., Requier-Desjardins, M., & Antona, M. (2001). A multi-agent model for describing transhumance in North Cameroon: Comparison of different rationality to develop a routine. *Journal of Economic Dynamics & Control, 25*(3-4), 527–559. doi:10.1016/S0165-1889(00)00035-X

Rounsevell, M. D. A., Pedroli, B., Erb, K. H., Gramberger, M., Busck, A. G., Haberl, H., ... Wolfslehner, B. (2012). Challenges of land system science. *Land Use Policy, 29*(4), 899–910. doi:10.1016/j.landusepol.2012.01.007

Rowley, A. (1996). Mixed-use Development: Ambiguous concept, simplistic analysis and wishful thinking? *Planning Practice and Research, 11*(1), 85–98. doi:10.1080/02697459650036477

Ruben, R., Moll, H., & Kuyvenhoven, A. (1998). Integrating agricultural research and policy analysis: Analytical framework and policy applications for bio-economic modeling. *Agricultural Systems, 58*(3), 331–349. doi:10.1016/S0308-521X(98)00034-1

Rubinowicz, P. (2000). Chaos and Geometric Order in Architecture and Design. *Journal for Geometry and Graphics, 4*(2), 197–207.

Ruiz, I., & Vargas-Silva, C. (2014). The Economics of Forced Migration. *The Journal of Development Studies, 49*(6), 772–784. doi:10.1080/00220388.2013.777707

Rutherford, G. N., Bebi, P., Edwards, P. J., & Zimmermann, E. N. (2008). Assessing land-use statistics to model land cover change in a mountainous landscape in the European Alps. *Ecological Modelling, 212*(3-4), 460–471. doi:10.1016/j.ecolmodel.2007.10.050

Şahbaz, Ö., & Hillier, B. (2007). The Story of the Crime: Functional, Temporal and Spatial Tendencies in Street Robbery. In *Proceedings of the 6th International Space Syntax Symposium.* Istanbul: ITU Faculty of Architecture.

Sailer, K., Pomeroy, R., Raheem, A., Budgen, A., & Lonsdale, N. (2012). The generative office building. In *Proceedings of the 8th International Space Syntax Symposium* (pp. 8010:01-8010:25). Santiago: Pontificia Universidad Católica.

Sailor, D. J. (2008). A green roof model for building energy simulation programs. *Energy and Buildings, 40*(8), 1466-1478.

Salama, S. A. (2006, December 14 - 20). Forgotten children. *Weekly Online.*

Sallis, J. F., Saelens, B. E., Frank, L. D., Conway, T. L., Slymen, D. J., Cain, K. L., . . . Kerr, J. (2009). Neighborhood built environment and income: Examining multiple health outcomes. *Social Science & Medicine (1982), 68*(7), 1285–1293. doi:10.1016/j.socscimed.2009.01.017

Salzano, E. (1997). Seven Aims for the Livable City. In *Making Cities Livable, International Making Cities Livable Conferences.* Gondolier Press.

Samardzic-Petrovic, M., Dragicevic, S., Bajat, B., & Kovacevic, M. (2015). Exploring the decision tree method for modeling urban land use change. *Geomatica*, *69*(3), 313–325. doi:10.5623/cig2015-305

Samardzic-Petrovic, M., Kovacevic, M., Bajat, B., & Dragicevic, S. (2017). Machine learning techniques for modeling short term land-use change. *International Journal of Geo-Information*, *6*(12), 387. doi:10.3390/ijgi6120387

Sande-Friedman, A. (2013). *The Four Times of Day by William Hogarth*. Retrieved from The Commononline: a modern sense of place: https://www.thecommononline.org/the-four-times-of-day-by-william-hogarth/

Sanders, L., Pumain, D., Mathian, H., Guerin-Pace, F., & Bura, S. (1997). SIMPOP: A multiagent system for the study of urbanism. *Environment & Planning B*, *24*(2), 287–305. doi:10.1068/b240287

Sargent, L. T. (2000). Utopian Traditions: Themes and Variations. In T. N. Schaer & G. C. Sargent (Eds.), *Utopia: The Search for the Ideal Society in the Western World* (pp. 8–17). New York: Oxford University Press in Association with The New York Public.

Sarpong, S., & Nabubie, I. B. (2015). Nuisance or Discerning? The social construction of street hawkers in Ghana. *Society and Business Review*, *10*(2), 102–117. doi:10.1108/SBR-03-2015-0005

Sasaki, K. (1987). A comparative static analysis of urban structure in the setting of endogenous income. *Journal of Urban Economics*, *22*(1), 53–72. doi:10.1016/0094-1190(87)90049-0

Savitch, H. (2010). What makes a great city great? An American perspective. *Cities (London, England)*, *27*(1), 42–49. doi:10.1016/j.cities.2009.11.012

Schaldach, R., & Koch, J. (2009). Conceptual design and implementation of a model for the integrated simulation of large-scale land-use systems. In Information technologies in environmental engineering. Environmental science and engineering, Part 2 (pp. 425-438). Academic Press. doi:10.1007/978-3-540-88351-7_32

Schappo, P., & Van Melik, R. (2017). Meeting on the marketplace: On the integrative potential of The Hague Market. *Journal of Urbanism*, *10*(3), 318–332. doi:10.1080/17549175.2016.1223741

Scheer, B. C. (2017). *The evolution of urban form: Typology for planners and architects / Brenda Case Scheer, AICP*. Abingdon, UK: Routledge. doi:10.4324/9781351179751

Schilders, P. (2010). *The Organic City: Method or Metaphor? The Meaning of 'Organic' in Architecture and Urban Planning* (M. Provoost, W. Vanstiphout, Eds., & B. Annable, Trans.). International New Town Institute.

Schneider, A. (2003). Decentralization: Conceptualization and Measurement. *Studies in Comparative International Development*, *38*(3), 32–56. doi:10.1007/BF02686198

Schwanen, T., Dijst, M., & Dieleman, F. M. (2004). Policies for Urban Form and their Impact on Travel: The Netherlands Experience. *Urban Studies (Edinburgh, Scotland)*, *41*(3), 579–603. doi:10.1080/0042098042000178690

Schwanen, T., & Mokhtarian, P. L. (2005). What affects commute mode choice: Neighborhood physical structure or preferences toward neighborhoods? *Journal of Transport Geography*, *13*(1), 83–99. doi:10.1016/j.jtrangeo.2004.11.001

Schwarz-v. Raumer, H., Jörg, J., & Alfiky, M. (2016). Respecting the Role of Agriculture for an Untegrated Landscape Development at the Urban-rural Fringe Using Geodesign Tools. *Journal of Digital Landscape Architecture*, (1–201), 327–336. doi:10.14627/537612037.This

Schwarz-v. Raumer, H.-G., & Stokman, A. (2011). GeoDesign-Approximations of a catchphrase. *Proceedings of Digital Landscape Architecture*, 106–115.

Schwarz-v. Raumer, H.-G., & Stokman, A. (2012). *GeoDesign : Herausforderungen an einen verständigen Umgang mit GIS Einleitung Wunsch und Wirklichkeit*. Academic Press.

Schwarz-v. Raumer, H.-G., & Stokman, A. (2013). GeoDesign : Herausforderungen an einen verständigen Umgang mit GIS Einleitung Wunsch und Wirklichkeit. In 18. Münchner Fortbildungsseminar Geoinformationssysteme (pp. 297–306). Academic Press.

Schwarz-v. Raumer, H.-G., & Stokman, A. (2014). Integrating Technology, Science and Creativity-A Challenge for Collaborative Settings in Geodesign. *Digital Landscape Architecture*, 15–26. Retrieved from http://dla2014.ethz.ch/talk_pdfs/DLA_2014_Keynote_Schwarz_v_Raumer.pdf

Seamon, D. (1993). The Life of the Place: A Phenomenological Commentary on Bill Hillier's Theory of Space Syntax. *Environmental & Architectural Phenomenology Newsletter*, *4*, 10–19.

Seamon, D. (2007). A Lived Hermetic of People and Place: Phenomenology and Space Syntax. In *Proceedings of the 6th International Space Syntax Symposium*. Istanbul: ITU Faculty of Architecture

Sebaa, A., Chick, F., Nouicer, A., & Tari, A. (2017). Research in big data warehousing using Hadoop. *Journal of Information Systems Engineering & Management*, *2*(2). doi:10.20897/jisem.201710

Séjourné, M. (2009). The history of informal settlements. In R. Kipper & M. Fischer (Eds.), *Cairo's informal areas between urban challenges and hidden potentials* (pp. 16–19). Cairo: GTZ Egypt.

Sen, J. (2013). Sustainable Urban Planning. New Delhi: The Energy and Resources Institute (TERI).

Serra, M., & Pinho, P. (2013). Tackling the structure of very large spatial systems: Space syntax and the analysis of metropolitan form. *Journal of Space Syntax*, *4*(2), 179–196.

Seto, K. C., Güneralp, B., & Hutyra, L. R. (2012). Global forecasts of urban expansion to 2030 and direct impacts on biodiversity and carbon pools. *Proceedings of the National Academy of Sciences of the United States of America*, *109*(40), 16083–16088. doi:10.1073/pnas.1211658109 PMID:22988086

Setsabi, S., & Leduka, R. C. (2008). The Politics of Street Trading in Maseru, Lesotho. *Urban Forum, 19*(3), 221-241.

Shaaban, K., Muley, D., & Elnashar, D. (2017). Temporal variation in walking behavior: An empirical study. *Case Studies on Transport Policy*, *5*(4), 671–680. doi:10.1016/j.cstp.2017.07.001

Shafiei, K. (2013). *The Fragmented Morphology of Spontaneous Settlements: The case study cities of Jeddah (Saudi Arabia) and Zahedan (Iran)* (Unpublished doctoral thesis). University of London.

Shalabi, K. (2008). *Saleh Heisa*. Cairo: Dar Al Shorouk.

Shamsuddin, S., Hassanb, N. R., & Bilyaminb, S. F. (2012). Walkable Environment in Increasing the Liveability of a City. *Social and Behavioral Sciences,* (50), 167 – 178.

Shawkat, Y., Khalil, O., & Zaazaa, A. (2015). *Parallel Urban Practice in Egypt 2015. Nairobi, Kenya: United Nations Human Settlements Programme*. UN-Habitat.

Shehayeb, D. (2009). Advantages of Living in Informal Areas. In R. Kipper & M. Fischer (Eds.), *Cairo's informal areas between urban challenges and hidden potentials* (pp. 34–43). Cairo: GTZ Egypt.

Shehayeb, D., & Eid, Y. (2007). Neighborhood Design and Community Building: A model of social interaction. In *The appropriate home: Can we design"appropriate" residential environments?* (pp. 129–144). Cairo: HBNRC.

Sheridan, P., Schroers, J. O., & Rommelfanger, E. (2007). GIS-based modeling of land use systems. In E. Koomen, J. Stillwell, A. Bakema, & H. J. Scholten (Eds.), *Modeling land use change progress and applications* (pp. 375–389). Dordrecht: Springer.

Sheshikala, M., Rao, D. R., & Prakash, R. V. (2016). Parallel Approach for Finding Co-location Pattern–A Map Reduce Framework. *Procedia Computer Science*, *89*, 341–348. doi:10.1016/j.procs.2016.06.081

Shooshtari, S. J., & Gholamalifard, M. (2015). Scenario-based land cover change modeling and its implications for landscape pattern analysis in the Neka Watershed, Iran. *Remote Sensing Applications: Society and Environment*, *1*, 1–19. doi:10.1016/j.rsase.2015.05.001

Shu, C. F., (2000). Housing layout and crime vulnerability. *Urban Design International, 5*(3-4), 177-188.

Sieverts, T. (1997). *Zwischenstadt zwischen Ort und Welt, Raum und Zeit, Stadt und Land.* Bauwelt Fundamente.

Siksna, A. (1997). The effects of block size and form in North American and Australian city centres. *Urban Morphology*, *1*, 19–33.

Silva, C. N. (2018). *New Approaches, Methods, and Tools in Urban E-Planning*. IGI Global. doi:10.4018/978-1-5225-5999-3

Silva, E. A., & Clarke, K. C. (2005). Complexity, emergence and cellular urban models: Lessons learned from applying Sleuth to two Portuguese Metropolitan Areas. *European Planning Studies*, *13*(1), 93–116. doi:10.1080/0965431042000312424

Simmonds, D., & Coombe, D. (2000). The Transport Implications of Alternative Urban Forms. In K. Williams, M. Jenks, & E. Burton (Eds.), *Achieving sustainable urban form* (pp. 121–130). London: E & FN Spon.

Sims, D. (2003). *Cairo, Egypt - understanding slums: case studies for the global report 2003.* UN Habitat. Available from: http://www.ucl.ac.uk/dpu-projects/Global_Report/pdfs/ Cairo.pdf

Sims, D. (2012). Understanding Cairo: The Logic of a City Out of Control. Cairo: The American University in Cairo Press.

Sims, D. (n.d.). *Library - analytical materials.* The Habitat International Coalition's specialized Member group: Housing and Land Rights Network (HIC-HLRN). Available at: http://www.landpedia.org/landdoc/Analytical_materials/Sims_Residential_Informality_in_Greater_Cairo. pdf

Singh, H., & Bawa, S. (2016). Spatial data analysis with ArcGIS and Map Reduce. *International Conference on Computing, Communication and Automation (ICCCA)*, 45-49. 10.1109/CCAA.2016.7813687

Sitte, C. (1945). *The Art of Building Cities: City Building According to Its Artistic Fundamentals.* New York: Reinhold Publishing Corporation.

Sitte, C. (1986). *City Planning According to Artistic Principles* (G. R. Collins & C. C. Collins, Trans.). Mineola, NY: Dover Publications, INC.

Sitte, C. (1996). *L'Art de bâtir la ville, l'urbanisme selon ses fondements artistiques.* Paris: Seuil Editions.

Smith, N., & Williams, P. (Eds.). (2013). *Gentrification of the City*. New York: Routledge. doi:10.4324/9781315889092

Snellen, D., Borgers, A., & Timmermans, H. (2016). Urban Form, Road Network Type, and Mode Choice for Frequently Conducted Activities: A Multilevel Analysis Using Quasi-Experimental Design Data. *Environment & Planning A*, *34*(7), 1207–1220. doi:10.1068/a349

Sohl, T. L., & Claggett, P. R. (2013). Clarity versus complexity: Land use modeling as a practical tool for decision-makers. *Journal of Environmental Management*, *129*, 235–243. doi:10.1016/j.jenvman.2013.07.027 PMID:23954777

Sohl, T. L., Sayler, K. L., Drummond, M. A., & Loveland, T. R. (2007). The FORE-SCE Model: A practical approach for projecting land cover change using scenario-based modeling. *Journal of Land Use Science*, *2*(2), 103–126. doi:10.1080/17474230701218202

sokan26july. (2018). *Residents of July 26 - Cairo Khedive (in Arabic).* Retrieved from www.facebook.com/sokan26july/: https://www.facebook.com/sokan26july/photos/pcb.207063756833259/207062963500005/?type=3&theater

Song, Y., Wang, X., Tan, Y., Wu, P., Sutrisna, M., Cheng, J., & Hampson, K. (2017). Trends and opportunities of BIM-GIS integration in the architecture, engineering and construction industry: A review from a spatio-temporal statistical perspective. *International Journal of Geo-Information, 6*, 397. doi:10.3390/ijgi6120397

Sousa, J. (2012, September). *Energy simulation software for buildings: review and comparison.* In International Workshop on Information Technology for Energy Applicatons-IT4Energy, Lisabo.

Southworth, F., Dale, V. H., & O'Neill, R. V. (1991). Contrasting patterns of land use in Rondonia, Brazil: Simulating the effects on carbon release. *International Social Science Journal*, *130*, 681–798.

Space Syntax Limited. (2004). *Agora Cities for People 2004, D3.1 Technology Platform Report*. London: Space Syntax Limited.

Spencer, H. (1872). *Principles of psychology* (Vol. 2). London: Williams and Norgate.

Srivanit, M., & Hokao, K. (2013). Evaluating the cooling effects of greening for improving the outdoor thermal environment at an institutional campus in the summer. *Building and Environment*, *66*, 158–172. doi:10.1016/j.buildenv.2013.04.012

Steadman, P. (2004). Developments in space syntax. *Environment and Planning. B, Planning & Design*, *31*(4), 483–486. doi:10.1068/b3104ed

Steiner, G. A. (1969). *Top Management Planning, Studies of the modern corporation.* New York: Macmillan.

Steinitz, C. (2010). Landscape Architecture into the 21st Century-Methods for Digital Techniques. In *Digital Landscape Architecture 2010* (pp. 2–26). Burner: Wichmann. Retrieved from http://www.kolleg.loel.hs-anhalt.de/landschaftsinformatik/fileadmin/user_upload/_temp_/2010/Proceedings/Buhmann_2-26.pdf

Steinitz, C. (2012). *A framework for geodesign : changing geography by design.* Esri. Retrieved from https://esripress.esri.com/display/index.cfm?fuseaction=display&websiteID=218&moduleID=0

Steinitz, C. (2016). *GeoDesign Dynamics - Carl Steinitz.* Coastal Regional Commision- Planning Department. Retrieved from https://www.youtube.com/watch?v=Jf_R4rB7MIQ&t=2021s

Steinitz, C. (2013). Beginnings of Geodesign: A Personal Historical Perspective. In *Geodesign: Past, Present, and Future* (pp. 4–14). Redlands: ESRI Press. Retrieved from http://www.esri.com/~/media/Files/Pdfs/library/ebooks/geodesign-past-present-future.pdf

Stevens, D., & Dragicevic, S. (2007). A GIS-based irregular cellular automata model of land-use change. *Environment & Planning B*, *34*(4), 708–724. doi:10.1068/b32098

Stewart Fotheringham, A., & Rogerson, P. A. (1993). GIS and spatial analytical problems. *International Journal of Geographical Information Systems*, *7*(1), 3–19. doi:10.1080/02693799308901936

Stock, W. G. (2011). Informational Cities: Analysis and Construction of Cities in the Knowledge Society. *Journal of the American Society for Information Science and Technology*, *62*(5), 963–986. doi:10.1002/asi.21506

Stolley, K. (2005). *The basics of sociology*. Westport, CT: Greenwood Press.

Straatman, B., White, R., & Engelen, G. (2004). Towards an automatic calibration procedure for constrained cellular automata. *Computers, Environment and Urban Systems*, *28*(1-2), 149–170. doi:10.1016/S0198-9715(02)00068-6

Strømann-Andersen, J., & Sattrup, P. A. (2011). The urban canyon and building energy use: Urban density versus daylight and passive solar gains. *Energy and Building*, *43*(8), 2011–2020. doi:10.1016/j.enbuild.2011.04.007

Stryker, B., & Nagati, O. (Eds.). (2016). Creative Cities: Re-Framing Downtown Cairo. Cairo: Cluster.

Sudhira, H. S. (2008). *Studies on Urban Sprawl and Spatial Planning Support System for Bangalore, India, PhD*. Bangalore: Centre for Sustainable Technologies and Department of Management Studies, Indian Institute of Science.

Sugumaran, R., & DeGroote, J. (2011). *Spatial Decision Support Systems - Principles and Practices*. CRC Press, Taylor & Francis Group.

Sui, D. Z. (1998). GIS-based urban modeling: Practices, problems and prospects. *International Journal of Geographical Information Science*, *12*(7), 651–671. doi:10.1080/136588198241581 PMID:12294534

Sun, B., & Robinson, D. T. (2018). Comparisons of statistical approaches for modeling land-use change. *Land (Basel)*, *7*(4), 144. doi:10.3390/land7040144

Sundquist, K., Eriksson, U., Kawakami, N., Skog, L., Ohlsson, H., & Arvidsson, D. (2011). Neighborhood walkability, physical activity, and walking behavior: The Swedish Neighborhood and Physical Activity (SNAP) study. *Social Science & Medicine (1982)*, *72*(8), 1266–1273. doi:10.1016/j.socscimed.2011.03.004

Susantia, R., Soetomoa, S., & Buchoria, I., & Brotosunaryo, P. (2016). Smart growth, smart city and density: In search of the appropriate indicator for residential density in Indonesia. *Social and Behavioral Sciences*, *227*, 194–201.

Swartz, S. (2012). Decentralization, Local Development and Women's Empowerment: Innovating Knowledge and Practice. *Universitas Forum*, *3*(1).

Tadamun. (2014). *Coming up short: Egyptian government approaches to informal areas*. The Tadamun Initiative. Available from: http://www.tadamun.info/

Tafirenyika, M. (Ed.). (2016). Africa's cities of the future. *Africa Renewal*, *30*(1).

Taher, B. (2008). *As Doha Said* (P. Daniel, Trans.). Cairo: The American University in Cairo Press.

Talen, E. (2003). Measuring urbanism: Issues in smart growth research. *Journal of Urban Design*, *8*(3), 195–215. doi:10.1080/1357480032000155141

Talen, E. (2006). Design That Enables Diversity: The Complications of a Planning Ideal. *Journal of Planning Literature*, *20*(3), 233–249. doi:10.1177/0885412205283104

Tandon, M., & Sehgal, V. (2017). Traditional Indian Religious Streets: A Spatial Study of the Streets of Mathura. *Frontiers of Architectural Research*, *6*(4), 469–479. doi:10.1016/j.foar.2017.10.001

Tangires, H. (2003). *Public Markets*. New York, NY: Norton.

Tang, T. Q., Yang, X. B., Wu, Y. H., Caccetta, L., & Huang, H. J. (2012). A multilane traffic flow model with lane width and the number of lanes. *Chinese Physics Letters*, *29*(9), 098903-1, 098903–098904. doi:10.1088/0256-307X/29/9/098903

Tan, R., Liu, Y., Zhou, K., Jiao, L., & Tang, W. (2015). A game-theory based agent-cellular model for use in urban growth simulation: A case study of the rapidly urbanizing Wuhan area of Central China. *Computers, Environment and Urban Systems*, *49*, 15–29. doi:10.1016/j.compenvurbsys.2014.09.001

Tao, W. (2013). Interdisciplinary Urban GIS for Smart Cities: Advancements and Opportunities. *Geo-Spatial Information Science*, *16*(1), 25–34. doi:10.1080/10095020.2013.774108

Tarasov, A., Cullen, C., & Delany, S. J. (2010). *Using crowdsourcing for labelling emotional speech assets.* W3C workshop on Emotion ML, Paris, France. doi:10.21427/D7RS4G

Tariq, T., & Arch, M. (2014). *An ENVI-met Simulation Study on Urban Open Spaces of Dhaka.* Academic Press.

Tavroulaki, I., Marcus, L., & Berghauser Pont, M. (2017). Representations of street networks in space syntax towards flexible maps and multiple graphs. *Proceeding of the 11th International Space Syntax Symposium.*

Taylor, J., & Song, L. (2016). Return to the Streets. *Cityscape: A Journal of Policy Development and Research, 18*(1), 71-88.

Tesfatsion, L. (2001). Introduction to the special issue on agent-based computational economics. *Journal of Economic Dynamics & Control*, *25*(3/4), 281–293. doi:10.1016/S0165-1889(00)00027-0

Thakur, R., & Sharma, M. (2009). Local Government Use of GIS in Comprehensive Planning. In J. D. Gatrell & R. R. Jensen (Eds.), Planning and Socioeconomic Applications (pp. 125–137). Springer. 13 doi:10.1007/978-1-4020-9642-6

The Commission on Human Rights and the General Assembly. (1998). *Guiding Principles on Internal Displacement, E/CN.4/1998/53/Add.2*. New York: United Nations.

The National Association of Regional Councils (NARC). (n.d.). *Livability Literature Review: A Synthesis of Current Practice.* Washington, DC: The National Association of Regional Councils (NARC).

The UN Refugee Agency. (2007). *Internally Displaced Persons: Questions and Answers.* Geneva, Switzerland: United Nations High Commissioner for Refugees (UNHCR). Retrieved from http://www.unhcr.org/basics/BASICS/405ef8c64

The World Bank. (2013, June 6). *Decentralization.* Retrieved from www.worldbank.org: http://www.worldbank.org/en/topic/communitydrivendevelopment/brief/Decentralization

Thorsson, S., Lindqvist, M., & Lindqvist, S. (2004). Thermal bioclimatic conditions and patterns of behaviour in an urban park in Göteborg, Sweden. *International Journal of Biometeorology*, *48*(3), 149–156. doi:10.100700484-003-0189-8 PMID:12955614

Throsby, D. (2006). The value of cultural heritage: What can economics tell us? In *Capturing the Public Value of Heritage: The Proceedings of the London Conference 25-26 January 2016*. London: English Heritage.

Timmer, V., & Seymoar, N.-K. (2005). *The livable city.* Vancouver: The World Urban Forum 2006, Working Group, International Centre for Sustainable Cities.

Tonkiss, F. (2013). *Cities by Design, The Social Life of Urban Form.* Cambridge, UK: Polity Press.

Topcu, M., Topcu, K., & Kabut, A. (2007), Movement economy dependent on urban design. In *Proceeding of the 6th International Space Syntax Symposium* (pp. 117-01: 117-06). Istanbul: Istanbul Teknik Üniversitesi.

Torgue, H. (2009). Composer le son des villes. WildProject, Ondes du monde: territoires sonores de l'écologie.

Torrens, M. P. (2000, November). *How Cellular Models of Urban Systems Work. (1. Theory).* Retrieved July 29, 2018, from http://discovery.ucl.ac.uk/1371/1/paper28.pdf

Torrisi-Steele, G., Wang, V. C., Sedivy-Benton, A., & Boden-McGill, C. J. (2015). Research Goes Digital: Some Methods, Frameworks, and Issues. *The Reference Librarian*, *56*(4), 239–258. doi:10.1080/02763877.2015.1057467

Townshend, T., & Lake, A. A. (2009). Obesogenic urban form: Theory, policy and practice. *Health & Place*, *15*(4), 909–916. doi:10.1016/j.healthplace.2008.12.002 PMID:19201641

Tung, A. M. (2001). *Preserving the world's great cities. The destruction and renewal of the historic metropolis*. New York: Clarkson Potter.

Turner, A. (2005). Could a road-centre line be an axial line in disguise? In *Proceeding of 5th International Space Syntax Symposium* (pp. 145-159). Delft: TU Delft, Tech. Press.

Turner, A. (2011). *UCL Depthmap: Spatial Network Analysis Software*. Retrieved from https://www.ucl.ac.uk/bartlett/architecture/research/space-syntax/depthmapx

Turner, A. (2001). Angular analysis. *Proceedings of the 3rd international symposium on space syntax*, 30-1.

Turner, A. (2003). Analysing the visual dynamics of spatial morphology. *Environment and Planning. B, Planning & Design*, 30657–30676.

Turner, A. (2003). Analyzing the Visual Dynamics of Spatial Morphology. *Environment and Planning. B, Planning & Design*, *30*(5), 657–676. doi:10.1068/b12962

Turner, A. (2004). *Depthmap 4: A Researcher's Handbook*. London: Bartlett School of Graduate Studies, UCL.

Turner, A. (2007). From axial to road-centre lines: A new representation for space syntax and a new model of route choice for transport network analysis. *Environment and Planning. B, Planning & Design*, *34*(3), 539–555. doi:10.1068/b32067

Turner, A., Doxa, M., O'Sullivan, D., & Penn, A. (2001). From isovists to visibility graphs: A methodology for the analysis of architectural space. *Environment and Planning. B, Planning & Design*, *28*(1), 103–121. doi:10.1068/b2684

Turner, A., & Penn, A. (1999). Making Isovists Syntactic: Isovist Integration Analysis. *Proceeding of the second space syntax symposium.*

Turner, A., Penn, A., & Hillier, B. (2005). An Algorithmic Definition of the Axial Map. *Environment and Planning. B, Planning & Design*, *32*(3), 425–444. doi:10.1068/b31097

Turner, B. L., Lambin, E. F., & Reenberg, A. (2007). The emergence of land change science for global environmental change and sustainability. *Proceedings of the National Academy of Sciences of the United States of America*, *104*(52), 20666–20671. doi:10.1073/pnas.0704119104 PMID:18093934

Turner, M. G., Wear, D. N., & Flamm, R. O. (1996). Land ownership and land-cover change in the Southern Appalachian Highlands and the Olympic Peninsula. *Ecological Applications*, *6*(4), 1150–1172. doi:10.2307/2269599

UN-Department of Economic and Social Affairs Statistics Division. (2017). *Principles and Recommendations for Population and Housing Censuses* (Vol. Revision 3). New York: United Nations. Retrieved from https://unstats.un.org/unsd/publication/seriesM/Series_M67Rev3en.pdf

UNESCO. (2013, August). *Warning: Looting of the Malawi National Museum in the Upper Egypt city of Minya*. Retrieved May 05, 2016, from http://www.unesco.org:http://www.unesco.org/new/en/culture/themes/illicit-traffic-of-cultural-property/emergency-actions/egypt/warning-looting-of-the-malawi-national-museum/

United Nations Economic Commission for Europe. (2017). *Country Profile of the Housing Sector: Kazakhstan*. New York: United Nation. Retrieved from http://www.unece.org/housing.html

United Nations General Assembly. (2016). *New Urban Agenda. United Nations Conference on Housing and Sustainable Urban Development (Habitat III)*, Quito, Ecuador.

United Nations High Commissioner for Refugees (UNHCR). (2015, June 18). *Worldwide displacement hits all-time high as war and persecution increase*. Retrieved from http://www.unhcr.org: http://www.unhcr.org/news/latest/2015/6/558193896/worldwide-displacement-hits-all-time-high-war-persecution-increase.html

United Nations Human Settlements Programme (UN-HABITAT). (2009). *International Guidelines on Decentralisation and Access to Basic Services for all*. Nairobi, Kenya: UN-HABITAT.

United Nations Human Settlements Programme. (2008). *The State of African City*. Nairobi, Kenya: UN-HABITAT.

United Nations Human Settlements Programme. (2016). *Urbanization and Development: Emerging Futures*. Nairobi, Kenya: UN-Habitat.

United Nations. (2005). *Decentralisation: Poverty Production, Reduction, Empowerment and Participation*. New York: United Nations. Retrieved from https://publicadministration.un.org/publications/content/PDFs/E-Library%20Archives/2005%20Decentralization_Poverty%20Reduction,%20Empowerment%20and%20Participation.pdf

Unlu-Yucesoy, E. (2013). Constructing the Marketplace: A Socio-Spatial Analysis of Past Marketplaces of Istanbul. *Built Environment*, *39*(2), 190–202. doi:10.2148/benv.39.2.190

Valbuena, D., Verburg, P. H., Bergt, A. K., & Ligtenberg, A. (2010). An agent-based approach to model land-use change at a regional scale. *Landscape Ecology*, *25*(2), 185–199. doi:10.100710980-009-9380-6

Valdez, V. C., Ruiz-Luna, A., & Berlanga-Robles, C. A. (2016). Effects of land use changes on ecosystem services value provided by coastal wetlands: Recent and future landscape scenarios. *Journal of Coastal Zone Management*, *19*(1), 1000418. doi:10.4172/jczm

Vale, D. S. (2013). Does commuting time tolerance impede sustainable urban mobility? Analysing the impacts on commuting behaviour as a result of workplace relocation to a mixed-use centre in Lisbon. *Journal of Transport Geography*, *32*, 38–48. doi:10.1016/j.jtrangeo.2013.08.003

Van de Coevering, P., & Schwanen, T. (2006). Re-evaluating the impact of urban form on travel patterns in Europe and North-America. *Transport Policy*, *13*(3), 229–239. doi:10.1016/j.tranpol.2005.10.001

Van Delden, H., & McDonald, G. (2010). *Towards the integration of economic and land use change models*. 5th International Congress on Environmental Modeling and Software, Ottawa, Ontoria, Canada.

Van Delden, H., van Vliet, J., Rutledge, D. T., & Kirkby, M. J. (2011). Comparison of scale and scaling issues in integrated land-use models for policy support. *Agriculture, Ecosystems & Environment*, *142*(1-2), 18–28. doi:10.1016/j.agee.2011.03.005

Van Keulen, H., Van Ittersum, M. K., & de Ridder, N. (2000). New approaches to land use planning. In R. P. Roetter, H. van Keulen, A. G. Laborte, C. T. Hoanh, & H. H. van Laar (Eds.), *Systems research for optimising future land use in South and Southeast Asia*. IRRI.

van Nes, A. (2011). The One- and Two-Dimentional Isovists Analysis in Space Syntax. In S. Nijhuis, R. Lammeren, & F. Hoeven (Eds.), *Exploring the Visual Landscape: Advances in Physiognomic Landscape Research in The Netherlands* (Vol. 2, pp. 163–183). Research in Urbanism Series.

Van Soesbergen, A. (2016). *A review of land use change models*. UNEP and John D. and Catherine T. Mac Arthur Foundation.

Vandezande, J., & Krygiel, E. (2015). Mastering Autodesk Revit Architecture 2016: Autodesk Official Press 1st. Alameda, CA: SYBEX.

Varga, M., Balogh, S., & Csukas, B. (2016). GIS based generation of dynamic hydrological and land patch simulation models for rural watershed areas. *Information Processing in Agriculture*, *3*(1), 1–16. doi:10.1016/j.inpa.2015.11.001

Varoudis, T. (2012). *DepthmapX Multi-Platform Spatial Network Analysis Software*. Retrieved from https://varoudis.github.io/depthmapX/

Vaughan, L. (1999). *Clustering, Segregation and the Ghetto: The Spatialisation of Jewish Settlement in Manchester and Leeds in the 19th Century* (Unpublished doctoral dissertation). University College London.

Vaughan, L. (2007). The Spatial Syntax of Urban Segregation. *Progress in Planning*, *67*(3), 199–294. doi:10.1016/j.progress.2007.03.001

Veldkamp, A., & Fresco, L. (1997). Exploring land use scenarios: An alternative approach based on actual land use. *Agricultural Systems*, *55*(1), 1–17. doi:10.1016/S0308-521X(95)00079-K

Veldkamp, A., & Fresco, L. O. (1996a). CLUE: A conceptual model to study the conversion of land use and its effects. *Ecological Modelling*, *85*(2-3), 253–270. doi:10.1016/0304-3800(94)00151-0

Veldkamp, A., & Fresco, L. O. (1996b). CLUE-CR: An integrated multiscale model to simulate land use change scenarios in Costa Rica. *Ecological Modelling*, *91*(1–3), 231–248. doi:10.1016/0304-3800(95)00158-1

Veldkamp, A., & Lambin, E. F. (2001). Predicting land use change. *Agriculture, Ecosystems & Environment*, *85*(1-3), 1–6. doi:10.1016/S0167-8809(01)00199-2

Verburg, P. H., Eickhout, B., & van Meijl, H. (2008). A multi-scale, multi-model approach for analysing the future dynamics of European land use. *The Annals of Regional Science*, *42*(1), 57–77. doi:10.100700168-007-0136-4

Verburg, P. H., & Overmars, K. (2007). Dynamic simulation of land-use changes trajectories with the Clue-S Model. In E. Koomen, J. Stillwell, A. Bakema, & H. J. Scholten (Eds.), *Modeling land-use change. The GeoJournal Library* (Vol. 90). Dordrecht: Springer.

Verburg, P. H., & Overmars, K. P. (2009). Combining top-down and bottom-up dynamics in land-use modeling: Exploring the future of abandoned farmlands in Europe with the Dyna-CLUE model. *Landscape Ecology*, *24*(9), 1167–1181. doi:10.100710980-009-9355-7

Verburg, P. H., Rounsevell, M. D. A., & Veldkamp, A. (2006b). Scenario-based studies of future land use in Europe. *Agriculture, Ecosystems & Environment*, *114*(1), 1–6. doi:10.1016/j.agee.2005.11.023

Verburg, P. H., Schot, P. P., Dijst, M. J., & Veldkamp, A. (2004). Land use change modeling: Current practice and research priorities. *GeoJournal*, *61*(4), 309–324. doi:10.100710708-004-4946-y

Verburg, P. H., Schulp, C. J. E., Witte, N., & Veldkamp, A. (2006a). A downscaling of land use change scenarios to assess the dynamics of European landscapes. *Agriculture, Ecosystems & Environment*, *114*(1), 39–56. doi:10.1016/j.agee.2005.11.024

Verburg, P. H., Veldkamp, A., de Koning, G. H. J., Kok, K., & Bouma, J. (1999). A spatial explicit allocation procedure for modeling the pattern of land use change based upon actual land use. *Ecological Modelling*, *116*(1), 45–61. doi:10.1016/S0304-3800(98)00156-2

Verburg, P., Soepboer, W., Veldkamp, A., Limpiada, R., Espaldon, V., & Mastura, S. S. A. (2002). Modeling the Spatial Dynamics of Regional Land Use: The CLUE-S Model. *Environmental Management*, *30*(3), 391–405. doi:10.100700267-002-2630-x PMID:12148073

Verhetsel, A., & Vanelslander, T. (2010). What location policy can bring to sustainable commuting: An empirical study in Brussels and Flanders, Belgium. *Journal of Transport Geography*, *18*(6), 691–701. doi:10.1016/j.jtrangeo.2009.11.003

Vernet, N., & Coste, A. (2017). Garden Cities of the 21st Century: A Sustainable Path to Suburban Reform. *Urban Planning*, *2*(4), 45–60. doi:10.17645/up.v2i4.1104

Victorian Competition and Efficiency Commission. (2008). *A state of liveability: An enquiry into enhancing Victoria's liveability*. Melbourne, Victoria: Victorian Competition and Efficiency Commission.

Vigarello, G. (2014). *Le Sentiment de soi. Histoire de la perception du corps. (XVIe - XXe siècle)*. Paris: Seuil Editions.

Villanueva, K., Knuiman, M., Nathan, A., Giles-Corti, B., Christian, H., Foster, S., & Bull, F. (2014). The impact of neighborhood walkability on walking: Does it differ across adult life stage and does neighborhood buffer size matter? *Health & Place*, *25*, 43–46. doi:10.1016/j.healthplace.2013.10.005 PMID:24239702

Vliet, J., White, R., & Dragicevic, S. (2009). Modeling urban growth using a variable grid cellular automaton. *Computers, Environment and Urban Systems*, *33*(1), 35–43. doi:10.1016/j.compenvurbsys.2008.06.006

Von Schonfeld, K. C., & Bertolini, L. (2017). Urban Streets: Epitomes of Planning Challenges and Opportunities at the Interface of Public Space and Mobility. *Cities (London, England)*, *68*, 48–55. doi:10.1016/j.cities.2017.04.012

Von Thunen, H. (1826). Der isolierte staat. In *Von Thunen's isolated state, 1968* (C. M. Warterberg, Trans.). Pergamon.

Vuchic, V. R. (1999). *Transportation for Livable Cities*. Center for Urban Policy Research.

Waddell, P., & Evans, J. (2002). UrbanSim: Modeling Urban Development for Land Use, Transportation and Environmental Planning. *Journal of the American Planning Association*, *68*(3), 297–314. doi:10.1080/01944360208976274

Wagner, D. F. (1997). Cellular automata and geographic information systems. *Environment & Planning B*, *24*(2), 219–234. doi:10.1068/b240219

Wakeman, R. (2016). *Practicing Utopia: An Intellectual History of the New Town Movement*. Chicago: University of Chicago Press. doi:10.7208/chicago/9780226346175.001.0001

Walsh, S. J., & Crews-Meyer, K. A. (Eds.). (2002). *Linking people, place, and policy. A GIScience approach*. Boston: Kluwer Academic Publishers. doi:10.1007/978-1-4615-0985-1

Walsh, S. J., Evans, T. P., Welsh, W. F., Entwisle, B., & Rindfuss, R. R. (1999). Scale dependent relationships between population and environment in Northeastern Thailand. *Photogrammetric Engineering and Remote Sensing*, *65*, 97–105.

Wang, M. (2012). *Extending Geographic Information systems to urban morphological analysis with a space syntax approach* (Unpublished M.Sc. thesis). University of Gävle (HIG).

Wang, X., & vom Hofe, R. (2007). *Research Methods in Urban and Regional Planning*. Tsinghua University Press, Beijing and Springer-Verlag GmbH Berlin Heidelberg.

Wang, J., Hou, L., Chong, H. Y., Liu, X., Wang, X., & Guo, J. (2014). *A cooperative system of GIS and BIM for traffic planning: A high-rise building case study. In Cooperative design, visualization, and engineering* (pp. 143–150). Springer.

Warnaby, G. (2009). Towards a Service-Dominant Place Marketing Logic. *Marketing Theory*, *9*(4), 403–423. doi:10.1177/1470593109346898

Wasft-Iskendrya. (2016). *Facebook page*. Retrieved on July 06, 2016, from https://www.facebook.com/groups/798916430159507/

Watson, S. (2009). The Magic of the Marketplace: Sociality in a Neglected Public Space. *Urban Studies (Edinburgh, Scotland)*, *46*(8), 1577–1591. doi:10.1177/0042098009105506

Wear, D. N., & Bolstad, P. (1998). Land-use changes in Southern Appalachian landscapes: Spatial analysis and forecast evaluation. *Ecosystems (New York, N.Y.)*, *1*(6), 575–594. doi:10.1007100219900052

Weber, B. (2004). Educating Multi-Disciplined Experts in GIS-Based Global Valuation Methods. *24th Annual Esri International User Conference*.

Webster, C. J. (1993). GIS and the scientific inputs to planning. Part 1: Description. *Environment & Planning B*, *20*(6), 615–748. doi:10.1068/b200709

Webster, C. J. (1994). GIS and the scientific inputs to planning. Part 2: Prediction and prescription. *Environment & Planning B*, *21*(2), 145–157. doi:10.1068/b210145

Wegener, M. (2004). Overview of land-use transport models. In D. A. Hensher & K. Button (Eds.), *Transport geography and spatial systems* (pp. 127–146). Kidlington: Pergamon/Elsevier. doi:10.1108/9781615832538-009

Wegener, M., Simmonds, D., & Waddell, P. (2011). *Equilibrium v. dynamics in urban modelling. In Applied Urban Modelling (AUM 2011) "Innovation in Urban Modelling*. Cambridge, UK: University of Cambridge.

Weick, K. (1979). *The Social Psychology of Organizing*. McGraw-Hill Humanities/Social Sciences/Languages.

Weiss, D., & Wurzel, U. (1998). *The Economics and Politics of Transition to an Open Market Economy: Egypt*. Paris, France: OECD. doi:10.1787/9789264163607-en

Wendel-Vos, G. C. W., Schuit, A. J., De Niet, R., Boshuizen, H. C., Saris, W. H. M., & Kromhout, D. (2004). Factors of the Physical Environment Associated with Walking and Bicycling. *Medicine and Science in Sports and Exercise*, *36*(4), 725–730. doi:10.1249/01.MSS.0000121955.03461.0A PMID:15064601

Westin, S. (2011). The life and form of the city: An interview with Bill Hillier. *Space and Culture*, *14*(2), 227–237. doi:10.1177/1206331211404391

Wheaton, W. (2004). Commuting, congestion, and employment dispersal in cities with mixed land use. *Journal of Urban Economics*, *55*(3), 417–438. doi:10.1016/j.jue.2003.12.004

Whitehand, J. W. R. (2001). British urban morphology: The Conzenian tradition. *Urban Morphology*, *5*(2), 103–109.

White, R. (2005). Modeling multi-scale processes in a cellular automata framework. In J. Portugali (Ed.), *Complex artificial environments. Verlag*. Springer.

White, R., & Engelen, G. (1993). Cellular automata and fractal urban form: A cellular modeling approach to the evolution of urban land use patterns. *Environment & Planning A*, *25*(8), 1175–1199. doi:10.1068/a251175

White, R., & Engelen, G. (1997). Cellular automata as the basis of integrated dynamic regional modeling. *Environment & Planning B*, *24*(2), 235–246. doi:10.1068/b240235

White, R., & Engelen, G. (2000). High-resolution integrated modeling of the spatial dynamics of urban and regional systems. *Computers, Environment and Urban Systems*, *24*(5), 383–400. doi:10.1016/S0198-9715(00)00012-0

White, R., Engelen, G., & Uijee, I. (1997). The use of constrained cellular automata for high-resolution modeling of urban land-use dynamics. *Environment & Planning B*, *24*(3), 323–343. doi:10.1068/b240323

Whyte, W. (1980). *The Social Life of Small Urban Spaces*. New York: Ingram.

Wicke, B., Verweij, P., Meijl, H. V., Vuuren, D. P. V., & Faaij, A. P. C. (2012). Indirect land use change: Review of existing models and strategies for mitigation. *Biofuels*, *3*(1), 87–100. doi:10.4155/bfs.11.154

Wiedmann, T., Lenzen, M., Turner, K., & Barrett, J. (2007). Examining the global environmental impact of regional consumption activities-Part 2: Review of input-output models for the assessment of environmental impacts embodied in trade. *Ecological Economics*, *61*(1), 15–26. doi:10.1016/j.ecolecon.2006.12.003

Wiet, G. (1964). *Cairo, City of Art and Commerce*. Norman, OK: University of Oklahoma Press.

Wilbanks, T. J., & Kates, R. W. (1999). Global changes in local places: How scale matters. *Climatic Change*, *43*(3), 601–628. doi:10.1023/A:1005418924748

Williams, K. (2000). Does Intensifying Cities Make them More Sustainable? In K. Williams, M. Jenks, & E. Burton (Eds.), *Achieving sustainable urban form* (pp. 30–45). London: E & FN Spon.

Williams, R. (1975). *The Country and the City*. Oxford University Press.

Wischmeier, W. H., & Smith, D. D. (1978). Predicting rainfall erosion losses - A Guide to Conservation Planning. *Agriculture handbook no. 537*. doi:10.1029/TR039i002p00285

Witzke, H. P., & Zintl, A. (2003). *CAPSIM-Complete documentation*. Bonn: Final Report to Eurostat.

Wong, N.-H., & Chen, Y. (2010). The Role of Urban Greenery in High-Density Cities. In E. Ng (Ed.), Designing high-density cities for social and environmental sustainability (pp. 227–262). Sterling, VA: Earthscan.

Wongtada, N. (2014). Street Vending Phenomena: A Literature Review and Research Agenda. *Thunderbird International Business Review*, *56*(1), 55–75. doi:10.1002/tie.21596

Wooldridge, M., & Jennings, N. (1995). Intelligent agents: Theory and practice. *The Knowledge Engineering Review*, *10*(2), 115–152. doi:10.1017/S0269888900008122

World Economic Forum. (2015). *Inspiring Future Cities & Urban Services Shaping the Future of Urban Development & Services Initiative*. Industry Agenda, Cologny/Geneva Switzerland.

World Economic Forum. (2017). *Migration and Its Impact*. Cologny: World Economic Forum. Retrieved from http://www3.weforum.org/docs/Migration_Impact_Cities_report_2017_low.pdf

Worrall, L. (1991). *Spatial Analysis and Spatial Policy Using Geographic Information Systems*. Belhaven Press.

Wright, D. J., & Montiel, L. M. (2007). *Divided they fall: hardship in America's cities and suburbs*. Albany, NY: The Nelson A. Rockefeller Institute of Government.

Wright, F. L. (1932). *The Disappearing City*. New York: Payson.

Wright, F. L. (1935). Broadacre City: A New Community Plan. *Architectural Record*, *77*(4), 345–349.

Wu, F., & Webster, C. J. (1998). Simulation of land development through the integration of cellular automata and multicriteria evaluation. *Environment & Planning B*, *25*(1), 103–126. doi:10.1068/b250103

Wu, Q., Li, H.-Q., Wang, R.-S., Paulussen, J., He, Y., Wang, M., ... Wang, Z. (2006). Monitoring and predicting land use change Beijing using remote sensing and GIS. *Landscape and Urban Planning*, *78*(4), 322–333. doi:10.1016/j.landurbplan.2005.10.002

Xu, X., Ding, L., Luo, H., & Ma, L. (2014). From building information modeling to city information modeling. *Journal of Information Technology in Construction*, *19*, 292–307.

Yang, J., Su, J., Chen, F., Xie, P., & Ge, Q. (2016). A local land use competition cellular automata model and its applications. *International Journal of Geo-Information*, *5*(7), 106. doi:10.3390/ijgi5070106

Yang, L. (2013). Research of urban thermal environment based on digital technologies. *Nature Environment and Pollution Technology*, *12*(4), 645.

Yatmo, Y. A. (2008). Street Vendors as 'Out of Place' Urban Elements. *Journal of Urban Design*, *13*(3), 387–402. doi:10.1080/13574800802320889

Yatmo, Y. A. (2009). Perception of Street Vendors as 'Out of Place' Urban Elements at Day-time and Night-time. *Journal of Environmental Psychology*, *29*(4), 467–476. doi:10.1016/j.jenvp.2009.08.001

Yeh, A. G. (1999). Urban Planning and GIS. *Geographical Information Systems, 2*(1), 877-888.

Young, E., & Hermanson, V. (2012). Livability Literature Review: A Synthesis of Current Practice. Washington, DC: The National Association of Regional Councils (NARC).

Yue, P., Baumann, P., Bugbee, K., & Jiang, L. (2015). Towards intelligent GIS Services. *Earth Science Informatics*, *8*(3), 463–481. doi:10.100712145-015-0229-z

Yue, P., & Jiang, L. (2014). BigGIS: How big data can shape next-generation GIS. In *Third International Conference on In Agro-geoinformatics (Agro-geoinformatics 2014).* IEEE. 10.1109/Agro-Geoinformatics.2014.6910649

Yun, Y. W., & Kim, Y. O. (2007). The Effect of Depth and Distance in Spatial Cognition. In *Proceedings of the 6th International Space Syntax Symposium.* Istanbul: ITU Faculty of Architecture.

Zacharias, J., Stathopoulos, T., & Wu, H. (2001). Microclimate and downtown open space activity. *Environment and Behavior*, *33*(2), 296–315. doi:10.1177/0013916501332008

Zairi, M. (2016). *L'ambiance comme enjeu politique dans l'espace public de Tunis lors du processus révolutionnaire* (Unpublished doctoral dissertation). Grenoble: PhD, ENSA, CRESSON.

Žák, Š. (2015). The identification of innovative research methods and techniques utilized in marketing research in the digital era. *Studia commercialia Bratislavensia, 8*(29), 139-152.

Zelinka, A., & Brennan, D. (2001). *Safescape: Creating Safer, More Livable Communities through Planning and Design.* Chicago: Planners Press.

Zhang, J. (2012). *Environmental quality assessment of outdoor public space in urban area: development of a Geoprocessing method.* Universität Stuttgart.

Zhao, L., Chen, L., Ranjan, R., Choo, K. K., & He, J. (2016). Geographical information system parallelization for spatial big data processing: A review. *Cluster Computing*, *19*(1), 139–152. doi:10.100710586-015-0512-2

Zhao, L., & Peng, Z.-R. (2012). LandSys: An agent-based Cellular Automata model of land use change developed for transportation analysis. *Journal of Transport Geography*, *25*, 35–49. doi:10.1016/j.jtrangeo.2012.07.006

Zhao, Y., & Zhu, Q. (2012). Evaluation on crowdsourcing research: Current status and future direction. *Information Systems Frontiers*, *12*(3), 1–18. doi:10.100710796-012-9350-4

Zhu, J., Zheng, Y., Carroll, A. L., & Aukema, B. H. (2008). Autologistic regression analysis of spatial-temporal binary data via Monte Carlo maximum likelihood. *Journal of Agricultural Biological & Environmental Statistics*, *13*(1), 84–98. doi:10.1198/108571108X273566

Zhu, Y., & Diao, M. (2016). The impacts of urban mass rapid transit lines on the density and mobility of high-income households: A case study of Singapore. *Transport Policy*, *51*, 70–80. doi:10.1016/j.tranpol.2016.03.013

Zimring, C., & Dalton, R. C. (2003). Linking Objective Measures of Space to Cognition and Action. *Environment and Behavior*, *35*(1), 3–16. doi:10.1177/0013916502238862

Zurek, M. B., & Henrichs, T. (2007). Linking scenarios across geographical scales in international environmental assessments. *Technological Forecasting and Social Change*, *74*(8), 1282–1295. doi:10.1016/j.techfore.2006.11.005

About the Contributors

Hisham Abusaada is a professor of architecture and urban design in the Housing and Building National Research Centre (HBRC), Egypt. During 1995-2015, he taught urban design course at universities in Saudia Arabia and Egypt. He is the author and co-authors of several scientific manuscripts and books in both Arabic and English published in scientific journals and conference proceedings. His area of interest is in urban design between theory and practice.

Abeer Elshater, has a PhD, and is an associate professor at Department of Urban Planning and Design, Faculty of Engineering, Ain Shams University (ASU), Cairo, Egypt, teaching and supervising multidisciplinary topics in the ideology of urban design. Her area of interest is in urban design disciplines. She has worked on some international research projects with international universities. Elshater is an author and co-author of five books as well as more than fifty scientific manuscripts in international journals and conference proceedings. Recently, she acts as an ambassador of the Regional Studies Association, United Kingdom.

* * *

Ahmed Hassan Abayazeed is an Egyptian architect. He received his Bachelor degree in Architecture in 2004 from The Faculty of Fine Arts at Alexandria University. In 2013, Abayazeed won a scholarship to study at a joint academic program (IUSD) hosted at the University of Stuttgart, Germany and Ain Shams University, Cairo where he earned his MSc double degree. His thesis topic concerns with residential urban spaces in Cairo's informal settlements. Abayazeed is working now as a senior urban designer at The National Organization for Urban Harmony (NOUH) where he participates in various projects covering a wide range of scope and complexity with special emphasis on urban design projects. Abayazeed's work and academic interests concern with public urban spaces, informal settlements, and urban design-related issues. He attended various workshops and summer schools. Also, he participated in international, local competitions and exhibitions and won several awards and certificates of merit.

Nancy M. Abd El-Moneim is an Egyptian Architect, holds a Ph.D. in Architecture from Cairo University in 2016 from where she got her Bachelor of Architecture (2000-2005). As for her master degree, she got it from the school of Economics at Catania University (2007-2008). She published her master thesis as a book with the same title. Her previous research includes issues related to informal settlements and governmental aspects. She worked, during the last four years for The Informal Settlement Development Fund (ISDF). Abd El-Moneim is interested in the process of resettlement projects and their effect on dwellers social and economic life's quality. Moreover, she is aware of the new challenges facing her geographical area and the MENA Region. Recently, she received a fund of the Urban Studies Foundation - Seminar Series Competition, 2018, which has a title "Gender Equity in Cities of the MENA Region". She has an experience in urban conservation she worked in many documentation and restoration projects all over Egypt. Currently, she is an assistant professor at the Arab Academy for Science Technology and Maritime Transport where she teaches classes in Urban Design, design studios, history of Architecture and graduation-project studio as well as postgraduate studies in her affiliated university.

Dalia Abdelfattah is an architect and researcher in urban affairs. Dalia is an assistant professor and a researcher in UN-HABITAT Programme of Planners for Climate Action. In 2018, she obtained her Ph.D. from Cairo University, Egypt. Her area of interest is in community design in Middle East and Northern Africa Region (MENA) with special focus on urban conservation, adaptation of multi-disciplinary approaches from environmental psychology. She has a number of scientific studies published in international journals and conference proceedings. Currently, Dalia works as a visiting lecturer at the Department of Architectural Engineering, Arab Academy for Science, Technology and Maritime Transport and the Architecture Engineering Technology (AET program) Faculty of Engineering, Cairo University.

Mougibelrahman Aboamer is a Ph.D. student at Grenoble School of Architecture (ENSA de GRENOBLE) University of Grenoble Alpes. He is also an assistant lecturer, Faculty of Engineering, Architectural Department, Misr International University (MIU). Egypt. Mougibelrahman gained his Master of Engineering in Architecture (MSc) from Faculty of Engineering, Cairo University for the topic "Architecture and Social Values through Political Paradigms". He is the author of the book "Meditations in the Lecture hall no. 3002" in Arabic (Tajaliat Modaraj 3002) Dar Leleit Press, Alexandria 2015. Mougibelrahman was also editor-in-chief of Architecture Wall Magazine, Cairo University-Faculty of Engineering, 2007-2008.

Samy Afifi is an assistant professor at the Urban Planning Department, Faculty of Engineering, Ain Shams University. He is a member of a team who is responsible for teaching many courses such as Theories and History of Urban Planning, Urban Economics, City Planning and Housing, and Computer Applications for Urban Planning. In professional practice, Samy is a co-founder for Urban Development Centre. After six years of experience in urban planning and design field, his elder colleague allowed him to participate in establishing their own firm in 2008. His career allowed him to act, discuss, negotiate and deal with a spectrum of people with different backgrounds and governmental organizations all over Egypt. His main interest is in tracing and analyzing dynamics during the evolution of traditional and contemporary urban settlements.

Abdalrahman Alashi is an architect. He enrolled in the Urban Planning Master Program at Istanbul Technical University (ITU), Department of Urban and Regional Planning, Turkey. He completed his BSc of Architecture in 2013 from Misr International University (MIU), Cairo. Abdalrahman is specialized and certified professional in software like Autodesk Revit Architecture 2015 and Autodesk Autocad 2015.

Mohammed Alfiky is a Ph.D. candidate at Stuttgart University, Researcher, urban planner, and GIS expert. Alfiky graduated from Department of Urban Planning and Design, Ain Shams University, Cairo in 2011. He obtained his M.Sc. in 2014 from Integrated Urbanism and Sustainable Design Programme

(IUSD) entitled "Community Based Security of Tenure: Initial Framework for Squatted Areas in Cairo". Since 2014, Alfiky was working for the IUSD-Cairo as researcher on the research projects entitled "Development Priorities in Informal Areas: Planning, Realization and Local Perceptions" DEVEPER in cooperation with the Habitat Unit-TU Berlin. Besides, he worked as a teaching assistant for the Integrated Case Study module (ICS) in the IUSD. In April 2015 till August 2015, he joined the Institute for Landscape Planning and Ecology (ILPÖ) in Stuttgart University for 4 months as a co-lecturer for Geo-design module for the IUSD. Since August 2015, he is back again for the IUSD-Cairo. In 2017, he co-founded Madyan Organization for Urban development with 8 other partners. He was the head of the Urban Planning team for Futuremakers project in Sohag Governorate with UNDP. Moreover, he was the head of the Urban Planning committee for the selection of candidates in the Urban Lab Camp (ULC) in el Gouna. Furthermore, he is a co-expert for GOPP for "urban development of Altawar' Axis" in Assalam district. Additionally, Alfiky has several publications regarding urban informality in Egypt.

Nada Alhakkak, PhD, is an associate professor in Computer Science and Information Technology. She is working at IT Faculty at Baghdad College of Economic Sciences Private University, Bagdad, Iraq.

Amany Ramadan Arisha is an Egyptian architect and an urban researcher, who holds a B.Sc. of Architectural Engineering and Environmental Design from Arab Academy for Science, Technology, and Maritime Transport. Currently, she is a part-time teaching assistant and M.Sc. candidate at the Department of Architecture Engineering and Environmental Design in the same institution where she received her B.Sc. In addition, she is a member of the research team that is co-organizing a series of seminars under the topic: "Gender Equity in Cities of the MENA Region: Women's Right to the City", funded by the Urban Studies Foundation. During the first seminar, she has been a keynote speaker in the field of Egyptian social housing projects and its compatibility with the specific socio-economic needs of women-headed families, entitled: "#She_Needs: Towards Effective Empowerment Programs that Unleash the Potentials of Female-Headed Households in Egypt". Her research interests include contested urbanism, spatial politics, social production of public space, the rhythm of place-temporality, and the role of urban design for socio-spatial transformations. She is investigating the effect of informality in terms of housing or employment on individuals' social behavior, as well as its impact on urban economies and spatial development; which is the underpinning motivation for her post-graduate research initiatives.

Arif Çagdaş Aydınoglu is a lecturer in urban information systems, geographic information science (GIS) and analysis techniques, geo-database management, sustainable land management, spatial data infrastructures, geo-web services and interoperability at the Department of Geomatics Engineering, Gebze Technical University, Turkey. His research interests include GIS, (national) spatial data infrastructure, interoperability and standards of geographic data, land management, urban/land GIS, and disaster management. He is currently vice-president of Gebze Technical University (GTU) and director of the Institute of Natural and Applied Sciences at GTU. Dr. Aydınoglu is professor at GTU since November 2017. He has published about 100 scientific papers, reports and journal articles and co-authored 3 books in Turkey and internationally.

Samaa R. Badawi is an assistant professor in architecture department, Mansoura University, Egypt. She holds a Philosophy of Doctorate degree in Urban Design from Mansoura University in 2013. She teaches urban design, landscape architecture, and architectural design studios. In 2013 she won the price of the best Ph.D. thesis at the university level. In 2016, she won the price of excellence in teaching from the same university she graduated from. She has a special interest in new urban design movements and smart cities. She participated in national and international conferences with research papers in the same field. She also worked as a reviewer for papers submitted to many conferences and journals. She, also, worked as a reviewer for research projects proposals submitted to Effat University Research Center. She also works as a supervisor and examiner of Urban Design Master Students.

Rahma Doheim is an assistant professor in Architecture. She received her Ph.D. in the Built Environment from the University of Ulster. Dr. Doheim holds a master's in safety management from the University of West Virginia, USA. Her research interests lie in smart cities, creativity and architectural pedagogy, LEED rating system, natural thermal and smoke ventilation, as well as CFD (FDS) modeling. Dr. Doheim is working at Assiut University, Egypt. Dr. Doheim has taught and practiced architecture in the USA, the Middle East, and the UK. She was a Senior Designer with BONIAN Architects, Egypt, and an Architect at the Herberger Centre for Design Excellence, Arizona State University, USA.

Alshimaa A. Farag is an assistant professor of Architecture. She holds a Philosophy of Doctorate degree in Urban Design from Zagazig University, Egypt. She is a LEED Accredited Professional in Neighborhood Development (LEED AP ND) and Home (LEED AP Home). She is currently teaching courses of Urban Design, Landscape Architecture, Research Method, and Architecture Design Studios at the College of Architecture and Design, Effat University. She is an EDRA member as she has special interests in human behaviors and their interrelationships with the built environment, especially in the Middle East and Northern Africa Region (MENA).

Omar Mohamed Galal is currently a Ph.D. candidate at the Technical University of Berlin, Germany. He got his Masters of Science degree in Architectural Engineering at Cairo University. His main research interest is related to the reinterpretation of Urban Form and Space within the context of current local and global challenges on the environmental level. Topics like energy efficiency, utilizing renewable energy within the urban settlements, and creatively addressing public transportation within the city are all related to his interest. He also has a vast experience in working with a diverse range of consultancy, real state developers and construction entities. Furthermore, he has participated in and won various architectural and urban design competitions. Currently, he is an assistant lecturer at Arab Academy for Science Technology and Maritime Transportation (AASTMT) where he participates in teaching Architectural and Urban Design classes.

Turgay Kerem Koramaz received the Ph.D. degree in Urban and Regional Planning with the dissertation about the perception of urban historic sites through computer-aided three-dimensional visualization techniques. He was visiting scholar in Royal Institute of Technology, Graduate School of Built Environment and Architecture – Stockholm (Sweden) in 2008. He has been working in Istanbul Technical University (ITU), Department of Urban and Regional Planning since 2000. He has publications about spatial analysis, internal migration in Turkey, urban conservation and renewal. He is teaching site conservation, information and communication technologies and research methods for urban design. He is involved in academic and professional activities about spatial analysis in urban planning, urban conservation, and regeneration.

Sana Layeb is a PhD student in architecture. She is affiliated to the Research Team on Ambiances (ERA) from the National School of Architecture (ENAU), Tunisia. She is interested on the altered perception in the urban and architectural space and the complexity of soundscapes.

Abdelbaseer Mohamed is a lecturer at Al-Azhar University, Urban Planning Department. He is also a former Carnegie fellow at American University in Washington, D.C. He recently completed a doctorate in Urban Design at Ain Shams University, Cairo and Cologne University of Applied Sciences in Germany.

Ahmed M. Refaat is an Assistant Professor in the Architecture Department, Assiut University, Egypt and Architecture Department, Effat University, KSA. He got his bachelor and master degrees from Assiut University. He got his Ph.D. from Manchester University in the field of Architecture and Urban Design. Academically, Ahmed has considerable experience in teaching and supervising under- and post-graduate students. His research work focuses on the morphology of the spatial structure of cities and the factors that propel their spatial transformations, and the achievement of walkable, livable and sustainable cities and neighborhoods. Using space syntax techniques, he conducted the spatial analysis of many cities and neighborhoods in Egypt, KSA, and Italy.

Hans-Georg Schwarz-v.Raumer (born 1959 in Triberg/Germany) studied Geography and Mathematics at University of Karlsruhe/Germany. Since 2003, he is Deputy Director of Institute of Landscape Planning and Ecology at Stuttgart University and since 2010 he is Head of Landscape Analysis and Modelling division attached to the Institute. His scientific fields cover applications of mathematics, modeling, and Geographical Information systems (GIS) in Geography and spatial planning. He participated in several projects on land, water and energy resource management for which he applied scenario technique, System Dynamics, spatially explicit land change modeling techniques and developed Multicriteria Decision Support tools. He is teaching GIS and quantitative methods, landscape planning, ecology and Environmental Impact Assessment in different international BSc. and MSc. programs at the University of Stuttgart.

Ahmad Toimah is an assistant professor at the Urban Planning Department, Faculty of Engineering, Ain Shams University. Toimah has about 17 years of academic and professional experience in the field of spatial planning and regional development. His academic experience includes an MSc thesis in "interaction between urban and rural settlements", a Ph.D. thesis in "urban-rural balance", a number of publications, and a teaching experience in housing, urban planning, and urban design courses. While professional experience includes participation in a wide range of projects including strategic planning, regional development, housing, urban design and capacity building for a number of governmental agencies, international organizations and private sector, and that was in cooperation with different universities and consulting firms in Egypt, United States, Germany, Canada, Saudi Arabia, and Libya.

Eda Ustaoglu has a PhD in Planning and Public Policy from the School of Architecture, Planning and Environmental Policy, University College Dublin, and until 2015 she worked at the European Commission-Joint Research Centre-Ispra, Italy as a scientific officer. She is currently working as a post-doc researcher at the Department of Geomatics Engineering at Gebze Technical University, Turkey. Her current research project focuses on developing a methodology for land-use modelling applications in selected case study areas in Istanbul. Over the past years, she has been involved in various projects, university studies and research activities in line with her graduate and post-graduate studies, which are directly related to the field of urban and environmental science. Her research interests include urban economics, transport policy and planning, transportation infrastructure appraisal, spatial economics and land-use research.

Index

www.ingramcontent.com/pod-product-compliance
Lightning Source LLC
LaVergne TN
LVHW080533300926
849162LV00007B/1595
* 9 7 8 1 5 2 2 5 9 2 3 8 9 *